Neuroanatomical Tract-Tracing Methods

Neuroanatomical Tract-Tracing Methods

Edited by

LENNART HEIMER

University of Virginia
Charlottesville, Virginia

and

MARTINE J. ROBARDS

University of California
Irvine, California

PLENUM PRESS • NEW YORK AND LONDON

Library of Congress Cataloging in Publication Data

Main entry under title:

Neuroanatomical tract-tracing methods.

Includes bibliographical references and index.
1. Brain—Research. 2. Neural circuitry. 3. Neuroanatomy. I. Heimer, Lennart. II. Robards, Martine J. [DNLM: 1. Nervous system—Anatomy and histology. 2. Neural pathways. 3. Cytological techniques. WL 102 N4945]

QM455.N46 599.04'8 80-20687

ISBN-13: 978-1-4613-3191-9 e-ISBN: 978-1-4613-3189-6

DOI: 10.1007/978-1-4613-3189-6 AACR1

Softcover reprint of the hardcover 1st edition 1981
A Division of Plenum Publishing Corporation
233 Spring Street, New York, N.Y. 10013

Contributors

George F. Alheid Department of Neurology
University of Virginia School of Medicine
Charlottesville, Virginia 22908

Georgia A. Bishop Department of Anatomy
Michigan State University
East Lansing, Michigan 48824

Theodor W. Blackstad Institute of Anatomy
University of Oslo
Oslo 1, Norway

Sven O. E. Ebbesson Department of Anatomy
University of Ponce
Ponce, Puerto Rico 00731

Stephen B. Edwards Department of Anatomy
University of Virginia School of Medicine
Charlottesville, Virginia 22908

Victor L. Friedrich, Jr. Laboratory of Neuromorphology
Department of Biobehavioral Sciences
University of Connecticut
Storrs, Connecticut 06268

Peter J. Hand
Laboratories of Anatomy,
Department of Animal Biology,
School of Veterinary Medicine, and
Institute of Neurological Sciences
University of Pennsylvania
Philadelphia, Pennsylvania 19104

Lennart Heimer
Department of Neurology
University of Virginia School of Medicine
Charlottesville, Virginia 22908

Anita Hendrickson
Department of Ophthalmology
University of Washington
Seattle, Washington 98195

Stephen T. Kitai
Department of Anatomy
Michigan State University
East Lansing, Michigan 48824

O. Eugene Millhouse
Department of Neurology
College of Medicine
University of Utah
Salt Lake City, Utah 84132

Robert Y. Moore
Department of Neurology
State University of New York at Stony Brook
Stony Brook, New York 11794

Enrico Mugnaini
Laboratory of Neuromorphology
Department of Biobehavioral Sciences
University of Connecticut
Storrs, Connecticut 06268

José S. de Olmos
Instituto de Investigación Medica Mercedes y Martin Ferreyra
Cordoba, Argentina

Melburn R. Park
Department of Anatomy
Michigan State University
East Lansing, Michigan 48824

Virginia M. Pickel
Laboratory of Neurobiology
Department of Neurology
Cornell University Medical College
New York, New York 10021

Oswald Steward Departments of Neurological Surgery and Physiology
University of Virginia School of Medicine
Charlottesville, Virginia 22908

Robert C. Switzer III Comparative Animal Research Laboratory
Oak Ridge, Tennessee 37830

W. Bruce Warr The Boys Town Institute for Communication Disorders in Children
Omaha, Nebraska 68131

Preface

Times of dramatic progress in brain research have often been correlated with the development of new and powerful techniques that have changed the kinds of questions one can ask. An historical example may illustrate the point. More than 50 years ago, Nissl studies (Ferraro, 1928) showed that extensive forebrain lesions resulted in chromatolysis and cell loss in the substantia nigra; thus, it was suggested that the substantia nigra gave rise to projections into the basal forebrain. In the late 1950s, another clue emerged, this time linking observations from the field of neuropathology with a discovery in experimental neuropharmacology (Carlsson, 1959a,b; Ehringer and Hornykiewicz, 1960). It had long been recognized that patients with Parkinson's disease suffered neuronal loss in the substantia nigra and that their symptoms were somehow related to striatal dysfunction. Thus, when fluorescent catecholamine assays were developed and combined with pharmacological and neuropathological studies of Parkinson's disease, the dopaminergic nature of the illness was shown. A bit later, Falck and Hillarp (Falck *et al.*, 1962) developed a fluorescent histochemical method to visualize monoamine-containing cells in the brain; this technique was soon applied to show that the rich dopaminergic terminal field in the striatum derived from neurons in the substantia nigra (Andén *et al.*, 1964). In the following decade, refinements in the histofluorescent method and the development of sensitive silver impregnation methods permitted a detailed light microscopic exploration of the dopaminergic nigrostriatal system. Along the way, the increased understanding of the nigrostriatal pathway—its anatomy, pathology, and pharmacology—led to the development of L-Dopa, the drug that revolutionized the treatment of parkinsonism. The elucidation of the nigrostriatal pathway continues with increasingly more sophisticated techniques including immunocytochemistry, HRP histochemistry, and electron microscopy (Moore *et al.*, 1971; Hökfelt *et al.*, 1973). Most recently, a combination of retrograde tract-tracing methods and immunohistochemical methods has been applied to the nigrostriatal pathway, allowing the histochemical identification of the transmitter in nigrostriatal neurons labeled by injection of a retrograde tracer in the neostriatum (Ljungdahl *et al.*, 1975; Hökfelt *et al.*, 1979).

This historical resume of one neural circuit serves to demonstrate the incremental nature of scientific inquiry, the step-by-step application of several generations' efforts, using a succession of analytical methods. Certainly, no single method could have elucidated this pathway alone.

In 1981, we find ourselves in a position the classical neuroanatomists would have envied. We have at our fingertips dozens of new and powerful techniques and a heritage of literature built on the older descriptive and experimental methods. To be sure, the expansion of neuroanatomical methods is far from over; each week seems to add another weapon to the technical arsenal. But now, more than ever before, new initiatives into the field of neuroscience are faced with difficult problems of choice. Which technique or combination of techniques should be used? How are the techniques applied? How shall the results be interpreted? This volume was prepared as an aswer to those questions.

The initial planning for this book took place several years ago at the University of Virginia School of Medicine in Charlottesville, where many of the contributors gathered for a two-day conference. Although the scope of the project has grown in the intervening years, the objective has remained the same: to produce a book that facilitates the choice of the most appropriate neuroanatomical method or combination of methods for studying the circuitry of the nervous system and that contains practical advice on how to use the most common tract-tracing methods in contemporary neuroscience. In general, each chapter provides the theoretical background of a method or set of methods in addition to specific procedures of application. In the appendices, detailed experimental protocols and comments on variations have been provided. Since many of the practical problems related to the day-to-day business of neuroanatomical research are common for most tract-tracing methods, we have devoted Chapter 1 to such general procedures. Of necessity, this chapter is presented a bit dogmatically, but its aim is to provide the young neuroscientist with a workable and coherent set of procedures such as one might acquire in a well-established neuroanatomy laboratory. The second and third chapters deal with lesion techniques and the injection of tracers into the brain. Here we have attempted to include many of the currently popular procedures and to give practical advice on how to apply them.

We wish to thank the Sloan Foundation, which, through its grant to the neuroscience program at the University of Virginia, supported the planning meeting for this project and sponsored the printing of the color plates. We appreciate the assistance of many colleagues who have read these chapters at various stages of their preparation and have offered their editorial advice. For excellent secretarial assistance, we would especially like to mention Mary

Staton, Inge Larsen, and Karin Wiedemann. Our interactions with the personnel at Plenum Press have been altogether pleasant, and we should like to thank Mr. Seymour Weingarten, Mr. Peter Strupp, and Mr. Leonard R. Pace for their patience and generous support.

Lennart Heimer
Martine RoBards

REFERENCES

Andén, N. E., Carlsson, A. Dahlström, A., Fuxe, K., Hillarp, N. A., Larsson, K., 1964, Demonstration and mapping out of nigro-neostriatal dopamine neurons, *Life Sci.* **3**:523–530.

Carlsson, A., 1959a, Detection and assay of dopamine, *Pharmacol. Rev.* **11**:300–304.

Carlsson, A., 1959b, The occurrence, distribution and physiological role of catecholamines in the nervous system, *Pharmacol. Rev.* **11**:490–493.

Ehringer, H., and Hornykiewicz, O., 1960, Verteilung von Noradrenalin und Dopamin (3-hydroxytyramin) in Gehirn des Menschen und ihr Verhalten bei Erkrankungen des Extrapyramidalen Systems, *Klin. Wochenschr.* **38**:1236–1239.

Falck, B., Hillarp, N. A., Thieme, G., and Torp, A., 1962, Fluorescence of catecholamines and related compounds condensed with formaldehyde, *J. Histochem. Cytochem.* **10**:348–354.

Ferraro, A., 1928, The connections of the pars suboculomotoria of the substantia nigra, *Arch. Neurol. Psychiatr.* **19**:177–180.

Hökfelt, T., Fuxe, K., and Goldstein, M., 1973, Immunohistochemical studies on monoamine-containing cell systems, *Brain Res.* **62**:461–469.

Hökfelt, T., Phillipson, O., Kuypers, H. C. J. M., Bentivoglio, M., Catsman-Berrevoets, C. E., and Dann, O., 1979, Tracing of transmitter histochemically identified neuron projections: Immunohistochemistry combined with fluorescent retrograde labeling, *Neuroscience Lett.* [*Suppl.*] **3**:341.

Ljungdahl, A., Hökfelt, T., Goldstein, M., and Park, D., 1975, Retrograde peroxidase tracing of neurons combined with transmitter histochemistry, *Brain Res.* **84**:313–319.

Moore, R. Y., Bhatnagar, R. K., and Heller, A., 1971, Anatomical and chemical studies of a nigro-neostriatal projection in the cat, *Brain Res.* **30**:119–135.

Contents

2 Methods for Selective, Restricted Lesion Placement in the Central Nervous System

Robert Y. Moore

3 Methods for Delivering Tracers

George F. Alheid, Stephen B. Edwards, Stephen T. Kitai, Melburn R. Park, and Robert C. Switzer III

4 Silver Methods for the Impregnation of Degenerating Axoplasm

José S. de Olmos, Sven O. E. Ebbesson, and Lennart Heimer

5 The Autoradiographic Tracing of Axonal Connections in the Central Nervous System

Stephen B. Edwards and Anita Hendrickson

6 Horseradish Peroxidase: The Basic Procedure

W. Bruce Warr, José S. de Olmos, and Lennart Heimer

7 Horseradish Peroxidase: Intracellular Staining of Neurons

Stephen T. Kitai and Georgia A. Bishop

8 Horseradish Peroxidase and Fluorescent Substances and Their Combination with Other Techniques

Oswald Steward

9 The Golgi Methods

O. Eugene Millhouse

12 Tract Tracing by Electron Microscopy of Golgi Preparations

Theodor W. Blackstad

13 Fluorescence Histochemical Methods: Neurotransmitter Histochemistry

Robert Y. Moore

14 Immunocytochemical Methods

Virginia M. Pickel

15 The 2-Deoxyglucose Method

Peter J. Hand

Experimental Neuroanatomy

1

General Approaches and Laboratory Procedures

THEODORE W. BLACKSTAD, LENNART HEIMER, and ENRICO MUGNAINI

I. INTRODUCTION

A great number of tract-tracing techniques are presently available for the exploration of nervous system organization—a bewildering array, perhaps, for the beginning neuroscientist. The intention of this volume is to facilitate the acquisition of the necessary skills and judgment to apply these methods correctly and to interpret their results wisely. But first, it is essential to deal with more basic topics: how to select among the available methods the most appropriate to apply to a problem of interest. Thus, we will begin by introducing the reader to the techniques available with some evaluation of when and how they may best be used. In addition, this chapter will deal with as-

THEODOR W. BLACKSTAD • Institute of Anatomy, University of Oslo, Oslo 1, Norway. LENNART HEIMER • Department of Neurology, University of Virginia School of Medicine, Charlottesville, Virginia 22908. ENRICO MUGNAINI • Laboratory of Neuromorphology, Department of Biobehavioral Sciences, University of Connecticut, Storrs, Connecticut 06268.

pects of neuroanatomical practice common to many, if not all, of the techniques dealt with in later chapters: the choice and handling of experimental animals, surgical methods, preparation of tissue sections, and analysis of results.

II. TRACT-TRACING METHODS

A. Types of Methods

The most popular techniques for studying the anatomy of the central nervous system fall into four main categories (Fig. 1).

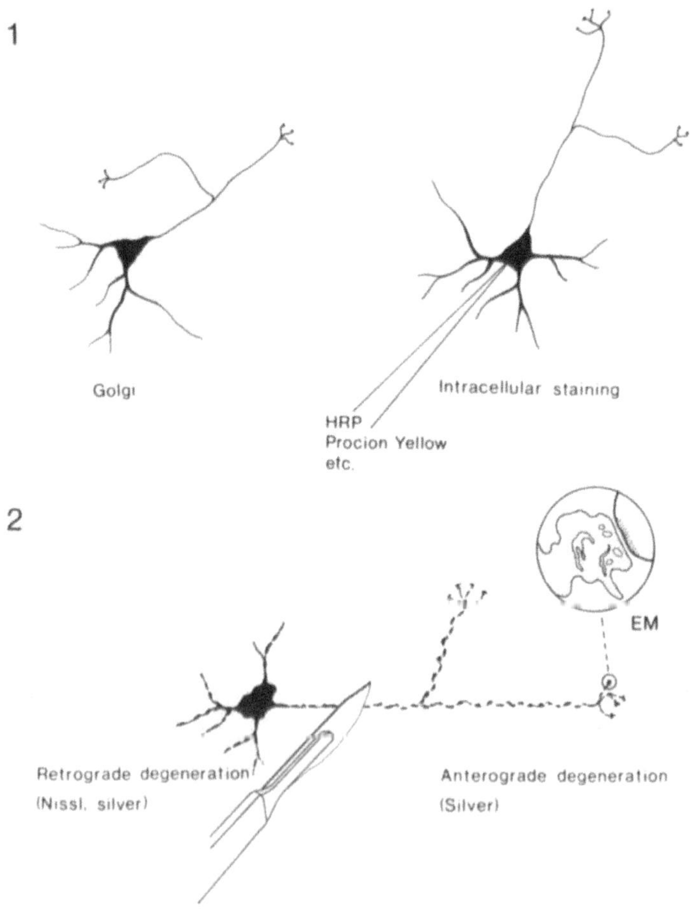

Figure 1. Different types of tract-tracing methods.

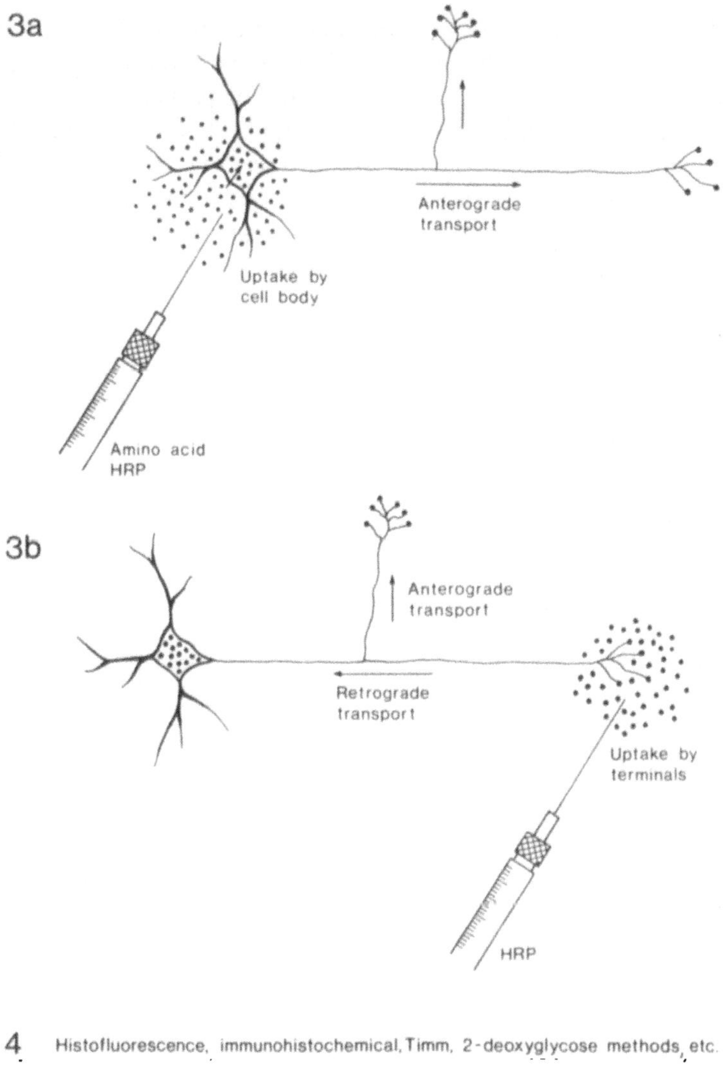

Figure 1. (*cont.*)

1. The first main category includes the Golgi methods and *in vivo* dye injection methods* for the study of normal neurons. These methods are especially useful when the immediate neighborhood relations of nerve cells are studied and when dendritic and axonal geometry are

**Dye injection methods* in this connection refer to any method in which a substance, including HRP, can be injected into individual neurons for the purpose of displaying the nerve cell body and its processes (Kater and Nicholsen, 1973).

of particular interest. The methods are discussed in Chapters 7 and 9.

2. The second main category includes methods that focus on degenerative and reactive changes in various parts of the neuron as a consequence of a lesion. Even the most distant parts of neurons involved in such a lesion will respond to the injury. These methods, therefore, usually permit connections to be traced far beyond the effective range of the first group. By using various stains for light microscopy or by subjecting the tissue to electron microscopy, one can visualize various aspects of the degenerative or reactive process. Classical examples of methods belonging to this group are the Nissl method for the demonstration of chromatolysis in the nerve cell body and the Marchi method for the staining of disintegrating myelin sheaths. Although the Nissl and Marchi methods may occasionally be useful for tracing central nervous connections, they have, to a large extent, been superseded by more recently developed tract-tracing methods, and they will, therefore, not be discussed in this volume. Silver methods for the impregnation of degenerating axons and axonal endings and electron microscopy of degenerating neural elements will be discussed in Chapters 4 and 11, respectively.
3. The third category includes tract-tracing methods that rely on a substance's incorporation into the neuron, its transport, and its subsequent localization in other parts of the neuron. Two methods will be discussed in Chapters 5, 6, and 8.
4. Several histochemical methods are used in the mapping of central nervous pathways. Examples of these are the monoamine histofluorescence methods (Chapter 13), the recently developed immunohistochemical techniques for the identification of various transmitter-related substances (Chapter 14), the Timm method for the demonstration of metals (Haug, 1973; Danscher and Zimmer, 1978), and the 2-deoxyglucose method (Chapter 15).

B. Choice of Method

Although we can offer no rigid prescription of how to conduct a tract-tracing study, we can list some factors that should influence the decision of where to begin and how to proceed. Once an area or a fiber system has been selected for study, published literature should be searched for information about the system. For example, descriptions based on the Golgi, Marchi, or Nauta methods may already exist. If so, substantial information regarding both the course and the general area of termination of the pathway may already be available, and one may wish to focus attention on its precise mode or locus of termination.

1. Termination of a Pathway

To study fiber terminations, many scientists start with a modern *silver impregnation method* for the tracing of degenerating axons including their end structures. The first experimental step may be to make a restricted surgical lesion in the area where the pathway originates (area of cell bodies). Alternatively, the axons composing the pathway may be transected surgically somewhere along their course. The animal is then allowed to survive for a postoperative interval between 1–7 days, depending on species and fiber system and selected to produce an optimal degenerative reaction for the silver impregnation. At the end of this interval, the central nervous system is fixed by vascular perfusion, and after a suitable period of fixation, the tissue is sectioned and impregnated with silver.

Most silver impregnation methods involve three main steps: an initial treatment in silver nitrate, a treatment in an ammoniacal silver solution, and a reduction of the silver in a reducing solution, usually a mixture of formalin, alcohol, and citric acid. By studying the silver-impregnated sections, the precise distribution of the degenerating nerve fibers and their terminals can be determined. This task is facilitated by comparing the silver-impregnated sections with adjacent sections stained by the Nissl method or another appropriate method. The region containing the lesion is often embedded and sectioned separately, particularly if it is remote from the area of termination. In any case, the location and the extent of the lesion are carefully evaluated, if necessary by comparison with material prepared to show cyto-, myelo-, or chemoarchitecture of the remaining area around the lesion.

Such a silver impregnation study usually provides detailed information on both the course of the fiber tracts emanating from the lesioned area and the sites of their termination. A serious weakness, however, is that fibers merely passing through, but not actually originating in, the area of the experimental lesion may have been interrupted by the lesion. There is also a risk that fibers outside the intended lesion may have been interrupted by the electrode track. If damage to such extraneous fibers does create problems of interpretation, then silver methods should be supplemented or replaced by *autoradiography.* In an autoradiographic tract-tracing study, a highly concentrated solution of radioactively labeled amino acid is injected into the area of presumed origin of the pathway. The value of this type of procedure is related to the fact that the labeled amino acid is taken up only by cell bodies in the area of injection and not by passing fibers. If a pathway has already been mapped by a silver method, the autoradiographic method can be more specifically devoted to the task of establishing to what extent, if any, the pathway originates in a particular area. The labeled tracer, taken up by the cell bodies in the vicinity of the injection, is transported anterogradely through the axons to their terminals. After a short postsurgical interval, usually less than 1 day, when most of the rapidly transported fraction of protein formed with the radioactive precursor has reached the axon terminals, the animal is anesthetized and its brain fixed by vascular perfusion. The brain is sectioned, and the sections mounted on microscope slides that are subsequently coated with a radiosensitive emulsion and exposed for a period of 2 to several weeks. The slides are then developed, fixed, and stained by the Nissl method. If the pathway originates in the area of injection, it will reveal itself in the sections through the presence of developed silver grains located over axons and axon terminals.

In the research strategy outlined above, silver impregnation of Wallerian degeneration precedes autoradiographic tracing. With the increasing use of the autoradiographic technique in many laboratories, one may wish to start with the autoradiographic technique rather than with the silver method. The choice of method also depends, to some degree, on the problem. For instance, the general distribution of the callosal fibers to an entire cerebral hemisphere is best studied with silver impregnation of degenerating fibers and terminals following complete callosal transection. However, if one wants to demonstrate pathways originating in the corpus striatum, the use of the autoradiographic technique following injections of labeled amino acid into the corpus striatum would be more appropriate because of the existence of many extrinsic fibers passing through this nuclear complex.

Beyond the experimental procedures outlined above, additional methods exist. The ultrastructural aspects of the termination of the pathway may be examined by *electron microscopy* (EM) alone or by EM in combination with degeneration methods or autoradiography. It should be pointed out, however, that EM autoradiography remains a specialized and time-consuming technique which should not be expected to yield reliable data without considerable training on the part of the researcher. The widespread availability of electron microscopic facilities, on the other hand, has made it possible to exploit the many advantages of EM analysis of anterograde degeneration without undue effort, and the technique can be successfully applied even by a neuroanatomist without years of training.

Fixation for electron microscopy requires more rigorous attention to the details of pH and osmolarity than that for light microscopy, but this should not deter a beginner from including an EM analysis as part of a tract-tracing study. Dehydrating and embedding tissue for electron microscopy can be done in a few hours, and the skills needed to prepare ultrathin sections and analyze them in the electron microscope may be acquired in a matter of weeks with the help of a competent instructor.

In principle, the light microscopic and EM study of a degenerating fiber tract are similar. An important difference, however, is that the optimal postoperative survival times are usually somewhat shorter in EM than in light microscopic studies. After fixation of the tissue by vascular perfusion, small slabs are excised from the regions where probable terminal zones are located. Following osmication and dehydration, the tissue is embedded in plastic and sectioned on an ultramicrotome. Orientation is obtained with the aid of 1- to 5-μm "semithin" sections which can be stained with different techniques. After appropriate trimming of the block, ultrathin sections are prepared for the identification of degenerating synaptic profiles, glial changes, and other signs accompanying anterograde axonal degeneration. In addition to verifying the existence of truly terminal structures belonging to the projection under investigation, the EM study can disclose the morphology of the synaptic contact (symmetric vs. asymmetric), the vesicle shape (round, flattened, or pleomorphic), and the identity of the postsynaptic element (dendrite, soma, or axon), as well as other features of the terminal area.

2. Cells of Origin of a Pathway

After it has been established that a specific projection exists, two important questions usually arise: (1) Where are the cells of origin of the pathway located? and (2) Which morphological characteristics do these cells display? To answer these questions, the horseradish peroxidase (HRP) method (Chapter 6) may be applied.* When this method is used, the enzyme is injected into the terminal area of the projection where it is taken up by the synaptic endings and transported in retrograde direction back to the cell bodies. After a short postoperative survival period, usually 12–48 hr, the tissue is fixed by vascular perfusion and sectioned on the microtome. The sections are incubated in a solution containing a chromogen, usually a benzidine derivative, which is oxidized to a colored product by oxygen released from hydrogen peroxide in the incubating medium. The cells that are labeled with the reaction product are those that give rise to the projection. After incubation for HRP, the tissue may be impregnated by the Golgi technique for further detailed characterization of the cells of origin (Chapters 9 and 12).

In this context, it should be recalled that HRP has a much wider field of application than mapping cells of origin of a pathway. Since locally injected HRP is taken up not only by axonal terminals but also by cell bodies and is subsequently transported in anterograde direction, HRP can be used for mapping the course and termination of central nervous pathways as well. Indeed, the use of a very sensitive HRP technique can demonstrate retrograde and anterograde transport in both the axon and collaterals of the same neuron (Chapter 8). HRP-containing terminals furthermore, can be investigated both with the light and electron microscope. Finally, since HRP can be injected intracellularly (Chapter 7), it can offer a spectacular view of neuronal morphology equaled only by an excellent Golgi impregnation.

This introductory review has emphasized the complementary nature of a number of various tract-tracing methods and the unique contribution of each to a thorough elucidation of a pathway. Certainly, not all tract-tracing problems will require such a multifaceted approach as advocated in the previous pages or the use of the methods in the order indicated above. Some

*Recently, several other substances have been introduced for similar tracing purposes (Chapter 8).

projects, furthermore, will require the use of other techniques. But under all circumstances, one should remember that it is better to attack a problem from many different directions than to rely on a single method.

III. PRACTICAL PROBLEMS

A. The Experimental Animal

1. Selection of Species

The selection of appropriate experimental animals is a critical step in any neuroanatomical study. If a problem can be solved as readily by using mice, rats, or some other inexpensive and easily obtained animal, then it is obviously advisable to choose one of these standard laboratory animals. Their virtues are several: rodents, in particular, are cheap, inexpensive to maintain, hardy in a laboratory environment, resistant to infection, and used by so many investigators in so many fields that a rich literature is available for inspiration and comparison.

Despite the many advantages of using the more common laboratory animals, certain systems are nevertheless best studied in other species in which they are particularly highly developed. The bat, for instance, may be especially well suited for the study of auditory mechanisms. In some cases, it is necessary to consider using an expensive and exotic primate, even if such an animal is difficult to obtain and may be difficult to handle and maintain. The most valid reason for selecting a primate is clear: in studies whose aim it is to specify as closely as possible the pattern of human neural connectivity, primates are the best choice.

In other cases, a species may recommend itself for less compelling reasons such as tradition. The use of the cat in studies on the visual and somatosensory systems is an example. While there is no *a priori* reason to assume that the cat's visual or somatosensory apparatus is particularly representative of either the mammalian or human condition, nevertheless an astounding quantity of research has been amassed on those systems in the cat. One may wish to choose the cat, then, because it offers a particularly good basis for further research in these fields.

Yet another reason to select particular animals is the desire to fit the results of one's study into an evolutionary or comparative scheme. For example, if there is interest in deducing how a particular nucleus in the primate may have evolved from the reptilian state, one would select for study carefully and rationally several species representing good living approximations of ancestral animals in the reptile-to-primate lineage: a lizard, an opossum, a tree shrew, a bush baby, and a rhesus macaque might constitute a defensible series for an evolutionary study. Needless to say, such an ambitious undertaking would require extensive animal facilities, major financial support, and a number of years to devote to its accomplishment.

2. Age of the Animal

Although the age of the animal is usually not a critical factor in most tract-tracing experiments, nevertheless, there are cases in which it may be the factor that determines the success or failure of an experiment. A classic example was the use of very young animals for the demonstration of cells of origin by the aid of the so-called modified Gudden method (Brodal, 1940). This method took advantage of the fact that retrograde degenerative changes in cell bodies following transection of their axons are more pronounced in young than in adult animals. But at the same time, this same susceptibility of young brains to degenerative changes is one of the reasons why juvenile animals are often avoided in anterograde degeneration studies. Transneuronal atrophy—a phenomenon rather rare in adults—often complicates the interpretation of degenerative changes occurring in response to an experimental lesion in young animals. Too old animals, on the other hand, should also be avoided in degeneration experiments because of the frequent occurrence of spontaneous degeneration (Chapter 4).

A situation where the use of young animals is advisable is when a primary lesion, after a certain survival time, is to be followed by a secondary lesion. This method, which is referred to as "chronic deafferentation" or "successive degeneration," is of special value when the terminal area of two overlapping pathways needs to be determined by the study of terminal degeneration. When the primary lesion is performed in young animals, the degenerational debris is usually removed within weeks, whereas in adult animals it may persist for months and complicate the interpretation of the terminal degeneration resulting from the secondary lesion. Although the introduction of autoradiographic mapping has made this approach less necessary for tract tracing, there will still remain specific cases in which the chronic deafferentation procedure is useful.

At present the chronic deafferentation procedure is, however, most typically used in studies of plastic changes in the nervous system following injury, the "sprouting" phenomenon. In this case, a first (priming) lesion is used to destroy the known afferents of a structure. After a time, a second lesion in another area of the brain is performed, and the brain is later subjected to histological analysis to see if the fibers destroyed in the second lesion have sprouted into the synaptic area vacated after the first lesion. The dramatic success of the successive degeneration method in this respect should, therefore, serve as a warning to investigators aiming to use the technique: chronic deafferentation may alter the connections of an area.

Age recommendations are furthermore appropriate if one considers the use of the HRP and Golgi techniques. The HRP technique works best in young animals, but most investigators intentionally select young adults for study simply because one can have greater certainty that the connections revealed will reflect the mature condition. Golgi impregnation is usually most successful in material from newborn animals, and most classical Golgi studies have been done on such material. Adult animals, however, are increasingly being used in order to obtain a picture of the definitive pattern.

3. Survival Time

The importance of the postoperative survival time of the animal was not always fully realized in the past. However, the advent of electron microscopy, modern silver methods, tritiated amino acid autoradiography, and HRP tracing have clearly demonstrated that the survival time is one of the most significant variables in tract-tracing experiments, and it is important to determine the optimal survival time in the early stage of each project.

If the method is based on degenerative phenomena, one should be aware that degenerating boutons are removed by glial elements in many areas of the brain within the first few days following the operation. Therefore, methods such as silver impregnation and electron microscopy, which reveal degenerating boutons, cannot be used to full advantage with excessively long survival times. It is also important to realize that the degeneration of axon terminals is usually faster than the degeneration of nonterminal parts of the axon. Therefore, whereas degenerating axon terminals can be demonstrated best in animals with 1–3 days' survival, degeneration of the parent fibers is demonstrated more clearly with longer survival times such as 3–5 days.

These guidelines relate to commonly used small laboratory mammals such as the rat and guinea pig. Other species may need different survival times. Cold-blooded animals, for instance, often require several weeks' survival time. Age, too, is a factor. Degeneration occurs faster in young animals, and axonal debris is more rapidly cleared. Additional factors influencing the choice of survival time in silver impregnation studies are discussed in Chapter 4.

Choice of survival time is equally important when connections are studied by means of the autoradiographic technique. Label in the terminal fields of many pathways can best be demonstrated with a survival time of 1 day or less, whereas the visualization of both parent fibers and terminals often needs several days' survival time. This may seem surprising. The reason that label is first demonstrated in larger quantity in the terminals is that the fast component of the axonal transport goes selectively to the axonal endings. Only later, when the slow transport mechanism has been given a chance to manifest itself, can the axon itself be readily demonstrated in many fiber systems. The optimal survival time in autoradiographic tract tracing is also related to the length of the fiber system studied as well as to such biological factors as the body temperature of the animal. It is highly advisable to determine the optimal survival time in each investigation empirically. This is true not only for the autoradiographic methods but also for the retrograde and anterograde transport techniques using HRP and other tracers.

4. Number of Animals

Although, at times, a limited number of successful experiments may provide a clear demonstration of a pathway, the most convincing results are derived from larger numbers of animals. In order to map in detail a single pathway in the brain, as many as 100 animals may be required. To study such a large number of animals may necessitate not only substantial financial investment but also self-discipline and tenacity on the part of the investigator. Needless to say, the species being studied will influence the number of animals used in a study. Whereas five to ten monkeys must often be accepted as sufficient, ten times that number of a small rodent would be realistic.

5. Anesthesia

Anesthesia relieves the animal of pain and prevents reflex responses during surgery, and it is an essential part of most operations. Correct application of the appropriate anesthetic agent can save valuable time and prevent serious complications during and after the operation. There are many pharmacological agents that can produce sedation and muscular relaxation in laboratory animals. To be efficient, however, the anesthetic agent should have some additional characteristics. It should not have an overly depressive effect on the respiratory and cardiovascular systems. It should be easy to administer, and it should permit a rapid induction of anesthesia as well as a relatively rapid recovery of the animal. Ideally, it should be stable, nonirritative, nonflammable, and nonexplosive. The barbiturates, which were introduced in the early part of this century, fulfill most of these requirements, and they are the most widely used anesthetic agents in animal surgery and are often used as induction agents in human surgery. However, many other anesthetic agents may be superior for special purposes. This subject will not be dealt with at length here. More specific information on the use of barbiturates as well as other agents for muscle relaxation and general anesthesia of different laboratory animals is given in the books *Veterinary Anesthesia* by Lumb and Jones (1973), *Textbook of Veterinary Anesthesia* by Soma (1971), and *Animal Anaesthesia* by Green (1979). A useful guide for drug dosage for use in laboratory animals has been written by Barnes and Eltherington (1973).

a. Barbiturates. The barbiturates, which are sodium salts of barbituric acid derivatives, are most often given as intraperitoneal or intravenous injections in animals. The intravenous route is generally safer because the effect can be closely observed during the injection. The intraperitoneal route, which is most often used in small laboratory animals, requires the injection of a predetermined dose. However, many factors besides the weight of the animal influence the effect of the anesthetic agent, and frequently the "average" dose is not ideal for a specific animal. For instance, aged, fat, or weak animals require less anesthetic than healthy muscular animals of the same weight.

The most commonly used barbiturate for surgery on laboratory animals is pentobarbital sodium (Nembutal®), commercially available as a 5 or 6% solution. In order to be able to inject a more accurate dose in small animals such as the rat, it is advisable to dilute one part of the anesthetic with one or two parts saline or water. The effective dose for dogs, cats, rabbits, guinea pigs, and rats is in the range of 25–40 mg/kg. However, since there are significant variations among different species and different strains of animals, most laboratories have usually devised their own charts of average doses for the animals they use. It is furthermore advisable to keep protocol on the effect of the anesthetic agent in each experimental animal.

b. Inhalation Anesthesia. We shall discuss here examples of two varieties of inhalation anesthetics: halogenated hydrocarbons and ether. While inhalation offers some advantages as a route of anesthetic administration, it also involves some risks and liabilities. Thus, careful evaluation is required before selecting this type of anesthesia for any experimental program.

Halogenated Hydrocarbons. The most commonly used anesthetics in this class are halothane (Fluothane*) and methoxyflurane (Metofane†). Halothane has higher vapor pressure than methoxyflurane, the latter requiring an active circulation device rather than just a vaporizer to dispense it. On the other hand, halothane is more expensive to use than methoxyflurane. Both require some vaporizing system, and, whereas some homemade devices may be sufficient, the least expensive commercial system costs over $1000. Neither is flammable. Taken together, their advantages as anesthetic agents are several: (1) since the anesthetic agent is rapidly eliminated through the lungs, the level of anesthesia can be controlled rather easily; (2) animals recover rapidly from the effects of the anesthetic; (3) animals may be anesthetized for long periods of time. However, disadvantages also are recognized: (1) since either a face mask or an endotracheal tube is required, mounting the animal in a standard stereotaxic head holder may be awkward; (2) while the slow recovery from anesthesia which barbiturates provide may at times be beneficial to postoperative animals in that their pain is temporarily relieved and they are not likely to injure themselves by getting up too soon after surgery, the opposite may be true with inhalation anesthetics; (3) finally, and probably the most important drawback to the use of halogenated

*USA: E. R. Squibb & Sons, 745 Fifth Avenue, New York, New York 10022. Europe: Fluthane ICI (Imperial Chemical Industries Ltd.) Pharmaceutical Division, Cheshire, England, and Halothan Farbwerke Hoechst, Frankfurt am Main, West Germany.

†Pitman-Moore Company, PO Box 344, Washington Crossing, New Jersey 08560.

hydrocarbons, a convincing body of evidence points to the danger to operating room personnel of prolonged or intermittent low-level exposure to these anesthetics in an improperly ventilated surgical facility; central lobular necrosis of the liver is caused by significant exposure to these substances. Exposed individuals are perhaps greater anesthetic risks should they ever need halothane in surgery as a patient.

Ether. Ether has for many years been a popular anesthetic agent primarily because it is fast-acting and efficient and has a wide margin of safety. However, it has three serious drawbacks: its tendency to irritate tissues, its flammability, and its enhancement of bleeding. Most laboratories are discontinuing the use of ether on a regular basis, and it is absolutely contraindicated in operations where electrocautery is used because of the danger of explosions. If ether is used, either alone or as a supplement to barbiturates, it is wise to give the animal an additional dose of atropine (about 0.1 ml of 0.1% per 100 g weight) to prevent excessive secretion in the respiratory passages.

Animals vary considerably in their response to an anesthetic agent as well as to surgery as a whole. Anesthetic emergencies, however, can be minimized by observing a few basic rules. Since even the short-acting barbiturates have a cumulative effect, it is easy to overdose when repeated doses are given. Therefore, it may be wise to anesthetize another animal rather than to give a lightly anesthetized animal a second dose. Make sure that only healthy animals are used. Even a trivial respiratory infection increases the risk of fluids accumulating in the airways with concomitant obstruction of breathing.

Anesthetic agents have a tendency to depress the respiratory and the cardiovascular systems, and it is important to pay special attention to the functions of these vital systems, both during the operation and in the postoperative stage. If the ventilation of the lungs is already somewhat impaired by the effect of the anesthetic, even a small obstruction of the airway may be fatal. If the breathing becomes shallow or irregular, ventilation should be assisted artificially. It is advisable to keep the animal's head in a natural position during artificial respiration and to pull out the tongue and free the mouth and nose of mucus. It is also useful to stimulate the respiratory centers with an analeptic agent, such as pentamethylene tetrozole (Metrazol®), which is usually administered intraperitoneally (30–50 mg/kg) in rats and guinea pigs and intravenously (10 mg/kg) in dogs.

c. Cardiac Arrest. Should cardiac arrest develop, it can sometimes be treated successfully provided immediate steps are taken to stimulate the heart and improve ventilation. The heart can be stimulated by compressing the chest at a rate of about 70 times per minute. In small animals like the rat, heart compression can be accomplished by pressing the heart between the thumb and the finger. After 10–15 heart compressions, the lungs should be inflated a couple of times by cupping the hands over the nose and mouth and blowing air into them. In larger animals, intubation and mechanical ventilation are usually required. The procedure should be repeated until heartbeat and circulation are clearly reestablished.

To facilitate the animal's recovery from anesthesia, one should perform the operation as smoothly and quickly as possible. Unnecessarily large incisions should be avoided, and bleeding should be stopped immediately. Since intrinsic thermoregulatory mechanisms are impaired during anesthesia, it is important to make sure that the animal's body temperature is near normal during and after the operation. The room temperature should be at least 20–22°C, and heating pad or heating lamp may be required to keep the animal warm. However, remember that the animal is also very sensitive to overheating.

6. Operative Approaches

It is beyond the scope of this book to discuss the countless surgical techniques and approaches conceivable in different experimental situations, and only a few general comments will be presented. Many tract-tracing studies are continuations of already published work, and it is often useful to start with modifications of surgical techniques used earlier. Do not hesitate to seek guidance of colleagues in the laboratory, but remember that no technique is so good that it cannot be improved. Indeed, this is an area where the investigator's ingenuity can play an important role.

The making of lesions in order to provoke reactions in neurons is an integral part of many experimental tract-tracing studies, and some of the most common lesion techniques are discussed in a systematic fashion in Chapter 2. Lesions can be made in different ways, depending on, among other factors, the region under study and the equipment available. As a rule, the simplest method that fulfills the requirements should be chosen. If the structure to be lesioned is located superficially and is immediately available to inspection, it is not necessary to use expensive equipment designed to make lesions in deeper brain structures. Such superficial lesions include transection of a superficial fiber bundle like the dorsal funiculus of the spinal cord or removal of cerebellar or cerebral cortex. Utensils suitable for superficial lesions can include needles, fine knives, scoop- or hooklike instruments, or suction pipettes, all of which are often adapted *ad hoc* to the particular task. The cutting edge of the knife should be sharp to secure efficient transection of brain matter as intended.

Heat is often used for the purpose of making superficial lesions. The lesioning instrument, flat or pointed as needed, may be heated by more or less sophisticated means ranging from an alcohol lamp to an electronically regulated apparatus. A well-known heat-lesion technique is Dusser de Barenne's "laminar coagulation" procedure, which is discussed in more detail in Chapter 2. The stereotaxic technique of making lesions in or injecting tracers into deep brain structures is also discussed in Chapter 2.

As already stated, tract-tracing techniques that rely on the incorporation of a tracer substance into the neuron and the subsequent anterograde or retrograde intraaxonal transport of this substance have become increasingly popular. As a consequence, the technique of delivering tracers to the nervous system has improved rapidly in recent years, and there are many published descriptions of extracellular and intracellular injection techniques using either a pressure system or iontophoresis. Some of the more commonly used injection techniques are described in Chapter 3.

Precise systematic mapping of the injection site or the lesion, including any surrounding region of necrosis resulting from vascular damage, is one of the most important aspects of tract-tracing research. A lesion or injection seldom proves to be exactly as desired. This apparent lack of success, however, may readily be turned into a positive result if the extent of the lesion or the injection site is determined precisely. Cases in which the electrode or the pipette barely missed the intended goal often constitute ideal controls that may be difficult to produce deliberately. Similarly, comparison of animals with varying degrees of involvement of an area will always permit valuable inferences. Therefore, instead of considering variability in placement of lesions or injections as failure, one should exploit the variability by meticulous mapping of each case. The literature abounds with examples of valuable latent information that has not been exploited because the experimental lesions were too crudely analyzed.

7. Surgery

a. The Surgical Wound. It is easier to close a wound nicely if the fur of the area has been cut short or shaved. When suturing a wound, one should ensure that epidermis is not folded into the wound, because healing only occurs between subepidermal surfaces. Dermis should face dermis precisely. The distance between the skin stitches must be small if the animal can reach the wound with a paw or if two or more animals are in the same cage and may chew at each others wounds. As rats are especially inclined to do this, each operated rat should be placed in an individual cage at least as long as it is affected by the anesthesia. Surgical wounds can also be closed by special wound clips.

b. Sterility and Antibiotics. Some animals, such as the rat, possess a great resistance against infections, and no precautions beyond general cleanliness are required during surgery. Other species, such as the rabbit or the cat, very easily attract infections of surgical sites and should be operated on under sterile or at least very clean conditions. To accomplish this, the field of operation is disinfected, instruments are sterilized, and antibiotics are given prophylactically. Some animals, like the monkey, are so expensive and so prone to infection that such measures should always be taken.

c. Bleeding and Hemostasis. Hemorrhage complicates nearly every surgical intervention. However, by studying the typical course of larger blood vessels in relation to the intended field of surgery before the experiment, an approach can be selected that causes minimal damage to these blood vessels. Bleeding from skin and musculature is generally unimportant for the later analysis of the CNS but should be avoided or stopped nevertheless, since it may weaken or even kill the animal. It may also obscure further operative access to the CNS. In most cases, moderate bleeding ceases spontaneously after a few minutes. If not, pinching with a cold or hot hemostatic forceps closes most vessels. A profusely bleeding cavity may be filled with hemostatic gel such as Gelfoam® or with a piece of fully or partially detached musculature which accelerates the spontaneous hemostasis. When approaching the skull, it is wise to separate skin, muscle, and periosteum from each other individually by blunt dissection in order to facilitate wound closing layer by layer after the operation. This creates a reasonably good barrier over the opening in the cranium. Deep layers should be sutured with as few loose stitches as possible in order to preserve their viability.

When surgery is close to a large dural sinus, accidental bleeding from the sinus is not uncommon. Such bleedings are often massive and have killed countless experimental animals. There are two efficient ways of preventing such bleedings. One, of course, is to entirely avoid contact with the sinus. The other is to free the sinus by careful dissection and then either pull it gently aside, thus giving room for the lesioning instrument, or ligate its end and remove the middle portion. These techniques are appropriate especially with regard to large sinuses such as the superior sagittal or the transverse sinus. It is feasible, even in small animals like the rat, to remove the skull over a dural sinus without rupturing the latter. This can be achieved by nibbling away the skull roof in tiny pieces with a small hemostatic forceps or a fine rongeur while working under a dissecting microscope.

d. The Craniotomy. It is important to remember that the friction of a rapidly rotating burr head used for drilling a hole through the skull may, in itself, evoke a heat lesion in the underlying cortex. To prevent this, the area of drilling can be irrigated with cold saline. If the skull roof is thin, as is the case with most small laboratory animals, a burr hole can easily be made by merely spinning the drill between the fingers. The slow rotation prevents overheating.

Pressure on the brain entails a certain risk of uncontrolled damage and should therefore be avoided if possible. When preparing to make an electrolytic lesion, for instance, one should make only a small hole in the cranium and the dura to allow access for the electrode. In the case of a superficial lesion, it is often useful to make the lesion with heat without even opening the dura. This also prevents spontaneous bleeding. In other instances, however, it may be advisable to make a wide opening in both the cranium and the dura in order to prevent brain tissue from being irreversibly destroyed by herniation through a small craniotomy opening. This is especially important in cases of massive subcortical lesions where the risk of brain edema is high.

e. Postoperative Care. The animal must be observed carefully after the operation both to ensure that the case will be as successful as possible from a purely scientific point of view and in order to assist the animal until it has recovered. Fortunately, the great majority of tract-tracing projects require limited surgical intervention, and the animal is usually able to do well on its own immediately after waking up from the anesthesia. Nevertheless, one should set aside some time for postoperative observation and care of the animal. The animal should be offered drinking water periodically as its level of consciousness rises. The postoperative care also includes removal of excrement, providing a clean layer of bedding under the animal, cleansing secretions away from nose or eyes, and offering friendly contact in general.

B. The Tissue

1. Fixation

Most staining procedures require an initial fixation of the brain tissue, and the quality of the fixation is of great importance for the final result. With the advent of electron microscopy and plastic-embedding techniques, it became possible to evaluate fully the effects of different fixatives on the preservation of tissues and the effects of such variables as mode of administration, osmolarity, and temperature of the perfusion medium. This knowledge is exploited in present-day fixation, even for many light microscopic procedures.

The fixative of choice, which varies for different histological procedures, is usually delivered through the vascular system in the deeply anesthetized animal or immediately after death. The cannula is most commonly inserted through the left ventricle into the ascending aorta, and the fluid is forced into the vascular system by hydrostatic pressure. Unless the spinal cord is being studied, it is advisable to clamp the descending aorta in order not to waste fixative on the lower part of the body. To allow for the escape of the blood and the fixative from the vascular system, an incision is usually made in the right atrium. Vascular perfusion is described in more detail in Chapter 10.

2. Removal of the Brain

Removing the brain from the skull without damage requires experience and special care. Although the brain usually can be removed successfully under simple visual guidance, a dissecting microscope, a jeweler's loupe, or a set of magnifying goggles (2.5–4 × magnification)* are valuable in carrying out the more delicate parts of the dissection such as removing the dura with-

*Dental magnifying loupe. Lactona Products Division, Morris Plains, New Jersey 07950.

out cutting the brain, clipping the cranial nerves, and leaving the hypophysis intact. Good illumination facilitates the dissection, and a well-ventilated work bench should be used to protect the experimenter against prolonged breathing of harmful aldehyde vapors.

There are two basic approaches for the removal of the brain: ventral and dorsal, and the beginner should familiarize himself with both, using animals not intended for scientific analysis. The approach of choice is determined by the preference of the investigator and by the location of the area under study. The ventral approach, often forgotten, is particularly helpful if the area of interest lies at the base of the brain.

Dissecting instruments, e.g., rongeurs, forceps, and scissors, should be selected to suit the specific animal. It is essential to keep cutting instruments sharp, and each investigator should have his own set of the more critical utensils. The brain must be moistened continually with fixative fluid during the removal. It is sometimes useful to let the perfused brain sit in the skull for a few hours or overnight before the dissection if such treatment is not contraindicated by the specific histological technique to be used. To facilitate the mounting of serial sections at a later stage, one side of the brain can be marked.

3. Embedding and Sectioning

Most tract-tracing procedures include embedding and sectioning of the tissue, and a survey of this subject may be useful for the beginner. There are many types of embedding and sectioning methods available. For some types of embedding, several sectioning methods are possible, and vice versa. The method chosen often depends on the specific project. Furthermore, each sectioning method requires a certain tissue consistency which is obtained by a specific pretreatment of the tissue. Thus, while many procedures are available, only the more usual combinations of pretreatment and sectioning methods will be mentioned here. We shall assume that the reader acquires familiarity elsewhere with the technical details of the various sectioning devices such as the rotary microtome for paraffin and thick plastic sections, the cryostat and conventional freezing microtome for frozen sections, the sliding microtome for several kinds of sections, the Vibratome and tissue chopper for soft and unfixed material, and the ultramicrotome for thin plastic sections.

The choice of sectioning procedure depends very much on the consistency of the material, the section thickness required by the analysis, and the histological staining procedure. Only with the high speed of cutting of a tissue chopper or sawing movement of the Vibratome can soft; unfixed tissue be sectioned without any type of support or hardening. Otherwise, it is always necessary to harden the tissue through freezing, fixation, or both. To provide further support for sectioning, the tissue can either be encased in celloidin, paraffin (Chapter 9), gelatin (Appendix, VI.A), agar (Chapter 12;

Appendix, V.C4), or egg yolk, or it can be fully embedded in paraffin, celloidin, or plastic. Embedding in plastic is described in Chapter 10. Protocols for paraffin and celloidin embedding can be found in most standard handbooks of histological techniques (e.g., Bancroft and Stevens, 1977; Gabe, 1976; Romeis, 1968; Thompson, 1963).

a. Frozen Sections. If freezing can be tolerated and sections of about 10–40 μm are required, as in most silver impregnation procedures, frozen sections are preferred. Previously fixed tissue is made sufficiently hard by the freezing to allow sections down to about 10 μm to be cut on one of the several types of freezing microtomes. Sometimes, especially if the brain is small and delicate, encasing in egg yolk (see Ebbesson, 1970, p. 157) or gelatin (Appendix, VI.A) is advisable. Encasing can also help to preserve the mutual relations among parts of larger brains that may become separated during the sectioning or further processing. This is apt to happen, for example, with cerebellar folia and hippocampus or when sections pass tangentially through the olfactory bulb or the collicles. Encasing of the specimen is also useful for left–right orientation of the specimen. Instead of making a cut in the brain itself, one can cut off a corner of the encasing material.

Içe crystal formation is a disturbing artifact that may give frozen sections a vacuolated appearance. This complication can be prevented by immersing the tissue in a 30% solution of sucrose or 20% DMSO in fixative or buffer before cutting. The tissue is ready for cutting when it has sunk to the bottom of the jar, usually after a few days for a rat brain or a brain fragment of similar size. To minimize ice crystal formation, it is important to freeze the tissue as rapidly as possible and to avoid thawing and refreezing during the cutting procedure. There are several models of the traditional freezing microtome on the market, and some types of microtomes can be adapted for freeze microtomy. With a simple freezing stage and the use of a mixture of dry ice and ethanol, a regular sliding microtome can easily be used for preparing frozen sections. A "step-by-step" instruction of a technique for cutting frozen sections on a sliding microtome is presented in Appendix, VI.B.

If thin sections of unfixed material are required, the cryostat is virtually indispensable. With it, complete series of sections, 10 μm thin or less, may be produced even from a fresh brain directly removed from the animal as required for certain enzyme histochemical procedures. This method of sectioning is also compatible with silver impregnation techniques and some other methods, thereby allowing the production of alternate series of sections stained with different methods for correlative analysis. Aldehyde-fixed material can also be sectioned with the cryostat. A few practical details regarding the cryostat and the preparation of cryostat sections are given in Section III.B.4c.

b. Paraffin and Celloidin Sections. Embedding in paraffin and celloidin are classical and useful procedures both for serial sectioning and for the study of fine detail at the light microscope level. Paraffin embedding is chosen if thin sections down to about 4 μm are required and if the staining method to be used will work after the tissue has been dehydrated and cleared as required by the embedding method. To ensure good cutting, it is essential to remove completely all water from the tissue by keeping the last change of the dehydrating agent, usually ethanol (analytical grade), absolutely free of water. This can be accomplished by storing the 100% (absolute) ethanol in a well-closed jar containing a bottom layer of white copper sulfate* covered by a filter paper. The specimen is placed on the filter paper. It is also important to remove the dehydrating agent completely, that is, to clear the block completely with chloroform or toluene before paraffin infiltration. Excessively long treatment in the clearing agent or in the hot paraffin will harden the block more than is desirable. The use of good embedding paraffin is important. If paraffin wax is heated well beyond its melting point before it is used for the first time, its embedding and cutting properties are improved. Excessive wax trimmed away from embedded blocks is placed in a funnel provided with a filter paper. The funnel is placed over the stock of melted paraffin kept in the oven, and the excessive wax is allowed to go back to the stock paraffin. It can later be used over and over again. Such old paraffin cuts better than fresh wax. A dull knife, wrong angle of knife clearance, or uncontrolled static electricity are frequent sources of difficulty in cutting complete series of uncompressed, whole paraffin sections; one should be able to do this routinely.

Celloidin (nitrocellulose) embedding is a more gentle method than paraffin embedding because heating of the tissue can be avoided. Furthermore, much larger sections can be prepared from celloidin-embedded material than from paraffin material. Celloidin sections, however, cannot be cut as thin as paraffin sections and can rarely be made less than 20–30 μm. Celloidin embedding, like paraffin embedding, permits Nissl and myelin staining. In contrast to paraffin embedding, celloidin embedding is appropriate also for Golgi-impregnated material. Celloidin embedding, however, is a slow procedure. After dehydration, the infiltration with nitrocellulose of increasing concentrations takes several days, and the further thickening of the last change of celloidin solution through evaporation must usually be protracted over as much as a couple of weeks in order to obtain a celloidin block of suitable firmness and with good cutting properties.

c. Plastic Sections. Plastic embedding is indispensible for the ultrathin sections needed for electron microscopy. Various types of resin are employed, epoxy and polyester resins being the most common. Further information on plastic embedding and ultramicrotomy is given in Chapter 10. It is worth remembering that resin sections of a thickness from about 0.5 μm to 200 μm can be prepared for light microscopy, and thin plastic sections are excellent for studying fine cytological details. Thinner sections are best obtained with glass knives, whereas thick sections can be cut with a heated

*Water-free copper sulfate is white; when it turns blue because of absorbed water, it is regenerated by heating in a stove or in a crucible on sand in a frying pan.

steel knife. Several light microscopic methods including the reduced silver technique (Heimer, 1970) have been modified for use on plastic sections. Other staining methods for plastic sections used for light microscopy have been described by several authors (e.g., Berkowitz *et al.,* 1968; see also Chapter 10).

d. Alternative Procedures. It is sometimes desirable to cut unfixed or fixed unembedded tissue without freezing. If 50- to 300-μm-thick sections can be tolerated, a very simple sectioning procedure, such as freehand cutting with a razor or razor blade with the tissue block glued to a base, can be used. To be able to do this, however, one must ensure that the tissue be of rather firm consistency. Essentially the same thing can be accomplished with the sliding microtome. More often the task is solved by the aid of a tissue chopper or a Vibratome supplied with a Schick® injector blade. The Vibratome permits preparation of the thinnest sections, down to about 10 μm. These methods are used primarily for histochemical purposes. If a tissue block lacks mechanical stability because it is too soft or too high, it can be supported through "encasing" in a shell of firmer substance as described in Chapter 12 (Appendix, V.C.4).

Although embedding and sectioning techniques are well-developed and standardized for the different tract-tracing methods, there will be instances when an unconventional approach proves useful. Sections for reduced silver methods are usually cut on the freezing microtome from formalin-fixed tissue. However, if the laboratory has a very good cryostat, it is certainly appropriate to use the cryostat for sectioning. If direct correlation between light and electron microscopy is needed, a reduced silver method may be applied directly to plastic sections of tissue fixed and embedded for electron microscopy (Chapter 11). Although frozen sections do not normally give optimal tissue quality when seen with the electron microscope, they may be preferred if required by other methods needed for direct correlation with the electron microscopic picture. In this context, it is worth mentioning that both autoradiography and HRP histochemistry for light as well as electron microscopy are feasible with tissue cut on the Vibratome or the freezing microtome.

In most studies it is important to section the material in various planes, usually frontal, sagittal, and horizontal planes. There are many reasons for cutting the brains in more than one plane. One is that relations that are difficult to interpret in one plane of sectioning may become quite obvious in a different plane of sectioning.

4. Collecting and Storing Sections

For analysis and reconstruction of a pathway or another brain structure, serial sections are indispensable. The methods of collecting and storing serial sections depend on the type of embedding.

a. Paraffin Sections. Such sections are made in long ribbons attached to the immobile knife of a microtome. A microtome with a moving knife is less appropriate for this purpose. After the ribbons are lifted over to trays, every second or third section is cut out with a scalpel; they are mounted on slides in serial order and neatly aligned. Alternate series for different methods may thus be mounted. The mounting of the sections can be done in the following way. Smear a thin layer of albumin–glycerol on the slide. Place the slide on a lukewarm heated plate large enough to hold several slides. Pour a layer of distilled water over the slide. Take one section at a time and place it on the water in the right order. With small sections and large slides, several rows of sections can be placed on the water pool. Place a moist water-absorbing cloth across one end of the slide and tip the slide to let most of the water run into the cloth. When this is correctly done, the sections do not float down to the cloth. The sections can now be precisely aligned with the aid of a needle attached to a shaft. The slide is left to dry on the heated plate and later in a stove at 30–45°C. The heated plate should be sufficiently hot to allow stretching of the sections but not so hot that the paraffin melts.

It is advisable to store the trays with extra sections. The storage temperature should be sufficiently low to prevent the sections from adhering to the tray, and the trays should be well covered to protect the sections against air currents that might cause derangement of their order. The mounting of paraffin sections is time-consuming, and in general it is advisable not to mount more sections than strictly necessary. On the other hand, storage of unmounted sections is risky and therefore impractical. A solution to this dilemma is to study the sections as soon as possible and quickly proceed either to mount further sections or to dispose of the extra sections.

b. Frozen Sections. Frozen sections should be collected in a manner that preserves the serial order. Although some investigators routinely sacrifice sections during sectioning (e.g., every two out of three or four out of five), at times it may be wise to keep all sections, especially if one expects to try difficult staining techniques and modifications. The sections are collected in groups in Petri dishes or in multicompartmental plastic trays, such as ice-cube trays, simple egg trays, or specially made trays.* Each group may contain a restricted number of adjacent sections, in which case a fully serial order is not achieved. It may, therefore, be better to place consecutive sections in an orderly fashion in different compartments: the first section in compartment No. 1, the second section in compartment No. 2, etc. Each compartment will then contain a complete series of evenly spaced sections. In the case of ten compartments, the first compartment will hold section Nos. 1, 11, 21, 31, etc. These sections are sufficiently far apart to be easily arranged in a serial fashion after final processing and mounting. Dish No. 2 will contain sections 2, 12, 22, 32, etc., each readily recognizable by comparison with the already aligned sections of compartment No. 1. Frozen sections

*Special "trans-boxes" are made by Vlchek Plastics Company, Middlefield, Ohio 44062, Item No. P914-2.

kept in formalin solution may be stored covered for weeks or months, preferably in a cool place. Staining properties, however, may deteriorate with time. Again, a quick decision as to whether more stained sections are needed is preferable to indecision and storage.

Processing of frozen sections is done either by transferring each section from one bath to the next or by transferring a batch of up to ten sections at a time. The second method is obviously more efficient and is usually the one used when the different parameters in the staining method have been established. An efficient way to transfer a batch of sections from one solution to another is to place them in a plastic cup or vial in which the bottom has been replaced by a nylon net or another type of solvent-resistent fine mesh net (i.e., nylon stocking). The net can be glued to the plastic by an appropriate adhesive such as RTV® silicon rubber* or melted onto it. This method works well for the silver methods as well as for the different HRP techniques. Frozen sections can also be mounted on coated slides prior to staining as, for instance, in the case of Nissl staining.

c. Cryostat Sections. These sections are cut individually and immediately thawed onto slides, thus enabling them to be processed in groups as are mounted paraffin sections. If the material is fixed, the sections should be mounted on slides coated with chrome alum–gelatin. There is little extra labor involved in preparing alternate series for different methods; consecutive sections are merely mounted on different slides, one set for each series. To prepare and store one or two extra series for reserve purposes is therefore quite realistic. Cryostat sections have the valuable property of staying untorn and undistorted through the entire staining process and are thus excellent for the purpose of three-dimensional reconstructions.

To base large amounts of experimental work on cryostat sectioning is worthwhile only if the cryostat satisfies a number of requirements. First, it must be able to make sufficiently large sections. A good cryostat should be able to produce flat sections 1 inch wide. Second, the user should be able to operate it for hours without becoming fatigued or developing ache in back and arms. Therefore, it must be possible to insert the hands into the microtome room horizontally at the front rather than at the top of the cryostat. Third, the microtome ought to be motor-driven and regulated with the knee or foot to free the user's hands for other purposes.† It is likely that the scientific output of many laboratories could be increased significantly by using cryostat sections for several types of staining of experimental material.

d. Celloidin Sections. Celloidin sections are usually sufficiently durable to withstand handling by a forceps. Thus, they can be processed individually or mounted on slides before processing. Unprocessed sections can be stored

*General Electric Co., Schenectady, New York.

†The cryostat "Duspiva," manufactured by W. Dittes in Heidelberg, West Germany, fully satisfies these requirements, especially when supplied with a "section stretcher" designed by Dr. F.-M. S. Haug, Anatomical Institute, Karl Johans Gate 47, Oslo, Norway. The model made by Ditte's, however, is quite expensive.

between paper sheets that are stacked and bundled together with a piece of string and stored in 80% alcohol. Designations can be written with a conventional lead pencil on the paper sheets, and individual sections can be retrieved later for staining as desired.

IV. ANALYSIS

A. Amount of Material

Energy and perseverance are needed to analyze a large number of sections from many experimental animals. It is usually wise to examine each brain in detail as soon as it is ready for examination. If a large number of brains are prepared without a continuous analysis of the results, one generally misses the opportunity to modify a less successful surgical approach or staining procedure or to complement the material when final processing parameters have been determined. As the analysis of the histological sections is often the most time-consuming part of an experiment, it is necessary to consider this point when planning ahead for research, teaching, and other activities. The preparation of two to four animals per week will often prove to be a normal "load" in tract-tracing experiments, especially if the methodological problems are to be given adequate attention.

B. Analytical Approach

When the sections from a series of experiments have become available, there is a real danger of becoming overwhelmed by the abundance of findings in the individual sections. Virtually every lesion or autoradiographic injection may bring into evidence several sites of termination, and an HRP injection in a specific area will almost always disclose several pathways to that area. But a thorough analysis does not necessarily mean that every positive finding should receive special attention. It is usually a better strategy to concentrate the attention on one specific pathway throughout the entire series of experimental animals. The work can then be more meticulous, and the results are easier to prepare for publication if they are related to a restricted topic rather than to a variety of different pathways or systems. One may even profit from analyzing one single feature in a specific fiber system at a time and studying that feature in great detail through the whole series of sections. Eventually a moment will be reached when the partial observations can be fitted into a total pattern, and a synthesizing conception of the pathway can be attained. Needless to say, there can be no strict rules on how to analyze sections. At times it may even be justified to investigate more than one pathway simultaneously.

Two subjects related to the analysis of the results obtained are of great importance: the study of the normal anatomy of the structures under investigation and procedures for mapping and reconstruction.

C. Normal Material

In the initial stage of almost any type of anatomical study, one should make an effort to acquire a good understanding of the normal anatomy and topography of the regions related to the study. Although an initial gross topographical orientation is often neglected, either because of ignorance of how to deal with the problem or because of a lack of traditions in the laboratory, it is one of the most important steps in a neuroanatomic study, especially for a beginner. This background information can be obtained with the aid of very simple methods such as gross anatomical dissection and brain cutting under a stereomicroscope. The brain dissection may give valuable information regarding the pathways. The study of brain slices can be facilitated by a simple instant thionin staining or by Mulligan's method (see Appendix, VI.C.1a,b). The preparation of schematic drawings and photographs of the appropriate structures at this point may be helpful for the further analysis and may even provide some useful basic illustrations for publication.

The initial orientation of normal anatomy should also include histological observations of relevant areas. For light microscopic studies, it is usually worthwhile to sacrifice one or more animals and prepare several series of sections stained with some of the classical staining methods, e.g., the Nissl method and one of the many myelin stains. A useful technique for orienting purposes in the context of tract tracing is the original Nauta (1950) method which impregnates a large number of normal nerve fibers. Two other techniques for the impregnation of normal fibers are described in the Appendix (Section VI.C.3), one for material fixed with unbuffered formalin, the other for cacodylate- or phosphate-buffered formalin. Obviously, the study of normal material is equally important in electron microscopic as in light microscopic investigations.

D. Mapping and Reconstruction

Mapping is mainly based on histological sections. The extent of a lesion, however, is sometimes recorded photographically before the embedding and sectioning procedures. Photographs of the whole brain are especially useful in the cases of superficial lesions or of a large brain, such as the monkey or human brain, which should be photographed from different angles before it is cut into slabs in one of the standard planes. The slabs are then stacked together and rephotographed so that the relationship of the plane of section to the whole brain is evident. The slabs can then be trimmed and their surfaces photographed before they are sectioned. Although a photograph of the whole brain is very useful if an experimental lesion is visible on the surface of the brain, an accurate drawing of the extent of the lesion is often sufficient for the record.

When mapping and reconstruction of a pathway is to be done on the basis of the histological sections, it is accepted practice to make drawings of regularly spaced sections with the help of an image-projection apparatus. In these drawings, the outline of the section, the major landmarks such as conspicuous fiber bundles, major nuclear groups, large blood vessels, etc. can be indicated. It may even be possible to indicate some of the salient features of the pathways being studied. The section is placed under the microscope, and the distribution of degenerating or labeled neuronal structures is recorded or mapped. Although in principle one may use a camera lucida (microscope drawing tube) instead of a projection apparatus as an alternative for making drawings from serial sections, this is in practice very slow and cumbersome except in cases where small areas are studied at high magnification.

To facilitate the mappings, X–Y recorders connected to potentiometers attached to the microscope stage are used in some laboratories. The movement of the recorder pen is graded with the power of the lens so that the investigator can record on the same map while scanning the sections both at low and high magnification. Some neuroanatomists who have an X–Y plotter prefer nevertheless to enter their observations manually in the drawings.

When an anatomical study is ready for publication, the results obtained are usually documented with line drawings and photographs. The quality of the illustrations is of crucial importance, especially as the prospective reader usually has no opportunity to inspect the original material. The prospect of producing first-rate illustrations is improved if the photographic recording of interesting data and striking results becomes an integral part of the continuing analysis.

It has been common practice in the past to illustrate a hodological paper with a series of line drawings in which the major findings are indicated and summarized with the help of various symbols and to restrict photographic illustrations to critical histological features. Although the drawings are usually a necessary and valuable part of the illustrative material, photographs represent a more immediate and exact documentation. It is therefore appropriate to encourage the use of photographs whenever practical. Many of the modern tract-tracing techniques produce sections of high contrast suitable for low-magnification photography (e.g., Fig. 4, Chapter 4 and Figs. 2A and 3, Chapter 6). A series of such photographs is likely to constitute more reliable documentation than a series of line drawings of the same structures, since the drawings may have a considerable element of subjectivity despite the experimenter's efforts to the contrary. To prepare successful photographs at low magnification necessitates special photographic equipment. Unfortunately, the high cost of printing halftone pictures usually limits the number of photographs in a paper, and line drawings will therefore con-

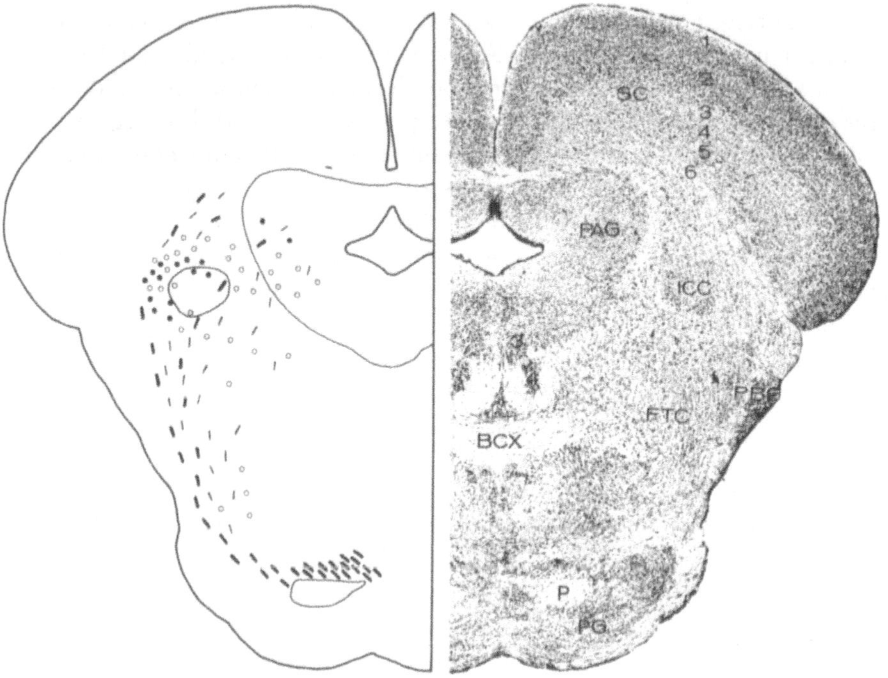

Figure 2. Mapping of degeneration in a brainstem section of the tree shrew. Broken lines, degenerating axons. Dots and circles, terminal degeneration (of afferent fibers from two different sources). [From Schroeder and Jane (1971) with permission from the Wistar Press.]

tinue to be an important part of the documentation in publications dealing with the experimental tracing of pathways. Line drawings should be produced with great care. Published drawings are often too small or too simplified to be sufficiently informative. For such reasons, misunderstandings are especially common if the interpretation and delineation of the brain areas under investigation vary among different authors. One way to solve the problem is to illustrate the data as shown in Fig. 2.

Three-dimensional reconstructions are of great help for understanding spatial relationships. Although such reconstructions can be done from serial sections stained in different ways, they are often prepared on the basis of Nissl-stained material. One may either prepare *solid* three-dimensional models and later make illustrations showing these or make *graphic* reconstructions, i.e., produce illustrations showing three-dimensional relations without first preparing a solid model.

When making reconstructions, the series of sections should be supplied with reference points in order to facilitate alignment of each section against the next. This can be done by cutting a notch along one side of the tissue or drilling a fine hole through frozen tissue before sectioning it in the cryostat. If the tissue is embedded, the embedding material can be marked in a similar fashion.

Solid three-dimensional reconstructions in wax, plastic, cardboard, or other material showing salient features of a studied region are extremely useful for orientation purposes. A mere turning of a solid model may replace a series of graphic reconstructions from different angles, and, on the basis of such a model, one may produce photographs and drawings both for recording and presentation of experimental results. It is our experience that time invested in such work is amply repaid by the advantages obtained.

We give no detailed instruction in this book on the production of three-dimensional models but refer the reader instead to the literature. Fortunately, the procedures are simple, and initiative and interest are more important than recipes. In principle, enlarged profiles of the structure to be reconstructed are made, cut out, and stacked suitably spaced with reference points in register to ensure proper alignment. If the drawings are enlarged n times (e.g., 150 ×), and every mth section (of d thickness) is used, the distance between the upper faces of adjacent plates is $d \times m \times n$, (e.g., $10 \mu\text{m} \times 5 \times 150 = 7500\ \mu\text{m}$, i.e., 7.5 mm). Any type of plate material that can be cut is usable. An important factor for the choice of plate type is the shape of the model: long, thin, or hollow models require a more rigid material than do solid, rounded structures for which soft wax or similar material is sufficient.

If a plate thickness precisely fitting the desired magnification is not available, the magnification may be adjusted to the available plate dimension, or the distance between thin plates may be adjusted with spacers glued between the plates.

Simple models may be made from cardboard suitably separated by spacers, the rest of the room between the profiles being filled with wax or other material. Useful models can be made from rigid, transparent plastic plates that are not glued together but joined by bolts through holes in the plates. Such models may be split temporarily for inspection of features in their interior.

If wax plates are used, the profiles to be reconstructed are drawn on very thin paper which is placed on the wax plates; the profiles are then cut out of the plates through the paper. Styrofoam® is a suitable material because it is light and can be cut with a heated needle. Whatever material has been used for the reconstructions, the surface of the model can be smoothed by filling the steps between the plate edges with wax or another suitable material.

When the profiles are drawn, interesting features such as limits between nuclei or cortical areas should be marked. The completed model then becomes a three-dimensional representation not only of the total structure but also of the relations among its individual components. A finished model should be labeled with magnification and a description of what it shows, and it should be mounted on a stand for protection.

Graphic reconstructions are described in more detail in the Appendix (Section VI.D). It is a basic requirement, even in this case, that sections be aligned against each other. However, if a high degree of accuracy is not required, it will often suffice to align neighboring sections by comparing their total outline and a number of details in the tissue. Plotting the lesion and the distribution of the pathway onto a simple kind of graphic reconstruction will be more valuable than having no reconstruction at all.

Graphic reconstructions permit research results to be presented in a more condensed form than does the illustration of only a single section. For example, the relations between lesions and associated areas of degeneration may be accommodated readily within a single figure. Graphic reconstructions based on alternating and differently stained series of sections may greatly increase the amount of information. For example, distribution of labeled axons may be related to nuclei and areas visible in Nissl-stained material.

V. THE NEUROANATOMICAL LABORATORY

This chapter has primarily been written for the younger neuroanatomist who normally does not make decisions related to the facilities of a laboratory. Within the frame of an already existing department, however, a younger scientist does have the possibility of constructing devices, suggesting rearrangements, and generally influencing the decisions of the responsible leader; therefore, we shall include some suggestions concerning the arrangement and use of a neuroanatomical laboratory.

A. Animal Quarters and Surgery

Animal quarters in biological research institutions are increasingly located in centrally placed facilities, supervised by professional personnel, and subject to governmental controls and regulations. In other places, however, the animal quarters are part of the individual laboratory; in such cases the condition of the facility is largely dependent on the persons using it. The active young neuroanatomist is likely to come there frequently, and it is reasonable to expect that he share in the efforts to keep it in good order. If animals are purchased from an outside dealer, one should not buy more at a time than needed. Keeping the number of animals as low as possible requires less work and provides a better opportunity to take care of the existing animals. It is also cheaper to let the dealer store and feed animals that are not earmarked for immediate use. For work with special or expensive animals, it may be better and cheaper to cooperate with a department that is already equipped to maintain such animals. It may often be possible to improve on the practical arrangements for feeding animals, changing their bedding materials, or washing cages. Automatic watering of the animals saves much work and is good for the animals, whereas flat dishes for water and food are regularly overturned. Install automatic on and off switches for the room lighting. In case of renovation, consider placing cages on transportable racks on wheels to save work and promote cleanliness. Minimizing the smell of excrement and urine is more a matter of adequate cleanliness than of strong ventilation.

A department engaged in neuroscience research must always have a room or an area reserved for *animal surgery*. It should have good general lighting and possibilities for additional strong illumination of the field of operation. The requirements for adequate lighting can be satisfied without buying the expensive types of lamps used in human surgery.

Misuse of surgical *instruments* is often a source of irritation, even in laboratories with strict discipline. It is absolutely necessary that each operator be responsible for immediately cleaning and placing in order the instruments used during an operation. Even if such a routine has been decided on, it is desirable that each user of the surgery room have his own set of the more critical instruments, such as scalpels, fine scissors, hemostatic and watchmaker's forceps, perfusion cannulae, injection needles and syringes, suture material and needles, expensive rongeurs, homemade special clamps, rubber gloves, and perhaps a few more items. Remarkably few instruments are needed in operations on small animals, and each person will soon discover what is needed for his or her specific type of work. Such personal equipment should be stored in a cabinet, drawer, or box that can be locked with a key. This will allow each user to know exactly in what state his instruments are. The extra expenses incurred in having individual sets of the more critical instruments are negligible compared to the advantages. Even with regard to equipment for perfusion, it may be practical that each user have some of the equipment to himself. A simple perfusion setup is described in Chapter 10.

B. Optical Equipment

Although some methods, such as fluorescence microscopy and electron microscopy, require specialized equipment, the major part of the work in neuromorphology is based on ordinary equipment available in most neuroscience laboratories.

The great usefulness of low-magnification equipment is often overlooked. Many operations involve manipulating small structures that can conveniently be observed with a simple hand lens or a set of magnifying goggles, thus freeing the hands of the surgeon. The next step of sophistication is a "magnifying lamp" consisting of a 10–15 cm wide magnifying lens with an attached circular fluorescent tube. Specially designed dissecting microscopes are very useful in a neuroanatomic laboratory. Such microscopes should be versatile with regard to magnifications, working distance, and illumination, and they should be capable of providing both incident and transmitted light.

Ideally, every neuronanatomist should have his own *research microscope*, allowing him to work independently of other students. The competent use of the microscope is more important than the construction, once the basic requirements of an adequate light source, condenser, and apertures have been satisfied. Too many investigators abuse their microscopes by not securing illumination according to the fundamental Köhler principle. As a rule, the expenses of equipment for polarization and phase-contrast microscopy can be avoided, since these methods are very rarely used by the tract-tracing neuroanatomist.

The *electron microscope* is an indispensible instrument in a modern department of anatomy or neuroscience, but this is not the place to deal with choice of type. The beginner must watch experienced colleagues before handling the microscope and must follow the established rules for the use of the instrument with absolute discipline, since deranging it in the least will stop or impede not only his own work but also that of others. Specimen preparation for electron microscopy can be efficiently carried out only relatively near the instrument itself; if a laboratory of neuroanatomy is without an electron microscope, the investigator should transfer his specimen preparation and EM darkroom work to an institution where an electron microscope and experienced people to guide him are available.

Earlier parts of this chapter have emphasized the need for making drawings of relevant structures in light microscopy by means of a *projection apparatus.* At least one projection apparatus is absolutely necessary in any laboratory devoted to tract tracing. When selecting a commercial model or constructing an instrument, one should take care that there be a very strong light source and a versatile condenser system, allowing sufficiently large image fields and magnifications. In addition, the lenses must not give distortions at the magnifications of interest. If *one* apparatus is to be used for both small and large magnifications, it must be possible to remove temporarily the ocular tube.

Equipment for *photomicrography* belongs among the most important optical items of the laboratory, but available options are too numerous to be discussed here. Large pictures at low magnifications make the greatest demands on the equipment. Still, it is a misunderstanding to think that modern expensive equipment is any guarantee of success. Looking at low-power photomicrographs made 50 or more years ago will demonstrate that it is the correct use of equipment and photographic material that is most important. Small pictures, particularly at higher magnifications, are more easily made and are often useful for recording and documentation during microscopy. To facilitate the work, it is practical to have a 35-mm camera mounted directly on the research microscope. Regarding *enlargers,* the lens quality is of paramount importance and should be thoroughly checked before the equipment is purchased.

C. Laboratory Precautions

1. Equipment Care

We strongly advise the beginner to become thoroughly familiar with the equipment in the laboratory, no matter who is officially in charge of servicing it. First, many equipment breakdowns arise from improper use; second, the majority of breakdowns are not difficult to prevent or to repair on the spot. Thus, the effort one invests in learning how an instrument works and how it should feel and sound when operated will result in fewer breakdowns and fewer and shorter interruptions to one's work. Furthermore, when a problem is solved immediately, the instrument is available to the next user without delay, and expensive, unwarranted work interruptions are prevented.

It goes without saying that a critical amount of discipline must go along with the curiosity; both qualities should be encouraged. There are a few good books and a great many instruction sets to help the industrious student. One, addressed to a variety of the instruments common in the histology laboratory, is written by W. Burrell (1978).

2. Regulations

Radioactive isotopes are used only in restricted areas of the laboratory that are strictly monitored and checked for radiation on a regular basis. Possible accidental spills should be reported to the local Radiation Safety agency and decontaminated immediately.

Users of radioisotopes should abstain from drinking, eating, smoking, or applying cosmetics or hand creams around these substances.

Some drugs used in the neuroanatomical laboratory, especially anesthetics, e.g., barbiturates and chloral hydrate, may be regulated by special government agencies in some countries and will then require careful bookkeeping and locked storage. Needless to say, these rules should be carefully respected. The credibility of a whole laboratory can easily be jeopardized by a single careless person. To reduce such bookkeeping one may replace regulated anesthetics with freely permitted ones such as urethane. The latter, however, should be used with care because it bears some carcinogenic properties.

Disposal of animal carcasses and chemical wastes should also conform to local regulations.

Finally, everyone should be aware that several agencies awarding research grants make the principal investigator or laboratory director responsible for compliance to strict rules. Failure to observe regulations may include the risk of loss of financial support or necessary licences.

3. Safety

Safety in the laboratory is a very important issue. Since many of the chemicals used in the neuroanatomical laboratory are explosive (gas mixtures), or toxic (solvents, osmium), may produce allergic reactions (aldehydes, resins), or involve an increased risk of developing cancer (benzidine, radioactive substances), care should be taken not to expose oneself and others unnecessarily to such substances.

Laboratory accidents do occasionally occur. Prevention, however, can be very effective, and there is no need for exaggerated anxiety. Again, awareness and discipline are important. Good ventilation, fume hoods, goggles, gloves, proper laboratory clothing, and, on special occasions, gas masks should be used whenever appropriate. Work surfaces should be kept free of clutter and cleaned regularly. Spills should be cleaned up immediately and thoroughly. It is also important to become familiar with local safety procedures, location of emergency showers, eye flushing equipment, and first aid kits.

Animal perfusions should be carried out in a hood or, if possible, on a well-ventilated, drained sink or table (see Chapter 10). The latter, not available commercially, has the advantage of being accessible from all sides, facilitating cooperation of assistants and demonstrations to onlookers. It consists of a stainless steel plate with an elevated rim and eccentrically located sink covered by a perforated plate onto which the animal is tied. The sink area is aspirated through side holes protected from the fluid spill. A powerful motor is necessary to provide sufficient aspiration. Venting is better done directly to the outside so that return of vapors is impossible.

VI. APPENDIX

A. Encasing of Tissue for Preparation of Frozen Sections

Many small brains require support during sectioning. This can be accomplished by embedding the tissue in paraffin or celloidin or by encasing it in various media appropriate to the sectioning procedure (see Chapter 12). Encasing in egg yolk or albumin–gelatin (Ebbesson, 1970; Crane and Goldman, 1979) has become quite popular in laboratories where frozen sections are needed for silver impregnation, HRP staining, autoradiography, or other conventional staining procedures.

Encasing in gelatin has several advantages. It affords a clear view of the brain, thus facilitating the initial orientation of the brain as well as the trimming of the block. Furthermore, in contrast to egg yolk and albumin–gelatin, it does not stain intensely with Nissl stains, thereby providing a more suitable background for photography. To avoid wrinkling of sections during mounting, it is advisable to pass them through an acetone, ammonia, acetate–buffer–ethanol sequence as described by Switzer (1979).

1. Encasing in Gelatin

To prevent detachment of the gelatin from the perimeter of the tissue, it is useful to remove the pia carefully from the brain using the rubber end of a pencil.

1. Slowly add 12 g of gelatin to a mixture of 88 ml distilled water and 26 g sucrose under constant stirring.
2. Incubate the mixture in a closed container at 60°C for 30 min or until the mixture is clear.
3. Remove the foamy layer from the surface.
4. Pour a little of the mixture into a cup and allow to harden in the refrigerator.
5. Place the brain in the solidified base of the cup in the desired orientation.
6. After the brain is firmly attached to the base, slowly pour the remainder of the mixture into the cup, avoiding the formation of bubbles.
7. Leave in the refrigerator until the gelatin has hardened.
8. Remove the casket and leave in a mixture of perfusion fixative and 30% sucrose for 2–3 days.
9. Cut immediately or change to 30% sucrose–buffer solution.

B. Preparation of Frozen Sections on a Sliding Microtome

The cutting of good frozen sections requires proper equipment and some training. Several microtome models are especially designed for the cutting of frozen sections. A cryostat is especially valuable if the sections are to be mounted immediately onto slides. The following technique for cutting frozen sections is based on the use of dry ice and a regular sliding microtome on which a freezing stage (Fig. 3) made of aluminum can be attached.

1. Pour 95% ethanol into the trough. The fluid should be about 0.5 cm deep.
2. Trim the tissue in order to make a flat base. This base should be blotted dry before it is fixed to the stage.
3. Break dry ice into small lumps and fill the trough with them.
4. Put embedding matrix* on the stage. Plane down (or smooth out) the frozen matrix with a knife that is reserved for this purpose only.
5. Wet the base of the tissue block with the embedding matrix and then put the block on the frozen matrix that covers the stage. If the brain block is larger than the base of the frozen matrix, add more matrix to support the part of the block that extends beyond the base.
6. When the bottom of the block begins to freeze, i.e., the tissue begins to turn white, add a small amount of powdered dry ice around the tissue. As more and more of the tissue turns white, add more layers of dry ice. In the end, the entire tissue block should be covered with dry ice.
7. Allow the block to become completely frozen. This usually takes 20–30 min for a rat brain.

*No. M-1 embedding matrix for frozen sections. Lipshaw Manufacturing Company, 7446 Central Avenue, Detroit, Michigan 48210.

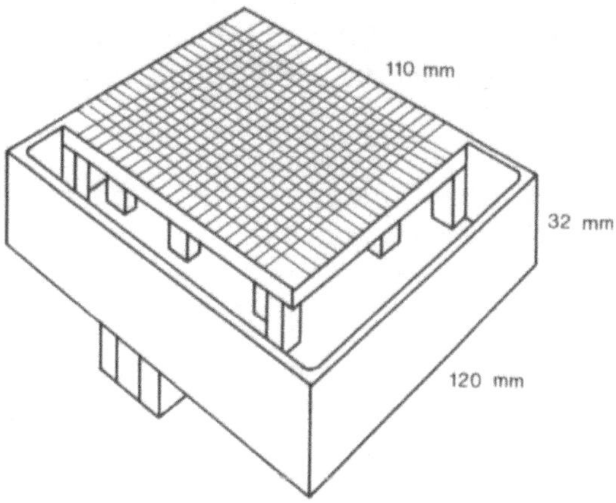

Figure 3. Freezing stage of aluminum.

8. Lubricate the sliding areas of the microtome and attach the knife.
9. Remove the ice powder and bring the block up to the level of the cutting edge of the knife. Adjust the thickness of the section and begin cutting sections. Draw the knife through the tissue with a continuous slow stroke. Maintain an even pressure on the knife.
10. Remove the sections from the knife with a small brush. Keep the knife edge free of the fluid that is on the brush.
11. Frequently lubricate the sliding surfaces of the microtome to ensure smooth cutting action.

C. Special Staining Procedures

Many conventional staining procedures can be used to great advantage as a complement to the specific tract-tracing techniques described in this book. Descriptions of such procedures can be found in most handbooks of histological technique (e.g., Bancroft and Stevens, 1977; Gabe, 1976; Romeis, 1968; Thompson, 1963).

In this section, only a few, mainly unpublished staining procedures will be described.

1. Staining of Cut Surfaces

a. The Mulligan Method (Mulligan, 1931). This is a method for the staining of cut surfaces of aldehyde-fixed brain, rendering gray matter bluish-black and white matter white. It is most often used on human brain material in dissection courses for medical students. In the slightly modified recipe given below, the volumes of the solutions have been adapted to smaller brains (rat, guinea pig). The basic principle of the method is that in a pretreatment the myelin is partly dissolved and smeared over the surface as a film that will protect the white matter against the dark stain produced in the last part of the procedure. If carefully done, this method can produce fine details detectable only in the dissecting microscope.

Stock Ingredients

(A) $CuSO_4 \cdot 5H_2O$	5% in hydrochloric acid, 1.25%, 50 ml
(B) Carbolic acid	20 g dissolved in 430 ml of water
(C) Iron alum	2%, 500 ml
(D) Tannic acid (powder)	5 portions, each 2 g

The stock solutions (A–C), which are made up with distilled water, can be kept in the refrigerator without deteriorating.

Staining Solutions

First solution	Mix 10 ml of A with 90 ml of B
Second solution	Solution C, unchanged (*quantum satis*)
Third solution	Dissolve 1 portion (2 g) of D in 100 ml of water

Staining

The surface to be stained must be cut with a sharp edge (razor blade, scalpel); a rough surface will give blurred staining. Keep the piece for 1 min each in the first, second, and third solutions, washing in distilled water for 1 min after each solution. The stained material can be kept in formalin; slices may be covered with gelatin and mounted in serial order on a glass plate and then kept in formalin. Counterstaining of Mulligan-stained slices with acid thionin is possible.

b. Acid Thionin Stain. This method, like the Mulligan method, works well on aldehyde-fixed material. The basic principle of this method is that the pH of a solution of thionin (or another thiazin dye) is kept so low that it will only stain the nucleic acids of the cell nuclei and Nissl substance. Cells in several layers under the surface may be discerned with the dissecting microscope, and a combined picture of gross features and cytoarchitecture is obtained.

Depending on the fixative used, or if it is desired to stain other tissue components for orientation purposes, the acidity of the stain may be varied somewhat. The following solutions will normally be appropriate.

Stock Solution

Ethanol (absolute)	60 ml
Distilled water	300 ml
Acetic acid (glacial)	10 ml
Thionin	4 g

Staining Solution

Stock solution	2 ml
Acetic acid (glacial)	1 ml
Distilled water	50 ml

Staining

Immerse pieces with cut surfaces in the stain for 1–2 min and then rinse in water. Stored pieces can be restained.

Pretreatment

To render white matter lightly brownish and to minimize staining of neuropil background, a pretreatment for 1 min with a solution containing equal parts of ethanol (95%), acetone, acetic acid, and water is often useful. Afterwards, rinse in water for 1–3 min and then stain.

2. Nissl Method for Free-Floating Sections

Staining. Dissolve 1 g cresyl violet acetate in 200 ml distilled water and adjust pH to 3.0–3.5 with glacial acetic acid. Well-fixed tissue should be washed in distilled water at least overnight and preferably for 48 hr.

Procedure

Reagent	*Time*	*Remarks*
Distilled water	At least 15 min	
1% HCl in 70% ethanol	5–15 min	
Staining solution	Until dark	Trials preferably
Tap water	Short wash	Change water after each section
Distilled water	5 min	Time not too essential
70% Ethanol	5 min	Shrink sections evenly
95% Ethanol	5 min	
95% Ethanol with 10% acetic acid (100:5)	Until darker areas begin to differentiate	When evaluating differentiation, take into account loss of stain in creosote
95% Ethanol	5 min	
Concentrated butanol	Until red background color begins to fade	
Creosote	Until sections soften	Stain fades in creosote
Mount on slides and coverslip		
100% Ethanol		Lift coverslip slightly several times
100% ethanol–xylene (1:1)	5 min	Lift coverslip slightly several times
Xylene	Short wash	Lift coverslip slightly several times
Mount and coverslip with Permount		

3. Silver Methods for Normal Axons

a. Nauta Silver Stain Modified by Vogt (1974). This recipe, which does not work well on tissue perfused for the cupric silver technique, is a slightly modified version developed by Blair Turner and Margaret Krapp for use on the monkey. However, it also works well on the rat and the guinea pig.

Perfuse with a 0.9% saline followed by 10% neutral formalin. Although Vogt suggests that best results are obtained if the uncut brains are stored in 10% formalin for 6 months, equally good results can be obtained with a 2–3 day fixation of the brain following perfusion provided the sections are transferred to the silver–pyridine solution the same day they are cut. Following the initial fixation, the brain is transferred to fixative containing 30% sucrose. When it has sunk to the bottom, it is cut at 20–30 μm. Sections may be either free-floating or mounted and air dried onto gelatin-coated glass slides.

Staining Procedure

1. Rinse sections for a total of 10 min in three changes of distilled water.
2. Place sections in a solution containing 8 volumes of 2.5% silver nitrate and 1 volume of pyridine. Keep the sections in this solution overnight. Put unmounted sections in small bottles containing this solution and then place on a tissue rotator.
3. Rinse sections for a total of 10 min in three changes of distilled water.
4. Transfer sections to a freshly prepared ammoniacal silver nitrate solution. Keep unmounted sections in this solution for about 1 min and carry through steps 4 and 5 individually with continuous motion. Keep mounted sections in the ammoniacal silver for approximately 8 min. Agitate several times.

   Ammoniacal silver nitrate:
   100 ml 4.5% Silver nitrate
   33 ml 95% Ethyl alcohol
   9 ml Concentrated ammonia
   Shake until solution becomes clear, then add 4 ml 4% sodium hydroxide, and shake again.

5. Without washing, place sections in two successive solutions of reducer until they turn a deep brown color (1–3 min). Agitate several times and allow most of the reduction to take place in the first reducer.

   Reducer:
   27 ml 10% Formalin
   90 ml Absolute ethyl alcohol
   27 ml 1% Citric acid
   Add distilled water to 1100 ml.

6. Rinse twice in distilled water.
7. Differentiate in modified Weigert's solution for 15–20 sec while gently agitating the solution. Continue differentiation until the fine fibers of the cortex are clearly seen. Fibers will appear black on a clear or light tan background. Loss of stained fibers will occur if the sections are kept too long in this solution. Differentiate unmounted sections individually in order to avoid folding and to ensure a consistent background color.

   Modified Weigert's solution (freshly made):
   1 g Potassium ferricyanide
   2 g Sodium borate
   per 100 ml distilled water

8. Rinse three times in distilled water.
9. Fix in 1% sodium thiosulfate for 1–2 min.
10. Rinse three times in distilled water.
11. Mount free-floating sections in gelatin–alcohol and blot or air dry.
12. Dehydrate and coverslip.

b. Silver Stain According to J. A. Spada, C. Diaz, and J. S. de Olmos. This method is specifically developed for material perfused according to the directions given in the cupric silver procedure (Chapter 4, Appendix, VII.C). The result obtained by the method is shown in Fig. 4.

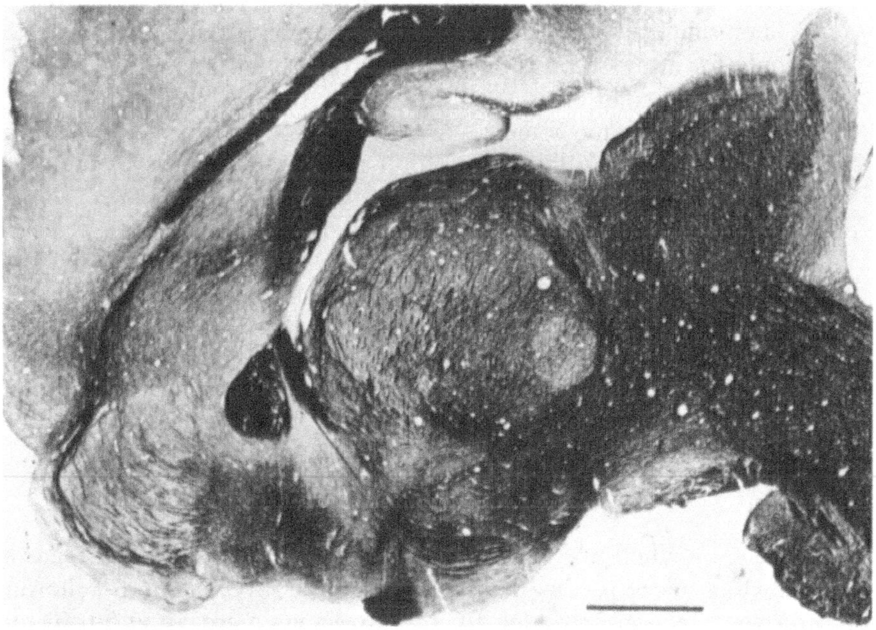

Figure 4. Sagittal section through the rat brain stained for normal axons according to the silver method by Spada, Diaz, and de Olmos. Bar, 2 mm.

1. Wash the sections in distilled water for 24 hr.
2. Place them in 10% $AgNO_3$ for 5 min.
3. Treat the sections in pure acetone until dehydration (shrinkage) is complete (10–30 sec).
4. Without washing, transfer the sections to an ammoniacal silver nitrate solution until rehydration is complete (tissue assumes original shape). This takes about 30 sec.

   Ammoniacal silver nitrate solution:
   9 ml 20% $AgNO_3$
   31 ml Distilled water
   20 ml Absolute alcohol
   2.4 ml Concentrated ammonia
   2.8 ml 2.5% NaOH
   Add the ingredients in this order.
5. Reduce the section for 5 min

   Reducer:
   400 ml Distilled water
   45 ml Absolute alcohol
   13.5 ml 1% Citric acid
   13.5 ml 10% Formalin
6. Wash in distilled water.
7. Transfer the sections to a 0.5% potassium ferricyanide solution until the tissue loses its red color and fibers can be seen.
8. Wash thoroughly in distilled water.
9. Return to step 4 and repeat the sequence 1–2 times for further fiber staining. Keep the sections in the ammoniacal silver solution only for a few seconds or until they turn black. Inability to remove all the red color from the tissue means that the sequence has been repeated once too often.
10. Fix in 1% sodium thiosulfate for 1 min.
11. Put in 95% alcohol.
12. Treat the sections twice in a solution of equal parts of 95% alcohol and acetone.
13. Put in creosote, mount, blot dry, and coverslip.

4. Myelin Method According to J. S. de Olmos

This is a modified Weigert technique especially designed for material fixed according to the prescriptions for the cupric silver method. Following fixation, frozen sections intended for the myelin stain are cut at 30–40 μm. The sections can be stored in the fixative for 2–60 days. The result obtained by the method is shown in Fig. 5, and a flow diagram of the method is presented in Fig. 6.

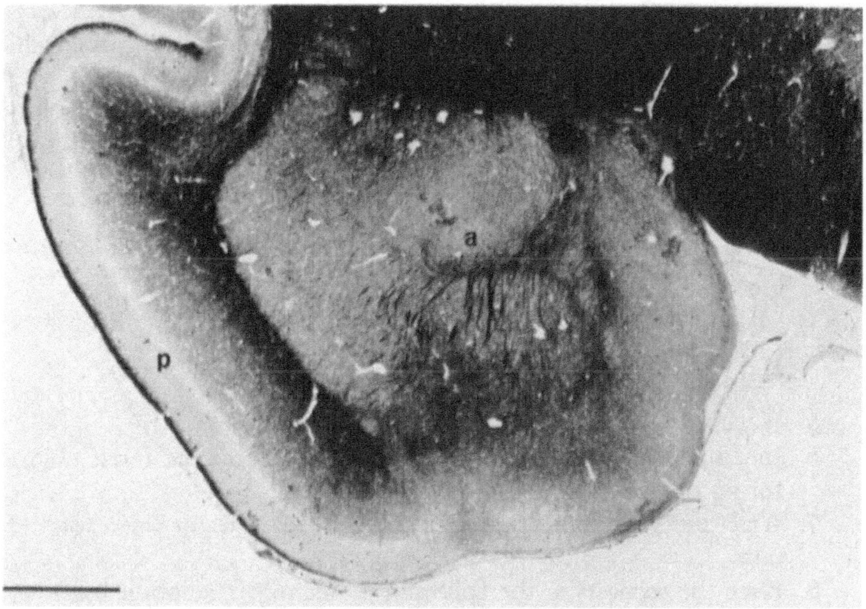

Figure 5. Frontal section through the amygdaloid body of a normal guinea pig stained with the myelin method by de Olmos. **a,** amygdaloid body; **p,** piriform cortex. Bar, 1 mm.

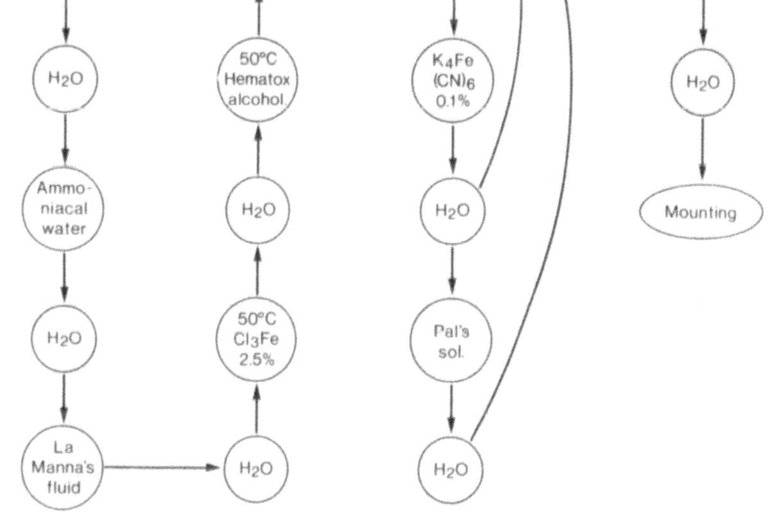

Figure 6. Flow diagram of the myelin stain by de Olmos.

Staining Procedure

1. Wash the sections at room temperature for 12–24 hr in distilled water.
2. Place the sections for 24–48 hr at room temperature in a solution of 8 drops of concentrated ammonium hydroxide in 100 ml of distilled water.
3. Wash thoroughly in distilled water.
4. Incubate for 24–40 hr at room temperature in La Manna's solution.

   La Manna's solution:
   9.5 g Potassium dichromate
   4.5 g Zinc chloride
   Distilled water, up to 100 ml.
5. Wash in distilled water.
6. Incubate the sections in a 2.5% aqueous solution of ferric chloride for 20 min at 50°C.
7. Wash the sections in two changes of distilled water for a total of 5 min.
8. Place the sections in the following hematoxylin solution for 1 hr at 50°C. Combine 1.5 ml 10% hematoxylin in absolute ethanol, 45 ml double distilled water, and 0.5 ml glacial acetic acid. Prepare the solution fresh every time and filter before use.
9. Wash in tap water for 5–10 min and proceed with the rest of the protocol at room temperature.
10. Differentiate in a solution of borax, 0.5%; sodium carbonate, 0.5%; and potassium ferricyanide, 0.5%. Bleach the sections until the major myelin bundles become visible.
11. Wash in distilled water.
12. Treat the sections in 0.1% potassium permanganate for 30–40 sec.
13. Wash in distilled water.
14. Repeat steps 10–13.
15. Treat the sections in fresh Pal's solution containing 1 g sodium or potassium sulfite and 1 g oxalic acid in 200 ml distilled water. Leave the sections in the solution for about 30 sec or until the myelin acquires a blue color. If sections are bleached too long, fine fibers will disappear.
16. Put the sections back in the solution of step 10 for 5min.
17. Wash twice in tap water.
18. Wash in tap water to which a few drops of lithium carbonate have been added.
19. Wash in distilled water.
20. Mount in a solution of equal parts of 2% gelatin and 60% ethanol.
21. Blot the sections carefully with filter paper and let dry in room temperature.
22. Dehydrate in 95% ethanol.
23. Clean the sections in creosote solution and coverslip.

D. Graphic Reconstructions

Theodor W. Blackstad

Four procedures for graphic reconstruction of anatomical structures will be mentioned. To facilitate comparison of these procedures, all illustrations will be based on the same material: a series of 33 equidistant sections through the fastigial nucleus of the cat.*

It is not possible to describe the reconstruction procedures fully in few words, and uninitiated readers are urged to familiarize themselves with the procedures by applying them to a few serial sections; 4–5 simple profiles will suffice.

It is appropriate to point out some morphological characteristics of the fastigial nucleus used as an example. The nucleus is somewhat elongated in the rostrocaudal direction. It is thicker anteriorly, taillike caudally. It resembles a tetrahedron, the top being the thinner caudal end, and the base the anterior end of the nucleus. Medially, the nucleus has a longitudinal keellike ridge where two relatively plane faces of the tetrahedron meet. These are the larger basal face and the smaller dorsal face. The lateral (dorsolateral) face, which opposes the nucleus interpositus, has a furrow in its rostral half. The dorsal face is bent: it is concave rostrally, convex caudally. The basal face is slightly concave in the caudal half of the nucleus and convex at the rostral end.

1. Vertical Projection

This method simply consists in copying the profiles, one at a time, onto a sheet of tracing paper placed over the profiles while checking that the reference points are made to coincide from one profile to the next. Copying the profiles with a pantograph allows a simultaneous change of magnification. Only those parts of each profile that are not covered by the profiles already copied are drawn. By starting from one end of the series, a picture of the brain structure seen from this end will result (Fig. 7). The drawing starting from the opposite end will show the structure from that end as a mirror image which can be reversed photographically. By starting the tracing from a larger profile in the middle, considering it to be an opening, and including the drawing profiles visible through this opening, a partial view of the interior relief of a structure can be obtained. Horizontal and oblique surfaces are well visualized in pairs of vertical projection reconstructions (Fig. 7; see also Haug *et al.*, 1971). Since this type of reconstruction can be made in a matter of minutes, and since no special equipment is needed, the method is recommended as an initial routine before other procedures of reconstruction are implemented.

*This series was prepared for other purposes by Dietrichs and Walberg (1979). The outlines of the nucleus (the *profiles*) with the reference points for alignment were drawn at a magnification of 17.5 ×. We are grateful for the permission to use this series here.

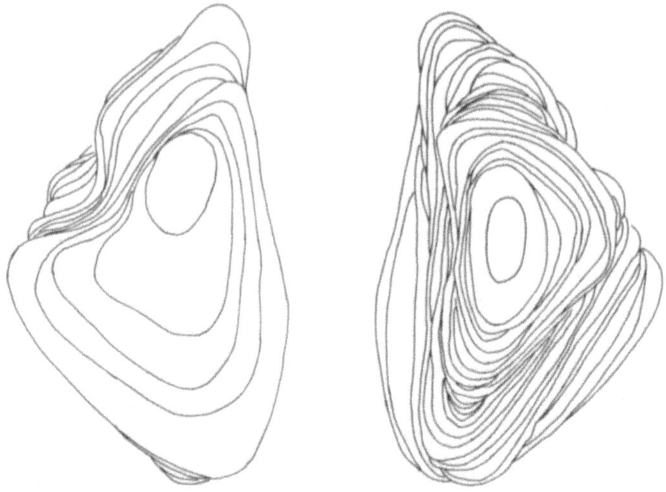

Figure 7. The fastigial nucleus reconstructed by vertical projection and seen from its rostral end (left) and its caudal end (right). The right drawing was reversed photographically (see text). All 33 sections were employed. Most of the morphological features described in the text are fairly well recognizable.

2. Lateral Projection

The sections (profiles) are placed, one at a time, under a sheet of tracing paper supplied with a set of parallel lines (one per section) perpendicular to the desired viewing direction, as shown in the upper left of Fig. 8. To avoid confusion, these lines will be called *transverse.* The reference points, marked off on the tracing paper, must coincide precisely with the reference points on each profile drawing. The transverse lines are separated by distances equal to the true distance between the sections × magnification. The edges and other interesting parts of each profile are projected onto the appropriate transverse line. When all sections have been treated as shown in Fig. 8, a drawing of the whole structure is made with the aid of the marks on the transverse lines, preferably while simultaneously consulting the original profile drawings. Limits between subnuclei, cortical areas, etc., may be established by drawing a continuous line through the individual marks corresponding to such features, or the surface relief including ridges and grooves may be indicated by imagining the reconstructed object to be illuminated from a selected angle and by entering varying degrees of shade (Fig. 8, right).

The lateral projection may, of course, be done towards any direction, and all sides of a structure may thus be displayed. In this sense, the method may be said to permit rotation but not tilt (see definitions, Section D.3 and with Fig. 9). Contrary to the results obtained by the vertical projection method, surfaces parallel to the plane of sectioning will be most poorly represented.

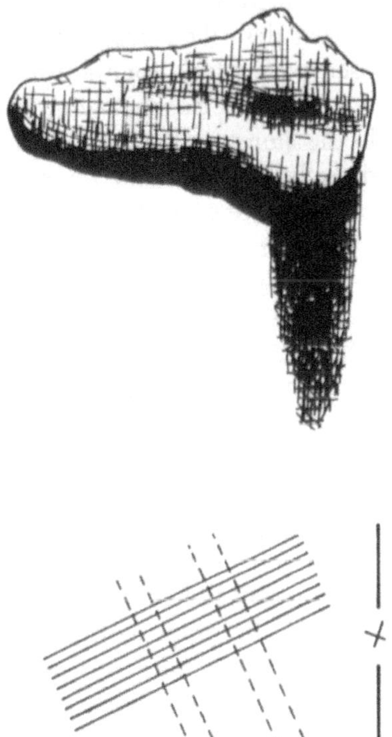

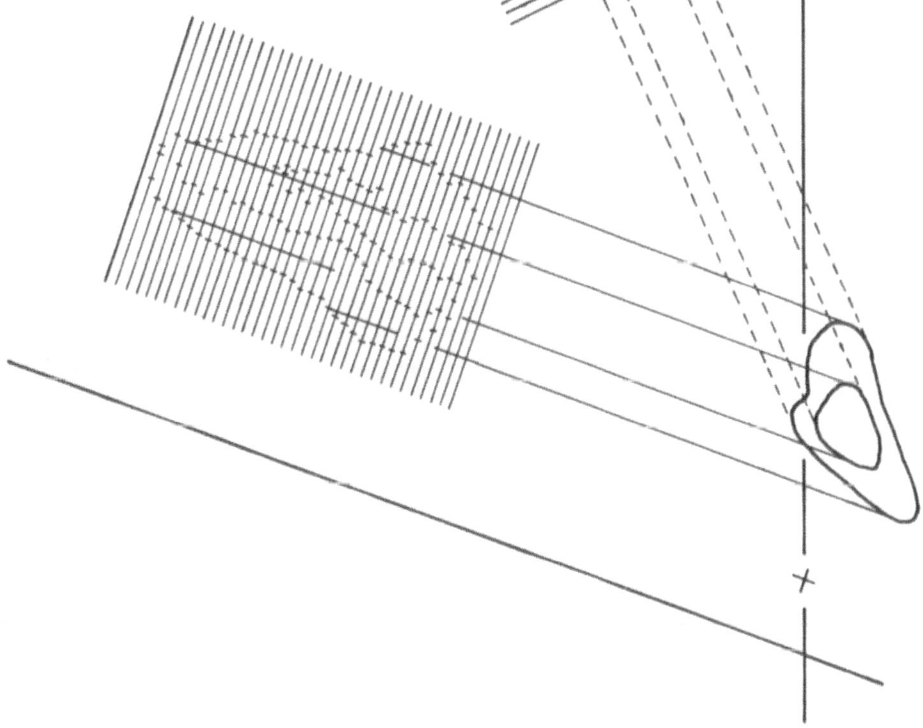

Figure 8. Lateral projection of the fastigial nucleus. The viewing direction (and, thus, the direction of projection) is parallel to the thick line along the left part of the figure. The two crosses are the reference points common to all profiles. The contour, the summit of convexities, and the depths of concavities have been projected onto the set of parallel (transverse) lines from all 33 sections. (The sections are numbered 2–34.) However, only two profiles—Nos. 4 and 22—and projection lines from four points (the contours of sections 4 and 22) have been drawn. One alternative direction of projection has been shown (broken lines toward the right with an incomplete set of transverse lines). Surface features are subjectively indicated by means of shadowing in the figure to the right. The caudal end of the fastigial nucleus is shown above, and the anterior end below. The dorsal surface is visible to the left in the reconstruction but is darkly shaded. The medial keel is along the leftmost edge of the reconstruction. The concavity in the major rostral part of the lateral surface is visible. The larger basal surface is hidden except for its transition to the lateral surface (right edge of reconstruction). The light is supposed to come from the right, parallel to the line through the reference points. It could equally well come from any other direction considered useful to emphasize relief characteristics.

A method for *unfolding* of surfaces makes use of a system of lines similar to that described here for reconstruction by lateral projection. It has been used particularly for the unfolding of curved cortical surfaces with delineation of areas, degeneration fields, etc., and numerous examples are available in the literature. Instead of projecting points along the profile circumference directly onto the transverse lines, the lengths between points of interest are measured, e.g., with a map measurer (curvimeter, space runner), and these distances are laid off on the appropriate transverse line. An arbitrary line on the surface of the brain structure is taken as the axis from which the distances are laid off to the right and to the left along the transverse lines. The axial line might be a relatively rectilinear boundary between two cortical areas. Finally, the points on the transverse lines are interconnected freehand, and a map of the unfolded surface is thus obtained. The dimensions will have correct proportions in the direction of the transverse lines (the plane of sectioning), but in the opposite direction, the curvedness of the surface must give distortions.

Two studies that include attempts to unfold heavily curved surfaces are those of Blackstad *et al.* (1970) and Swanson *et al.* (1978), both dealing with parts of the hippocampal region. In both cases, pliable sheets were applied over curved surfaces of three-dimensional models. After being appropriately marked, the sheets were detached and flattened out as far as possible. Nieuwenhuys (1974) described a purely graphic method for displaying the nuclear masses of the brain stem.

3. Tilted Projection

As seen immediately when a three-dimensional solid reconstruction (model in wax or other material) is available, the most informative viewing angle is often somewhere between a horizontal (lateral) and a vertical direction. It is possible to mimic inspection of a tilted solid model by applying elementary principles of geometry and projection. This can be done in many ways, one of which is shown in Fig. 9.† This method, which does not produce perspective, can be performed with pencil, compass, and ruler (see below).

*Tilting** a profile implies reducing all intraprofile distances (lengths, l) in the direction of the tilt to $l \cdot \cos v$, v being the tilt angle. All intraprofile lengths at right angle to the tilt direction are to remain unchanged. The

*Tilt occurs when a vertical axis through a plane of sectioning is caused to deviate from the vertical orientation (Fig. 10, middle and right). However, in the text and figures presented here, only tilt toward plain left or right is considered, and *direction of tilt,* therefore, means to the left or right. When a vertical axis is tilted, the sections (profiles), which must be imagined as forming a stack (Figs. 10–12), will be inclined to the same degree (indicated by the tilt angle); this inclination will also be called tilt. *Rotation* is here meant to signify rotation of sections (profiles) on an axis perpendicular to the plane of sectioning; this axis will be called vertical (Fig. 10, left). When a reconstruction is made involving both rotation and tilt, first the stack of profiles is rotated; thereupon the rotated stack is tilted.

†Another paper-and-pencil method, and the topic of reconstruction in general, are dealt with by K. Peter (pp. 1994–2020) in Krause (1927).

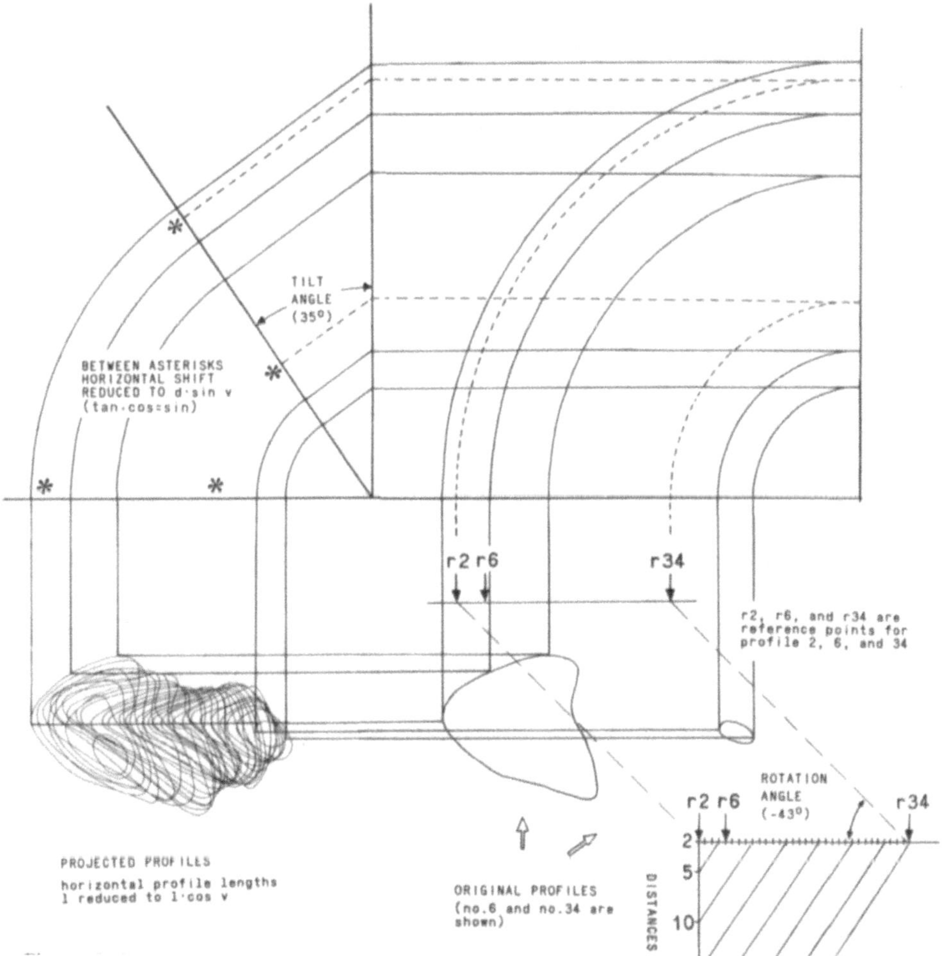

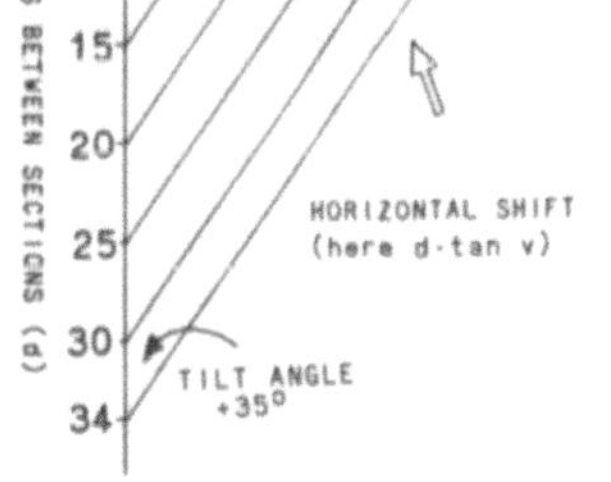

Figure 9. Reconstruction of the fastigial nucleus with rotation and tilt. To the lower right, geometric aid for transforming the distance between the sections (× magnification), d, laid off on the vertical axis, to d·tan v, marked on the horizontal axis. For the sake of clarity, only eight sections have been marked on the vertical axis. The acute angle adjacent to the vertical axis is the tilt angle v (here, +35°). $d \cdot \tan v$, upon projection, will become $d \cdot \sin v$; broken lines and asterisks illustrate this in the case of the distance between profiles 2 and 34, represented by reference points r2 and r34 (one of each). When an individual profile is to be projected, it is placed so that its reference points come to lie in the correct position to ensure proper horizontal shift of the projected profiles. Points r2, r6, and r34 indicate where the two reference points (corresponding to the two crosses in Fig. 8) associated with profiles 2, 6, and 34 are when each of these profiles are projected with a tilt of 35°. However, it is not necessary that a reference point be placed on the horizontal coordinate axis itself as is done here. For further explanations, see the labeling in the figure, and the main text.

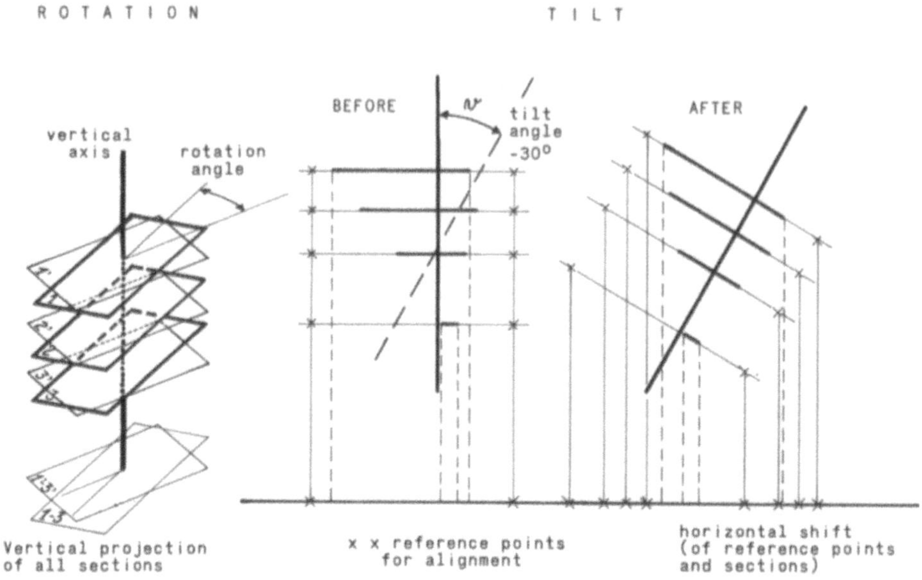

Figure 10. Illustration defining rotation, tilt, etc. For additional explanations, see footnote, p. 46.

projection lines in Fig. 9 make clear how a projected profile is thus produced from the original profile and drawn at a suitable distance from the latter.* For clarity, only projection lines from a few (five) points are shown.

When the stack of profiles is imagined tilted, the projected profiles are shifted in the horizontal direction as indicated in Fig. 10. Lower profiles move to the right when the tilt angle is positive, i.e., counterclockwise, and to the left when the angle is negative. This shift, which increases with the distance (d) between the sections and with the tilt angle, is $d \cdot \sin v$.

In order to ensure that the projected profile is shifted for the correct distance (Fig. 9), the original profile is moved a distance of $d \cdot \tan v$ before being projected. It is not necessary to compute $d \cdot \tan v$, since it is found geometrically as shown in the lower right of Fig. 9. Thereafter, it is used directly for the shift of the reference points and thereby of the original profile itself. When carried through the projection system, the length $d \cdot \tan v$ will be reduced to $d \cdot \tan v \cdot \cos v = d \cdot \sin v$. In other words, when the profiles are projected according to the method shown in Fig. 9, their mutual distances after projection, i.e., in the three dimensional graphic reconstruction, will be $d \cdot \sin v$ as required.

Practical Execution

1. Place a large sheet of paper on a drawing board with adjustable rulers. Draw a coordinate system like the one shown in the lower right

*Laying off the intraprofile lengths on the vertical line to the upper right in Fig. 9 instead of immediately swinging them up to the left is done to give the same left-to-right sequence of projected profiles as of original profiles.

of Fig. 9. Lay off the distances between the sections to be used (× magnification) along the left vertical axis and draw lines up to the horizontal axis at an angle equal to the desired tilt of the reconstruction. (For a negative angle the line must go upwards to the left in contrast to the situation in Fig. 9.)

2. Draw lines (in ink) corresponding to the thick horizontal line in the middle of Fig. 9 (a "baseline") and the two vertical lines upward from it. From where the left vertical line meets the base line draw a line upward to the left at an angle equal to the absolute value of the selected tilt angle.
3. Place the first profile somewhere on the sheet approximately as shown for the profile in Fig. 9 after having selected an appropriate rotational (azimuthal) orientation. Mark off on the large sheet the reference points (two or more) accompanying the profile. Connect one reference point with the origin of the coordinate system (corresponding to r2–r2 in Fig. 9).
4. With a ruler, project a number of points from the original profile up to the base line. Then lay these points off on the right vertical line with the compass, project them over to the left vertical line, and drop perpendiculars from them onto the oblique line defined by the unsigned tilt angle. Finally, using the compass, transfer the points forming the ends of the perpendiculars to the base line and draw vertical lines from there downward on the sheet.
5. Draw horizontal lines toward the left from the points selected for projection in the original profile; where these lines cut the vertical projection lines just mentioned, the desired points of the projected profile are found. Draw all projection lines very lightly with a sharp pencil. Transfer enough points to allow them to be smoothly connected freehand with a pencil. Ink in the final contour and rub out all auxiliary pencil projection lines to ready the large sheet for the processing of a new profile. It is wise to have tracing paper taped over the area where the graphic reconstruction will appear, since after completion of the whole drawing, this will make any desired reversal of the picture an easy task.
6. Place the next profile on the large sheet. (The profiles with associated reference points must have been drawn on tracing paper.) Determine the position of the profile by placing its reference points as far to the right or left of the reference points of the previous profile as indicated in the coordinate system. This might be similar to the r2–r6 shift in Fig. 9. Transfer points as described in 4 and 5. Continue until all profiles have been projected.
7. The final drawing can now be made. In Fig. 9, all 33 projected profiles are fully shown. While actually carrying out a projection, one may omit points that will become hidden (Fig. 11). In that case, it is feasible to make the final reconstruction in one single stage, drawing in ink the nonhidden parts of each profile as soon as it is ready.

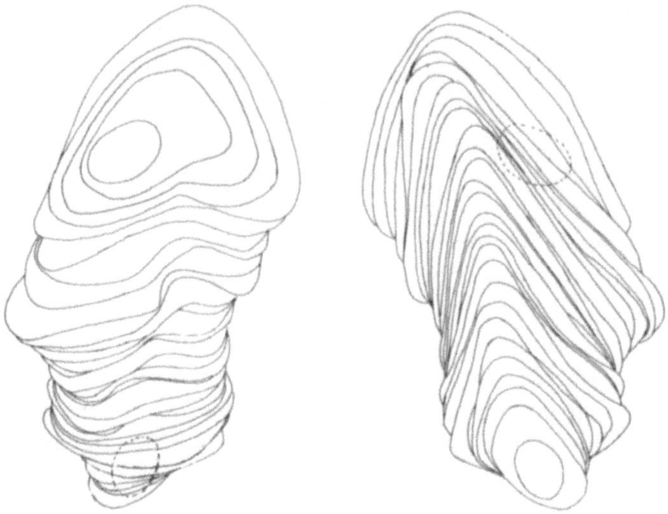

Figure 11. Graphic reconstruction of the fastigial nucleus prepared as described in Fig. 9. Accordingly, the stack of sections (profiles) has first been rotated 43° counterclockwise from the position in which the line through the two reference points is horizontal as in Fig. 8. The stack has been tilted 35° to the left. The left figure was made by first tracing section No. 2 (the most anterior section), then the following sections, omitting hidden parts. The lateral surface has been exposed to view (cf. Fig. 7, left). The right figure was similarly made by starting at profile 34 and reversing photographically the mirror image then resulting. It shows the opposite aspect, invisible in the left figure. Here, the medial keel and the basal and dorsal surfaces are visible. Other surface characteristics (concavities, convexities, bends) mentioned above, are visible as well. To facilitate orientation, the last hidden profile (the end of the nucleus) has been indicated with a broken line in each of the reconstructions, also the figures have been turned about 90° counterclockwise from the original positon (Fig. 9).

Clearly, it is necessary to start with the profile selected to form the uppermost level of the reconstruction. However, it is recommended that the line-hiding procedure be postponed until after all profiles have been projected, since this offers the opportunity of making an additional reconstruction as seen from the opposite end, without further projection work (Fig. 11). It is also possible to reconstruct only parts of the full structure, possibly separated to permit inspection of the cut surface at a specific level, etc.

The method described here permits inspection of a brain structure from all angles, since the applicable −90° to +90° will cover all possible degrees of tilt, and the original profiles, which are rotated 43° in Fig. 9, can be placed in any rotational orientation.

The time needed for this method of reconstruction varies with the number and the degree of complexity of the profiles. The magnification of the original profiles should not be made too small, since this will increase the need for time-consuming exactness in drawing and tracing. Projecting one profile may roughly require from 5 to 20 min, and a full reconstruction as shown in Fig. 9 can be completed in from ½ to 2 days.

Figure 12. Example of computer-aided three-dimensional reconstruction based on the same series of profiles from the fastigial nucleus as illustrated in Figs. 7–9 and 11. (For a further reconstruction of the same nucleus with different angels of rotation and tilt, see Dietrichs and Walberg, 1979). Polygonal shape of profiles results from the limited number of coordinates collected with the digitizer along each profile. To create a prettier impression, freehand smoothing of the profiles could have been done, or more coordinates have been collected. The degree of rotation is close to that used in Figs. 9 and 11. The degree of tilt is 35° in the right plot, i.e., as in Figs. 9 and 11, while in the left plot it is 28°. The two figures therefore constitute a stereo pair. Reconstruction time for one drawing on a Tektronix storage screen (digitizing and plotting excluded) is about 1 min. New angles of rotation, tilt, scale, etc. can be selected in few seconds. In this stereo pair, some of the surface relief features seen in the other illustrations can be checked.

Mechanical equipment for reconstruction involving tilting with associated horizontal shift has been described by Forster (1968) who introduced the word "Perspektomat." It has also been used in neuroanatomy by Hjorth-Simonsen (1972). It is worth using if available (most likely in a department of geography). The speed of reconstruction using this pantographlike apparatus is, at least in the author's experience, not very different from the method described here.

4. Computer-Aided Reconstruction

A computer with associated terminal screen and plotter is potentially a powerful tool for all kinds of three-dimensional reconstruction. Equipment and programs for doing such work are likely to become available in many departments of morphology and neuroscience in the near future, and we suggest that this alternative be then preferred to both the mechanical reconstruction principle and the pencil-and-paper method just described. See Fig. 12 for an example of reconstructions of the fastigial nucleus performed by the author's computer program.*

*The author may be contacted concerning the possibilities of copying his computer program for three-dimensional reconstructions. This program, which includes hidden-line procedures, does not produce perspective.

REFERENCES

Bancroft, J. D., and Stevens, A., 1977, *Theory and Practice of Histological Techniques,* Churchill Livingstone, New York.

Barnes, C. D., and Eltherington, L. G., 1973, *Drug Dosage in Laboratory Animals, Second edition,* University of California Press, Berkeley.

Berkowitz, L. R., Fiorello, O., Kruger, L., and Maxwell, D. S., 1968, Selective staining of nervous tissue for light microscopy following preparation for electron microscopy, *J. Histochem. Cytochem.* **16:**808–814.

Blackstad, T. W., Brink, K., Hem, J., and Jeune, B., 1970, Distribution of hippocampal mossy fibers in the rat. An experimental study with silver impregnation methods, *J. Comp. Neurol.* **138:**433–449.

Brodal, A., 1940, Modification of Gudden method for study of cerebral localization, *Arch. Neurol. Psychiat. (Chic.)* **43:**46–58.

Burrell, W., 1978, *A Comprehensive Handbook for General and Applied Microscopy,* Halsted Press, New York.

Crane, A. M., and Goldman, P. S., 1979, An improved method for embedding brain tissue in albumin–gelatin, *Stain Technol.* **54:**71–75.

Danscher, G., and Zimmer, J., 1978, An improved Timm sulphide silver method for light and electron microscopic localization of heavy metals in biological tissues, *Histochemistry* **55:**27–40.

Dietrichs, E., and Walberg, F., 1979, The cerebellar cortinuclear and nucleocortical projections in the cat as studied with anterograde and retrograde transport of horseradish peroxidase I. The paramedian lobule, *Anat. Embryol.* **158:**13–39.

Ebbesson, S. O. E., 1970, The selective silver-impregnation of degenerating axons and their synaptic endings in nonmammalian species, in: *Contemporary Research Methods in Neuroanatomy* (W. J. H. Nauta and S. O. E. Ebbesson, eds.), p. 157, Springer-Verlag, New York.

Forster, F., 1968, Der Perspektomat P-40. Ein selbstzeichnender Apparat für Parallelperspektive, *Neue Zürcher Z.* Fernausgabe **51:**49–51.

Gabe, M., 1976, *Histological Techniques,* Masson, Paris, and Springer-Verlag, New York.

Green, C. J., 1979, *Animal Anaesthesia,* Laboratory Animals Handbooks 8, Laboratory Animals Ltd., London.

Haug, F.-M. Š., 1973, Heavy metals in the brain. A light microscope study of the rat with Timm's sulphide silver method. Methodological considerations and cytological and regional staining patterns, *Adv. Anat. Embryol. Cell. Biol.* **47**(4)**:**1–71.

Haug, F.-M. Š., Blackstad, T. W., Simonsen, A. H., and Zimmer, J., 1971, Timm's sulfide silver reaction for zinc during experimental anterograde degeneration of hippocampal mossy fibers, *J. Comp. Neurol.* **142:**23–31.

Heimer, L., 1970, Bridging the gap between light and electron microscopy in the experimental tracing of fiber connections, in: *Contemporary Research Methods in Neuroanatomy* (W. J. H. Nauta and S. O. E. Ebbesson, eds.), pp. 162–172, Springer-Verlag, New York.

Hjorth-Simonsen, A., 1972, Projection of the lateral part of the entorhinal area to the hippocampus and fascia dentata, *J. Comp. Neurol.* **146:**219–231.

Kater, S. B., and Nicholson, C., 1973, *Intracellular Staining in Neurobiology,* Springer-Verlag, Berlin and New York.

Krause, R., 1927, *Erzyklopädie der Mikroskopischen Technik,* Vol. III, Urban & Schwarzenberg, Berlin and Vienna.

Lumb, W. V., and Jones, E. W., 1973, *Veterinary Anesthesia,* Lea & Febiger, Philadelphia.

Mulligan, J. H., 1931, A method of staining the brain for macroscopic study, *J. Anat.* **65:**468–472.

Nauta, W. J. H., 1950, Über die sogenannte terminale Degeneration im Zentralnervensystem und ihre Darstellung durch Silberimprägnation, *Schweiz. Arch. Neurol. Psychiatr.* **66:**353–376.

Nieuwenhuys, R., 1974, Topological analysis of the brain stem: A general introduction, *J. Comp. Neurol.* **156:**255–276.

Romeis, B., 1968, *Mikroskopische Technik,* R. Oldenbourg, Munich and Vienna.

Schroeder, D. M., and Jane, J. A., 1971, Projection of dorsal column nuclei and spinal cord to brainstem and thalamus in the tree shrew, *Tupaia glis, J. Comp. Neurol.* **142:**309–350.

Soma, L. R., 1971, *Textbook of Veterinary Anesthesia,* Williams & Wilkins, Baltimore.

Swanson, L. W., Wyss, J. M, and Cowan, W. M., 1978, An autoradiographic study of the organization of intrahippocampal association pathways in the rat, *J. Comp. Neurol.* **181:**681–715.

Switzer, R. C., 1979, Wrinkle-free mounting of sections from gelatin-embedded brains, *Neurosci. Lett.* **12:**103–105.

Thompson, S. W., 1963, *Selected Histochemical and Histopathological Methods,* Charles C. Thomas, Springfield, Illinois.

Vogt, B. A., 1974, A reduced silver stain for normal axons in the central nervous system, *Physiol. Behav.* **13:**837–840.

Methods for Selective, Restricted Lesion Placement in the Central Nervous System

2

ROBERT Y. MOORE

I. INTRODUCTION

Experimentally induced lesions of specific parts of the brain have been used extensively in two contexts. The first is in anatomical investigations designed to elucidate the pathways of the CNS with the aid of tract-tracing techniques such as the silver methods (see Chapter 4). Second, experimentally produced brain lesions have also been used widely as a means of studying the functions of specific brain regions: if a certain population of neurons or a pathway is destroyed, there is usually a loss or alteration of the function normally mediated by that structure. The purpose of this chapter is to outline some of the techniques available for experimental production of CNS lesions in animals. It should not be forgotten, however, that analysis of the human brain following spontaneously occurring lesions can also provide valuable information concerning pathways and function.

ROBERT Y. MOORE • Department of Neurology, State University of New York at Stony Brook, Stony Brook, New York 11794.

Brain lesions can be produced in different ways. A simple method is to use a knife or another mechanical device. Brain tissue can also be destroyed by other means such as electricity, heat, aspiration, or by the use of neurotoxins. One circumstance that greatly influences the choice of method is the location of the structure to be destroyed. Needless to say, it is easier to make a lesion in a superficial part of the brain than in a deep brain structure. Fortunately, however, a stereotaxic instrument for making subcortical lesions with great precision was developed more than 70 years ago (Horsley and Clarke, 1908). The stereotaxic instrument is one of the most important pieces of equipment in a modern neuroscience laboratory; it is used widely, not only in tract-tracing experiments but also in a variety of physiological and behavioral studies. A description of the stereotaxic technique, therefore, will precede the review of different methods for making brain lesions.

II. STEREOTAXIC TECHNIQUE

A. Theoretical Background

With this technique, the animal's head is held in a rigid instrument with fixation of the head at either three or four points. With the exception of some adaptations for the rabbit, two points that are universally used are the bony external auditory meatuses on each side. Two bars are inserted into these meatuses until firmly fixed and are attached to the frame of the stereotaxic apparatus. The second fixation points for stereotaxic apparatuses for larger animals (e.g., cat, monkey) are the infraorbital ridges. An appropriately fitted metal bar is placed over the lowest point of the infraorbital ridge on each side, and the head is held firmly against these by another bar that fixes the position of the palate or the mandible. In smaller species (such as rat, guinea pig, or hamster), the infraorbital ridge does not provide an adequate support, and the head is held rostrally by placing the incisors over a bar and firmly fixing the palate directly behind the incisors against the bar (Fig. 1). In each case, the situation is the same, the head is held firmly in a rigid structure at two fixed and reproducible points. Consequently, this fixation of one rigid structure, the cranium, within another rigid structure, the stereotaxic apparatus, will place the brain in a reproducible position. That is, presuming that the brain has a relatively constant anatomical relationship with the cranium, a constant fixation of the cranium within the stereotaxic instrument will give the brain a fixed relationship to the instrument. This is the basis of stereotaxic technique, and it should be noted that success of the technique theoretically requires only a constant relationship between the brain and the bony auditory meatuses and the infraorbital ridges or palate. In some work on the rat and other small mammals, stereotaxic coordinates are given in distances from the bregma, midline, and top of the skull. Although we prefer to use the interaural line as a referent for small animals as well as for large ones, fairly precise stereotaxic localization can apparently

Figure 1. Stereotaxic apparatus for small animals (Kopf Instrument Co., Tujunga, California). The ear bars (A) are attached to specially fitted ear plugs (B) which aid in obtaining optimal positioning of the animal in the instrument. Rostral fixation of the head is attained by placing the incisors over the incisor bar (C). The instrument shown is fitted with a holder (D) containing a microknife.

be obtained by using the bregma as the zero referent (Whishaw *et al.*, 1977). In either case, skull angle should be kept constant, and this is usually done by placing bregma and lambda on the same horizontal plane. A large number of stereotaxic instruments are now available commercially that are suitable for use with either large or small animals, and several companies produce instruments that can be adapted for a wide variety of species.

B. The Stereotaxic Atlas

With the head of the animal fixed in this position, it is possible to construct a three-dimensional grid, a stereotaxic atlas, that will allow the accurate production of a lesion or placing of an electrode in any area of the brain. The techniques for constructing an atlas have been reviewed many times (see Clarke, 1920; Olszewski, 1952; Bleier, 1961) and will not be considered extensively here. The basis of the methodology is to form reference planes. The original reference planes employed by Horsley and Clarke (1908) were as follows: a basal horizontal zero plane determined by the interaural line (between the bony auditory meatuses) and by a line passing through the lowest point on the lower margin of the orbit on each side; a sagittal zero plane crossing through the cranium in the midsagittal plane; and a vertical zero plane passing through the interaural line and perpendicular to the basal horizontal zero plane (Fig. 2). All of the planes are oriented at right angles to one another. Because the basal, horizontal zero plane is almost entirely below the base of the brain in the cat and monkey, it is customary, in atlases for these and other large species, to use a horizontal plane 10 mm above the basal horizontal plane.

Reference planes for atlases to be used with smaller animals, however, are much more variable, and the user should become thoroughly familiar, first, with his instrument and, second, with the stereotaxic atlas he is employing. For example, rat atlases are available in which the rostral fixation point, the tooth bar, is placed 2.5 mm below the interaural line (Hurt *et al.*, 1968), 2 mm above the interaural line, or 5 mm above the interaural line (deGroot, 1959). In addition, the horizontal zero plane may be identified arbitrarily at the level of the interaural line (Hurt *et al.*, 1971) or at some distance, such as 5 mm, above that line (deGroot, 1959). The relationship of the brain to these reference planes is shown for one rat atlas (Fig. 2). A list of stereotaxic atlases is given in the Appendix (VI.A–D).

Although the stereotaxic technique affords the best opportunity to place discrete lesions reproducibly in subcortical structures, the application of the technique is beset with many problems and pitfalls that must be carefully avoided. Many of these problems relate to the stereotaxic atlas. Most atlases are constructed using animals of a fairly uniform age, size, and, on occasion, sex. However, if animals of a different age or size are used, the relationship between the brain and its bony encasement also may differ, giving inaccurate lesion placement if proper precautions are not taken. Often a stereotaxic

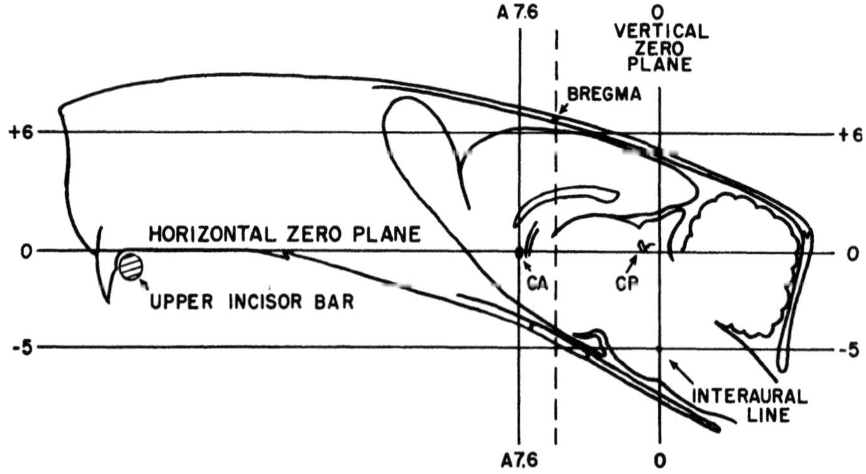

Figure 2. Midsaggittal section through the head of the rat illustrating the construction of horizontal and vertical stereotaxic planes. [Reproduced with permission from deGroot (1959).]

atlas may be inaccurate because it was produced from very few, or even one brain. This leads to two sources of inaccuracy in lesion placement. The brain used may not be representative, or there may be sufficient variation among individuals of the species for coordinates for a particular structure to vary considerably. The latter problem is demonstrated quite clearly for the cat by Bleier (1961). In addition, some atlases are anatomically of poor quality, and structures may be mislabeled or the photographs too poor to allow visual verification of the labeling. A further problem may be that the atlas does not adequately take into account problems such as shrinkage and distortion of the tissue during histological preparation (see Olszewski, 1952). Another factor that may produce problems is that the size of the region to be destroyed is not necessarily constant from animal to animal. This is rarely a problem in small animals, but it may necessitate an adjustment of stereotaxic coordinates and lesion size in larger species where the brain size may vary significantly from individual to individual.

C. Verification of the Coordinates

Considering these different problems, it is usually better to obtain one's own precise coordinates rather than to rely on a specific atlas for the final coordinates. This usually is done in one of two ways. The first method is to take several atlases and select coordinates from them. Following a series of practice lesions,* a final set of coordinates can be selected. These coordinates are then checked in additional animals by histological, confirmation of the lesion placement. This procedure confers the additional benefit of allowing a check of lesion or injection size by testing different parameters. A more accurate method is to determine stereotaxic coordinates by directly visualizing the structure in which one intends to place a lesion. This is accomplished by decapitating an anesthetized animal, placing the skull in the stereotaxic machine, removing the occipital, parietal or frontal bones, and opening the dura. Part of the brain is aspirated, or slices are cut through the brain until the desired structure is uncovered. An electrode can then be placed in the structure, and the coordinates are obtained directly. As with the first method, though, it is always advantageous to place some practice lesions before embarking on a major experiment. In cases where the region to be lesioned is not distinctly separated from adjacent structures, the method may not be satisfactory. Recently, a very precise method for obtaining stereotaxic coordinates in small animals has been reported (Dixon *et al.*, 1978; Marchand and Riley, 1979). We have employed this method and find it superior to those outlined above, but it has a minor disadvantage of requiring some modification of the stereotaxic instrument.

The use of optimal stereotaxic technique can only be achieved by prac-

*The best results are obtained if the animals used for these experiments are of the same size as those for which the coordinates are developed.

tice, but success in lesion placement can be obtained by the use of different stereotaxic procedures. However, there are certain basic requirements for a successful application such as the proper selection of animal, use of a good atlas, availability of a stable, well-made stereotaxic instrument, verification of the atlas coordinates, and proper surgical technique. The latter includes proper placement of the head in the stereotaxic machine. With regard to the rat, which is the most commonly used laboratory animal, correct placement can be achieved in the following way. Select one supplier of an inbred strain of albino or hooded rats and use only female animals at a selected weight. Since weight and age correlate well in the young female rat, this implies a given age. Obviously in situations where males are preferred, the same procedure should be followed.

The animals are anesthetized, and a small slit is made with scissors in the skin at the base of pinna to expose the external auditory meatus. Ear plugs are inserted into the external auditory canal until the tympanic membrane is ruptured. The rupture often can be heard or felt, and it is attended, even in anesthetized animals, by a reflex retraction of the eye. When the ear plug is securely situated on each side, the animal's head is lifted up to the level of the ear bars, and these are consecutively inserted into the ear plugs, maintaining pressure to assure that the ear bars will designate the interaural line. Some stereotaxic instruments do not use separate ear plugs and, in these, the ear bars are inserted directly into the external auditory canal. The head is then centered in the machine. At this point the top of the head should be flat, and there should be little or no lateral play or head movement if the jaw is grasped and moved vertically up and down. The incisors are then placed over the tooth bar, and the tooth bar is retracted until the back of the incisors are firmly against it. The nose bar is then lowered until the palate is held firmly against the incisor bar. At this point the head should feel quite solidly held in the apparatus. It cannot be emphasize too strongly that proper placement of the head in the instrument is essential to successful stereotaxic technique.

III. NONSELECTIVE LESION TECHNIQUES

A. Mechanical Lesions

1. Aspiration

Central nervous system tissue is sufficiently soft to be removed by aspiration under vacuum without difficulty. Commercial machines for this purpose are available from surgical supply houses.* More simply, aspiration can be achieved by the aid of a standard vacuum pump or laboratory vacuum

*Metal aspirators for human neurosurgery are generally of such large gauge as to render them useless in work on smaller animals.

line that is attached by vacuum tubing to a trap which is, in turn, attached by another piece of vacuum tubing to the aspirator. The aspirator can be produced by heating glass tubing (about 5–6 mm in diameter) over a Bunsen burner until it is soft and then stretching it, with whatever bend is desirable, so that a gradually diminishing diameter is produced. The tubing is then broken at the desired diameter after scoring it with a diamond pencil. This produces an even break which is necessary for well-controlled aspiration lesions. For example, a diameter of 1 mm or more might be ideal for aspirating monkey cortex, whereas a much smaller diameter would be appropriate for work on the mouse or rat.

Aspiration lesions are most readily applied to external structures, especially cerebral cortex (e.g., Walker, 1938; Rose and Woolsey, 1943, 1949), but they have also been made in a number of subcortical structures such as hippocampus and septum (e.g., Moore, 1964), superior colliculus (Schneider, 1970), and corpus callosum (Myers, 1956), to mention a few. The technique for cortex is exemplary for all areas. With the head held securely in a stereotaxic or some alternative stable head holder, the area of interest is exposed by removal of bone and reflection of the dura. The exact limits of the cortex to be removed are noted. These may be indicated by sulci in a convoluted cortex, or they may be marked by dots of India ink on a lissencephalic brain. A small incision in the pia is made with a thin, pointed scalpel blade, and the cortex is removed by subpial aspiration. If this is carried out under an operating microscope, the lesion can be carried down to the border between cortex and white matter quite precisely. The size of the aspirator and the force of the vacuum required for a given lesion in a particular species can only be determined empirically. Obviously, the technique is crude, but it should be emphasized that with proper application and, particularly, with use of an operating microscope, very precise and reproducible lesions of cortical and some subcortical structures can be made. Aspiration lesions in subcortical structures will obviously necessitate some disruption of overlying cortex. Therefore, in instances where this is undesirable, stereotaxic placement of electrolytic or other forms of lesion would be more appropriate.

2. Knife Methods

The knife methods provide another type of mechanical lesion technique that is simple and inexpensive and primarily employed to transect fiber bundles. It is possible, for example, to selectively transect a fiber pathway such as the lateral lemniscus with a very fine knife (Goldberg and Moore, 1967), but it is also possible to use the knife method to transect the entire neuraxis at some critical point as in the classical studies of Bremer (1935). Each of these examples represents an extreme of the application of a knife method with the lesion made under direct visual guidance and without mechanical placement. Thus, in each case the methodology is simple: the structure(s) to be transected are identified visually, and the appropriate knife cut is made freehand. Recently, a number of studies* have been reported that have combined the knife method with stereotaxic technique to provide more controlled and reproducible lesions, particularly in deep structures (Voloschin *et al.*, 1968; Sclafini and Grossman, 1969; Gold *et al.*, 1973; Moore, 1978). However, knife lesions of subcortical structures have a limited application in tract-tracing experiments because they tend to cause massive incidental damage. They are employed only when the desired effect is to eliminate a particular fiber system totally without regard to effects on adjacent systems.

B. Injected, Nonselective Toxins

A very large number of studies have been reported over the past 75 years in which CNS lesions have been produced by the injection of a wide variety of toxic materials. Excluding the selective neurotoxins, which will be covered later in this chapter, this method has proven the least satisfactory one for reliable production of CNS lesions. The major problem is that the diffusion of the injected material is often unpredictable, thus making it difficult to control lesion size and location. However, small and apparently well-controlled chemical lesions have been produced by Groeneweger and Voogd (1977) who injected sodium cyanide iontophoretically through a glass micropipette (Knook, 1969). The chemical lesion method is reviewed extensively by Carpenter and Whittier (1952) and Routtenberg (1972), and the interested reader is referred to their papers for further critical details. Some further comments on this matter will arise in the discussion of selective neurotoxins (Section V).

C. Alterations of Cerebral Vasculature

Experimental lesions can also be made by selective interruption of cerebral vasculature, that is, by the production of an experimental cerebrovascular accident. This requires exact knowledge of the distribution of the cerebral vessels in the animal to be studied, their reproducibility, and the extent of collateral vascularization available. In some animals, for example in the cat, it is very difficult to produce cerebral infarction with ligature of major cerebral vessels, whereas in the gerbil, quite standard and reproducible lesions can be produced by occlusion of the middle cerebral artery. This method has little application at present, but there is a growing literature implicating monamines in some of the phenomena related to the sequelae

*The principal stimulus for these studies appears to have been initiated by the paper of Halasz and Pupp (1965) demonstrating that it was possible to totally deafferent the hypothalamus and subsequent studies indicating that such lesions are a powerful tool for the analysis of neuroendocrine function (Halasz, 1969).

of vascular accidents, and some very interesting studies have been reported using the method (e.g., Robinson *et al.*, 1975).

D. Radioisotopic and Related Methods

Two methods have been employed for the production of restricted CNS lesions using ionizing radiation. The first of these depends upon the localized application of a short-half-life isotope to the CNS (Schulman, 1964), and the second method, described by Malis *et al.* (1960), depends on the radiation by high energy particles generated from a cyclotron. Although the size of the lesion can be accurately predicted from the dose of the radiation, these procedures for making lesions are of limited value in tract-tracing experiments for obvious reasons. They require an appropriate source of high energy particles, and they are technically cumbersome.

E. Ultrasound Lesions

In the late 1950s and early 1960s, one group of investigators developed the technique of producing highly localized CNS lesions by focusing several beams of ultrasound upon a single, stereotaxically determined target within the brain (see Fry *et al.*, 1964, for review). With this method, each individual beam produces insufficient tissue disruption to injure neurons. However, at the point where several beams intersect, a lesion is produced. With precise control of several beams, it is possible to place very localized lesions deep in the brain without incidental damage to any overlying structures. The method has been employed in several behavioral and anatomical (e.g., Fry and Cowan, 1972; Pandya *et al.*, 1973) studies and has provided very useful data. Unfortunately, the method is somewhat cumbersome and also too expensive to be established routinely.

F. Thermal Lesions

1. Lesions Produced by Heat

The use of heat to destroy neural tissue is one of the most widely applied methods for making discrete ablations of neural tissue. The method in general use is to apply radiofrequency current (Aronow, 1960) through an insulated electrode stereotaxically placed in the desired brain region. Low power continuous wave radiofrequency current is passed between the brain electrode and an "indifferent" electrode placed in the muscles of the neck beneath the skin of the back or in some other innocuous area. The intervening medium acts as a leaky dielectric so that heat is generated by dissipation of electrical energy within the tissue. The diameter of the lesion is roughly a linear function of the current; that is, of milliamperes of radiofrequency current. Several radiofrequency current-producing devices are commercially available and relatively inexpensive.* The most elaborate allow very precise control of lesion size because the brain electrode contains a small thermistor within its tip, and the radiofrequency current is adjusted to produce an experimentally determined temperature at the electrode tip for a duration of 1 min. The maximal lesion is produced at 1 min, and further passage of current does not increase the size of the lesion because of the limits over which the heat is dissipated at a given temperature. Instruments with a thermistor within the electrode have one drawback. Because the electrode is nearly a millimeter in diameter, it produces a substantial electrode track with considerable tissue damage, and it is not possible to reliably produce lesions 1 mm or less in diameter. This technique, therefore, may not be acceptable in studies using small animals such as rats, mice, or hamsters, but it is quite useful for larger animals where the electrode size is not important and one rarely wishes to make lesions smaller than a millimeter.

With any of the radiofrequency lesion-producing devices, one can employ a small diameter metal electrode, insulated except for the tip, and pass radiofrequency current of known strength for an established time that will produce lesions of the size desired. The lesion parameters, i.e., the strength and duration of the current, should be carefully determined for each area to be studied, as they vary depending upon the composition of the tissue. Radiofrequency lesions are becoming increasingly more popular because they tend to be quite uniform for a given current, are usually approximately spherical in shape, and do not present problems of either deposition of metal or release of gas during production of the lesion as do electrolytic lesions (see Section III.G, below). The progression of histologic events following a radiofrequency lesion does not differ significantly from that to be described for electrolytic lesions in the following section.

Another heat lesion technique that is quite useful for superficial structures is the so-called "laminar thermocoagulation method." In this technique, heat is usually applied to the exposed surface of the brain† by means of a 0.5-mm-thick silver plate attached to an ordinary soldering iron. By applying a proper voltage from a Variac, the plate is heated to about 70°C. This temperature is determined by placing a piece of Wood's metal with a melting temperature between 68–72°C on the silver plate. When the plate has reached the desired temperature, it is brought into flat contact with the brain. By standardizing the number of seconds of contact, the depth of the lesion can be relatively well controlled. An advantage of this method is that the lesion is usually confined to the heated area without the added complication of tissue necrosis deep to the intended lesion. The reason for this

*Radiofrequency current instruments can be obtained from several companies including Grass Instruments (Quincy, Massachusetts) and Kopf Instruments (Tujunga, California).

†The silver plate can be applied directly to the intact dura in order to avoid accidental damage of the brain tissue. In animals with a thin calvarium, laminar heat lesions can even be produced by applying heated saline directly to the skull.

seems to be that a silver plate of 70°C produces just enough heat to destroy the nervous elements but not quite enough to disrupt the vascular supply. This feature is especially useful when lesions are aimed at superficial structures at the base of the brain, such as the olfactory tubercle, where many blood vessels perforate the brain substance.

2. Cryogenic Lesions

Cooling of nervous tissue can be used for the production of brain lesions (Cooper, 1962), but the method has never become popular for two reasons. First, the equipment necessary is cumbersome, and the probe required to make the cold lesion is quite large. Consequently, the method is primarily applicable in situations where only a limited number of lesions are required in a reasonable time and where the species has a large brain in which the cold probe does not produce unacceptable incidental damage. Second, the lesions produced are more difficult to control than electrolytic or radiofrequency lesions, which can be produced with a minimum of equipment and expense and are quite applicable to small animals. The cold technique, however, does have one aspect that can be extremely useful, particularly in human surgery but also in some animal experiments. During the process of cooling, there is a period in which the tissue has lost function but, if warmed, will again function normally. This permits gauging the effect of a lesion before it has been made permanent.

G. Electrolytic Lesions

The passage of direct current from the tip of an insulated metal electrode has, for many years, been the most popular method of producing discrete lesions within the CNS. The mechanisms of lesion production with direct current is based on the consequences of ionic flow in inducing an irreversible alteration of cell metabolism and, hence, cell death. Ionic flow is measured in terms of the product of amperes of current applied times the duration in seconds, the unit being coulombs. In general, the size of a lesion is a direct function of the coulombs of current applied. This varies considerably, however, with the type of metal used for the electrode and whether anodal or cathodal current is used.

Lesions produced by anodal current are more reproducible, both in size and shape, than those made by cathodal current. However, anodal lesions do produce deposition of metal in the lesion site. This is substantially greater with some metals than with others, but all, including platinum, deposit some metal during passage. This has generally been viewed to be of little significance, but in some situations (Velasco, 1972), the metal used rather than lesion parameters seems to have been responsible for the functional effect of the lesion. Cathodal lesions, on the other hand, tend to release gas during current passage, and this creates two technical problems. First, the release of gas results in a frequent loss of electrode contact with the tissue and, consequently, the duration of current passage is often uncertain. Second, cathodal lesions are quite variable in size for a constant current and duration of passage, and they also tend to be unpredictably irregular in size. For these reasons cathodal lesions are rarely employed experimentally. In a recent study by Cox *et al.* (1978), however, it was noticed that "cathodal lesions could be made quite regular by decreasing the intensity of the current." Rowland (1966) has recommended a bidirectional current lesion-producing method that does not deposit metal or produce gas, but it has never come into popular use.

IV. EVALUATION OF THE ELECTROLYTIC LESION

The most extensive recent discussion of the mechanism of lesion production of anodal current is provided by Gold (1975). Some aspects of his discussion warrant repetition as they are of considerable significance. Of all the materials now available, only platinum–iridium (90%–10%) and platinum-clad tungsten have been found to be free of corrosion and, at the same time, to offer sufficiently low electrical resistance to be of practical use. However, platinum-containing electrodes are seldom used in experimental work. The commonly used stainless steel or "nichrome" electrode has markedly different properties in lesion production. Gold found that, at identical current parameters, slow iron anode lesions were 1.3–3 times as large as stainless steel anode lesions and these, in turn, were 5–10 times as large as platinum–iridium anode lesions. Consequently, the investigator must be cognizant of the properties of the electrode used.

The progression of histologic events transpiring after placement of a small, anodal electrolytic lesion has been described in detail by Wolf and DiCara (1969). Within an hour after the lesion is made, a central core of necrotic debris is evident. This core is surrounded by a narrow ring of concentrated cellular material representing glial cells, lysed neurons, blood vessels, and some extravasated blood cells. External to this there is a narrow zone containing a few tissue elements, and this is surrounded by a broad ring of tissue containing cells that have been killed by the current but do not yet show pathological changes in the usual cell stains. One day after the lesion is made, the cellular debris is largely removed from the central portion of the lesion, and there is migration of phagocytic cells into the area of the lesion. In the broad area that did not show changes at 1 hr, there is generalized neuronal pathology ranging from lytic changes to complete neuron loss. At 1 week the entire region surrounding the central cavity is filled with glial cells, phagocytic microglia, and astrocytes, and, at 2 weeks,a clear process of walling-off of the lesion by gliosis is apparent. Over the remaining period of the study, a total of 16 weeks, two processes were evident. First, the area of the lesion is cleaned up by reparative processes of phagocytosis, leaving an empty central cavity surrounded by a narrow rim of reactive as-

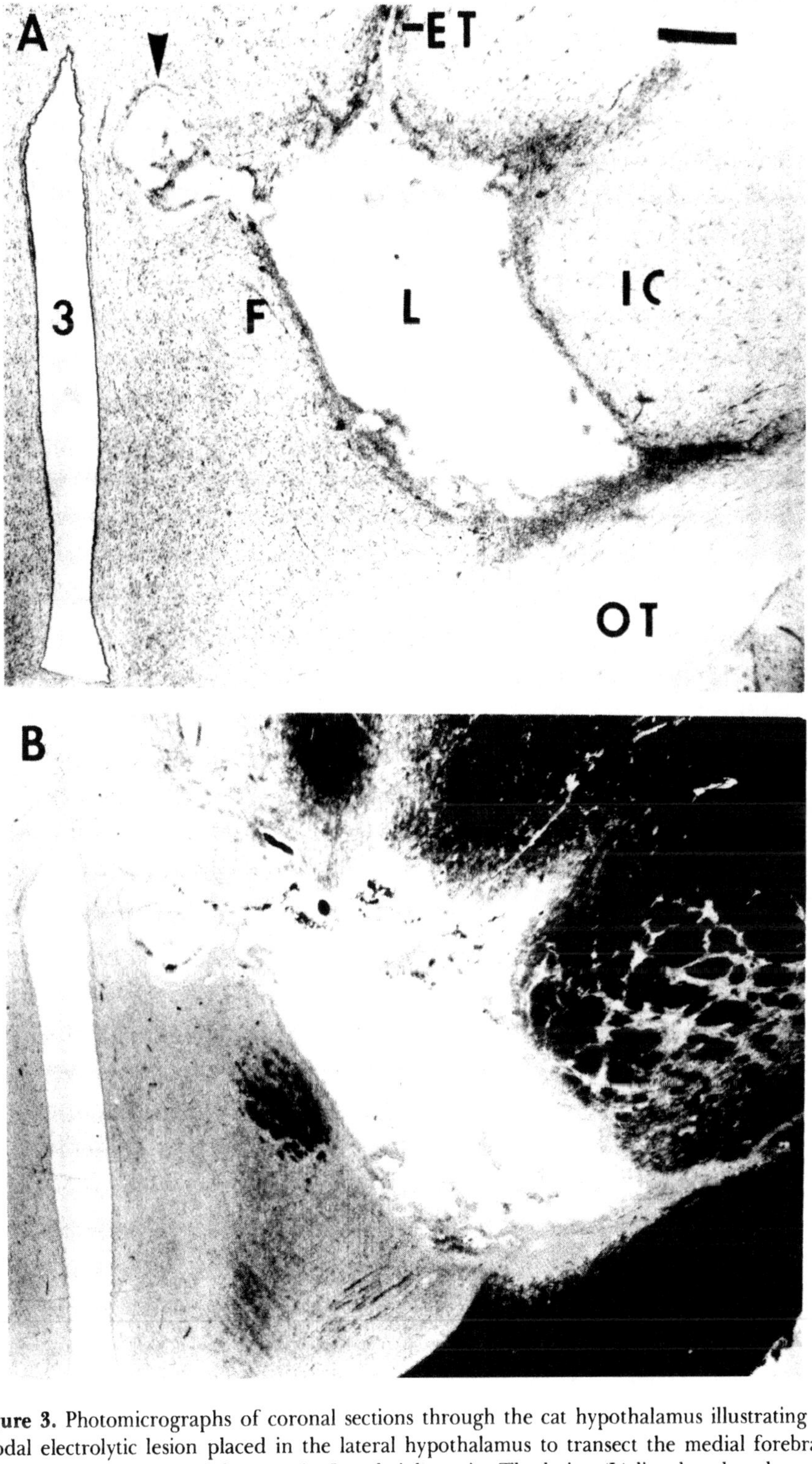

Figure 3. Photomicrographs of coronal sections through the cat hypothalamus illustrating an anodal electrolytic lesion placed in the lateral hypothalamus to transect the medial forebrain bundle and the nigrostriatal tract. (A) Cresyl violet stain. The lesion (L) lies dorsal to the optic tract (OT) and extends from the fornix (F) into the internal capsule (IC) and is represented by a large cavity with an extension (arrow) medially toward the third ventricle (3). The electrode tract (ET) is evident dorsally. (B) Woelcke stain for myelinated fibers, section adjacent to that shown in (A). Bar, 500 μm.

trocytes. The activity of these cells appears to continue well beyond the immediate postoperative recovery period, however, as between 4 and 16 weeks, there is a small but significant shrinkage in the total size of the lesion in Wolf and DiCara's (1969) study. Thus, the process of change they observed is as follows. Between 1 hr to 1 day after the lesion, apparent lesion size increases by as much as 400% (range 100% to 400%). This probably reflects two factors: the difficulty of assessing the exact boundaries of the lesion in the outer zone at 1 hr and the presence of some edema surrounding the lesion at 1 day. The apparent lesion size is rather constant during the next several weeks. As most of the animals used in tract-tracing experiments are allowed to survive for a period of somewhere between 1 and 14 days, the determination of lesion size does not usually create problems. This is especially true if the extent of the lesion is evaluated in silver-impregnated material where the tissue affected by the lesion is characteristically argyrophilic. Subsequently, there is a marked shrinkage of the lesion, and this is characteristic of many electrolytic lesions. In some lesions the shrinkage leads to marked tissue distortion which may have functional consequences.

Two further points of importance emerge from Wolf and DiCara's (1969) careful study. First, the size and shape of the lesion, given a constant current and time, is different in different areas of the brain. Second, the

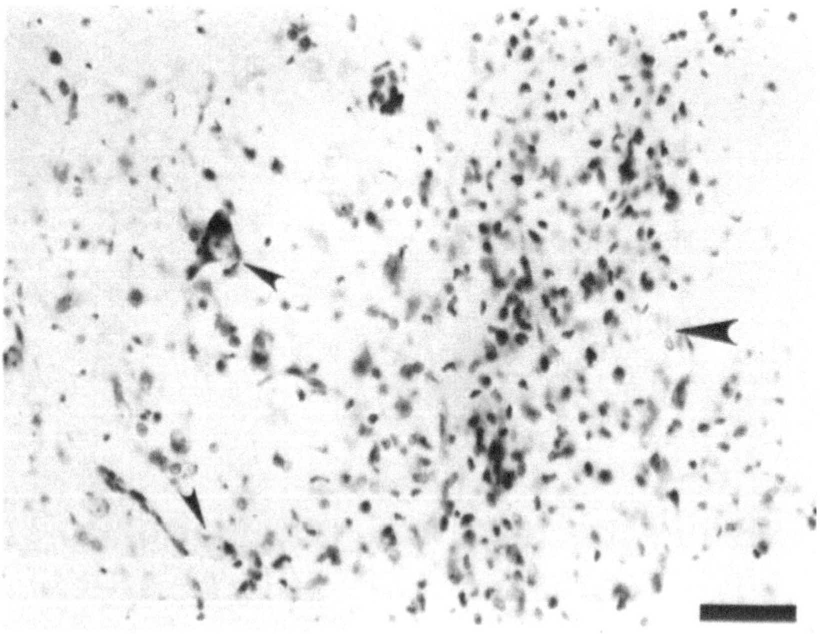

Figure 4. Photomicrograph from the section shown in Fig. 3A demonstrating the appearance of the border of the lesion. This is composed of a thin, dense glial scar (large arrow) with the immediately adjacent tissue having a normal histology and containing intact neurons (arrows). Bar, 35 μm.

amount of postoperative shrinkage of the lesions varies from area to area. It is evident from our own studies that two lesions may differ significantly in this respect. In Fig. 3, a lesion destroying the medial forebrain bundle including the nigrostriatal pathway is illustrated. This lesion, which was produced with anodal current 30 days before sacrifice of the animal, has resulted in little shrinkage or distortion of the tissue. It is represented by a central cavity with a very narrow rim of gliosis (Fig. 4), adjacent to which are completely normal appearing nerve cells. The lack of distortion is particularly evident in the myelin-stained section (Fig. 3). In contrast to this, another lesion produced by heat (radiofrequency lesion) in the locus coeruleus (Fig. 5) produces a progressive scarring with shrinkage and distortion of tissue (Fig. 6). These lesions emphasize the variability of electrolytic lesions, both as a function of method of production and as a function of the time interval following production of the lesion. The time factor is also important for evaluation of radiofrequency lesions, and these may show the same shrinkage and distortion as electrolytic lesions.

The interpretation of lesion effects in a functional context is a matter of considerable importance in many experimental studies. However, the subject is beyond the scope of this book, and the interested reader is referred to a recent discussion by the present author (Moore, 1978).

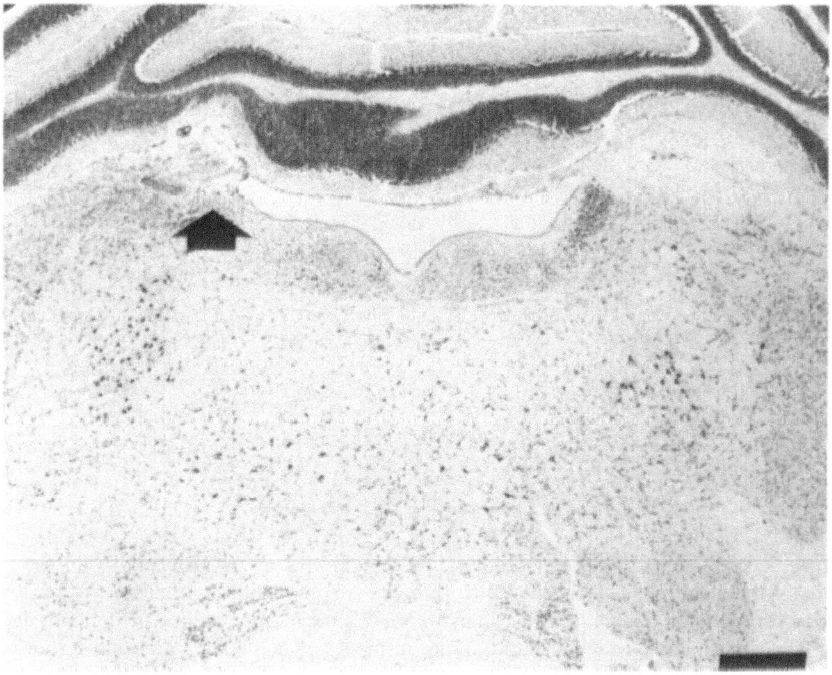

Figure 5. Photomicrograph of a unilateral locus coeruleus lesion in the rat produced by cathodal electrolytic current. The area of the lesion on the left is designated by an arrow. There is marked scarring, shrinkage, and distortion. Bar, 400 μm.

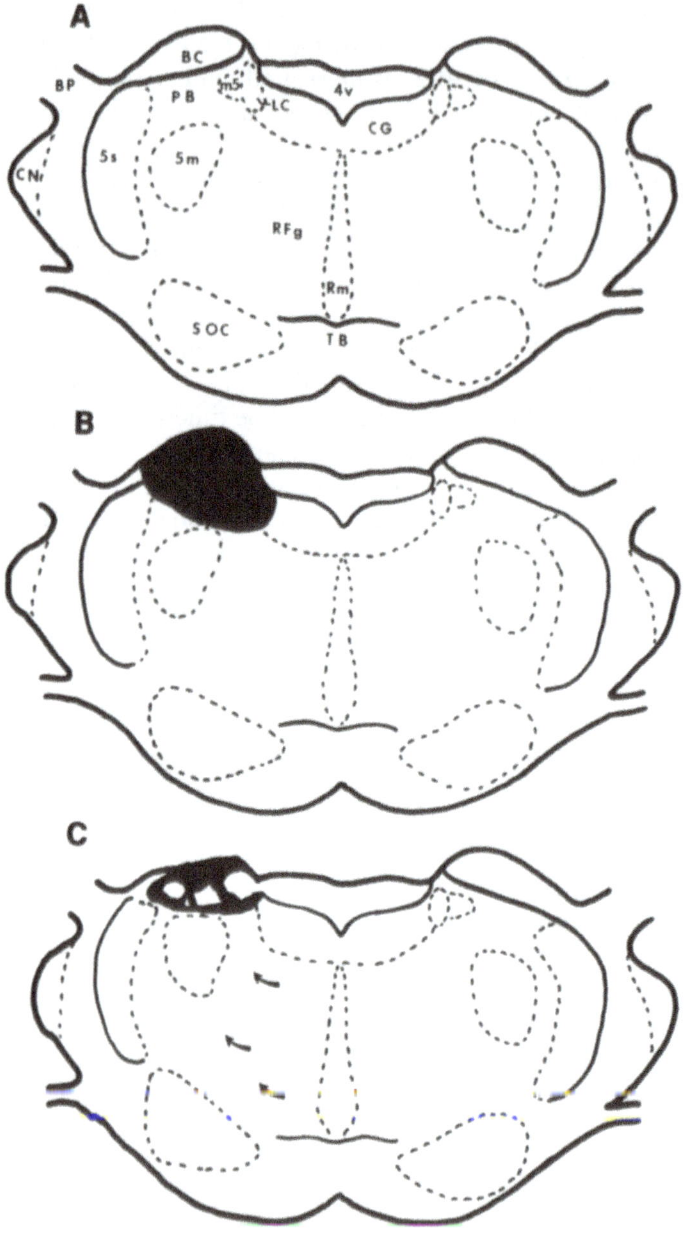

Figure 6. Diagrammatic representation of the sequence of events that follows an electrolytic lesion of the locus coeruleus produced by cathodal current as shown in Fig. 5. A drawing of a section through a normal brain is shown in (A). In (B), the location of an acute lesion is depicted by the blacked-in area. In (C), the appearance of the lesion at 30 days or beyond is shown. The blacked-in area represents glial scar tissue within and surrounding the lesion. The white patches within this area represent cavitary areas in the lesion. The arrows designate the direction of the distortion of brainstem structures that has taken place as the scarring occurs. In the section from which the diagram was made, the motor and sensory trigeminal nuclei and the adjacent

V. SELECTIVE LESION TECHNIQUES

The selectivity of some compounds as neurotoxins has long been discussed, particularly in the clinical literature, but it is only recently that selective neurotoxins have come into widespread experimental use. An early example of the experimental use of selective neurotoxins is the elegant study of Richter (1945) demonstrating selective destruction of the striatal complex with chronic carbon disulfide administration. More recently Purpura and Gonzalez-Monteagudo (1960) showed selective effects of methoxypyridines on hippocampal pyramidal cells. However, the greatest experimental interest has been generated by the development of 6-hydroxydopa, 6-hydroxydopamine, and related compounds (see de Champlain and Nadeau, 1971; Kostrzewa and Jacobowitz, 1974; Breese, 1975, for reviews), the dihydroxytryptamines (see Bjorklund *et al.*, 1973, 1974, 1975; Baumgarten *et al.*, 1975 for reviews), and kainic acid and other glutamate derivatives (see McGeer *et al.*, 1978a for review). These reviews summarize the history of the development of these compounds and their modes of action as well as their usage and effects, and most of this will not be considered further here. Rather, we shall concentrate on two aspects of the use of these selective neurotoxins, the method of application and the specificity of their effects in producing lesions in the CNS. For some time this has been a highly controversial field, but it appears now that some reasonably definitive statements can be made about the use of these compounds to produce selective lesions in the CNS.

A. Kainic Acid and Glutamate Derivatives

During the past 5 years there has been increasing interest in the use of kainic acid and other compounds related to glutamate as specific neurotoxins. The background for this information concerning the origin and chemistry of the compounds and a review of recent literature have been provided by McGeer *et al.* (1978a) and will not be repeated in detail here. The most provocative early work came from the studies of Olney and his associates (Olney, 1969a,b; Olney *et al.*, 1974) who demonstrated that glutamate produced retinal degeneration in neonatal mice, that monosodium glutamate produced discrete hypothalamic lesions in mice, and that kainic acid, a glutamate analogue obtained by extraction from the seaweed, *Digenea simplex*, is a much more powerful neurotoxin than glutamate itself. Subsequent work by a number of investigators, but particularly by Olney and co-workers, by

---

tegmentum were distorted. Abbreviations: 4v, fourth ventricle; 5s, trigeminal sensory nucleus; 5m, trigeminal motor nucleus; BC, brachium conjunctivum; BP, brachium pontis; CG, central gray; CN, cochlear nucleus; LC, locus coeruleus; m5, mesencephalic trigeminal nucleus; PB, parabrachial nucleus; RFg, gigantocellular reticular nucleus; Rm, nucleus raphe magnus; SOC, superior olivary complex; TB, trapezoid body.

Coyle and co-workers, and by the McGeers and their associates, have provided a large amount of information on the neurotoxic action of kainic acid.

It was noted early by McGeer and McGeer (1976) and by Coyle and Schwarcz (1976) that intrastriatal injection of kainic acid produced cell loss with axonal sparing in a manner that produced biochemical abnormalities similar to those observed in Huntington's disease. Further anatomical and biochemical aspects of his phenomenon have been rapidly elucidated (e.g., Schwarcz and Coyle, 1977; Coyle *et al.*, 1978; Hattori and McGeer, 1977). Herndon and Coyle (1977) demonstrated selective destruction of cerebellar neurons by kainic acid, and Nadler *et al.* (1978) demonstrated that intraventricular kainic acid injection produced highly specific destruction of neuronal groups in the hippocampal formation. Since the introduction of kainic acid as a selective neurotoxin, four major questions have emerged. First, what is the mechanism of action of kainic acid that produces selective neuronal destruction? Second, how selective are lesions produced by kainic acid? Third, if kainic acid is a selective neurotoxin for neuronal perikarya and dendrites but spares axons of passage, is it further selective in affecting only certain groups or classes of neurons? Fourth, what are the optimal procedures for the utilization of kainic acid as a selective neurotoxin?

1. Mechanism of Action

The neurotoxic potency of glutamate analogues and derivatives of kainic acid was examined in retina and neostriatum (Schwarcz *et al.*, 1978). Kainic acid was found to have greater neurotoxicity than any other compound studied, and a correlation was observed between the neurotoxic and neuroexcitatory effects of glutamate analogues. In other studies, Biziere and Coyle (1978) demonstrated that decortication, which removes the major glutamate innervation of neostriatum, greatly attenuates the neurotoxic action of kainic acid injected into neostriatum. This suggests that glutamate receptors are not the site of action of the kainic acid neurotoxicity. This is further supported by the observation that injection of kainic acid and glutamate into the neostriatum denervated of its neocortical input shows some cell destruction, indicating a partial restoration of the kainic acid effect. This suggests that the neuron must be activated by glutamate in order to be affected by kainic acid. Further support for this view derives from the work of Campochiaro and Coyle (1978) who elucidated the developmental time course of the neurotoxic effects of intrastriatal kainic acid. When injected into the neostriatum of 7-day-old rats, kainic acid has little or no effect. By 3 weeks, however, intrastriatal kainic acid produces neuronal loss similar to that observed in the adult. This interval conforms to the time during which the neocortical glutamine-producing innervation to neostriatum develops, further suggesting that activation by this pathway is necessary for the neurotoxic effect of kainic acid. Thus, it would appear that kainic acid acts by an excitatory driving of neurons with glutamate innervation to exhaustion and death through a mechanism that, at present, has not been elucidated. That is, neurons seem to require a glutamate innervation to be affected by kainic acid, and the effect is apparently dependent upon excitation of the neurons. However, the mechanism by which kainic acid produces such a marked and destructive, excitation is unknown. That neurons must be excitable to be susceptible to the neurotoxic effect of kainic acid is further indicated by the fact that neurotoxic effects are extremely difficult to obtain when the animal is deeply anesthetized with barbiturate (see Section V.A.3 below).

2. Selectivity

Most of the evidence obtained to the present has clearly indicated that kainic acid is a selective neurotoxin for dendrites and neuronal perikarya, sparing axons of passage and axon terminals. Evidence for this has been obtained for cerebellum (Herndon and Coyle, 1977) and neostriatum (Coyle *et al.*, 1978). We also have obtained evidence to support this view. Injections of kainic acid into lateral hypothalamus, which produce massive cell loss (Fig. 7), do not appear to affect axons of the medial forebrain bundle or of the ascending catecholamine neuron systems (Peterson and Moore, 1980). Mason and Fibiger (1979) have reported recently that injections of kainic acid into the midbrain tegmentum in the region of the ascending dorsal noradrenergic bundle arising from the locus coeruleus produce an approximately 25% depletion of hippocampal and cortical norepinephrine content. The authors suggest that kainic acid may cause some damage to axons of passage, but it is not clear from their studies that the effect does not result from axonal damage consequent to the glial reaction. Nevertheless, this aspect of the selectivity of the lesion remains open for further investigation. Our studies (Peterson and Moore, 1980) have not provided evidence for axonal damage (Fig. 8).

Similarly, it is not entirely clear at present whether any central neurons would be totally resistant to the neurotoxic effects of kainic acid. Nadler *et al.* (1978) have shown that there is a differential susceptibility to kainic acid in the hippocampal formation but, with sufficiently high doses of the toxin administered intraventricularly, all hippocampal neurons degenerate. In our own studies (Peterson and Moore, 1980), we have found most neurons of the dorsal thalamus and medial hypothalamus to be remarkably resistant to even high doses of kainic acid. All of our studies have been carried out with doses in the general range reported in the literature, so it is not possible to state that the neurons are absolutely resistant. Nevertheless, from all studies available, it would appear that there is a clear hierarchy of susceptibility to kainic acid neurotoxicity and that some neurons may be totally resistant to any specific effects of the neurotoxin, regardless of whether or not they receive a glutamate-containing innervation.

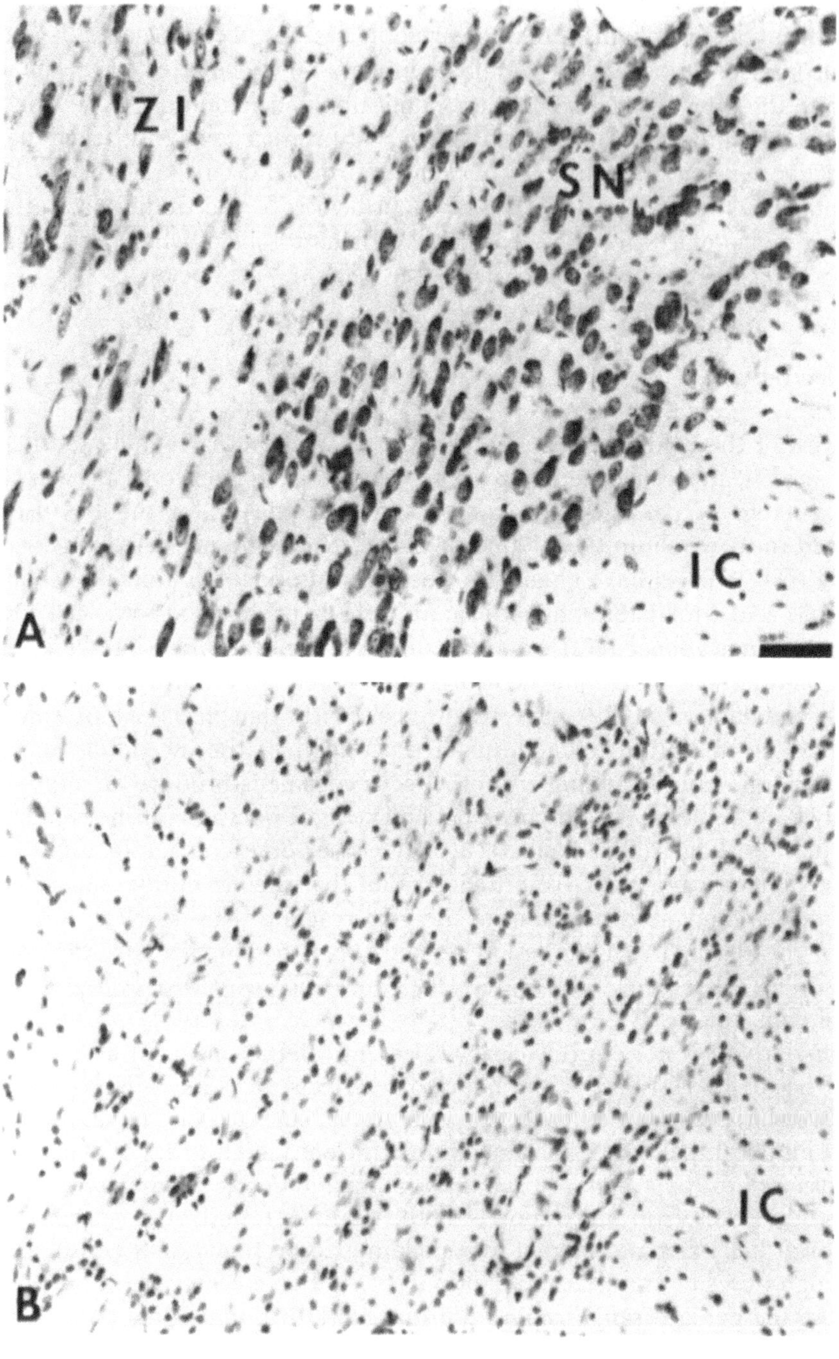

Figure 7. Photomicrograph of a cresyl-violet-stained coronal section through the subthalamic nuclei from an animal that had a unilateral injection of kainic acid into the lateral hypothalamus. (A) Contralateral to the injection site showing intact neurons of the subthalamic nucleus (SN) and zona incerta (ZI). (B) Ipsilateral side (negative reversed for comparison) showing marked neuronal loss in both subthalamic nucleus and zona incerta lying bove the internal capsule (IC). Bar, 100 μm.

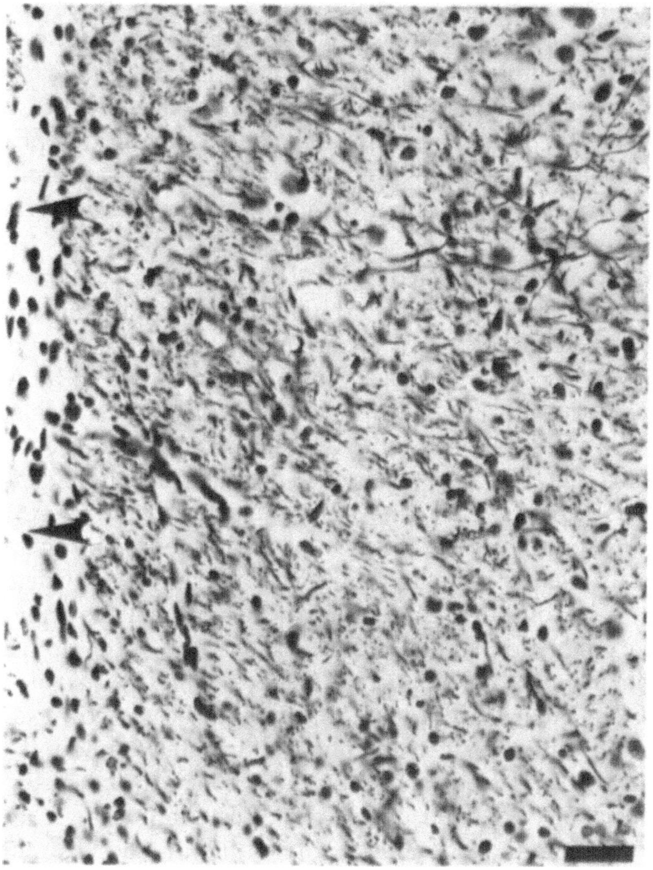

Figure 8. Photomicrograph of a coronal section through the lateral hypothalamus immediately adjacent to the end of the tract (arrows) produced by a needle through which a kainic acid injection was made. Numerous intact axons are present immediately adjacent to the site of injection. Bodian stain. Bar, 100 μm.

3. Procedures for Administration

Kainic acid has been used with three routes of administration. The first is intravenous. The toxin does not pass the blood–brain barrier and, consequently, only structures outside the blood–brain barrier will be affected by this route of administration. A second route of administration is intraventricular injection. This has been used by Nadler *et al.* (1978) in their studies of the hippocampal formation and appears particularly useful for producing selective lesions in this structure. Intraventricular administration of kainic acid is attendant by toxic symptoms which are dose related. In particular, the toxin produces prolonged convulsions that may be fatal. Nadler *et al.* (1978) state that at doses of kainic acid greater than about 5 nmoles injected intraventricularly over a 30-min period, most rats will be killed. In addition, the more rapidly the drug is administered, the lower is the dose required to achieve a fatal outcome. These investigators found that injection of 2.35 nmoles of kainic acid into each lateral ventricle over a 30-min period reliably destroyed all CA3 and CA4 pyramidal cells and, occasionally, CA1 pyramidal cells. For most purposes, intracerebral injection appears to be the optimal route. As with the neurotoxic monoamines, intracerebral injection of kainic acid results in a region of nonspecific damage surrounded by a zone of selective neurotoxicity.

Several factors should be considered when using the intracerebral route of administration. First, the concentration of kainic acid and speed of injection will play marked roles in determining its effects. McGeer *et al.* (1978b) report that injection of 2.5 nmoles of kainic acid in 1 μl of buffer produces greater effects with more rapid injection than with slow injection. The problem of diffusion differs significantly from that complicating the use of neurotoxic monoamines. In our experience, kainic acid injected into brain parenchyma in amounts as small as 1 nmole in 0.5 μl may diffuse and demonstrate neurotoxic effects for as far as 2–3 mm from the site of injection. This may even occur with neurotoxic effects being evident at the site of injection, a surrounding zone of apparent normality as demonstrated by histological examination, and a second, distant zone of cell destruction. Consequently, in the use of kainic acid, the possibility of distant effects must always be taken into consideration, and careful histological examination of material made. A further consideration is that the compound may spread into the ventricular system and cause effects at a distance by that route. Another source of undesired neurotoxic effects can occur from leakage of even very small amounts of kainic acid from the injection cannula.

Two further considerations deserve mention. First, anesthesia has a marked effect on the neurotoxic action of kainic acid. Animals anesthetized with pentobarbital or other barbiturates require much larger doses of kainic acid in the desired area to produce cell destruction than animals anesthetized with, for example, halothane. In general, it is advisable to utilize the smallest possible dose of kainic acid in animals anesthetized with halothane or some other anesthetic agent that does not produce marked and long-lasting central nervous system depression. Second, the investigator must be cognizant of other problems with the use of kainic acid that may markedly alter the results obtained. McGeer *et al.* (1978b) report that consistent results can be obtained with aliquots of a 10–20 μmoles/ml solution stored at −20°C and thawed immediately before use. Solutions stored at 4°C or at low concentration appear to deteriorate and are inconsistent in effect. Further, there has been variability from batch to batch of kainic acid, with some batches obtained in our laboratory having much less effect per dose than others, and some being totally without effect. For that reason, the investigator should carefully test each batch for neurotoxic action before using it in an extensive experiment.

For the investigator who wishes to initiate use of kainic acid, the following protocol should be helpful. Kainic acid should be dissolved in physiological saline or a phosphate buffer at physiological pH in a concentration of approximately 5 nmoles per 1 μl of solution. This amount of solution should be injected into the neostriatum over 1–5 min, the animal sacrificed 1–5 days later, and the neostriatum examined histologically for evidence of neuronal destruction. This amount of kainic acid should produce considerable neuronal destruction. If this is obtained, the investigator can then proceed to the region of his interest. If this is a small nucleus, the volume of solution should be reduced, and the amount of kainic acid can be determined empirically. Usually this should fall in the range of 1–5 nmoles. It should be injected through a small cannula such as a 32 gauge needle attached to a microsyringe. For optimal results, the injection should be carried out over a 1- to 5-min period with the animal anesthetized with halothane or very light ether. This protocol will reduce variability in the technique to a minimum but certainly will not eliminate it.

Although kainic acid has the potential of becoming a very powerful tool for production of selective lesions in the central nervous system, the search for a more suitable substance continues. Recently ibotenic acid, which is less toxic than kainic acid, has been shown to produce kainatelike lesions (Schwarcz *et al.*, 1978). The work with these compounds has just begun, and it is not surprising to find that their application is beset with technical problems and uncertainties. However, a chemical compound that would have the extraordinary ability to destroy nerve cell bodies but not axons and could be easily injected and contained within a restricted area of the central nervous system would have a profound impact on anatomical and behavioral studies of nervous system organization.

B. Neurotoxic Catecholamine and Indolamine Derivatives

A large number of chemical analogues of the catecholamines have been described to have neurotoxic effects (Breese, 1975) but, at present, only 6-hydroxydopamine (6-OH-DA) and 6-hydroxydopa (6-OH-DOPA) are in general experimental use. 6-OH-DA can be administered parenterally, intraventricularly, and by intracerebral injection. Parenteral administration in the adult animal has significant, but modest, effects on norepinephrine content in some areas of the CNS but none in others (Sachs and Jonsson, 1973). In the newborn animal, prior to the development of a blood–brain barrier for 6-OH-DA (Sachs, 1973), parenterally administered 6-OH-DA produces marked decreases in telencephalic norepinephrine content (Singh and de Champlain, 1972; Jonsson *et al.*, 1974) with subsequent increases in brainstem amine content. Injection of 6-OH-DA into the adult animal either intracisternally or intraventricularly has differing effects on brain catecholamines depending upon dosage, schedule of administration, and the concomitant use of other pharmacologic agents that affect catecholamine metabolism. The effects appear to be relatively specific on a biochemical level; levels of catecholamine are reduced, whereas serotonin and other putative neurotransmitter levels appear unaffected (Breese, 1975; Sachs *et al.*, 1973). On a morphological level, intracisternal injection of 6-OH-DA produces degeneration of known dopamine neuron pathways demonstrable by silver impregnation techniques and cell loss in the substantia nigra and ventral tegmental area (Hedreen and Chalmers, 1972). Yet, despite the fact that norepinephrine levels were markedly reduced, Hedreen and Chalmers (1972) could not find any indication of degeneration in norepinephrine pathways or terminal areas. This probably represents a technical problem related to the silver impregnation methods used. In addition, it should be noted that little nonspecific degeneration was noted. Both degeneration of norepinephrine neurons and nonspecific changes occur in some circumstances, however, as Descarries and Saucier (1972) have reported total loss of the locus coeruleus and a generalized hydrocephalus after intraventricular 6-OH-DA administration.

6-OH-DOPA has the advantage that it can be determined either parenterally (Jacobowitz and Kostrzewa, 1971, Sachs *et al.*, 1973) or by intraventricular injection (Richardson and Jacobowitz, 1973). Since 6-OH-DOPA passes the blood–brain barrier, either route of administration will affect central catecholamine neurons. Higher doses are required for parenteral administration, and the agent will also affect, in this case, peripheral as well as central adrenergic neurons; this may be of importance in analysis of functional effects. Since the drug reaches the brain via its vascular supply, however, its distribution within the CNS will be relatively uniform. In contrast to this, intracisternal or intraventricular injection will produce greatest effects near the ventricular and pial surfaces. Consequently, the route of administration will be determined by the requirements of the particular experimental design. Specific dose schedules and other details concerning administration of 6-OH-DA or 6-OH-DOPA can be obtained from references cited by Kostrzewa and Jacobowitz (1974) and Breese (1975).

The application of 6-OH-DA or 6-OH-DOPA, either parenterally or by intracisternal or intraventricular injection, is advantageous in studies where widespread effects on CNS catecholamine neuron systems are desired. However, because of the widespread effects of intraventricular administration of 6-OH-DA, Ungerstedt (1971) introduced direct, intracerebral injection of a drug to produce restricted lesions intended principally to affect only a catecholamine neuron system in the region of the injection. Although this methodology has been employed extensively since that time, it also has become the subject of substantial controversy (see below). The general procedure is as follows (see Ungerstedt, 1971, for details). The animal is held in a stereotaxic machine with an appropriate device (e.g., microliter syringe) attached for injecting the 6-OH-DA. Dosages usually range from 1–2 μg 6-OH-DA to 10 μg (calculated as the base), although much higher doses have been used. The drug is injected in either water or some isotonic solution containing small amounts of ascorbic acid to prevent autooxidation. The volumes of solution used usually range from 0.25 to 4 μl. There are a number of factors that determine the character of the final lesion produced. First, the size of the injection needle will determine the amount of direct nonspecific, mechanical damage produced (Fig. 9). Second, the speed of injection will determine similar effects. Injections of material at a rate of greater than 1 μl in 1–2 min appear to produce more nonspecific tissue damage and less diffusion than do more slowly applied injections. From our own studies we would recommend injection rates in the range of 1 μl every 5 min to obtain minimal local tissue injury and maximum diffusion. Third, the distance over which injected 6-OH-DA diffuses has clearly been shown to be a function of the volume of material injected.

The major controversy regarding intracerebral 6-OH-DA injection has arisen over whether the effects are specific and, if so, the extent to which they are specific (Hökfelt and Ungerstedt, 1973; Sotelo *et al.*, 1973; Agid *et al.*, 1973; Butcher *et al.*, 1974; Simon *et al.*, 1974; Bloom, 1975; Poirier, 1975; Javoy, 1975; Butcher *et al.*, 1975; Butcher, 1975; Stahl *et al.*, 1975). At the present time most authors agree that, beyond the acute period, a lesion induced by the injection of 6-OH-DA into the cerebral substance has three or four more or less distinctive zones. The first of these is the mechanical damage, noted above, produced by passage of the injection needle. This is usually surrounded, particularly in older lesions, by a second zone of reactive astrocytic gliosis, the usual response of CNS to a lesion of this sort. The

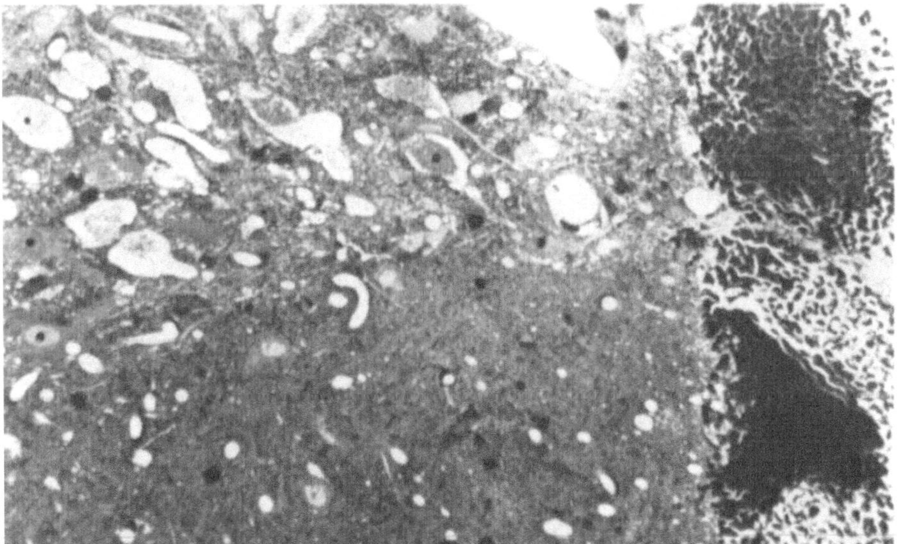

Figure 9. Photomicrograph of substantia nigra 24 hr after injection of 6-OH-DA. The hemorrhagic needle tract is present on the right extending through the pars compacta into pars reticulata. Many degenerating neurons are evident in the pars compacta. Bar, 200 μm. [Courtesy of Dr. Tomas Hökfelt.]

third zone is the region within the injection site where all, or nearly all, neuronal elements appear to be affected. The size of this zone appears to be a function of the concentration of 6-OH-DA injected. There is a general consensus that 6-OH-DA in sufficient concentration is a potent nonselective neurotoxin, and such effects would appear to account, for example, for the observations reported by Poirier *et al.* (1972). It has been recommended that the dose not exceed 4 μg of 6-OH-DA in 2 μl of solution to minimize such nonselective toxic effects. The zone of nonselective effects grades into a third zone which, in turn, grades imperceptibly into noninvolved neural tissue. In the third zone (Hökfelt and Ungerstedt, 1973; Sotelo *et al.*, 1973), there is selective degeneration of some neuronal constituents with other, adjacent neuronal structures remaining entirely intact (Figs. 10 and 11). The extent of this zone is dependent upon the distance over which the 6-OH-DA diffuses. As noted above, this is a function of the volume and rate of injection. Most authors interpret the morphological changes in this zone as reflecting selective changes in catecholamine neurons because of the specific

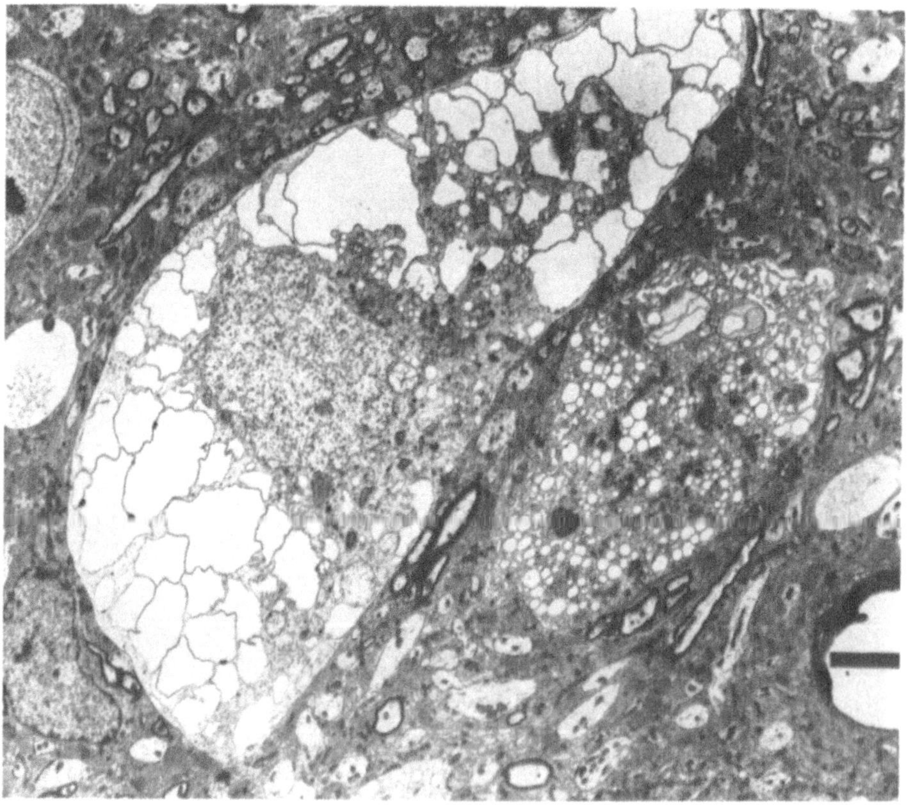

Figure 10. Electron micrograph demonstrating degenerating neuron and large dendrite in substantia nigra pars compacta 24 hr after 6-OH-DA injection. Bar, 250 μm. [Courtesy of Dr. Tomas Hökfelt.]

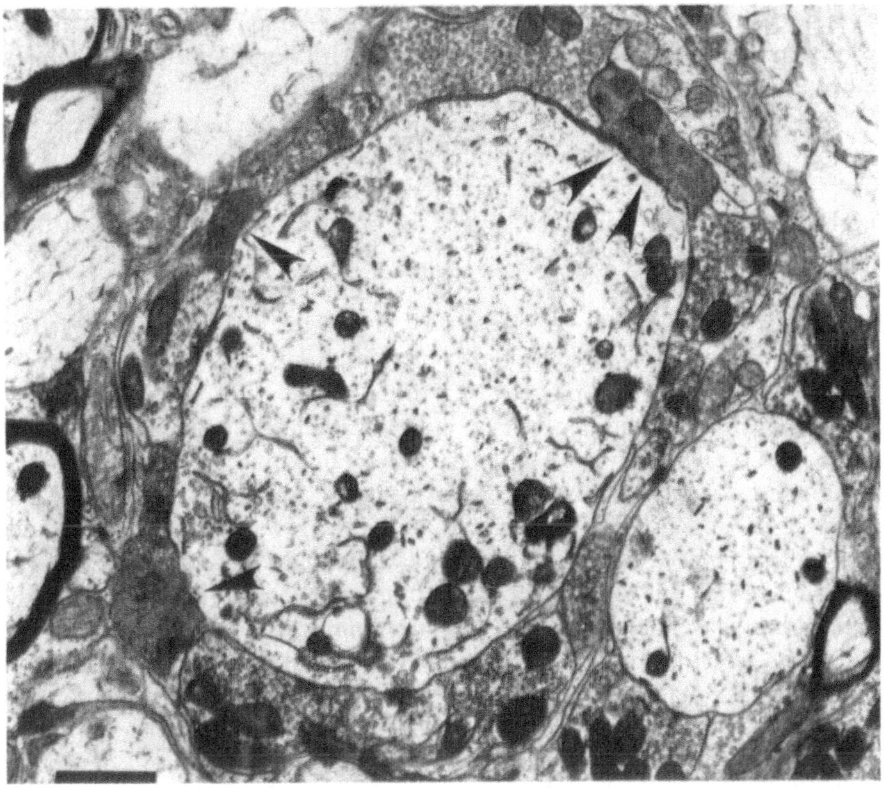

Figure 11. Electron micrograph showing degenerating axon terminals (arrows) along large dendrite in the motor trigeminal nucleus 24 hr after injection of 6-OH-DA. Bar, 1 μm.

uptake mechanisms present over the entire cell surface of these neurons in the CNS. Certainly, there is ample evidence that both central neuronal perikarya, axons, and axon terminals are destroyed. The susceptibility of terminals is predictable on the basis of their well-known uptake mechanisms for catecholamines. The susceptibility of axons is more surprising, but these are undoubtedly unmyelinated axons that, because of their highly collateralized organization and their availability to contents of the extracellular space, may also be differentially susceptible to 6-OH-DA. Certainly it is commonplace to see catecholamine accumulations in axons traversing the brainstem at a distance from the site of nonselective damage with a 6-OH-DA injection.

Under ideal circumstances, an intracerebral injection of 6-OH-DA will produce a small area of damage caused by the injection cannula. This will be surrounded by a thin glial scar and an area of nonspecific cellular pathology. The extent of this will be a function largely of the concentration of 6-OH-DA injected and the rate of injection. Surrounding this will be a zone of selective neuronal damage to catecholamine neurons and this, too, will be a function of the concentration, volume, and rate of injection of the 6-OH-DA. For each experiment, these variables must be determined empirically in order to produce the optimal selective lesion with a minimum of mechanical and nonselective effects.

Following the introduction of the neurotoxic catecholamines there was a search for similar compounds that would affect indolamine neuron systems. The obvious choice, a hydroxylated serotonin, 5,6-dihyroxytryptamine (5,6-DHT), proved to produce significant and long-lasting effects on brain serotonin levels (Baumgarten *et al.*, 1971). Subsequently, a wide variety of such compounds has been produced and studied (Baumgarten *et al.*, 1975; Björklund *et al.*, 1975), and a great deal is known of the structure–activity relationships of these hydroxylated tryptamines with respect to neurotoxicity. As Breese (1975) has pointed out, these compounds have limitations in applicability. When administered by intraventricular injection, 75 μg (calculated as the base) of 5,6-DHT appears to be a maximum dose but produces only modest changes in CNS serotonin content, at least in most areas. The introduction of 5,7-dihydroxytryptamine (5,7-DHT) has proven advantageous in this respect. When administered alone, 5,7-DHT produces not only significant reductions in CNS serotonin but a marked reduction in CNS norepinephrine as well (Baumgarten and Lachenmayer, 1972; Björklund *et al.*, 1974). The selectivity of 5,7-DHT can be improved with the use of desipramine to block uptake by norepinephrine neurons, so that very marked decreases in CNS serotonin can be produced without alteration of norepinephrine levels (Björklund *et al.*, 1975). The application of neurotoxic indolamines has markedly advanced our understanding of the organization and response to injury of central serotonin neuron systems (Björklund *et al.*, 1973, 1974; Baumgarten *et al.*, 1975). Much of the work on these compounds has employed intraventricular injection. It is a common observation (Breese, 1975; Halaris *et al.*, 1977; Moore *et al.*, 1978) that hydrocephalus, ependymal staining of the ventricular surfaces, and other evidences of nonspecific effects are commonplace in the brains of animals receiving such injections. Nevertheless, we have seen a number of brains in which 75 μg of 5,6-DHT was placed intraventricularly without such changes and have carried out a study of the pattern of neuronal degeneration in such brains (Moore, *unpublished observations*) using the Fink–Heimer silver impregnation method. Degenerating axons and terminals are present at both 2 and 4 days after the intraventricular administration of 75 μg 5,6-DHT. The distribution of these conforms in large part to the known distribution of serotonin neuron systems in the rat brain, and there is no evidence for massive, nonselective degeneration. Indeed, save for what appears to be more than the expected periventricular degeneration and some degenerating axons in the medial lemniscus, the degeneration appears quite selective. It is of interest that the proximal axons and cell bodies within the raphe nuclei showed no changes in these brains.

The neurotoxic indolamines have also been used for intracerebral injection (Björklund *et al.*, 1973, 1974), usually in doses of 4 μg in 4 μl of solution, but some parametric studies have been performed (Björklund *et al.*, 1973). As with the 6-OH-DA injections, there is an area of central damage produced by the cannula. This is surrounded by an area of nonselective damage and a final zone which is pigmented but appears instact save for the serotonin neurons. The observations of Björklund *et al.* (1973) indicate that this is a promising technique, but it has not been applied extensively, and no ultrastructural studies have been performed on the region of the injection site to allow comparisons with the 6-OH-DA studies. The dihydroxytryptamines have come into general use, and it is evident that they provide a powerful tool for producing selective lesions of CNS indolamine neuron systems.

VI. THE INTERPRETATION OF LESION EFFECTS

In the introduction to this chapter it was noted that CNS lesions have been used extensively to analyze the structure and function of the nervous system. The basis for study of the effects of lesions is the general neurological dictum that specialization of structure implies specialization of function. The purpose of this chapter has been to provide a critical presentation of lesion techniques in order that investigators may employ them most effectively in the analysis of the organization of the nervous system. This provides at least a portion of the basis for analysis of function. Here, the technical problems of lesion placement, selection of the most appropriate means of making the lesion, and determination of what has been ablated remain the same as for purely morphological studies. Interpretation of functional deficits represents a much more difficult problem and will not be considered here (see Brodal, 1973; Schoenfeld and Hamilton, 1977; Moore, 1978, for reviews).

VII. APPENDIX: STEREOTAXIC ATLASES

The atlases listed here are selected from several sources including the author's personal experience and extended lists of stereotaxic atlases that can be found in the following publications:

Heimer, L., and Lohman, H. M., 1976, Anatomical methods for tracing connections in the central nervous system, in: *Handbook of Psychobiology* (M. Gazzaniga and C. Blakemore, eds.), pp. 73–105, Academic Press, New York.

Stereotaxic Atlases—An Annotated List, Brain Information Service, University of California at Los Angeles, Los Angeles, California 90024.

Stereotaxic Atlases, David Kopf Instruments, 7324 Elmo Street, Tujunga, California 91042.

The atlases listed below are for groups of animals in common laboratory use. Those marked with an asterisk (*) are those the author has personally used and found to be anatomically sound and reasonably reliable within the limitations of stereotaxic technique discussed in the text.

A. Rat Brain Atlases

*deGroot, J., 1959, The rat forebrain in stereotaxic coordinates, *Tr. R. Neth. Acad. Sci.* **52:**1–40.

*deGroot, J., 1959, The rat hypothalamus in sterotaxic coordinates, *J. Comp. Neurol.* **113:**389–400.

Fifkova, E., and Marsala, J., 1960, Stereotaxic atlas for the rat, in: *Electrophysiological Methods in Biological Research* (J. Bures, N. Petran, and J. Zachar, eds.), pp. 444–453, Academic Press, New York.

*Hurt, G. A., Hanaway, J., and Netsky, M. G., 1971, Stereotaxic atlas of the mesencephalon in the albino rat, *Confinia Neurologica* **33:**93–115.

König, J. F. R., and Klippel, R. A., 1963, *The Rat Brain—Stereotaxic Atlas*, Williams & Wilkins, Baltimore.

*Pellegrino, L. J., Pellegrino, A. S., and Cushman, A. J., 1979, *A Stereotaxic Atlas of the Rat Brain,* 2nd edition, Plenum Press, New York.

*Sherwood, N. M., and Timiras, P. S., 1970, *A Stereotaxic Atlas of the Developing Rat Brain*, University of California Press, Berkeley.

*Szentagothai, J., 1962, Cytoarchitectonic atlas of the rat brain in Horsley–Clarke coordinates, in: *Hypothalamic Control of the Anterior Pituitary*, pp. 20–37, Publishing House of the Hungarian Academy of Sciences, Budapest.

B. Cat Brain Atlases

*Berman, A. R., 1968, *The Brain Stem of the Cat: A Cytoarchitectonic Atlas with Stereotaxic Coordinates*, University of Wisconsin Press, Madison.

*Bleier, R., 1961, *The Hypothalmus of the Cat—A Cytoarchitecture Atlas in Horsley–Clarke Coordinate System*, Johns Hopkins University Press, Baltimore.

Fifkova, E., and Marsala, J., 1960, Stereotaxic atlas for the cat, in: *Electrophysiological Methods in Biological Research* (J. Bures, N. Petran, and J. Zachar, eds.) pp. 426–43, Academic Press, New York.

*Jasper, H. H., and Ajmone-Marsan, C., 1954, *A Stereotaxic Atlas of the Diencephalon of the Cat*, National Research Council, Ottawa, Canada.

Snider, R. S., and Niemer, W. T., 1961, *A Stereotaxic Atlas of the Cat Brain*, University of Chicago Press, Chicago.

Verhaart, W. J. C., 1964, *Stereotaxic Atlas of the Brain Stem of the Cat*, Davis, Philadelphia.

C. Primate Brain Atlases

Davis, R., and Huffman, R. D., 1968, *A Stereotaxic Atlas of the Brain of the Baboon (Papio)*, University of Texas Press, Austin.

Delucchi, M. R., Dennis, B. J., and Adey, W. R., 1965, *A Stereotaxic Atlas of the Chimpanzee Brain*, University of California Press, Berkeley.

*Emmers, R., and Akert, K., 1962, *A Stereotaxic Atlas of the Brain of the Squirrel Monkey*, University of Wisconsin Press, Madison.

Gergen, J. A., MacLean, P. D., 1962, *A Stereotaxic Atlas of the Squirrel Monkey's Brain*, U.S. Public Health Service Publication No. 933, Washington.

Kusama, T., and Mabuchi, M., 1970, *Stereotaxic Atlas of the Brain of Macaca Fuscata*, University Park Press, Baltimore.

Manocha, S., Shantha, T. R., and Bourne, G. H., 1968, *A Stereotaxic Atlas of the Brain of the Cebus Monkey (Cebus apella)*, Clarendon Press, Oxford.

*Olszewski, J., 1952, *The Thalamus of the Macaca Mulatta*, Karger, Basel.
Shantha, T. R., Manocha, S. L., and Bourne, G. H., 1968, *A Stereotaxic Atlas of the Java Monkey Brain (Macaca Irus)*, Williams & Wilkins, Baltimore.

D. Dog Brain Atlases

Dua-Sharma, S., Sharma, S., and Jacobs, H. L., 1970, *The Canine Brain in Stereotaxic Coordinates*, MIT Press, Cambridge.
Lim, R. K. S., Liu, C., and Moffitt, R., 1960, *A Stereotaxic Atlas of the Dog's Brain*, Charles C. Thomas, Springfield, Illinois.

ACKNOWLEDGMENTS. This review represents a revision of one previously published in the *Handbook of Psychopharmacology, Volume 9* (L. Iversen, S. Iversen, and S. H. Snyder, eds.), Plenum Press, New York, 1978. The preparation of the chapter and some of the work reported in it were supported by USPHS Grants NS 16303, NS 16304, National Institutes of Health, and by Grant MH 34202 from the National Institute of Mental Health.

REFERENCES

Agid, Y., Javoy, F. Glowinski, J., Bouvet, D., and Sotelo, C., 1973, Injection of 6-hydroxydopamine into the substantia nigra of the rat. II. Diffusion and specificity, *Brain Res.* **58**:291–301.

Aronow, S., 1960, The use of radiofrequency power in making lesions in the brain, *J. Neurosurg.* **17**:431–438.

Baumgarten, H. G., and Lachenmayer, I., 1972, 5,7-Dihydroxytryptamine improvement in chemical lesioning of indoleamine neurons in the mammalian brain, *Z. Zellforsch.* **135**:399–414.

Baumgarten, H. G., Björklund, A., Lachenmayer, I., Nobin, A., and Steveni, U., 1971, Long-lasting selective depletion of brain serotonin by 5,6-dihydroxytryptamine, *Acta Physiol. Scand.* [*Suppl.*] **373**:1–15.

Baumgarten, H. G., Björklund, A., Nobin, A., Rosengten, T., and Schlossberger, H. G., 1975, Neurotoxicity of hydroxylated tryptamines: Structure–activity relationships. I. Long-term effects on monoamine content and fluorescence morphology of central monoamine neurons, *Acta Physiol. Scand.* [*Suppl.*] **429**:1–27.

Biziere, K., and Coyle, J. T., 1978, Influence of cortico-striatal afferents on striatal kainic acid neurotoxicity, *Neurosci. Lett.* **8**:303–310.

Björklund, A., Nobin, A., and Stenevi, U., 1973, The use of neurotoxic dihydrotryptamines as tools for morphological studies and localized lesioning of central indolamine neurons, *Z. Zellforsch.* **145**:479–501.

Björklund, A., Baumgarten, H. G., and Nobin, A., 1974, Chemical lesioning of central monoamine axons by means of 5,6-dihydroxytryptamine and 5,7-dihydroxytryptamine, *Adv. Biochem. Psychopharmacol.* **10**:13–33.

Björklund, A., Horn, A. S., Baumgarten, H. G., Nobin, A., and Schlossberger, H. G., 1975. Neurotoxicity of hydroxlated tryptamines: Structure–activity relationships. 2. *In vitro* studies on monoamine uptake inhibition and uptake impairment, *Acta Physiol. Scand.* [*Suppl.*] **429**:29–60.

Bleier, R., 1961, *The Hypothalamus of the Cat—A Cytoarchitecture Atlas in Horsley–Clarke Coordinate System*, Johns Hopkins University Press, Baltimore.

Bloom, F. E., 1975, Monoaminergic neurotoxins: Are they selective?, *J. Neural Transm.* **37**:183–187.

Breese, G. R., 1975, Chemical and immunochemical lesions by specific neurotoxic substances and antisera, in: *Handbook of Psychopharmacology, Vol. 1* (L. L. Iversen, S. D. Iversen, and S. H. Snyder, eds.), pp. 137–189, Plenum Press, New York.

Bremer, F., 1935, Cerveau isolé et physiologie du commeil, *C. C. Seances Soc. Biol. Paris* **118:**1235–1242.

Brodal, A., 1973, Self-observations and neuro-anatomical considerations after a stroke, *Brain* **96:**675–694.

Butcher, L. L., 1975, Degenerative processes after punctate intracerbral administration of 6-hydroxydopamine, *J. Neural Transm.* **37:**189–208.

Butcher, L. L., Eastgate, S. M., and Hodge, G. K. 1974, Evidence that punctate intracerebral administration of 6-hydroxydopamine fails to produce selective neuronal degeneration, *Naunyn-Schmiedebergs Arch. Pharmacol.* **285:**31–70.

Butcher, L. L., Hodge, G. K., and Schaffer, J. C., 1975, Degenerative processes after intraventricular infusion of 6-hydroxydopamine, in: *Chemical Tools in Monoamine Research* (G. Jonsson, T. Malmfors, and C. Sachs, eds.), Elsevier/North Holland, Amsterdam.

Campochiaro, P., and Coyle, J. T., 1978, Ontogenetic development of kainate neurotoxicity: Correlates with glutamatergic innervation., *Proc. Natl. Acad. Sci. USA.* **75:**2025–2029.

Carpenter, M. B., and Whittier, J. R., 1952, Study of methods for producing experimental lesions in the central nervous system with special reference to stereotaxic technique, *J. Comp. Neurol.* **97:**73–132.

Clarke, R. H., 1920, *Investigation of the Central Nervous System,* Johns Hopkins Hospital Reports, Special Vol. 1, pp. 1–160, Lord Baltimore Press, Baltimore.

Cooper, J. S., 1962, A cryogenic method for physiologic inhibition and production of lesions in the brain, *J. Neurosurg.* **19:**853–855.

Cox, T. R., Gisi, J. M., and Wolf, G., 1978, Deposit-free electrolytic brain lesions: Methodologic and histologic observations, *Physiol. Behav.* **18:**967–973.

Coyle, J. T., and Schwarcz, R., 1976, Model for Huntington's chorea: Lesion of striatal neurons with kainic acid, *Nature* **263:**244–246.

Coyle, J. T., Molliver, M. E., and Kuhar, M. J., 1978, *In situ* injection of kainic acid: A new method for selectively lesioning neuronal cell bodies while sparing axons of passage, *J. Comp. Neurol.* **180:**301–323.

de Champlain, J., and Nadeau, R., 1971, 6-hydroxydopamine, 6-hydroxydopa and degeneration of adrenergic nerves, *Fed. Proc.* **39:**877–885.

deGroot, J., 1959, the rat hypothalamus in stereotaxic coordinates, *J. Comp. Neurol.* **113:**389–400.

Descarries, L., and Saucier, G., 1972, Disappearance of the locus coeruleus of the rat after intraventricular 6-hydroxydopamine, *Brain Res.* **37:**310–316.

Dixon, R. G., Riley, J. N., and Marchand, E. R., 1978, Simple template method for targeting brain loci in small animal stereotaxy, *Brain Res. Bull.* **3:**721–722.

Fry, F. J., and Cowan, W. M., 1972, A study of retrograde cell degeneration in the lateral mammillary nucleus of the cat, with special reference to the role of axonal branching in the preservation of the cell, *J. Com. Neurol.* **144:**1–24.

Fry, W. J., Fry, F. J., Malck, R., and Paukau, J. W., 1964, Quantitative neuroanatomic studies implemented by ultrasonic lesions—Mammillary nuclei and associated complex of cat brain, *J. Acoust. Soc. Am.* **36:**1795–1835.

Gold R. M., 1975, Anodal electrolytic brain lesions: How current and electrode mental influence lesion size and hyperphagiosity, *Physiol. Behav.* **14:**624–632.

Gold, R. M., Kapatos, G., and Carey, R. J., 1973, A retracting wire knife for stereotaxic brain surgery made from a microliter syringe, *Physiol. Behav.* **10:**813–815.

Goldberg, J. M., and Moore, R. Y., 1967, Ascending projections of the lateral lemniscus in the cat and monkey, *J. Comp. Neurol.* **129:**143–156.

Groeneweger, H. J., and Voogd, Y., 1977, The parasagittal zonation within the olivocerebellar projection. I. Climbing fiber distribution in the vermis of cat cerebellum, *J. Comp. Neurol.* **174:**417–488.

Halaris, A. E., Jones, B. E., and Moore, R. Y., 1976, Axonal transport in serotonin neurons of the midbrain raphe, *Brain Res.* **107:**555–574.

Halasz, B., 1969, The endocrine effects of isolation of the hypothalamus from the rest of the brain, in: *Frontiers in Neuroendocrinology* (W. F. Ganong and L. Martini, eds.), pp. 307–342, Oxford University Press, New York.

Halasz, B., and Pupp, L., 1965, Hormone secretion of the anterior pituitary gland after physical interruption of all nervous pathways to the hypophysiotrophic area, *Endocrinology* **77**:553–562.

Hattori, T., and McGeer, E. G., 1977, Fine structural changes in the rat striatum after local injections of kainic acid, *Brain Res.* **129**:174–180.

Hedreen, J. C., and Chalmers, J. P., 1972, Neuronal degeneration in rat brain induced by 6-hydroxydopamine: A histological and biochemical study, *Brain Res.* **47**:1–36.

Herndon, R., and Coyle, J. T., 1977, Selective destruction of neurons by a transmitter agonist, *Science* **198**:71–72.

Hökfelt, T., and Ungerstedt, U., 1973, Specificity of 6-hydroxydropamine-induced degeneration of central monoamine neurons: An electron and fluorescence microscopic study with special references to intracerebral injection on the nigrostriatal dopamine system, *Brain Res.* **60**:269–297.

Horsley, V., and Clarke, R. H., 1908, The structure and functions of the cerebellum examined by a new method, *Brain* **31**:45–124.

Hurt, G. A., Hanaway, J., and Netsky, M. A., 1971, Stereotaxic atlas of the midbrain of the albino rat, *Confin. Neurol.* **33**:93–115.

Jacobowitz, D., and Kostrzewa, R., 1971, Selective action of 6-hydroxydopa on noradrenergic terminals: Mapping of preterminal axon of the brain, *Life Sci.* **10**:1329–1342.

Javoy, F., 1975, Partial selectivity of 6-OHDA induced neuronal degeneration after intratissular injection in the brain with special reference to the nigro-striatal dopamine system, *J. Neural Transm.* **37**:219–227.

Jonsson, G., Pycock, C., Fuxe, K., and Sachs, C., 1974, Changes in the development of central noradrenaline neurons following neonatal administration of 6-hydroxydopamine, *J. Neurochem.* **22**:419–426.

Knook, H. L., 1969, Stereotaxic lesioning with micro-electrodes, *Psychiatr. Neurol. Neurochir.* **72**:61–64.

Kostrzewa, R. M., and Jacobowitz, D. M., 1974, Pharmacological actions of 6-hydroxydopamine, *Pharmacol. Rev.* **26**:199–288.

Malis, L. I., Beaker, C. P., Kruger, L., and Rose, J. E., 1960, Effects of heavy ionizing monoenergetic particles on cerebral cortex. I. Production of laminar lesions and dosimetric considerations, *J. Comp. Neurol.* **115**:219–242.

Marchand, E. R., and Riley, J. N., 1979, A self-centering head holder system for small animal stereotaxy, *Brain Res. Bull.* **4**:141–143.

Mason, S. T., and Fibiger, H. C., 1979, On the specificity of kainic acid, *Science* **204**:1339–1341.

McGeer, E. G., and McGeer, P. L., 1976, Duplication of biochemical changes of Huntington's chorea by intrastriatal injections of glutamic and kainic acid, *Nature* **263**:517–519.

McGeer, E. G., Olney, J. W., and McGeer, P. L., 1978a, *Kainic Acid as a Tool in Neurobiology*, Raven Press, New York.

McGeer, P. L., McGeer, E. G., and Hattori, T., 1978b, Kainic acid as a tool in neurobiology, in: *Kainic Acid as a Tool in Neurobiology* (E. G. McGeer, J. W. Olney, and P. L. McGeer, eds.) pp. 123–138, Raven Press, New York.

Moore, R. Y., 1964, The effect of some rhinencephalic lesions on retention of a conditioned avoidance response in the cat, *J. Comp. Physiol. Psychol.* **57**:65–71.

Moore, R. Y., 1978, Surgical and chemical lesion techniques, in: *Handbook of Psychopharmacology, Vol. IX* (L. Iversen, S. Iversen, and S. H. Snyder, eds.) pp. 1–39, Plenum Press, New York.

Moore, R. Y., Halaris, A. E., and Jones, B. E., 1978, Serotonin neurons of the midbrain raphe: Ascending projections, *J. Comp. Neurol.* **180**:417–438.

Mullan, S., Harper, P. V., Tani, E., Vailata, G., and Lathrop, K. A., 1963, A nuclear needle for use in neurosurgery, *J. Neurosurg.* **20**:940–947.

Myers, R. E., 1956, Function of the corpus callosum in interocular transfer, *Brain* **79**:358–363.

Nadler, J. V., Perry, B. W., and Cotman, C. W., 1978, Preferential vulnerability of hippocampus to intraventricular kainic acid, in: *Kainic Acid as a Tool in Neurobiology* (E. G. McGeer, J. W. Olney, and P. L. McGeer, eds.), pp. 219–238, Raven Press, New York.

Olney, J. W., 1969a, Glutamate-induced retinal degeneration in neonatal mice. Electron microscopy of the evolving lesion, *J. Neuropathol. Exp. Neurol.* **28:**455–474.

Olney, J. W., 1969b, Brain lesions, obesity and other disturbances in mice treated with monosodium glutamate, *Science* **164:**719.

Olney, J. W., Rhee, V., and Ho, O. L., 1974, Kainic acid: A powerful neurotoxic analogue of glutamate, *Brain Res.* **77:**507–512.

Olszewski, J., 1952, *The Thalamus of the Macaca mulatta,* Karger, Basel.

Pandya, D. N., Karol, E. A., and Lele, P. P., 1973, The distribution of the anterior commissure in the squirrel monkey, *Brain Res.* **49:**177–180.

Peterson, G. M., and Moore, R. Y., 1980, Selective effects of kainic acid on diencephalic neurons, *Brain Res.* **202:**165–182.

Poirier, L. H., 1975, Histopathological changes associated with the intracerebral injection of 6-hydroxydopamine (6-OH-DA) and peroxide (H_2O_2) in the cat and rat, *J. Neurol. Transm.* **37:** 209–218.

Poirier, L. H., Langelier, P., Roberge, A., Boucher, R., and Kitsikis, A., 1972, Non-specific histopathological changes induced by the intracerebral injection of 6-hydroxydopamine (6-OH-DA), *J. Neurol. Sci.* **16:**401–416.

Purpura, D. P., and Gonzalez-Monteagudo, O., 1960, Acute effects of methoxypyridine on hippocampal and blade neurons: An experimental study of "special pathoclisis" in the cerebral cortex, *J. Neuropathol. Exp. Neurol.* **19:**421–431.

Richardson, J. S., and Jacobowitz, D., 1973, Depletion of brain norepinephrine by intraventricular injection of 6-hydroxydopa: A biochemical, histochemical and behavioral study in rats, *Brain Res.* **58:**17–133.

Richter, R. B., 1945, Degeneration of the basal ganglia in monkeys from chronic carbon disulfide poisoning, *J. Neuropathol. Exp. Neurol.* **4:**324–353.

Robinson, R. G., Shoemaker, W. J., Schlumpf, M., Valk, T., and Bloom, F., 1975, Effect of experimental cerebral infarction in rat brain on catecholamines and behavior, *Nature* **255:**332.

Rose, J. E., and Woolsey, C. N., 1943, A study of the thalamocortical connections in the rabbit, *Bull. Johns Hopkins Hosp.* **73:**65–128.

Rose, J. E., and Woolsey, C. N., 1949, The relations of thalamic connections, cellular structure and evocable electrical activity in the auditory region of the cat, *J. Comp. Neurol.* **91:**441–466.

Routtenberg, A., 1972, Intracranial injection and behavior: A critical review, *Behav. Biol.* **7:**601–641.

Rowland, V., 1966, Stereotaxic techniques and the production of lesions, in: *Neuroendocrinology, Vol. 1* (L. Martini and W. F. Ganong, eds.), pp. 107–132, Academic Press, New York.

Sachs, C., 1973, Development of the blood–brain barrier for 6-hydroxydopamine, *J. Neurochem.* **19:**1561–1575.

Sachs, C., and Jonsson, G., 1975, Mechanisms of action of 6-hydroxydopamine, *Biochem. Pharmacol.* **24:**1–8.

Sachs, C., Jonsson, G., and Fuxe, K., 1973, Mapping of central noradrenaline pathways with 6-hydroxy-dopa, *Brain Res.* **63:**249–261.

Schneider, G. E., 1970, Mechanisms of functional recovery following lesions of visual cortex or superior colliculus in neonate and adult hamster, *Brain Behav. Evol.* **3:**295–323.

Schoenfeld, T. A., and Hamilton, L. W., 1977, Secondary brain changes following lesions: A new paradigm for lesion experimentation, *Physiol. Behav.* **18:**951–957.

Schulman, S., 1964, Impaired delayed response from thalamic lesions, *Arch. Neurol.* **11:**477–499.

Schwarcz, R., and Coyle, J. T., 1977, Striatal lesions with kainic acid: Neurochemical characteristics, *Brain Res.* **127:**235–249.

Schwarcz, R., Scholz, D., and Coyle, J. T., 1978, Structure–activity relations for the neurotoxicity of kainic acid derivatives and glutamate analogues, *Neuropharmacology* **17:**145–151.

Sclafini, A., and Grossman, S. P., 1969, Hyperphagia produced by knife-cuts between the medial and lateral hypothalamus in the rat, *Physiol. Behav.* **4**:533–539.

Simon, H., Le Moal, M., Galey, D., and Cardo, B., 1974, Selective degeneration of central dopaminergic systems after injection of 6-hydroxydopamine in the ventral mesencephalic tegmentum of the rat: Demonstration by the Fink–Heimer stain, *Exp. Brain Res.* **20**:375–384.

Singh, B., and De Champlain, J., 1972, Altered ontogenesis of central noradrenergic neurons following neonatal treatment with 6-hydroxydopamine, *Brain Res.* **48**:432–437.

Sotelo, C., Javoy, F., Agid, Y., and Glowinski, J., 1973, Injection of 6-hydroxydopamine in the substantia nigra of the rat. I. Morphological study, *Brain Res.* **58**:269–290.

Stahl, S. M., Daniels, A. C., Derda, D., and Spehlmann, R., 1975, Injection of 6-hydroxydopamine and hydrogen peroxide into the substantia nigra and lateral ventricle of the cat: Specific and nonspecific effects on striatal amines, *J. Neurochem.* **24**:165–172.

Ungerstedt, U., 1971, Stereotaxic mapping of the monoamine pathways in the rat brain, *Acta Physiol. Scand.* **82** (*Suppl.* 367):1–48.

Velasco, M. E., 1972, Opposite effects of platinum and stainless-steel lesions of the amygdala on gonadotrophin secretion, *Neuroendocrinology* **10**:301–308.

Voloschin, L., Joseph, S. A., and Knigge, K. M., 1968, Endocrine function in male rats following complete and partial isolations of the hypothalamo-pituitary unit, *Neuroendocrinology* **3**:387–397.

Walker, A. E., 1938, *The Primate Thalamus*, University of Chicago Press, Chicago.

Whishaw, T. Q., Cioe, J. D. D., Previsich, N., and Kolb, B., 1977, The variability of the interaural line vs. the stability of bregma in rat stereotaxic surgery, *Physiol. Behav.* **19**:719–722.

Wolf, G., and DiCara, L. V., 1969, Progressive morphologic changes in electrolytic brain lesions, *Exp. Neurol.* **23**:529–537.

Methods for Delivering Tracers

3

GEORGE F. ALHEID, STEPHEN B. EDWARDS, STEPHEN T. KITAI, MELBURN R. PARK, and ROBERT C. SWITZER III

I. INTRODUCTION

The delivery of chemical tracers into the central nervous system is often an essential, although sometimes problematic aspect of modern tract-tracing experiments. In this chapter we shall concentrate on the most common methods of tracer delivery: small injections of a solution, either using a micro syringe or through glass micropipettes, or extracellular of intracellular iontophoresis. It should be kept in mind, however, that other methods of delivering tracers may be more suitable for certain types of problems. Tracers have been injected into the periphery—for instance, into muscle in order to label motoneurons in the spinal cord or brain stem (Kristenson, 1970; Kristenson *et al.,* 1971; Kristenson and Olsson, 1973)—and they have been applied to the cut or crushed ends of peripheral nerves (Kristenson, 1975; Iles and Mulloney, 1971; Tweedle, 1978) and deposited in naturally closed structures such as the eye (LaVail and LaVail, 1972, 1974). In cases

GEORGE F. ALHEID • Department of Neurology, University of Virginia School of Medicine, Charlottesville, Virginia 22908. STEPHEN B. EDWARDS • Department of Anatomy, University of Virginia School of Medicine, Charlottesville, Virginia 22908. STEPHEN T. KITAI and MELBURN R. PARK • Department of Anatomy, Michigan State University, East Lansing, Michigan 48824. ROBERT C. SWITZER III • Comparative Animal Research Laboratory, Oak Ridge, Tennessee 37830.

where diffusion from the site of deposition is not a serious limitation, tracers may be applied directly to the surface of the brain, either as a liquid or by means of absorbant pellets or wicks that have been soaked in the tracer solution (Arbuthnott, 1969; Descarries and Lapierre, 1973; Held and Young, 1969; Weiss and Holland, 1967).

Many different carriers have been used to limit the diffusion of chemicals injected directly into the brain (Myers, 1974). These include substances such as agar, cocoa butter, paraffin, gum arabic, and silicone rubber (Silastic®). While the most recent methods of liquid or electrophoretic injections allow controlled localized deposition, the use of solid or semisolid carriers is an alternative means of discrete tracer placement (Davis and Agranoff, 1977; Kawamura *et al.,* 1978; Griffin *et al.,* 1979).

Experimenters have developed increasingly sophisticated methods of controlling injection volume and delivery rate of tracer solutions in order to minimize damage to the neuropil in the area of injection. This could be of critical importance for tracers that are also taken up by damaged axons passing through the injection site. Additionally, injection of HRP through a chronically implanted guide cannula has been reported to minimize labeling of axons of passage when compared with acute injections (Wakefield and Shonnard, 1979).

II. PRESSURE INJECTION

The use of pressure, either hydraulic or pneumatic, is by far the most common means of extracellular injection of tracers, and various schemes have been developed for the control of minute liquid injections. In the following section, detailed descriptions are given for pressure injection via microsyringe or micropipette. The use of a microsyringe represents perhaps the simplest and most popular tracer delivery method. The particular microsyringe technique described below (Edwards and Shalna, 1974) has been successfully used in numerous anatomical investigations. Alternatively, micropipettes either coupled to a microsyringe or directly to a pressure source, while allowing smaller injections, require somewhat more complex procedures.

A. Microsyringe Injection

Tracer injections of varying sizes may be made simply and reliably using a 1, 5, or 10 μm Hamilton fixed-needle syringe. These syringes are supplied with 26-gauge needles, 5 cm in length, and specially drawn to a 114-μm diameter at the tip. The needles have sufficient rigidity to resist displacement during entry but may be too large to permit labeling of a highly restricted area of tissue. In general, smaller needles (e.g., 4 cm long 30-gauge needles) produce smaller injection sites with less tissue damage and without adverse deviation during advancement through the brain. Upon special request to the Hamilton Company, Reno, Nevada, microliter syringes can be obtained with fixed needles of nearly any length, gauge, and tip style.

The syringe must be secured firmly in an apparatus that offers controlled advancement of the syringe plunger. This can be accomplished by placing the syringe plunger in line with a 25-mm micrometer drive system (Fig. 1). If injections are to be stereotaxically placed, the injection unit must be attached to a stereotaxic carrier. It is good precaution to sterilize and thoroughly dry the syringe before use in order to maintain the sterility of the tracer. After filling the syringe, the outside of the needle should be rinsed with sterile water or saline. If there is concern about leakage from the needle tip during advancement of the needle through the brain, the plunger may be raised in order to withdraw the column of fluid 0.1–0.2 μl. Once the needle tip has been placed at the desired coordinates, the injection is made manually at a slow and constant rate. Rates of around 0.05 μl/min are commonly used. Measurement of the amount injected should always be made by watching the fall of the meniscus in the syringe barrel rather than the advancement of the plunger inasmuch as the plunger can advance without a corresponding movement of the meniscus. After completion of the injection, the syringe needle should be left in place for 15–30 min in order to reduce the chance of the substance spreading along the needle track when the syringe is withdrawn.

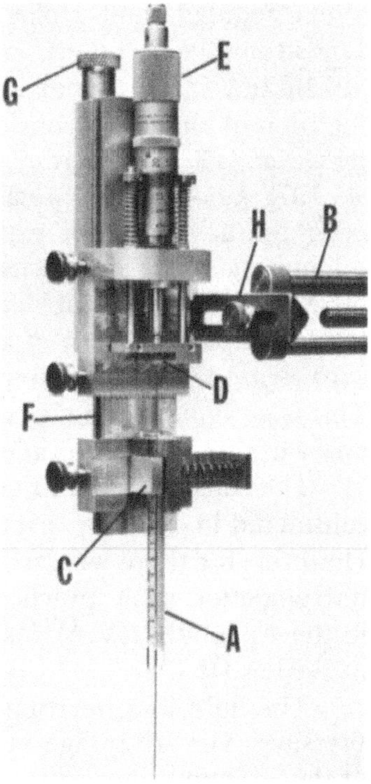

Figure 1. Microinjector shown with a Hamilton 701N 10 μl syringe (A) and attached to a David Kopf electrode carrier (B). C, syringe barrel pressure clamp; D, grooved plate for plunger button; E, 25-mm micrometer head; F, vertical adjustment plate; G, vertical adjustment knob, H, triangular attachment bar.

The size of the actual injection site depends on many factors including the diameter of the needle, the volume injected, the rate of injection, and the nature of the tissue into which the injection is made. Reducing the volume and rate of injection will produce smaller injection sites and will also minimize the tissue damage produced by hydrostatic pressure. The spread of the tracer tends to be restricted in dense fiber systems. It is possible that the compactness of cell bodies may also influence the spread of injected substances.

Hydraulic pressure systems have also been designed to drive the plungers of microsyringes (e.g., Schubert and Holländer, 1975; Price *et al.*, 1977). While mechanically more complex, these systems have the advantage of isolating the ejecting force from the frame holding the stereotaxically implanted syringe, thus reducing the possibility of damage from inadvertent movements of the mechanism in response to the fluid delivery. Such systems also provide good control over the time course of the injection when the hydraulic system is driven by a standard infusion pump.

B. Micropipette Injection

Drawn glass micropipettes have been used for both extracellular and intracellular injection of tracers. While more commonly used with iontophoresis (see Section III below), they have also been coupled with microsyringe or other injection systems for extracellular delivery of tracers (Lasek *et al.*, 1968; Schubert and Holländer, 1975; Price *et al.*, 1977; Woodward and Lindström, 1977; Eide *et al.*, 1976; Guglielmotti, 1979; Kelly and Gilbert, 1975), and various schemes for pressure delivery have been used for the injection of single neurons (e.g., Remler *et al.*, 1968; Kater *et al.*, 1973; Pitman *et al.*, 1972; Sakai *et al.*, 1978; Källström and Lindström, 1978; Koike *et al.*, 1972; Chowdhury, 1969; McCamen *et al.*, 1977). Drawn glass pipettes offer the advantage of extremely small shank and tip diameter, thereby minimizing tissue destruction by the pipette. This, however, is somewhat offset by a tendency for micropipettes to become plugged with tissue or by the tracer itself. A second advantage of micropipettes is their ability to record neuronal activity from the site at which the tracer is being deposited. This is an optional part of extracellular injections but is, of course, mandatory when one wishes to inject single cells in the central nervous system.

The construction and use of micropipettes in neurophysiology has been considered in detail by several authors (Nastuk, 1964; LaVallee *et al.*, 1969). However, for those who are contemplating their first involvement with glass micropipettes, some practical details of pipette preparation are considered in the Appendix (IV.A). Aspects of physiological recording are considered in Section III.C.2.

The fluid in a micropipette may be ejected by hydraulic or pneumatic pressure. The advantage of using hydraulic pressure is that any movement of the essentially incompressible driving fluid results in the movement of the same volume of the injection liquid. This advantage disappears if air bubbles are present in the system. Bubbles of air may be compressed, and insufficiently rigid tubing may expand, resulting in inaccurate volume measurements. Furthermore, once movement of the tracer solution has begun, a slightly lower pressure is required for continued movement, so that expansion of compressed air bubbles can result in a sudden uncontrolled ejection of fluid from the micropipette tip. Because of reduced pressure requirements at larger diameters, this is less of a problem for tip diameters over 50 μm. Pneumatic pressure may be conveniently used if provision is made for good pressure regulation and for the rapid release of the driving pressure as soon as the desired volume of fluid is ejected.

1. Extracellular Injection

In the following section, the procedure and the different components needed for extracellular micropipette pressure injections are described. The method has been successfully used by one of us (R. C. Switzer III) and is a modification of a system used by Dr. Glenn Northcutt (University of Michigan). An effort has been made to use components that are commercially available. The assemblage consists of a tapered glass micropipette, a holder for carrying the micropipette on a micromanipulator, thick-walled plastic tubing including appropriately sized connectors and T-fittings, a vacuum source, a low-pressure gas regulator, and an air supply (Fig. 2). Particular aspects of these components are treated in the following paragraphs.

a. Micropipette Holder and Micropipettes. Commercial pipette holders are available for pressure injections as well as for recording or iontophoresis. The pipette holder, model EH-2S,* has a standard bore or can be bored to a specific diameter upon request. This holder is provided with a side port for application of pressure, while the pin on top can be used to attach the holder to a micromanipulator as well as to provide electrical contact for the Ag/AgCl internal electrode. The pipette is held and sealed by a rubber compression gasket. Connection to this holder may be made using the conventional tapered hubs designed for hypodermic needles.

It is convenient to use calibrated 1- to 5-μl glass pipettes† for construction of the micropipettes used with this system. This allows for visual determination of the amount injected with or without the use of a dissecting microscope. The distance between the 1 μl marks on these tubes can be divided into suitable divisions by attaching a transparent scale to the shaft of the micropipette. For injections accurate to ±5 nl, a strip of Chartpak® dotted line can be used as a scale. The pipettes have greater wall thickness and

*W-P Instruments Inc., P.O. Box 3110, New Haven, Connecticut 06575.

†Corning No. 7099U-5, Corning Glass Works, Science Products Division, Corning, New York 14830, or Gold Seal Micropet No. 4614, Clay–Adams, Division of Becton–Dickinson, Parsippany, New Jersey 07054.

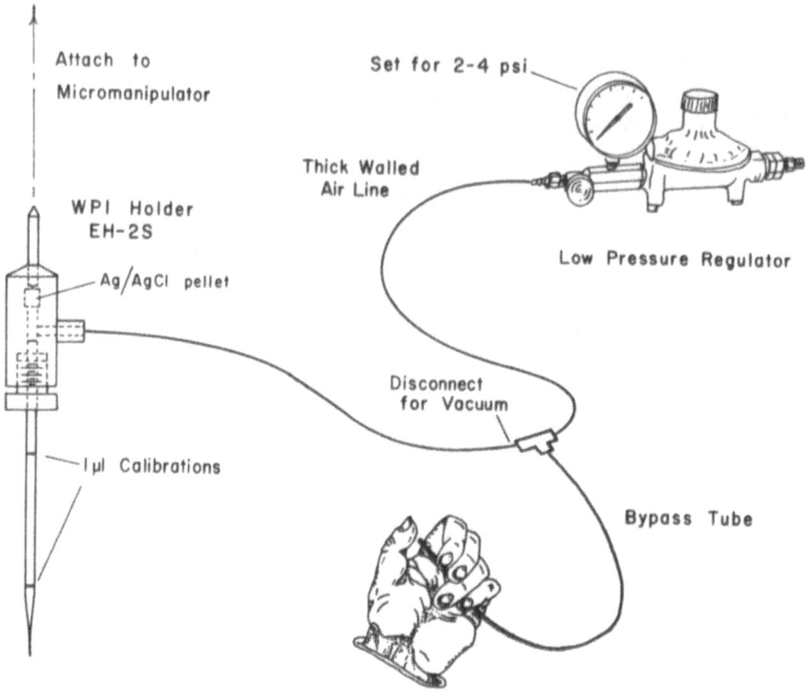

Figure 2. Schematic figure of a pneumatic drive for fluid in a micropipette. The WPI micropipette holder is provided with a side port for entry of air pressure and a pin at the top for attachment to a micromanipulator. The pin also provides electrical contact to an internal Ag/AgCl pellet that permits electrical recordings or a means for iontophoresis of fluid in the pipette. Another WPI holder (EH-2R) is provided with the pin on the side and the port on top. Low air pressure is supplied via a low-pressure gas regulator. A bypass tube between the regulator and the micropipette allows pressure to the micropipette to be controlled manually. A connector at the bypass tube junction provides a point for coupling to a vacuum line, so that the fluid can be drawn into the micropipette.

thereby greater strength than that found in conventional micropipette glass tubing. In general, it is recommended that the micropipette tip be beveled by a suitable grinding procedure (see, Appendix, IV.A.4). This eliminates or minimizes plugging of the tip, especially with the larger tip diameters used for extracellular tracer injections. The micropipette is easily filled by connecting the air line to a vacuum source: 1–2 µl of fluid can be drawn into the micropipette within 5–10 min, even for tips of 5–10 µm in diameter.

b. Pressure Delivery. The tubing used to deliver air pressure to the micropipette is an important consideration, especially if the amounts to be injected are less than 10 nl. Thick-walled tubing such as the Tygon® brand* is recommended for nanoliter amounts, while a sufficient length of flexible hypodermic metal tubing† may be necessary for critical applications. A pres-

*Tygon, Norton Plastics and Synthetics, P.O. Box 350, Akron, Ohio 44309.
†Hamilton, Inc., P.O. Box 10030, Reno, Nevada 89510.

sure regulator that allows accurate maintenance of pressure 1–15 lb/in.^{2} (*ca.* 7–100 kPa) is necessary.*

For injections of 10 nl or less, the system described by McCaman *et al.* (1977) may be useful. For a given tip size and pressure, the amount ejected can be regulated by the length of time the air pressure is on. The system requires a higher pressure source (4–100 lb/in.^{2} or 27–690 kPa) and electrical control of duration by means of a quick-acting solenoid valve.

c. Calibration. To calibrate the system, the diameter, *D*, of the droplet can be measured with a microscope and the volume determined by $V = \pi D^3/6$. For droplets less than 10 nl, aqueous solution will quickly evaporate and usually creep up the shaft, thereby preventing measurement. In this range, calibration might be made with a variation of one of the assay techniques described in the Appendix (IV.B). Alternatively, measurement of diameters of ejected aqueous liquid drops can be made while the micropipette is submerged in a fluid such as paraffin oil.

d. Ejection Rate. With the arrangement shown in Fig. 2, the pressure is set at 2–4 lb/in.^{2} (ca. 13–28 kPa). The flow valve on the regulator is set to allow air to flow at a slow rate. To eject fluid, the bypass tube is closed manually or with a suitable mechanical device if desired. It was found that with a tip size of 15–30 μm a sequence of 1 sec closed and 29 sec open yields a good delivery rate for a 10% horseradish peroxidase solution.

Spread of the ejected fluid from the micropipette tip and up along the shaft of the pipette is the primary consideration in determining the rate of delivery. Damage and spread are minimal if a rate no greater than 20–40 nl/min is employed. However, it is difficult to fix a rate that is applicable for every substance and all areas of the brain. The investigator must determine empirically the optimal parameters for a given experiment.

Measures for cleaning the micropipette shaft just before insertion, insertion rate, time in place before ejection, time in place after completion of ejection, and withdrawal as discussed for microsyringe injections are also applicable to micropipettes and should be followed.

Just before insertion, the pressure can be applied to the pipette and the tip observed with an operating microscope to determine if the fluid is, in fact, coming out. Often the tip will plug slightly because of evaporation. This is quickly remedied by lowering the micropipette so that the tip briefly touches tissue fluid and then raising the pipette slightly and applying air pressure. A small bead should quickly appear at the tip and may even creep up the shaft slightly. If no droplet appears, the sequence can be repeated. If this fails, then another micropipette should be selected from the storage box and filled. The frequency of tip plugging may be reduced by keeping the tip in the solvent or in saline after filling. Alternatively, parafin oil may

*Two regulators with suitable range are (1) Life-O-Gen, Cambridge, Maryland 21613, No. 1000-B10-5030-5532, 0-15 (Fisher 10-599-28); and (2) Low Pressure Gas Regulator Model 70 Matheson, 1/2–10 (Curtin Matheson Scientific 163–030), Box 1546, Houston, Texas 77001. The latter regulator is shown in Fig. 2.

be drawn into the micropipette after the tracer liquid and ejected immediately prior to use.

2. Intracellular Injection

Pneumatic or hydraulic pressure has been used in the past to inject chemicals into relatively large neurons in invertebrates, and recently this approach has been applied to small mammalian neurons as well. The advantage of pressure techniques, as opposed to iontophoresis, is that neuron filling may be accomplished more rapidly, and tracers that are difficult to iontophorese may be readily injected. A disadvantage, however, is that a certain volume of tracer solution must be injected within the cell, and this may lead to some structural distortion. Nevertheless, recent studies aided by intracellular pressure injections of horseradish peroxidase have given encouraging results (Källström and Lindström, 1978; Sakai *et al.*, 1978; Ahlsén *et al.*, 1978).

The main difference between the system described for extracellular injections (Section II.B.1) and that used for intracellular injections is that smaller tip diameters are generally required for intracellular studies (0.5–1.0 μm). This, in turn, may require significantly higher pressure than that required for the larger, extracellular micropipettes. Of course, there is the additional necessity for physiological amplification and display in order to find and impale neurons prior to injection.

Sakai *et al.* (1978) injected HRP into cortical cells of cats using a solenoid-gated, pneumatic pressure system similar to that described by McCaman *et al.* (1977). The microelectrode holder was connected to a source of regulated air pressure (60–90 lb/in.^{2} or 415–620 kPa) that was gated by means of a solenoid valve. By monitoring the output of injection fluid under a microscope, they estimated the output rate at 1 μm^3/sec. After successful penetration of a neuron and subsequent recording, pressure was applied for 1–10 sec. Successful injections were invariably accompanied by a sudden decrease in electrode resistance.

Källström and Lindström (1978) used pressure to inject cells and interneurons in the lateral geniculate of cats. In their experiments, hydraulic pressure was generated by a screw-driven generating device and delivered through paraffin oil in thick-walled polyethylene tubing to a sealed electrode holder. For very high pressure, they describe a system for locking the microelectrode in the holder to prevent movement along the vertical axis. The reported pressure required for initiation of the injection was quite variable, perhaps because of tissue obstruction of the tip. Initiation of tracer (HRP) flow in their experiment was usually accompanied by a decrease in amplitude of the action potential and of postsynaptic potentials as well as a fall in the membrane potential. Injections were terminated after 10–60 sec by opening a screw valve that immediately released the pressure.

III. IONTOPHORETIC INJECTION

The terms "electrophoresis" and "iontophoresis" are used interchangeably in the current neuroscientific literature to describe the use of electrical current to eject substances from a micropipette, but both terms originated outside neuroscience. Electrophoresis, "the movement of suspended particles through a fluid under the action of an electromotive force applied to electrodes in contact with the suspension," is widely known as a technique of macromolecular chemistry quite distinct from drug or tracer electrophoresis from a microelectrode. Iontophoresis was originally defined as the "introduction of drugs through intact skin by the transfer of ions effected by means of the application of a direct electric current." Perhaps because no other field currently uses it, "iontophoresis" is becoming the term more often used in neuroscience.

A. General Considerations

1. Tracer Charge and Current

The technique of iontophoresis permits charged molecules of a chemical or tracer to be ejected by establishing a suitable voltage gradient between the tracer solution in a micropipette and the tissue in which the pipette has been implanted. The tracer must be charged and soluble. The polarity of the current used for iontophoresis depends upon whether the predominant charge on the substance to be deposited is mainly positive or negative. The amount of tracer ejected is proportional to the total charge (current × time) delivered by a current source. In practice, most experimenters have kept the current constant and varied the duration of injection in order to vary the amount of tracer delivered. Since direct currents often cause a change in the resistance of liquid-filled microelectrodes, a source of constant current has been used to drive the iontophoresis of tracers (see Section III.C). Direct currents may cause rapid impedance buildup in the micropipette which can lead to blocking when the capability of the current source is exceeded. Pulsed direct currents help prevent this buildup, but if blocking does occur, a brief current reversal may help to clear the micropipette.

It should be kept in mind that the rate of tracer deposition does not depend on the current alone, but also on parameters such as the concentration of other similarly charged ionic species in the solution and the degree of ionization of the tracer. If, for instance, positive direct current is used, movement of any other positive ion in the tracer solution will contribute to the current that is monitored, so that identical currents may not deposit the same amount of tracer even when identical tracer concentrations are used. It is also profitable to manipulate the pH of the solution to insure optimal ionization of the tracer.

Connection between the current source and the micropipette is easily made by inserting a fine gauge wire into the solution in the shaft of the pipette. Gold wire, silver wire, chloridized silver wire, platinum–iridium, and stainless steel have been used for this purpose. Pure gold is useful since it is quite malleable, and electrical connections may require some bending of the connecting wire. A second wire is connected to exposed muscle tissue to provide the reference electrode.

Alternatively, electrical connection may be made with a commercially available electrode holder. In the type made by WPI,* the holder is filled with a chloride-containing solution prior to insertion of the micropipette. The electrical connection is through the liquid to a Ag/AgCl electrode embedded in the holder.

2. Passive Diffusion

Diffusion of tracers from the micropipette is a problem that is proportional to the tip diameter. A low current opposite in polarity to that used for iontophoresis may be used as a retaining current while inserting and removing the micropipette. Many anatomical studies aided by extracellular iontophoresis have been performed without the use of retaining currents. It is interesting, however, that passive diffusion has been used in at least two tract-tracing experiments to deposit [^{3}H]amino acids (Jones *et al.*, 1978; Simon *et al.*, 1979). The tip of the micropipettes in these instances ranged from 20–50 μm, which is comparable to the sizes used for extracellular iontophoretic tracer delivery.

3. *In Vitro* Assays

Despite successful passage of current, unsatisfactory deposition of tracer sometimes occurs, even though normally successful parameters are used for tracer concentration, current, tip size, etc. Several researchers (Schubert *et al.*, 1972; Schubert and Holländer, 1975; Spencer *et al.*, 1978) have described *in vitro* microassays for testing the quantitative release of tracers prior to extracellular iontophoresis. Such tests apparently dramatically improve the success rate of *in vivo* injection by screening out faulty electrodes (see Appendix, IV.B).

B. Extracellular Injection

Iontophoresis of tracers into the extracellular space may be accomplished with a minimum of equipment if physiological recording is not required. Basically, it requires a micropipette of appropriate tip size mounted

*W-P Instruments Inc., P.O. Box 3110, New Haven, Connecticut 06575.

in a suitable micromanipulator and connected to a current source (see Section III.C.2).

1. Horseradish Peroxidase

For iontophoresis of HRP, pulsed positive direct current has been most commonly used. Actual frequencies have varied from 5 to 0.07 Hz. Most investigators have used currents that range from 1 to 4 μA with "on times" that vary from less than 5 min to more than 30 min. In some cases as much as 10 μA has been used (Abols and Basbaum, 1979). However, care should be taken at higher current levels, since the combination of high current and small tip diameters may cause current density high enough to disrupt the tissue. Luiten (1975), for instance, noted that iontophoresis of HRP with current levels of 1–2 μA often resulted in tissue damage, possibly to axons of passage, and suggested that lower current levels might alleviate this effect. Phillipson (1979), on the other hand, found that although HRP was deposited, no transport occurred at current levels below 1 μA. Quite different solvents and species were used by these experimenters, but these reports indicate that the experimenter should determine the minimal current level consistent with transport in his preparation.

A wide variety of solvents has been used for extracellular iontophoresis of HRP. Graybiel and Devor (1974) used a 4% solution of HRP in Tris buffer at pH 8.6, and many investigators have successfully used this solvent. The concentration of HRP, however, has varied considerably among different laboratories. Other solvents, including water (Simon *et al.*, 1979), dilute saline (0.01 M NaCl) (Bunney and Aghajanian, 1976; Aghajanian and Wang, 1977; Gallager and Pert, 1978; Cedarbaum and Aghajanian, 1978; Edwards and Henkel, 1978; Henkel and Edwards, 1978), 0.1 M NaOH (Phillipson, 1978, 1979), and 0.5 M KCl (Ottersen and Ben-Ari, 1978a,b, 1979), have also been used. Lynch *et al.* (1974) used low concentration of HRP in 2 M NaCl for anterograde filling of neurons. In a recent review of this work, however, Spencer *et al.* (1978) report that 0.1–0.2 M NaCl is more practical, since HRP may precipitate when using the higher molarity. They also suggest that the use of phosphate buffer (pH 6.2) may facilitate neuron filling.

Since most of the transportable enzyme activity of HRP (Sigma type VI) is attributable to the basic isoenzyme C (Chapter 6), Bunt *et al.* (1976) have suggested that buffering the HRP at a pH lower than 8.6 may aid its iontophoresis. However, no systematic study seems to have resolved this question. Successful HRP iontophoresis and transport from buffered solutions with pH 8.2 to 7.4 have been reported (Luiten, 1975; Luiten and Van der Pers, 1977; Kawamura and Diamond, 1978; RoBards, 1979; Maekawa and Takeda, 1979; Abols and Basbaum, 1979).

The HRP concentration, usually 10–50%, does not appear to be critical. The higher concentrations, however, may be quite viscous and therefore more difficult to handle and more likely to plug the pipette. Tip diameters for HRP iontophoresis have ranged from 1 to 70 μm, but tip sizes of less than 30 μm are most frequent.

2. Tritiated Amino Acids

Tritiated glycine, leucine, and proline have been iontophoresed using positive direct currents between 0.1 and 2.0 μA for periods of 2–15 min. Tip diameters for the micropipettes have ranged from 3 to 20 μm, and the concentrated amino acids were dissolved in dilute acid, i.e., 0.01 M acetic acid or 0.01 M HCl (Schubert *et al.*, 1972; Norgren, 1976; Herkenham, 1978, 1979; Beckstead, 1979a,b; Herkenham and Nauta, 1979).

C. Intracellular Injection

In this section we shall briefly discuss requirements for intracellular recording and injection. A comprehensive treatment of the electronics and physiology involved is beyond the scope of this chapter. A more complete discussion of electrophysiological methods may be found in Nastuk (1964), Brown *et al.* (1973), Thompson and Patterson (1973, 1974), and Bures *et al.* (1976). Intracellular injection of HRP is described in Chapter 7 of this book, whereas intracellular staining techniques in general have recently been discussed by Kater and Nicholson (1973) and Tweedle (1978).

1. Surgery

Successful impalement and injection of vertebrate neurons will be possible only if detailed attention is paid to surgical techniques. It is important to use an adequate (massive) stereotaxic frame, preferably designed for intracellular work and well-isolated from mechanical vibration. It is also essential to maintain good control of anesthesia and to control the movements of the animal (e.g., from respiration, stimulus-induced movements, or even from nearby blood vessels). In addition, some type of microdrive may be necessary for advancing the micropipette. Commercially available devices* allow steps in the range of 1 to 4 μm; this is quite compatible with the requirements of intracellular recording and injection.

2. Recording and Current Injection

The problems involved in recording and iontophoretic injection of tracers are not unlike those met during drug iontophoresis through a high-

*Burleigh Instruments Inc., Burleigh Park, Fishers, New York 14453; David Kopf Instruments, 7324 Elmo Street, Tujunga, California 91042; Narishige Scientific Instruments, 27-9 lab, Minami-Karasuyama, 4-Chome, Setagaya-Ku, Tokyo 157, Japan.

resistance micropipette, and many of the problems have been solved by earlier workers.

a. Recording and Display. A practical recording and injection setup requires a sensitive preamplifier suitable for biological potentials, some form of amplification and display and, optionally, a method of permanently recording these potentials. A current source and a current monitor are required for iontophoresis. The current can be measured directly with a sensitive galvanometer or by measuring the voltage drop across a known, small resistance in series with the electrode. Potentials are readily amplified and displayed by means of an oscilloscope, while permanent recording may be obtained by photographing the oscilloscope display. More sophisticated recording may be obtained by use of multichannel FM tape recorders and/or by signal conditioning with digital methods for computer storage and analysis.

b. Stimulus Isolation. Frequently, a voltage or a current source is used to stimulate a pathway or nucleus. In this case the stimulating electrode must be isolated from ground to prevent interaction with the normally grounded circuit of the biological amplifier. It is possible to use a battery-driven stimulator that is not grounded or to use any of the various commercial stimulus isolation units* that have been developed for this purpose.

c. Operational Amplifiers. Relatively inexpensive physiological amplifiers are commercially available, and the purchase of such equipment is a reasonable investment for the investigator with long-range interests in these techniques. If the experimenter has some familiarity with modern integrated circuits, high-impedance amplifiers and current sources can be constructed quite simply by the use of standard operational amplifiers (op-amps) with a field effect transistor (FET) op-amp as an input stage. Op-amps are quite versatile and useful components for construction of physiological recording and stimulating equipment, and examples of their use have been discussed in detail by Brown *et al.* (1973), Hoenig and Payne (1973), and Bures *et al.* (1976).

d. Current Sources. An approximate constant current source uses a battery and a series resistance. To create a practical circuit, the following components are added: a voltage-dividing potentiometer to control the amplitude and a relay to disconnect the amplifier from the electrode during current injection. Such a circuit is drawn in Fig. 3, A and B. In practice, the particular range of recording electrode resistances needed for intracellular recording places limitations on the circuit, making its performance less than ideal. Nonetheless, the errors introduced are still small, and this circuit remains quite useful.

*W-P Instruments Inc., P.O. Box 3110, New Haven, Connecticut 06575; Grass Instrument Co., 101 Old Colony Avenue, Quincy, Massachusetts 02169.

The behavior of the circuit is described by Ohm's Law, $I = (V_b - V_c)/(R_s + R_e)$, where V_b is the driving voltage determined by the battery and potentiometer, and V_c is the potential of the cell which is also the potential at the tip of the electrode. Therefore, the difference $V_b - V_c$ is the driving force responsible for current injection. R_s is the value of the series resistance, and R_e that of the electrode resistance. The current is constant only insofar as fluctuations in R_e and V_c are not large in relation to R_s and V_b, respectively. R_s is therefore chosen to be much larger than R_e, and V_b much larger than any changes in V_c that might be expected to occur. There are, unfortunately, practical limitations. With the largest valued resistor available at 1000 MΩ, if, for example, R_e changes 20 MΩ during current injection, as might be expected, then the total error from this source is only 2%. If, however, the electrode goes into a high-resistance state (blocking) and increases to 500 MΩ, then only two-thirds of the desired current will actually be passing through the electrode. This is a drawback, but, on the other hand, the circuit of Fig. 3B is always stable. Electrode blocking cannot lead to oscillations and possible cell injury, as is the case for some active (electronic) current sources. Returning to the example, in order to inject 5 nA, a V_b of just 5 V is needed. As the cell voltage fluctuates 0.1 V, as it can be expected to do during an action potential, the driving voltage $V_b - V_c$ is changed only 2%. This is quite tolerable.

In the case of extracellular iontophoresis of HRP, where substantially higher currents (1–4 μA) have been used, voltage requirements with this circuit can become excessive despite the fact that lower impedance electrodes

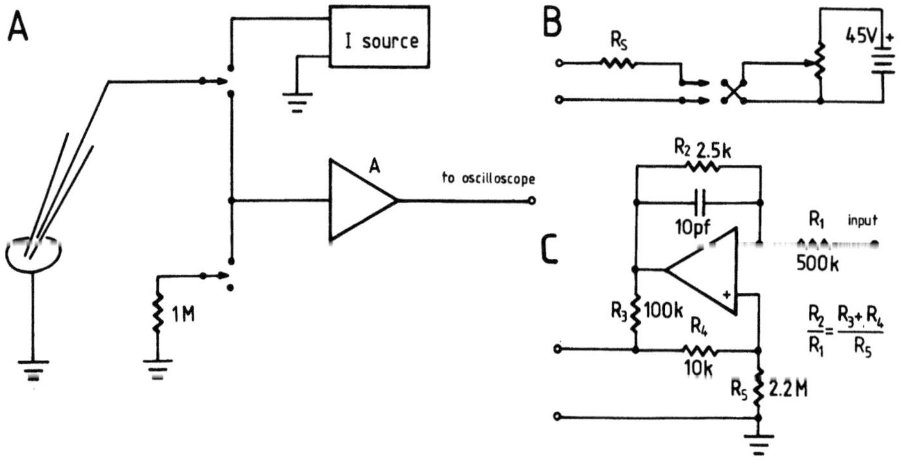

Figure 3. System of injecting current using constant-current sources. (A) diagrams the relationship of a current source to the recording amplifier, indicated as A, the preparation, and the microelectrode. A switch or relay is used to disconnect and ground the amplifier A from the microelectrode and to connect the current source for current injection. Schematics for two current sources are shown. (B) employs a battery and series resistor. (C) is a Howland current pump, an electronic constant-current source using an operational amplifier.

are used. For example, 1 μA driven through a 10 MΩ electrode (R_e) and a 90 MΩ series resistance (R_s) would require a 100 V battery or alternative voltage source. This would represent ±2% current regulation if R_e varied ±20%. However, it is easily seen that most of the voltage is required because of R_s, and this load can quickly become an impractical requirement if the same regulation is attempted with higher impedance electrodes, and/or if higher currents are required.

The circuits described in the remainder of this section are all true current sources. They have in common the use of feedback networks to provide a precisely controlled current through the load despite variations over a wide range of V_c. For these circuits, R_e is only a factor in the voltage drop across the microelectrode ($V_e = IR_e$) and contributes to the voltage at the upper end of the microelectrode, $V_{in} = V_c + IR_e$. A constant-current amplifier is accurate only if the maximum voltage it can deliver, called its compliance, is greater than V_{in}. As long as this condition is met, it accurately passes a regulated amount of current through the microelectrode. The higher the compliance of the current amplifier, the higher R_e it can tolerate. It is for this reason that current sources with compliances of ±100 V have been designed and used.

The *Howland current pump* has been used extensively in iontophoresis, and its use in a practical circuit has been described by Geller and Woodward (1972). A high-voltage operational amplifier is used to obtain high compliance. A circuit tailored for the injection of HRP is portrayed in Fig. 3C. The resistors R_1 and R_2 determine the gain of the amplifier and therefore the amount of current injected as a function of command voltage. This particular Howland current pump is calibrated to deliver 50 nA per command volt. Positive feedback through R_3 and the voltage divider network R_1 and R_5 maintains the constant current, measured as a voltage drop across R_s, even though changes in V_c and R_e occur.

This circuit, though widely used, is nonetheless far from ideal. To function properly, the condition $R_2/R_1 = (R_3 + R_4)/R_5$ must be precisely met. Because of its use of positive feedback, the circuit can become unstable. Transients in the power supply or in the input or output can lead to high-frequency oscillations that swing over the full range of supply voltage and are nearly always fatal to the penetrated cell. Such oscillations can occur if the electrode suddenly blocks. A further limitation of the Howland current pump is that it does not compensate for capacitive losses that exist between a high-resistance microelectrode and the surrounding tissue. A square-wave pulse delivered to the microelectrode barrel is always shaped by the time constant of the electrode resistance and any shunt capacitance parallel to that resistance. If the experimenter bears these difficulties in mind, he can make good use of this system of constant-current injection.

e. Active Bridge Preamplifiers. None of the circuits discussed so far allow the monitoring of membrane potential during current injection. Therefore, it is impossible to know whether the intracellular penetration is

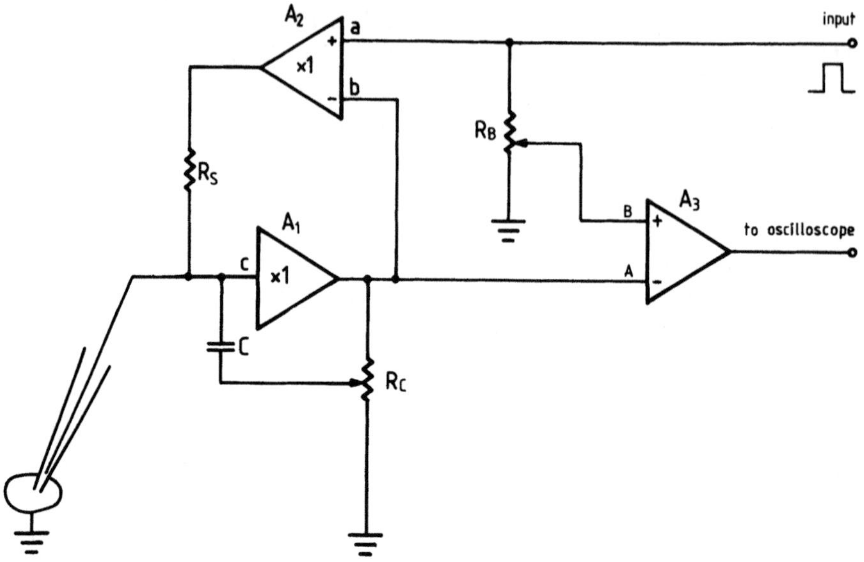

Figure 4. Block diagram of a microelectrode preamplifier with active bridge. The circuit portrayed here serves both to record voltage and to inject current through the microelectrode. It represents, in simplified form, the basic elements common to nearly all biological amplifiers employed for intracellular work.

being maintained as current is being injected. Many recording amplifiers, however, allow for the passing of current through the recording micropipette. Excellent results have been obtained using HRP-filled microelectrodes with resistances less than 100 MΩ. In this range, the ±15 V compliance of most commercially available biological amplifiers with current injection capabilities is adequate. High-compliance (at least ±100 V) preamplifiers specifically for microiontophoresis (Dreyer and Peper, 1974) have been designed to overcome the capacitive losses observed with other current injection systems.

The essential elements of a biological amplifier with an active bridge are portrayed in Fig. 4. Nearly all commercially available amplifiers for intracellular work are similarly designed. The block diagram consists of a unity-gain input amplifier, A1; a unity-gain amplifier, A2, for current injection; and a differential amplifier, A3, for output. A1 has the high input impedance necessary for microelectrode work and is the counterpart of amplifier A in Fig. 3A. All microelectrode amplifiers require a form of capacitance neutralization to compensate for the high-frequency components of the intracellular signal that are lost through shunt capacitances. Such capacitance, for instance, exists between the shaft of a microelectrode and surrounding tissue. In the circuit of Fig. 4, capacitance neutralization operates through the control, R_c, which returns a portion of the intracellular voltage signal through capacitor C to the microelectrode. C has a very small value, insuring that only high-frequency signal components participate in this positive feedback loop.

The key element for current injection is the closed loop formed by A_1, A_2, and R_s. The gains of A_1 and A_2 are precisely adjusted so that the total gain of that loop is precisely one. Two conditions become important in order to understand the functioning of this loop. First, as long as there is no command to inject current, i.e., the voltage at a is 0, any signal appearing at the electrode is duplicated by the output of A_2 at the upper end of R_s. Second, with identical voltages at both ends of R_s, no current flows through R_s, and, therfore, none flows through the electrode. Nor are the input characteristics of A_1 in any way degraded despite the connection of R_s whose resistance is many times lower than the input impedance of A_1 (Fein, 1966). When the current command input a receives a signal, it adds to b, and A_2 produces a voltage difference across R_s precisely equal to the voltage at a (V_a). A current $I = V_a/R_s$ will then flow through both R_s and the electrode. This current is constant despite potential changes at c such as would occur if R_e or V_c change. The remainder of the circuit, the differential amplifier A_3 and the bridge balance potentiometer R_B, is only for the convenience of reducing the artifact resulting from the voltage drop across R_e. R_B is adjusted to subtract a fraction of the current command input from the output signal. That fraction, BV_a, corresponds to the voltage IR_e that an electrode of constant R_e would contribute to the signal. The term bridge balance remains from the older technique of using a Wheatstone bridge to pass current through the microelectrode. The bridge-balance setting of a modern intracellular amplifier with an active bridge does not affect the amount of current being injected.

While current is being injected, the input of the amplifier registers a voltage that is the sum of the intracellular potential, V_c, and the voltage drop across the electrode caused by the injected current, $V_{in} = V_c + IR_e$. If the experimenter follows the commonly used procedure of using pulses of current to inject tracer substances (Chapter 7, Table I), he can conveniently monitor both electrode blocking and the quality of the penetration. During a pulse of injected current, a voltage excursion to the limit of amplifier compliance indicates that the electrode resistance is too high for the amount of current being injected. The amplitude of the current being injected should be reduced. If the high R_e is the result of blocking, then momentarily lowering the current is often sufficient to unblock the electrode. Between pulses, the current is zero, and an error-free measure of V_c can be made. This is the time to verify that membrane potential is being maintained.

IV. APPENDIX

A. Microelectrode Preparation

1. Capillary Tubing

Glass capillary tubing for manufacturing micropipettes is available from a number of distributors. Tubes come in various diameters, ranging from less than 1 mm to over 3 mm outside diameter. While smooth-bore tubing is available, tubes with a filament running down the inner wall have become popular, since they facilitate the filling of the microelectrode tip by promoting capillary action. If smooth capillary tubing is used, it is quite useful to insert glass fibers (preferably of the same type of glass) before pulling to facilitate later filling (Tasaki *et al.*, 1968). In addition to single capillary tubing, multichannel capillaries are commercially available. Some of these consist of several capillaries fused together, others, so-called "theta" or "quad" tube capillaries, contain multiple channels created by internal dividing walls. As indicated in the section on pressure injections, precalibrated micropipette capillaries may be used for extracellular injections. Most commonly used at present is borosilicate glass which is hard, strong, and, once drawn, may often be used repeatedly. Spencer *et al.* (1978) have suggested that in some instances interactions may occur between the tracer and the glass micropipette such that blockage or precipitation of the tracer (HRP) could occur. The particular source of the interference was not firmly established, but it should be kept in mind that this is a possible source of trouble that may require sampling of glass capillaries from various suppliers.

2. Cleaning of Capillary Tubing

Clean, uncontaminated tubing should be used for the manufacture of micropipettes in iontophoretic experiments. Kater *et al.* (1973), for example, suggest cleaning the capillaries in hot (80°C) concentrated H_2SO_4 followed by 3 rinses in 80°C filtered double distilled H_2O, and subsequent drying in a dust-free oven. Capillaries should then be stored in a suitable dust-free container prior to pulling.

For less critical applications, capillaries may be individually cleaned using tubing connected to a vacuum source such as a water-driven venturi. If building-supplied vacuum is used, a catch bottle must be used to insure that liquid is not drawn into the house system. If the connective tubing is chosen so that it just fits snugly on the outside of the capillary, the vacuum may be used to draw various solvents through the glass capillary at a rapid rate. If acetone is followed by distilled water and subsequently by isopropyl alcohol, the rapid evaporation of the alcohol with continued vacuum will allow the capillary to be used immediately for micropipette construction. With moderate tip sizes (tips >10 μm internal diameter), this cleaning procedure may be repeated on "pulled" micropipettes just prior to filling.

3. Pulling Micropipettes

The essentials of obtaining acceptable tapers on micropipette tubes depend on (1) the concentric placement of the tube in the heating coil, (2) the temperature of the coil, and (3) the strength of the pull and, in certain machines, the sequencing of different strengths of pull.

The tissue to be penetrated and the depth and size of tissue mass dictate how long the taper should be. For injections that penetrate 3 mm or less, short, stubby micropipettes are sufficient, whereas depths greater than this require a longer taper if tissue damage is to be avoided near the surface. When depths of 20 mm or more are used, deviation of the tip in the course of penetration becomes a problem. To prevent deviation, a concentric guide tube can be used (Graybiel and Devor, 1974).

Horizontal pipette pullers have variable strength weak and strong pull stages, allowing one to obtain the precise length and degree of taper necessary for a variety of conditions. The most common vertical puller employs gravity for the weak pull state. With a little practice, adequate pipettes can be obtained with any of the pullers, although to obtain the finest tips recommended for intracellular recording and injections, some authors recommend a Livingston-type puller or a version of a gas-cooled pipette puller (Brown and Flaming, 1977). The former produces an extra long taper used by many to obtain ultrafine tips, whereas the latter allows the pulling of fine pipette tips with larger taper angles, thus producing a somewhat more sturdy microelectrode tip geometry.

4. Beveling Tips

The regular beveled surface produced by sharpening techniques eleminates or minimizes plugging of the electrode tip by the tissue and facilitates micropipette passage through the tissue.

Once the right taper has been selected, the most suitable tip size must be determined. The tip is then beveled, using either a special beveling device or a mechanical bumping method. In the latter technique, the electrode tip is touched against some solid object under microscopic observation until appropriate tip diameter and bevel are obtained (Fig. 1, Chapter 1). Electrodes for intracellular penetration with tip diameters 0.5–1.0 μm and D.C. resistance values ranging between 30 and 60 MΩ can be obtained in this manner. Very fine beveled tips can also be obtained by the jet stream method (Ogden *et al.*, 1978) or a recent modification by Corson *et al.* (1979). These methods simply rely beveling of the micropipette tip by a suspension of fine grinding powder in a moving liquid, usually saline. While the mechanical bumping method or the liquid beveling techniques are simple and inexpensive to use, they may not be suitable for all experiments. One of the beveling techniques based on a wobble-free rotating surface embedded with grinding compound may be more suitable (Barrett and Whitlock, 1973; Brown and Flaming, 1974, 1975, 1979; Shaw and Lee, 1973).

A fast, easy, and reliable method of beveling tips is used in one of our laboratories (R. C. Switzer III). It was developed by Dr. David Watkins and is a modification of a method used in Dr. Ford Ebner's laboratory at Brown University. A circle of Flexigrit®,* an abrasive-coated polyester available in

*RDC Industries, 21st and Clearfield Streets, Philadelphia, Pennsylvania 19131.

hobby shops, is attached with spray contact adhesive to a disk that is held in the head stock of a minilathe. The micropipette to be beveled is attached to a holder that is mounted on the cross slide of the lathe. The pipette is positioned at *ca.* 45° to the plane of the disk with the tip pointing in the direction of rotation of the disk. With the aid of a small spotlight, the pipette is observed as it is advanced toward the disk. The tip is brought into contact with the disk and advanced further until the taper of the tip is seen to bend. Only 10–15 sec is necessary to bevel the micropipette using an 8-μm grit abrasive disk.

It is convenient to store unfilled micropipettes by attaching them to glass microscope slides with adhesive tape. Following inspection under a microscope, the diameters of the micropipette tips can be recorded on the slides which can then be stored in a slide box.

5. Micropipette Size

Micropipette size may be specified in terms of tip diameter (μm). Tips of >1 μm diameter (preferably internal) may be measured by observing the tip under a microscope with a calibrated reticle. However, as the limit of resolution for the light microscope is around 0.25 μm, it is difficult to measure tip sizes below 0.5–1 μm. Another way to estimate the size of the micropipette is to determine its impedance (MΩ), especially since electrode impedance is proportional to the tip diameter for a given filling solution. Impedance is often measured with A.C. or pulsed currents, since use of direct current may result in tip junction potentials in the micropipette with resulting errors in the impedance measurement.

6. Filling Micropipettes

Various methods have been devised for filling micropipettes. As with all procedures involving micropipettes, satisfactory filling and ejection of tracers will be enhanced if care is taken to use clean pipettes and clean solutions. It may be quite helpful to filter the tracer solution or the solvents used to carry the tracer.

Glass fibers inserted into capillary tubing prior to pulling of micropipettes facilitates the filling of the tip (Tasaki *et al.*, 1968). Such micropipettes are easily filled by injecting the tracer solution into the back of the pulled pipette using a fine gauge needle. The needle should be inserted up to the shoulder of the micropipette, and the solution slowly injected while withdrawing the needle. Care should be taken to avoid introducing air bubbles into the pipette. When dealing with expensive tracer solutions, it is preferable to use a microliter syringe for pipette filling so that small amounts of tracer material may be conveniently handled. With filament-filled tubing, the tip of the micropipette should rapidly become filled by capillary action.

When using more viscous solutions of tracers such as HRP for iontophoresis, it may be advisable to fill just the tip of the micropipette with the solvent, then fill the remainder of the pipette with the tracer solution, and allow the tip to fill by diffusion (Graybiel and Devor, 1974). When using pipettes containing glass fibers, the tip may be filled with dilute tracer or solvent solutions simply by holding the barrel of the pipette in the solution for 1 to 2 min.

If bubbles occur in the barrel of the pipette, they may be dislodged by tapping the side of the pipette or by means of a fine gauge wire used to tease them out. Bubbles in the very tip of the micropipette are more difficult to get rid of. However, if they do occur, an attempt to force them out can be made by connecting the barrel of the micropipette to a pressure source (see Section II.B) by means of snugly fitting polyethylene tubing. By applying pressure, one may be able to fill the tip with the tracer solution and force the bubble out, in which case one should be able to observe a fine drop of solution forming at the tip of the pipette. The connecting tubing may then be cut off proximal to the micropipette and the end removed. This avoids creating any vacuum that might reintroduce air into the pipette when the tubing is removed. The success of this procedure will be roughly proportional to the tip diameter of the pipette. As suggested in Section II, care should be taken to rinse off the pipette before it is inserted into the tissue.

If smooth-bore capillary tubing is used, micropipette filling can be accomplished by a method that involves filling the electrodes first with alcohol by gently boiling at reduced pressure. This solution can then be replaced over a few days by diffusion, first in distilled water and then in the tracer solution.

Once the micropipette has been filled, it is practical to keep the tip immersed in the solvent used for the tracer or in saline to prevent blockage of the tip by evaporation. If the pipette is to be stored for any length of time, it may be kept in a humid environment with the tip submerged in the solvent in order to minimize effects of evaporation in the barrel of the micropipette. Alternatively, the micropipette may be stored with the tip in paraffin oil, and a drop of paraffin oil may be placed in the barrel of the pipette to prevent evaporation.

B. Iontophoresis Assays

One recurring problem with iontophoretic application of tracers is assuring a successful deposit of tracer with each surgery. Quantification of chemical release by iontophoresis has been a serious problem in pharmacological investigations, and, in general, drug release is judged to be proportional to current. In such investigations, the experimenter may also monitor the neuronal response to the ionophore. In tract-tracing studies, however, a certain amount of the tracer must be injected in order to insure transport and labeling which occur in the absence of any clear neuronal response. The problem was illustrated in a recent study by Norgren (1976) who reported that no labeling was observed in 50% of iontophoretic injections although current and electrode impedance appeared normal.

One way to assure a successful deposit is to pretest the microelectrode for adequate tracer ejection. Schubert and colleagues (Schubert *et al.*, 1972; Schubert and Holländer, 1975) have suggested that the release of [^{3}H]amino acid from the micropipette should be checked preoperatively by iontophoresis, e.g., 100 nA for 5 min into a drop of saline, and subsequent measurement in a scintillation counter. They report that adequate tissue labeling can be expected if the electrodes measure 15 μm at the tip and pass about 4000–8000 counts per min *in vitro*.

A similar test for HRP release was devised by Spencer *et al.* (1978) as a variation of the peroxidase *o*-dianisidine-HCl assay. In this case, the color reaction with HRP, H_2O_2, and the chromogen *o*-dianisidine is used to judge HRP iontophoresis into a drop of saline. A test solution is made by adding 50 μl of 3% H_2O_2 plus 50 μl of *o*-dianisidine (7.5 mg/ml) to 6.0 ml of 0.01 M phosphate buffer (pH 6.0). When a drop of this solution is added to a drop of saline into which HRP has been injected, color should develop within 60 sec if sufficient HRP has been injected. "Good" HRP release results in a purple color that changes to a clear straw brown. Care should be taken to minimize air exposure to the tip of the micropipette in order to minimize plugging.

One of us (G. F. Alheid) has recently attempted to develop an HRP assay using tetramethyl benzidine (TMB) which has become a popular chromogen because of its sensitivity and apparent noncarcinogenic properties (Holland *et al.*, 1974). The stock solution (A) consists of 1 ml 0.05 M acetate buffer (pH 3.5), 1 ml 0.2 M sodium nitroprusside, 1 ml 1% H_2O_2, and 10 ml H_2O. A drop of saline into which HRP has been injected is transferred to a piece of filter paper. One drop of solution (A) is added to the saline drop followed by one drop of a TMB solution (1 mg/ml in absolute ethyl alcohol). The affected part of the filter paper will become blue–green in color if HRP has been injected. It is useful to simultaneously spot check a drop of pure saline as a check for inadvertant contamination. This assay and that described by Spencer *et al.* (1978) can be adaptable to quantitative colorimetric measurements with a spectrophotometer. Spencer *et al.* suggest the use of a 50-μl microcuvette for such a test.

It should be pointed out that quantification of both the [^{3}H]amino acid and the HRP-labeling techniques might be quite useful. If, for instance, the micropipette is known to pass a given number of units/min per μA, injection current and duration can be adjusted to produce equal deposits in successive animals, or when a variety of injection sizes are required, they can be regularly increased in unit size. This, in turn, might aid in understanding some of the quantitative relationships among tracer injection parameters, deposit size, and transport.

When using HRP solutions of mixed isoenzymes, one should keep in mind that the transportable HRP only represents a proportion in the *in vitro* color reaction, since not all isoenzymes are equally transported (Chapter 6, and Bunt and Haschke, 1976, 1978).

ACKNOWLEDGMENTS. This work was supported in part by Grants NS 11254 to S. B. Edwards and NS 14866 to S. T. Kitai and M. R. Park. R. C. Switzer III was a staff scientist at NIMH and G. F. Alheid was supported by NIMH Fellowship 7F32 MH 07677-01.

REFERENCES

Abols, I. A., and Basbaum, A. I., 1979, The posterior pretectal nucleus: Evidence for a direct projection to the inferior olive of the cat, *Neurosci. Lett.* **13:**111–116.

Aghajanian, G. K., and Wang, R. Y., 1977, Habenular and other midbrain afferents demonstrated by a modified retrograde tracing technique, *Brain Res.* **122:**229–242.

Ahlsén, G., Lindström, S., and Sybirska, E., 1978, Subcortical axon collaterals of principal cells in the lateral geniculate body of the cat, *Brain Res.* **156:**106–109.

Arbuthnott, G. W., 1969, Noradrenaline uptake into cerebral cortex: A histochemical study, *J. Neurochem.* **16:**1599–1604.

Barrett, J., and Whitlock, D. G., 1973, Technique for believing glass microelectrodes, in: *Intracellular Staining in Neurobiology* (S. B. Kater and C. Nicholsen, eds.), pp. 297–299, Springer-Verlag, Berlin.

Beckstead, R. M., 1979a, An autoradiographic examination of corticocortical and subcortical projections of the mediodorsal projection (prefrontal) cortex in the rat, *J. Comp. Neurol.* **184:**43–62.

Beckstead, R. M., 1979b, Convergent prefrontal and nigral projections to the striatum of the rat, *Neurosci. Lett.* **123:**59–64.

Brown, K. T., and Flaming, D. G., 1974, Beveling of fine micropipette electrodes by a rapid precision method, *Science* **185:**693–695.

Brown, K. T., and Flaming, D. G., 1975, Instrumentation and technique for beveling fine micropipette electrodes, *Brain Res.* **86:**172–180.

Brown, K. T., and Flaming, D. G., 1977, New microelectrode techniques for intracellular work in small cells, *Neuroscience* **2:**813–827.

Brown, K. T., and Flaming, D. G. 1979, Technique for precision beveling of relatively large micropipettes, *J. Neurosci. Methods* **1:**25–34.

Brown, P. B., Maxfield, B. W., and Moraff, H., 1973, *Electronics for Neurobiologists,* M.I.T. Press, Cambridge, Massachusetts.

Bunney, B. S., and Aghajanian, G. K., 1976, The precise localization of nigral afferents in the rat as determined by a retrograde tracing technique, *Brain Res.* **117:**423–435.

Bunt, A., and Haschke, R. H., 1978, Features of foreign proteins affecting their retrograde transport in axons of the visual system, *J. Neurocytol.* **7:**665–678.

Bunt, A. H., Haschke, R. H., Lund, R. D., and Calkins, D. F., 1976, Factors affecting retrograde axonal transport of horseradish peroxidase in the visual system, *Brain Res.* **102:**152–155.

Bures, J., Buresova, O., and Huston, J., 1976, *Techniques and Basic Experiments for the Study of Brain and Behavior*, Elsevier, Amsterdam.

Cedarbaum, J. M., and Aghajanian, G. K., 1978, Afferent projections to the rat locus coeruleus as determined by a retrograde tracing technique, *J. Comp. Neurol.* **178:**1–16.

Chowdhury, T. K., 1969, Techniques of intracellular injection, in *Glass Microelectrodes* (M. LaVallee, O. F. Schanne, and N. C. Herbert, eds.), pp. 404–423, John Wiley & Sons, New York.

Corson, D. W., Goodman, S., and Fein, A., 1979, An adaptation of the jet stream micropipette beveler, *Science* **205:**1302.

Davis, R. E., and Agranoff, B. W., 1977, Microimplantation of ^{3}H-proline on a single bead of ion-exchange resin, *Brain Res.* **124:**341–346.

Descarries, L., and Lapierre, Y., 1973, Noradrenergic axon terminals in the cerebral cortex of rat. I. Radioautographic visualization after topical application of *d,l*-^{3}H-norepinephrine, *Brain Res.* **51:**141–160.

Dreyer, F., and Peper, K., 1974, Iontophoretic applications of acetylcholine: Advantages of high resistance micropipettes in connection with an electronic current pump, *Pfluegers Arch.* **348:**263–272.

Edwards, S. B., and Henkel, C. K., 1978, Superior colliculus connections with the extraocular motor nuclei in the cat, *J. Comp. Neurol.* **179:**451–468.

Edwards, S. B., and Shalna, E. J., 1974, Microinjector for use in the autoradiographic neuroanatomical tracing method, *Pharmacol. Biochem. Behav.* **2:**111–113.

Eide, E., Illert, M., and Tanaka, R., 1976, Injection of horseradish peroxidase solution in calibrated volumes (nanolitre) into the spinal cord, *Neurosci. Lett.* **2:**51–56.

Fein, H., 1966, Passing current through recording glass micro-pipette electrodes, *IEEE Trans. Biomed. Eng.* **13:**211–212.

Gallager, D. W., and Pert, A., 1978, Afferents to brain stem nuclei (brain stem raphe, nucleus reticularis pontis caudalis and nucleus giganto cellularis) in the rat as demonstrated by microiontophoretically applied horseradish peroxidase, *Brain Res.* **144:**257–275.

Geller, H. M., and Woodward, D. J., 1972, An improved constant current source for micro-iontophoretic drug application studies, *Electroencephalogr. Clin. Neurophysiol.* **33:**430–432.

Graybiel, A. M., and Devor, M. A., 1974, A microelectrode delivery technique for use with horseradish peroxidase, *Brain Res.* **68:**167–173.

Griffin, G., Watkins, L. R., and Mayer, D. V., 1979, HRP pellets and slow release gels: Two new techniques for greater localization and sensitivity, *Brain Res.* **168:**595–601.

Guglielmotti, V., 1979, A thermally controlled microinjector, *Stain Technol.* **54:**151–157.

Held, J., and Young, I. J., 1969, A comparative study of the somato-axonal flow of protein in the feline hypoglossal and vagus nerves, *Exp. Brain Res.* **8:**150–162.

Henkel, C. K., and Edwards, S. B., 1978, The superior colliculus control of pinna movements in the cat: Possible anatomical connections, *J. Comp. Neurol.* **182:**763–776.

Herkenham, M., 1978, The connections of the nucleus reuniens thalami: Evidence for a direct thalamo-hippocampal pathway in the rat, *J. Comp. Neurol.* **177:**589–610.

Herkenham, M., 1979, The afferent and efferent connections of the ventromedial thalamic nucleus in the rat, *J. Comp. Neurol.* **183:**487–518.

Herkenham, M., and Nauta, W. J. H., 1979, Efferent connections of the habenula nuclei in the rat, *J. Comp. Neurol.* **187:** 19–48.

Hoenig, S. A., and F. L. Payne, 1973, *How to Build and Use Electronic Devices without Frustration, Panic, Mountains of Money, or an Engineering Degree,* Little, Brown & Co., Boston.

Holland, V. R., Saunders, B. C., Rose, F. L., and Walpole, A. L., 1974, A safer substitute for benzidine in detection of blood, *Tetrahedron* **30:**3299–3302.

Iles, J. F., and Mulloney, B., 1971, Procion yellow staining of cockroach motor neurons without the use of microelectrodes, *Brain Res.* **30:**397–400.

Jones, E. G., Coulter, J. D., and Hendry, S. H. C., 1978, Intracortical connectivity of architectonic fields in the somatic sensory, motor and parietal cortex of monkeys, *J. Comp. Neurol.* **181:**291–348.

Källström, Y., and Lindström, S., 1978, Simple device for pressure injections of horseradish peroxidase into small central neurons, *Brain Res.* **156:**102–105.

Kater, S. B., and Nicholson, C., 1973, *Intracellular Staining in Neurobiology*, Springer-Verlag, Berlin and New York.

Kater, S., Nicholson, C., and Davis, W. J., 1973, Guide to intracellular staining techniques, in: *Intracellular Staining in Neurobiology* (S. B. Kater and C. Nicholson, eds.), pp. 307–325, Springer-Verlag, Berlin and New York.

Kawamura, S., and Diamond, I. T., 1978, The laminar origin of descending projections from the cortex to the thalamus in Topaia glis, *Brain Res.* **153:**333–339.

Kawamura, S., Fukushima, N., Hattori, S., and Tashiro, T., 1978, A ventral lateral geniculate nucleus projection to the dorsal thalamus and the midbrain in the cat, *Exp. Brain Res.* **34:**95–106.

Kelly, J. P., and Gilbert, C. D., 1975, The projections of different morphological types of ganglion cells in the cat retina, *J. Comp. Neurol.* **163:**65–80.

Koike, H., Eisenstadt, M., and Schwartz, J. H., 1972, Axonal transport of newly synthesized acetylcholine in an identified neuron of aplysia, *Brain Res.* **37:**152–159.

Kristenson, K., 1970, Transport of fluorescent protein tracer in peripheral nerves, *Acta Neuropathol. (Berlin)* **16:**293–300.

Kristenson, K., 1975, Retrograde axonal transport of protein tracers, in: *The Use of Axonal Transport for Studies of Neuronal Connectivity* (W. M. Cowan and M. Cuénod, eds.), Elsevier, Amsterdam.

Kristenson, K., and Olsson, Y., 1973, Diffusion pathways and retrograde axonal transport of protein tracers in peripheral nerves, *Prog. Neurobiol.* **1:**85–109.

Kristenson, K., Olsson, Y., and Sjöstrand, J., 1971, Axonal uptake and retrograde transport of exogenous proteins in the hypoglossal nerve, *Brain Res.* **32:**399–406.

Lasek, R., Joeseph, B. S., and Whitlock, D. G., 1968, Evaluation of a radiautographic neuroanatomical tracing method, *Brain Res.* **8:**319–336.

LaVail, J. H., and LaVail, M. M., 1972, Retrograde axonal transport in the central nervous system, *Science* **176:**1416–1417.

LaVail, J. H., and LaVail, M. M., 1974, The retrograde intraaxonal transport of horseradish peroxidase in the chick visual system: A light and electron microscope study, *J. Comp. Neurol.* **157:**303–358.

LaVallee, M., Schanne, O. F., and Herbert, N. C., 1969, *Glass Microelectrodes*, John Wiley & Sons, New York.

Luiten, P. G. M., 1975, The horseradish peroxidase technique applied to the teleostan nervous system, *Brain Res.* **89:**181–186.

Luiten, P. G. M., and Van der Pers, J. N. C., 1977, The connection of the trigeminal and facial motor nuclei in the brain stem of the carp (*Cyprinus carpio* l.) as revealed by anterograde and retrograde transport of horseradish peroxidase, *J. Comp. Neurol.* **174:**575–590.

Lynch, G., Deadwyler, S., and Gall, C., 1974, Labeling of central nervous system neurons with extracellular recording microelectrodes, *Brain Res.* **66:**337–341.

Maekawa, K., and Takeda, T., 1979, Origin of descending afferents to the rostral part of the dorsal cap of inferior olive which transfers contralateral optic activities to the flocculus. A horseradish peroxidase study, *Brain Res.* **172:**393–405.

McCaman, R. E., McKenna, D. G., and Ono, J. K., 1977, A pressure system for intracellular and extracellular ejections of picoliter volumes, *Brain Res.* **136:**141–147.

Myers, R. D., 1974, *Handbook of Drug and Chemical Stimulation of the Brain*, Van Nostrand Reinhold Co., London.

Nastuk, W. L., 1964, *Physical Techniques in Biological Research, Vol. V. Electrophysiological Methods, Part A,* Academic Press, New York.

Norgren, R., 1976, Taste pathways to hypothalamus and amygdala, *J. Comp. Neurol.* **166:**17–30.

Ogden, T. E., Citron, M. C., and Pierantoni, R., 1978, The jet stream microbeveler: An inexpensive way to bevel ultrafine glass micropipettes, *Science* **201:**469.

Ottersen, O. P., and Ben-Ari, Y., 1978a, Pontine and mesencephalic afferents to the central nucleus of the amygdala of the rat, *Neurosci. Lett.* **8:**329–334.

Ottersen, O. P., and Ben-Ari, Y., 1978b, Demonstration of a heavy projection of midline thalamic neurons upon the lateral nucleus of the amygadala of the rat, *Neurosci. Lett.* **9:**147–152.

Ottersen, O. P., and Ben-Ari, Y., 1979, Afferent connections to the amygdaloid complex of the rat and cat. I. Projections from the thalamus, *J. Comp. Neurol.* **187:**401–424.

Phillipson, O. T., 1978, Afferent projections to A10 dopaminergic neurons in the rat as shown by the retrograde transport of horseradish peroxidase, *Neurosci. Lett.* **9:**353–359.

Phillipson, O. T., 1979, Afferent projection to the ventral tegmental area of Tsai and interfascicular nucleus: A horseradish peroxidase study in the rat, *J. Comp. Neurol.* **187:**117–144.

Pitman, R. M., Tweedle, C. D., and Cohen, M. J., 1972, Branching of central neurons: Intracellular cobalt injections for light and electron microscopy, *Science* **176:**412–414.

Price, P., Fisher, A. W. F., and Redstone, P., 1977, A simple apparatus for injection of small (nanolitre) volumes of horseradish-peroxidase, *Neurosci. Lett.* **6:**21–25.

Remler, M. P., Selverston, A. I., and Kennedy, D., 1968, Lateral giant fibers of crayfish: Location of somata by dye injection, *Science* **162:**281–283.

RoBards, M. J., 1979, Somatic neurons in the brainstem and neocortex projecting to the external nucleus of the inferior colliculus: An anatomical study in the opossum, *J. Comp. Neurol.* **184:**547–566.

Sakai, M., Sakai, H., and Woody, C. D., 1978, Intracellular staining of cortical neurons by pressure microinjections and recovery by core biopsy, *Exp. Neurol.* **58:**138–144.

Schubert, P., and Holländer, H., 1975, Methods for delivery of tracers to the central nervous system, in: *The Use of Axonal Transport for Studies of Neuronal Connectivity* (W. M. Cowan and M. Cuénod, eds.), Elsevier, Amsterdam.

Schubert, P., Kreutzberg, G. W., and Lux, H. D., 1972, Use of microelectrophoresis in the autoradiographic demonstration of fiber projections, *Brain Res.* **39:**274–277.

Shaw, M. L., and Lee, D. R., 1973, Micropipette sharpener with audio and hydraulic readouts, *J. Appl. Physiol.* **34:**523–524.

Simon, A., LeMoal, M., and Calas, A., 1979, Efferents and afferents of the ventral tegmental A-10 region studied after local injection of ^{3}H-leucine and horseradish peroxidase, *Brain Res.* **178:**17–40.

Spencer, H. J., Jones, R. K., and Lynch, G., 1978, The use of somatougal transport of horseradish peroxidase for tract tracing and cell labeling, in: *Methods in Physiological Psychology, Vol. II, Neuroanatomical Research Techniques* (R. T. Robertson, ed.), Academic Press, New York and London.

Tasaki, I., Tsukahara, Y., Ito, S., Wayner, M. J., and Yu, W. Y., 1968, A simple, direct and rapid method for filling microelectrodes, *Physiol. Behav.* **3:**1009–1010.

Thompson, R. F., and Patterson, M. M., 1973, *Methods in Physiological Psychology, Vol. 1. Bioelectric Recording Techniques, Part A, Cellular Processes and Brain Potentials,* Academic Press, New York and London.

Thompson, R. F., and Patterson, M. M., 1974, *Methods in Physiological Psychology, Vol.* I. *Bioelectric Recording Techniques, Part C, Receptor and Effector Processes,* Academic Press, New York and London.

Tweedle, C. D., 1978, Single cell staining techniques, in: *Methods in Physiological Psychology, Vol. II, Neuroanatomical Research Techniques* (R. T. Robertson, ed.), Academic Press, New York and London.

Wakefield, C., and Shonnard, N., 1979, Observations of HRP labeling following injection through a chronically implanted cannula. A method to avoid diffusion into injured fibers, *Brain Res.* **168:**221–226.

Weiss, P., and Holland, Y., 1967, Neuronal dynamics and axonal flow. II. The olfactory nerve as a model test object, *Proc. Natl. Acad. Sci. USA* **57:**259–264.

Woodward, W. R., and Lindström, S. H., 1977, A potential screening technique for neurotransmitters in the CNS: Model studies in the cat spinal cord, *Brain Res.* **137:**37–52.

Silver Methods for the Impregnation of Degenerating Axoplasm

4

JOSÉ S. de OLMOS, SVEN O. E. EBBESSON, and LENNART HEIMER

I. INTRODUCTION

The reduced silver methods were introduced in the beginning of this century (Ramón y Cajal, 1904; Bielschowsky, 1904), and they were initially appreciated primarily for their remarkable capacity to reveal normal histology and fiber architecture of the brain. Unfortunately, attempts to trace specific fiber systems using these methods were generally unsuccessful, because fiber bundles from a selected neuronal cell group could not be followed with any certainty through the complex meshwork of axons arising from other parts of the nervous system. In the 1930s, neuroanatomists (e.g., Hoff, 1932a,b; Glees

JOSÉ S. de OLMOS • Instituto de Investigación Medica Mercedes y Martin Ferreyra, Cordoba, Argentina. SVEN O. E. EBBESSON • Department of Anatomy, University of Ponce, Ponce, Puerto Rico 00731. LENNART HEIMER • Department of Neurology, University of Virginia School of Medicine, Charlottesville, Virginia 22908.

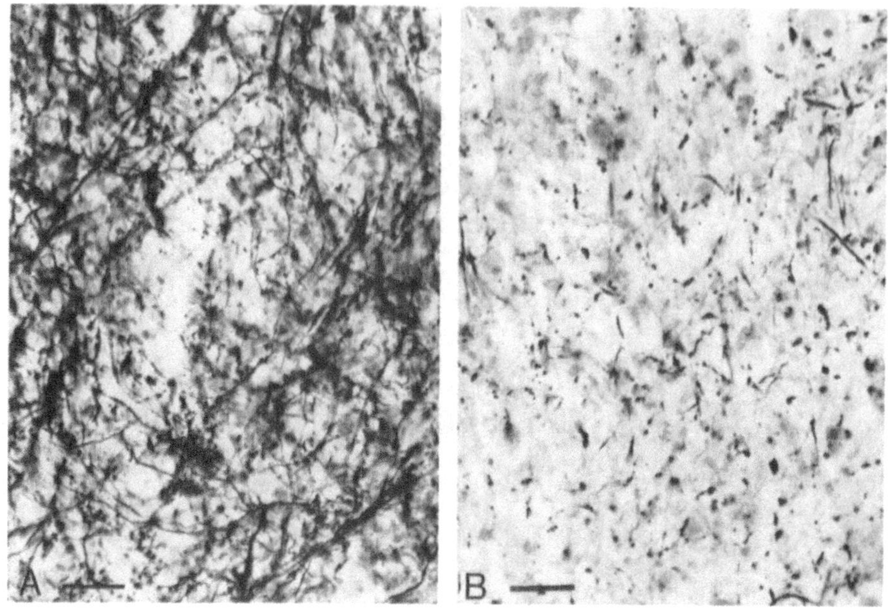

Figure 1. (A) Photomicrograph of a Glees-impregnated section demonstrating degeneration in the mammillary body of the rat following transection of the fornix 3 days before sacrifice of the animal. The degenerating terminals and fibers are partly camouflaged by the impregnation of the normal fibers. Bar, 20 μm. (B) Same case as in (A) but stained with the Nauta–Laidlaw method. The suppression of the argyrophilia of normal fibers enhances the study of degenerating fibers. Bar, 20 μm.

and Le Gros Clark, 1941; Glees, 1946) discovered that the capacity of silver stains to identify axons in some stage of Wallerian degeneration* could be utilized to "mark" a system of interest, thus facilitating its recognition as it meandered through the nervous system. Thus, an area of the nervous system could be subjected to an experimental lesion, and, after an appropriate survival interval, the animal could be sacrificed, and its brain processed by a silver method to reveal degenerating axons en route to their destination. Although degenerating axons could usually be distinguished from normal ones in such preparations, evaluating this histological material was tedious, and the results often inaccurate or incomplete. A major breakthrough occurred when Nauta and his colleagues (Nauta, 1950; Nauta and Gygax, 1951, 1954; Nauta and Ryna, 1952) first developed a silver stain that was *selective* for *degenerating* neural elements. Compare Figs. 1A and 1B for the

*In 1850, Augustus Waller demonstrated that a nerve fiber that is separated from its cell body undergoes a gradual disintegration referred to as direct Wallerian degeneration. This important phenomenon has formed the basis for several tract-tracing techniques including the famous Marchi method (Marchi and Algeri, 1885) for the staining of degenerated myelin. However, compared to the silver methods that stain degenerated axoplasm rather than the degenerated myelin sheaths, the use of the Marchi method is severely limited by the fact that many nerve fibers are poorly myelineated or unmyelinated. (For a definition of "indirect Wallerian degeneration," see Section VI.B.)

results obtained with nonselective (Glees) and selective (Nauta–Gygax) silver stains. Unfortunately, the family of Nauta stains was more effective at staining axons than axon terminals. Thus, projection pathways could often be traced only to their approximate site of termination. Naturally, for some systems, the information provided by the Nauta method was quite adequate; for others, major problems remained unsolved. The development of a Nauta-type stain, the Fink–Heimer (1967) technique, capable of impregnating degenerating terminals permitted a more complete demonstration of degenerating pathways. Since then, this procedure has been modified repeatedly to facilitate its use in various systems and species.

There is little doubt that many of the modern silver methods are capable of impregnating a significant number of synaptic endstructures. Convincing evidence comes especially from electron microscope studies of previously silver-impregnated sections (Heimer and Peters, 1968; Walberg, 1971). These ultrastructural studies have also shown that terminal segments of degenerating axons contribute in varying degrees to the light microscopic picture of so-called "terminal degeneration."* As the problem of defining the termination areas of a fiber system is of fundamental importance in experimental silver impregnation studies, this subject will be discussed in more detail in Section V.B.

II. THEORETICAL CONSIDERATIONS

A. When to Use the Silver Methods

The methods for selective silver impregnation of degenerating fibers and terminals are most suitable when the main emphasis of a study is to determine the efferent projections of an area or the distribution of a fiber tract. However, to what extent a silver impregnation study will be able to elucidate a pathway and its distribution areas depends greatly on the general anatomy and topography of the fiber system.

Especially valuable results can be expected in the study of efferent projections of superficially located structures such as the cerebral cortex in which lesions can be relatively easily made. This situation is illustrated in Fig. 2. A lesion of neurons in A, for instance, results in the degeneration of axons originating in A and projecting to B and C. The result would be similar if a lesion were made in the fiber tract between A and B. This approach is likely to be especially valuable in cases where the axons of a particular set of widely distributed neurons form a compact fiber bundle that can be transected conveniently. Typical examples of such fiber systems are the ascending and descending spinal pathways and the various commissural systems such as the corpus callosum.

If, on the other hand, the lesion is made in B, the axons originating in

*The term "terminal degeneration" refers to the degenerative changes occurring in the terminal axon arborizations and their synaptic endings (Nauta, 1950).

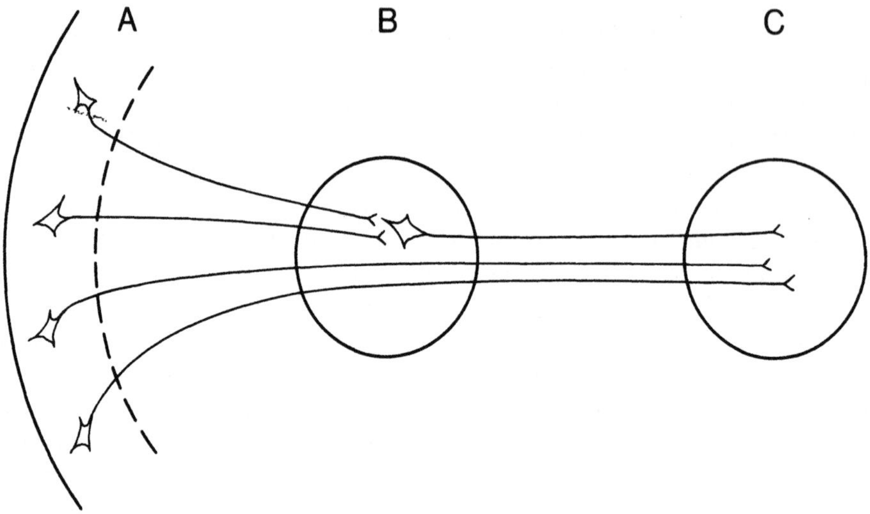

Figure 2. A schematic illustration of common situations in which the silver impregnation method can be used. See text for further explantion.

B and terminating in C will degenerate. It is important to realize, however, that the axons that originate in A and pass through B on their way to C will also be interrupted. A certain number of the degenerating axons entering C will therefore represent such "fibers of passage" that have their cells of origin in A rather than in B. The potential destruction of "fibers of passage" is a very important consideration in any type of experimental silver impregnation study, especially when the lesion is made in a subcortical structure.

Any deliberation on the usefulness of the silver methods should include some consideration of how the experimental lesion can best be provoked. It is especially important to realize that, besides the classical surgical fiber transections and electrolytic lesions, there are other alternatives. For example, the laminar coagulation technique, which is described in Chapter 2 (Section III.F.1), can be used for the tracing of pathways originating in superficial structures. It has also been used with success for the study of intracortical connections (Nauta *et al.*, 1973). Some chemical compounds that have a toxic effect on specific neuronal populations can be used to provoke selective axonal degeneration within those systems. 6-Hydroxydopamine, for instance, produces degeneration of dopamine pathways which can then be traced by the aid of a silver method (Hedreen and Chalmers, 1972), and selective degeneration of climbing fibers can be provoked by injections of 3-acetylpyridine (Desclin and Escubi, 1974). A detailed discussion of the many possibilities for producing lesions in the central nervous system is presented in Chapter 2.

When considering the choice of method, one should remember that the silver methods are often able to disclose morphological characteristics of the axons or the terminals in a given pathway. As a result, it may be possible to obtain information that has physiological relevance (Guillery and Oberdorfer, 1977). In order to take full advantage of this ability, however, one must be prepared to select the most appropriate survival time on the basis of a series of experiments with different postoperative intervals. Another unconventional application of the experimental silver methods has recently been suggested by Leonard (1973, 1975) who impregnated the Wallerian degeneration following experimental lesions in the developing visual and olfactory pathways in the hamster. As one might have expected, Leonard found that the onset and duration of the argyrophilia varies according to the age of the animal, and her hypothesis is that the appearance of a special kind of long-lasting argyrophilia can be correlated with the functional maturation of the system.

B. The Choice of Silver Method

Since the chemicophysical basis for silver impregnation is largely unknown, and since we do not know how variations in quality of the different chemicals affect the result, the methods are characterized by a certain capriciousness. A recipe that has given good results in some laboratories may be unsuccessful in others, thereby prompting repeated modifications of the original recipe. As a consequence, the literature contains a great number of modifications, and it may seem difficult for a beginner to choose the most appropriate recipe. Although most of the more recently published methods may serve the purpose if they are applied with some finesse, we have been most consistently successful with the Nauta–Laidlaw method for the staining of axonal degeneration and with the Fink–Heimer and the cupric silver methods for the impregnation of terminal degeneration (see Sections X.B–C).

III. PRACTICAL ASPECTS

A. Postoperative Survival Time

Providing appropriate steps for making the experimental lesion have been taken, the next consideration concerns the choice of postoperative survival time, which is one of the most important variables in experimental silver impregnation studies. Actually, the successful impregnation of "terminal degeneration" by the modern variants of the Nauta Gygax methods seems to be related as much to the use of shorter postoperative survival times as to histotechnical modifications.

The degeneration of a fiber system is a progressive process that eventually results in the complete disintegration of the entire system. However, the process proceeds at different rates in different parts of the system, and the optimal survival time, therefore, depends to a certain extent on whether the trajectory of a projection system or its synaptic field is the main focus of interest.

1. Synaptic Fields

A dramatic argyrophilic response may have developed in degenerating terminals by a couple of days following the operation. The degenerating stem fibers, however, may need a few more days to develop a response whereby they can be recognized clearly as degenerating axons. That the optimal postoperative survival time for the impregnation of degenerating synaptic endstructures can be as short as 2 or 3 days was demonstrated in one of the very first experimental degeneration studies aided by a reduced silver technique (Hoff, 1932a,b). Indeed, the terminal endstructures of some fiber systems may have disappeared completely by the time of optimal argyrophilic response in the parent axons. The reason for this seems to be that in many parts of the nervous system, degenerating terminal fibers and boutons are rapidly engulfed by glia, and the glial reaction may completely remove all signs of terminal degeneration within less than a week. Some examples from the literature will serve to illustrate the importance of this subject.

In a paper describing the dorsal root distribution to the substantia gelatinosa of the rat, Heimer and Wall (1968) made the following comments:

> The optimal survival time for the light microscopic demonstration of terminal degeneration in the substantia gelatinosa seems to fall between 1 and 3 days. This remarkably short time indicates a high rate of degeneration of the axonal parts in this region of the spinal cord. If the spinal cord is examined 1 week after dorsal root section—a time interval often used routinely for the demonstration of anterograde degeneration in mammals—little evidence of dorsal root connections with the substantia gelatinosa is left, at least at the light microscopic level.

Another interesting example is provided by the climbing fiber terminals in the cerebellum. For many years, degenerating climbing fibers were considered to be generally refractory to impregnation by silver techniques. However, when the postoperative survival time of the animal was given special attention, it became obvious that the excessively long survival times used in the past were responsible for the negative results. By shortening the survival time to 1 or 2 days, degenerating climbing fiber terminals could be impregnated with silver in several species including the rabbit (Mejia, 1976), pigeon (Freedman *et al.*, 1977), and cat (Groenewegen and Woogd, 1977).

As indicated above, it is common experience that degenerating synaptic endstructures of many fiber systems are best demonstrated with relatively short survival times following the experimental lesion. This may be somewhat surprising considering the many reports that have favored a somatofugal spread of the Wallerian degeneration (see review by Joseph, 1973). On the other hand, several electron microscopic studies have provided convincing evidence that the latency of degeneration in the synaptic endstructures is as short as or even shorter than in corresponding axons (e.g., Heimer,

1972; Donat and Wisniewski, 1973; Demêmes *et al.*, 1974; Morest and Baptiste, 1975). This would certainly be in harmony with the results obtained in the vast majority of silver impregnation studies. However, there may well be exceptions to this rule, and if the synaptic fields of a specific fiber system are difficult to display with short survival times, longer times should be tried.

One should keep in mind that the rate of degeneration seems to be dependent on axon thickness (Ramón y Cajal, 1928; Van Crevel and Verhaard, 1963a,b) as well as on the distance from the experimental lesion (Donat and Wisniewski, 1973; Demêmes *et al.*, 1974; Vaccarezza *et al.*, 1970). This may explain why the survival time for the optimal argyrophilic response may vary in different projection fields of a given fiber system (Schneider, 1968).

2. Parent Stem Fibers

A complete study of a fiber system includes the tracing of the course of the pathway as well as the mapping of its projection fields. The tracing of the degenerating stem fibers can often be done best if the postoperative survival is long enough to allow for a clear-cut fragmentation of the axons. For this to happen, a survival time of 5–10 days may be needed in many of the commonly used laboratory animals. In other words, the optimal survival time for the tracing of degenerating fibers is usually somewhat longer than the optimal time for demonstrating terminal degeneration.

Although a postoperative survival time between 1 and 2 weeks is generally suitable for the tracing of degenerating axons in most mammals, there are occasions in which such long survival times may lead to serious misinterpretations. This may happen if the degenerating axons are part of a diffusely organized fiber system, particularly if the plane of sectioning is more or less perpendicular to the course of the fibers. The argyrophilic particles may then be arranged in such disorder that the pattern of degeneration gives the false impression of representing terminal degeneration rather than degenerating stem fibers. Although a change of plane of sectioning may occasionally facilitate the interpretation, this problem can usually be solved only by decreasing the survival time.

3. Other Factors

It should be clear from the previous account that the optimal survival time is very much dependent on the part of the degenerating fiber system one wants to display. The choice of survival time, however, is dependent on many other factors including the species and the age of the animal. Nonmammalian species, for instance, usually require considerably longer survival time than do mammals (Ebbesson, 1970), and the optimal time for adult animals is generally somewhat longer than for young animals (Cook *et al.*, 1974; Saavedra *et al.*, 1973). The environmental temperature is also of importance (Monckleberg and Bethe, 1899; Merzbacher, 1903), especially in poikilothermic animals, which must be kept at a relatively constant temperature in order to avoid unnecessary fluctuations in the argyrophilic response (Gamble *et al.*, 1957; Ebbesson, 1970). Finally, one should keep in mind that the rate of degeneration does vary in different fiber systems and even in different populations of fibers within the same system (e.g., Lamothe, 1977).

4. Conclusion

Certainly, there is no easy answer to the question of the optimal survival time. Therefore, it is advisable to set aside some time at the beginning of a project to determine the optimal time on an empirical basis. In other words, unless the relationship between the argyrophilic response and the postoperative survival time is known from previous studies, it is important to sacrifice a series of animals at different postoperative intervals in order to determine the optimal survival times for different parts of the system. As a general guide, one may say that the optimal survival time for the demonstration of axon terminals in small laboratory animals such as rat and guinea pig is usually 1–3 days. Although the nonterminal parts of many degenerating pathways are also well-impregnated with such short times, degenerating fibers are often more clearly demonstrated in animals that have survived the operation for 3–5 days. The optimal survival time in the cat is usually a few days longer: 3–5 days for terminal degeneration and 5–9 days for fiber degeneration. In the monkey, both terminal degeneration and degenerating stem fibers can usually be studied with a survival time of 5–9 days.

B. Fixation and Sectioning

Fixation is usually initiated by perfusion with isotonic saline followed by the fixative. The technique of vascular perfusion is described in detail in Chapter 10. For most silver impregnation procedures, a 10% formalin solution is recommended, and the amount of fixative is dependent on the size of the animal. The volume needed for a 150-g rat, for instance, is about 200–300 ml, while a liter or more may be required for an adult cat. Following the perfusion, the brain is carefully removed from the skull and stored in fixative for one to several weeks before it is sectioned, usually on a freezing microtome.

Although much attention is paid to the composition of the fixative solution used for preparing tissue for electron microscopy, it is probably fair to say that the fixation process has received a rather perfunctory treatment in most descriptions of silver impregnation. One exception is the fixation procedure recommended for the cupric silver method (Section X.C). That

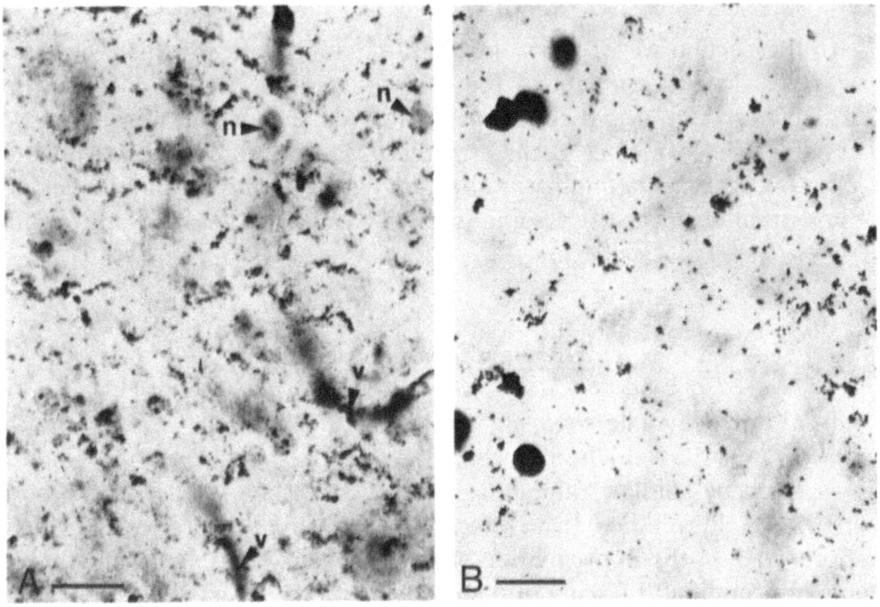

Figure 3. (A) Silver precipitate on the surface of a section that was cut with a dull knife from a block that was poorly fixed. Except for the dense staining of vessels (v) and a few cell nuclei (n), all of the remaining densities are silver precipitate. A similar artifact is often obtained if the freezing of the block is insufficient. Cupric silver method. Bar, 20 μm. (B) Surface precipitate resulting from contaminated silver solutions. The large round black profiles represent red blood cells. Fink–Heimer, Procedure II. Bar, 20 μm.

the fixation procedure may have an effect on the results obtained with silver impregnation techniques was recently demonstrated by Groenewegen and Woogd (1977). They showed that the silver impregnation of cerebellar climbing fiber terminals improved if the formalin-fixed material was post-fixed in glutaraldehyde.

Although special silver impregnation procedures have been developed both for paraffin- and plastic-embedded material (e.g., Guillery *et al.*, 1961; Heimer, 1970b) most procedures seem to work best on frozen sections, usually 25–40 μm thick. To avoid the formation of ice crystals and disruption of the tissue during the freezing procedure, the brain is soaked in a 30% sucrose solution made up in fixative or buffer. The tissue is kept in the sucrose solution until it sinks to the bottom of the jar, usually after a couple of days for a rat brain or a piece of similar size.

How to prepare and collect frozen sections is discussed in Chapter 1 and will not be repeated here. However, attention should be drawn to the fact that much valuable material has been ruined by the use of a dull knife, since any irregularity on the surface of the cut section collects silver and produces disturbing artifacts (Fig. 3A).

IV. GENERAL CHARACTERISTICS OF THE SILVER METHODS

In this section we will consider three methods with which we have had reasonably long experience, the Nauta–Laidlaw (Nauta, 1957), the Fink–Heimer (1967), and the cupric silver (de Olmos and Ingram, 1972) methods. Detailed recipes of these methods are presented in the Appendix (Sections X.A–C). Attention will also be given to some recently developed methods that seem to offer some distinct advantages (Desclin and Escubi, 1975; Gallyas *et al.*, 1980a–c).

A. The Nauta–Laidlaw Method

In 1956, Chambers *et al.* proposed a modified version of the Nauta–Gygax (1954) method in which the ammoniacal silver nitrate solution was replaced by Laidlaw's ammoniacal silver carbonate solution. This modification, like the method from which it is derived, is characterized by the "suppression" of the impregnation of normal fibers. This, in turn, enhances the identification and tracing of degenerating nerve fibers. The suppression of normal fibers is obtained by a pretreatment of the frozen sections in a solution containing a heavy metal compound, phosphomolybdic acid, followed by an oxidant, potassium permanganate. This is followed by a decoloration in a mixture of hydroquinone and oxalic acid. The sections are then impregnated in a silver nitrate solution and subsequently treated in Laidlaw's ammoniacal silver carbonate solution before being reduced in an alcoholic formalin solution acidified by citric acid.

Degenerating axons are easily identified in Nauta–Laidlaw sections provided that important variables such as survival time, fixation, and extent of oxidation are optimal. The degenerating axons appear beaded and fragmented, and they are, in general, easy to trace against a yellow-brown background (Fig. 1B). If properly handled, the technique also gives a more or less detailed picture of the degenerating axon arborization, including a variable number of degenerating axon terminals.

Several modifications of the Nauta–Laidlaw method have been developed. In one of these (Ebbesson and Rubinson, 1969), phosphomolybdic acid has been replaced by uranyl nitrate, and the oxidation of the sections is accomplished by a prolonged treatment in a weak potassium permanganate solution. Although this procedure was developed especially for nonmammalian species, it also works well in mammals. Another Nauta–Laidlaw method especially suited for the human brain (Section X.D.5b) has been developed by Albrecht and Fernstrom (1959).

Although the Nauta–Laidlaw method quickly became the method of choice in experimental neuroanatomy, it was gradually realized that the suppression of normal fiber impregnation is often obtained at the expense of a considerable reduction in the impregnation of the terminal axon arbor.

This prompted the development of the Fink–Heimer and the cupric silver methods.

B. The Fink–Heimer Method

This method consists of two variants that are principally similar to the original Nauta–Gygax (1951) procedure. However, they differ from it by (1) the application of potassium permanganate without pretreatment of the sections in phosphomolybdic acid and (2) subsequent impregnation in a mixture of silver nitrate and uranyl nitrate (Procedure I) or in a mixture of silver nitrate and pyridine following a pretreatment in uranyl nitrate (Procedure II).

The pretreatment in potassium permanganate alone (Procedure I) or in potassium permanganate followed by uranyl nitrate (Procedure II) is of importance not only for the suppression of the argyrophilia of normal axons but also for the impregnation of degenerating axoplasm. As with the original Nauta–Gygax (1954) procedure, the impregnation is completed in an ammoniacal silver nitrate solution rather than in Laidlaw's solution. Reduction of the silver is obtained by using the same acidic reducer as in the Nauta–Gygax method.

Both versions of the Fink–Heimer method impregnate degenerating axons black against a background of normal axons, the impregnation of which is suppressed to various degrees. The most striking quality of the Fink–Heimer technique is the impregnation of a large number of more or less spheroidal particles varying in diameter between 0.5 and 2 μm (Fig. 4). As shown by Heimer and Peters (1968), these particles represent degenerating terminal axon ramifications and their endstructures, together often referred to as "terminal degeneration." It should be mentioned that the suppressive Nauta–Gygax method may, at least in some fiber systems, produce a picture of "terminal degeneration" that is very similar to that obtained in Fink–Heimer sections (Guillery and Ralston, 1964; Lund and Westrum, 1966; Heimer and Ekholm, 1967; Walberg, 1971, 1972).

In general, Procedure I yields a better selective impregnation of degenerating axons and a better definition of normal neuronal cell bodies. Procedure II is generally characterized by an ill-defined and lightly stained background that allows for good counterstaining with the Nissl technique. A modification of the second procedure, published by Wiitanen in 1969, has been used with success especially on sections of the monkey brain, and another modification adapted to cryostat sections has been published by Hjorth-Simonsen (1970). A simplified silver impregnation technique was published by Eager in 1970. In the Eager technique, the oxidation in potassium permanganate as well as the impregnation in silver nitrate are deleted. The other steps in the Eager method are basically similar to those in the Fink–Heimer Procedure II.

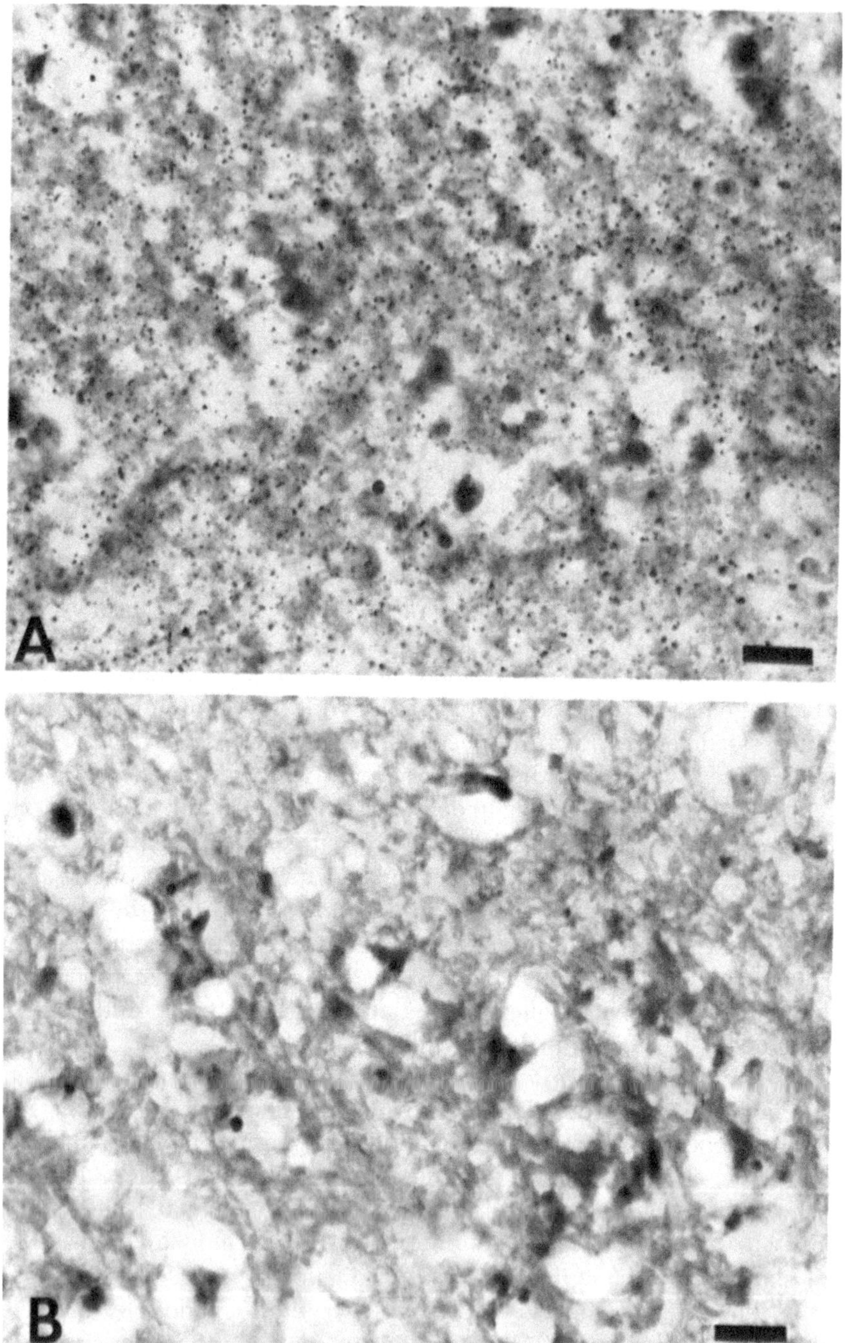

Figure 4. Substantia nigra lesion producing terminal degeneration in the cat caudate nucleus. (A) Lesion side. (B) Control side. Numerous degenerating terminals are evident as black dots on the lesion side. Fink–Heimer stain. Bar, 20 μm.

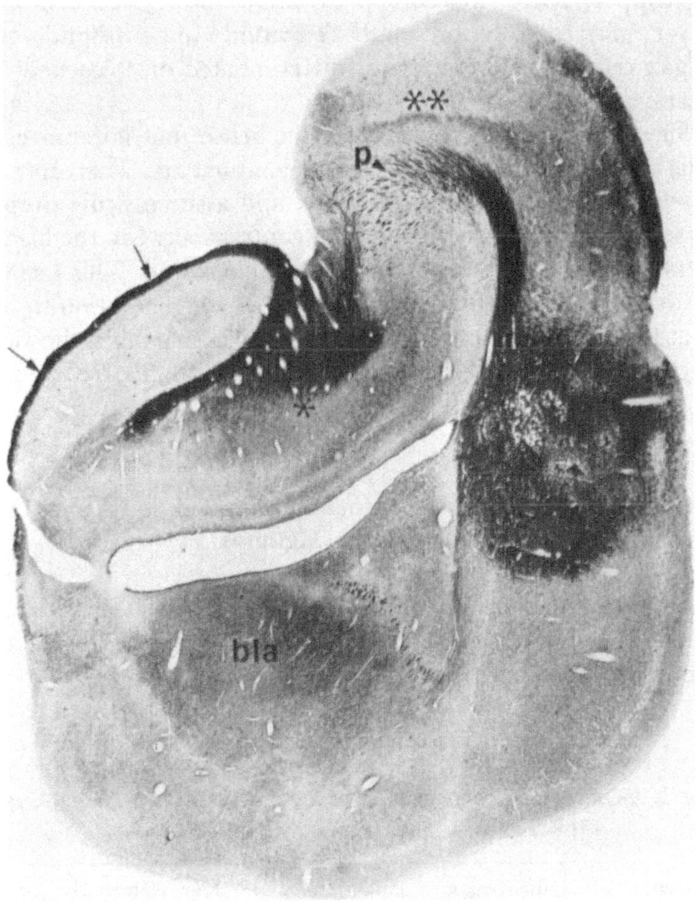

Figure 5. Degeneration in the perforant path system of the guinea pig demonstrated by the aid of the cupric silver method. The clearly demarcated lesion (round black area on right side of figure) is located in the caudal part of the primary olfactory cortex including the ventral part of the entorhinal area. The degenerating fiber bundles forming the perforant path (p) are passing through the subiculum to their termination in the hippocampus (*) and the dentate gyrus (small arrows). Note the additional fields of terminal degeneration (**) in the entorhinal area (ea) and the basolateral amygdaloid body (bla). Bar, 10 μm. Four days survival time.

C. The Cupric Silver Method

Although the cupric silver method* produces basically the same picture of terminal degeneration as the Fink–Heimer procedures, it differs from them in three important ways. First, the method is critically dependent on fixation in a slightly alkaline (pH 7.2–7.4) formalin solution. Second, the

*The first results obtained by this type of procedure were published in an abstract (de Olmos, 1966), and the first description of the procedure appeared 3 years later (de Olmos, 1969).

primary impregnation is carried out in an alcoholic silver–pyridine solution containing copper. Third, following a secondary impregnation in an ammoniacal silver nitrate solution and reduction in a slightly modified Nauta–Gygax reducer, the material is differentiated or "bleached" with potassium ferricyanide.

The "bleaching" plays a key role in the sense that it removes most of the metallic silver from the normal tissue components. Therefore, there is hardly any impregnation of normal fibers, and a successfully prepared cupric silver section provides an outstanding contrast between the black degenerating nerve fibers and the lightly stained background. This facilitates the tracing of degenerating pathways and allows the degeneration to be visualized even at low magnification (Fig. 5). Particularly striking is the richness of morphological detail rendered by this technique (Figs. 6B, 7B, and 8B).

D. Comparison among the Nauta–Laidlaw, Fink–Heimer, and Cupric Silver Methods

A summary of the impregnation characteristics of the three methods just discussed is presented in Table I. The table reflects the results obtained by the methods under optimal conditions.

Table I. General Impregnation Characteristics of the Nauta–Gygax, Fink–Heimer, and Cupric Silver Methods[a]

Tissue stained	Nauta–Gygax	Fink–Heimer I	Fink–Heimer II	Cupric silver
Reticular fibers	−	+	++	−
Endothelial nuclei	±	++	+	−
Neuronal soma	++	++	+	±
Neuronal nucleus	+	+	±	−
Neuronal nucleoli	+ (brown)	++ (brown)	+ (brown)	++ (black)
Chromatin bodies	+	++	±	++
Glia nuclei	± (brown)	±	± (brown)	++ (black)
Normal fibers	+	+	+	±
Degenerating axons of passage	++	+	+	++
Degenerating axon ramifications	+	++	++	++
"Terminal degeneration"	±	++	++	++
Degenerating cell bodies and dendrites	±	+	+	++

[a]Symbols used: ±, scarce or erratic impregnation; +, moderate impregnation; ++, strong impregnation.

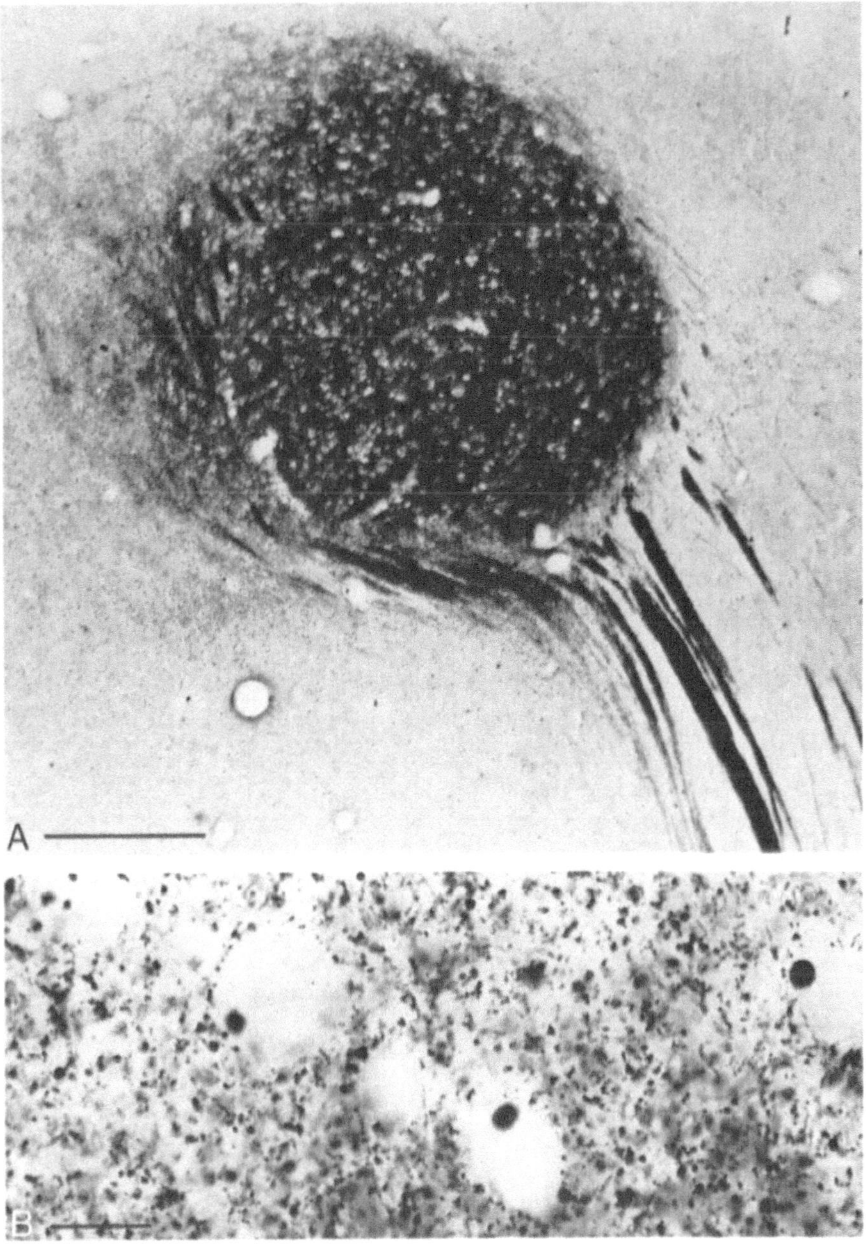

Figure 6. (A) Massive terminal degeneration in the central amygdaloid nucleus of the guinea pig following a lesion in the piriform–amygdaloid transition area 4 days before sacrifice of the animal. The degenerating piriform–amygdaloid fiber bundle (lower right corner) could be easily followed in continuity from the lesion to the terminal field in the central amygdaloid nucleus. Cupric silver method. Bar, 20 μm. (B) High-power photograph from the field of terminal degeneration in the central amygdaloid nucleus. The large black particles in the clear spaces (unstained cell bodies) represent nucleoli. Bar, 10 μm.

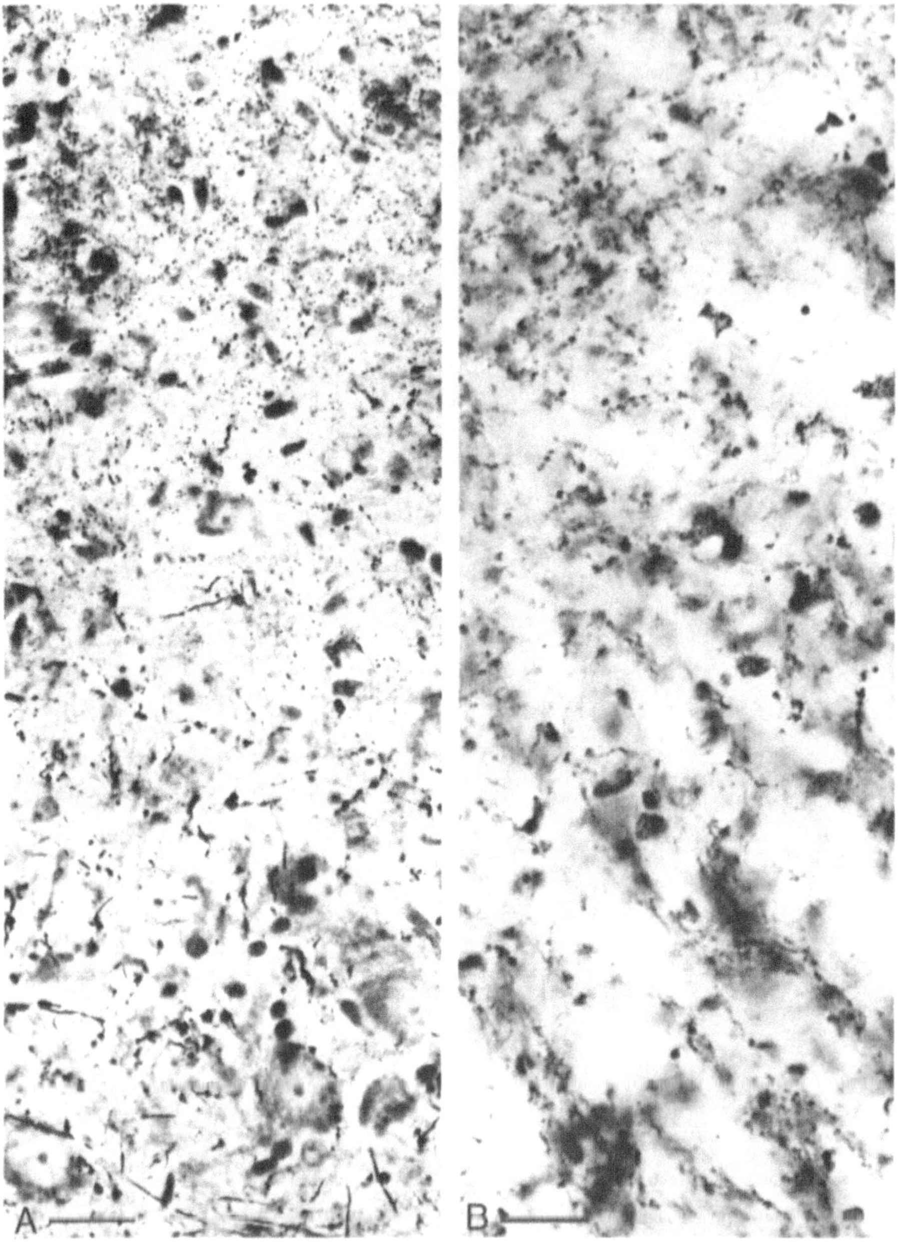

Figure 7. (A) Terminal degeneration in the ventral part of the globus pallidus (upper half of figure) of the rat following a laminar heat lesion in the olfactory tubercle. Degenerating fibers in the region of the medial forebrain bundle are seen in the lower half of the picture. Fink–Heimer method, Procedure II. Two days survival time. Bar, 20 μm. (B) Terminal degeneration in the longitudinal association system (upper half of figure) and the mossy fiber system (lower half of figure) of the guinea pig hippocampus. Arrows point to large mossy fiber terminals. Cupric silver method. Four days survival time. Bar, 20 μm.

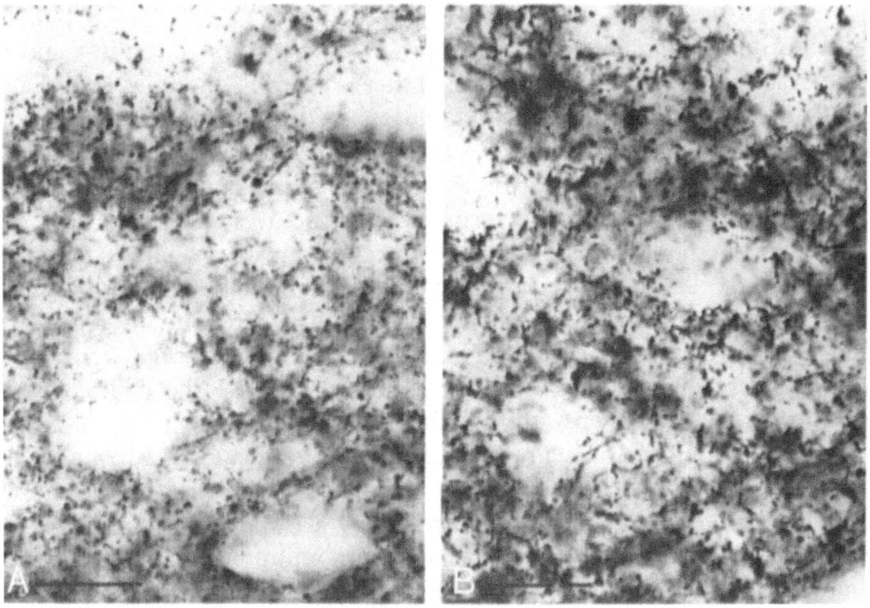

Figure 8. (A) Terminal degeneration in the anterior thalamic nucleus of the guinea pig following a lesion of the corticothalamic fiber system in the internal capsule. The argyrophilic pattern has a dustlike character, and there have, therefore, been some doubts about the nature of the degeneration (see text). Fink–Heimer method, Procedure II. Survival time, 4 days. Bar, 10 μm. (B) Same as in A but impregnated with the cupric silver method. The high sensitivity of this procedure has apparently made it possible to impregnate many fragmented degenerating preterminal fibers, and the picture is more reminiscent of terminal degeneration. Bar, 10 μm.

E. Other Silver Methods

1. The Silver Method of Desclin and Escubi

Desclin and Escubi (1975) recently published a silver procedure that is a combination of the Fink–Heimer procedures and the cupric silver method. In the method by Desclin and Escubi, the sections are oxidized in a solution of potassium permanganate. Following bleaching in a mixture of oxalic acid and hydroquinone, the sections are pretreated in a cobalt or lead nitrate solution rather than in uranyl nitrate. The impregnation takes place in a cupric–silver lactate mixture, and the sections are then treated in the ordinary sequence of ammoniacal silver nitrate solution followed by a Nauta–Gygax reducer.

As we have had no experience with the method developed by Desclin and Escubi, we hesitate to compare it with the other silver methods. It should be mentioned, however, that Desclin and Escubi consistently managed to impregnate degenerating climbing fibers.

2. Silver Methods of Gallyas *et al.*

In order to stabilize the silver impregnation procedure, Gallyas and his collaborators (Gallyas *et al.*, 1980a–c) have recently modified the Eager (1970) method by paying special attention to the physicochemical mechanism involved. Their work has resulted in two apparently highly selective and reliable procedures, one for the impregnation of degenerating axons and another for the staining of terminal endstructures and lysosomes.

In the method for the selective impregnation of axonal degeneration (Gallyas *et al.*, 1980b), the sections are pretreated in alkaline hydroxylamine, impregnated in a ferric silver nitrate solution, and subsequently reduced in a physical developer. The method results in a specific impregnation of predominantly myelinated degenerating axons which have the typical beaded appearance (Fig. 9A) seen with the other silver methods.

The method for the staining of degenerating terminals is equally simple. The formalin-fixed sections are first treated in a sodium hydroxide–ammonium nitrate mixture. They are then silver impregnated, washed in an alcoholic sodium carbonate–ammonium nitrate solution for the removal of unbound silver, and subsequently reduced in a modified Nauta–Gygax reducer. Like the method for axonal degeneration, this method is characterized by low background staining and results in a high-contrast image of the argyrophilic particles representing degenerating terminals and lysosomes (Fig. 9B). According to the authors, the technique is reliable and is characterized by great sensitivity for localizing terminal degeneration. In a recent review article, Zaborszky and co-workers indicate that catecholaminergic terminals can be impregnated more easily with the method of Gallyas and his collaborators than with the Fink–Heimer technique (Zaborszky *et al.*, 1979).

V. INTERPRETATION OF DEGENERATING FIBERS AND TERMINAL DEGENERATION

A. Axonal Degeneration

The picture of axonal degeneration is familiar to most neuroanatomists, and it has been described and illustrated by countless authors, sometimes in great detail (e.g., Glees and Nauta, 1955). Therefore, only the most significant aspects of the axonal degeneration will be described here.

The first manifestation of the process of axon degeneration in silver-impregnated material is usually restricted to an increase in argyrophilia of the affected fibers. In the next stage, beading of the fibers and an increased number of fusiform swellings appear (Fig. 10A). The beading becomes accentuated with time until, at a later stage, the fibers appear fragmented (Fig.

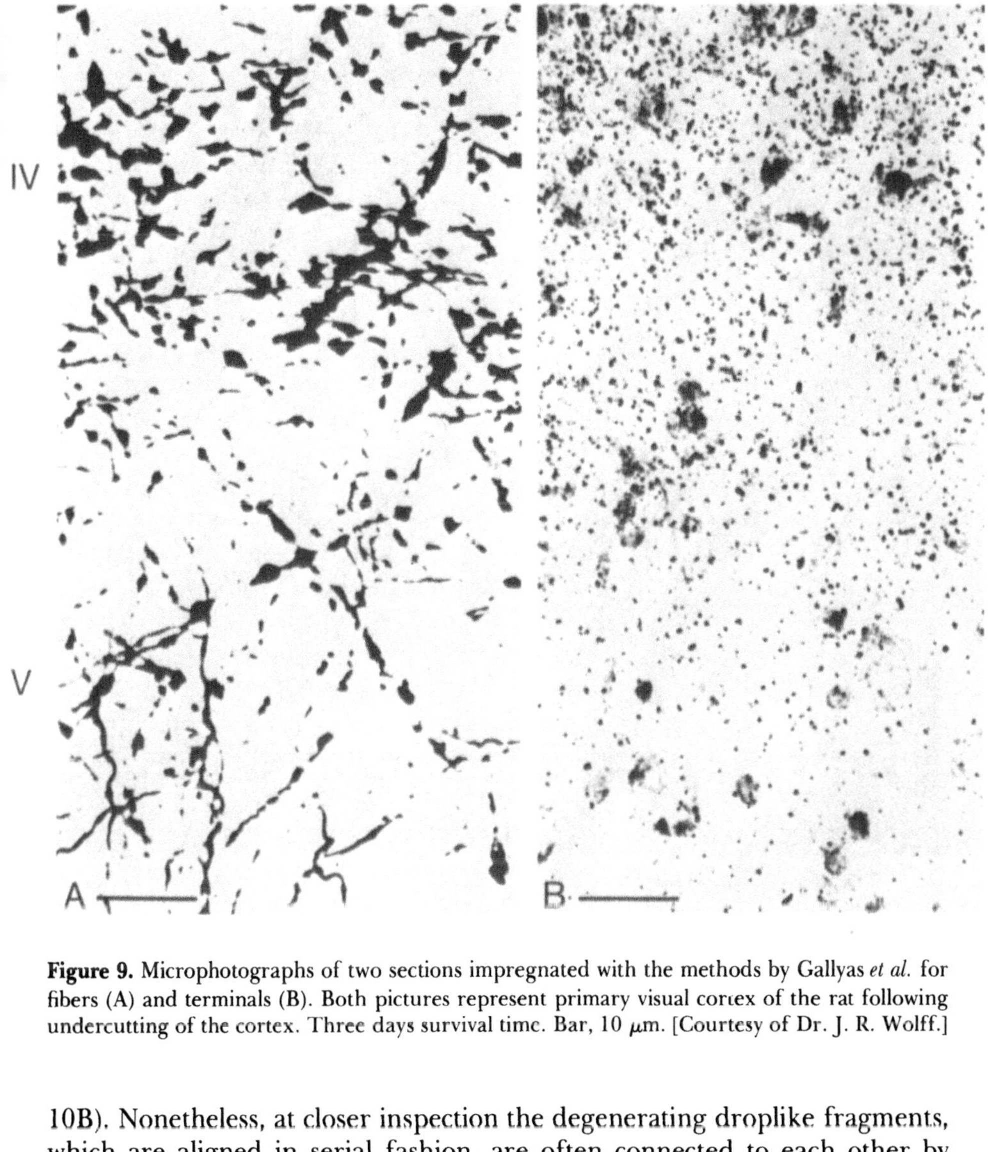

Figure 9. Microphotographs of two sections impregnated with the methods by Gallyas *et al.* for fibers (A) and terminals (B). Both pictures represent primary visual cortex of the rat following undercutting of the cortex. Three days survival time. Bar, 10 μm. [Courtesy of Dr. J. R. Wolff.]

10B). Nonetheless, at closer inspection the degenerating droplike fragments, which are aligned in serial fashion, are often connected to each other by more or less visible bridges composed of very small argyrophilic granules. Finally, a granular disintegration of the different fragments take place, and if the animal is allowed to survive long enough, the degeneration products disappear altogether.

It is important to keep in mind that even normal nerve fibers can assume a beaded appearance under certain conditions, causing problems of interpretation (compare Fig. 10A with Fig. 11A). Thus, caution should be exercised in interpreting beaded axons as degenerating, unless longer survival times permit the same axons to be visualized in later, more fragmented stages of degeneration.

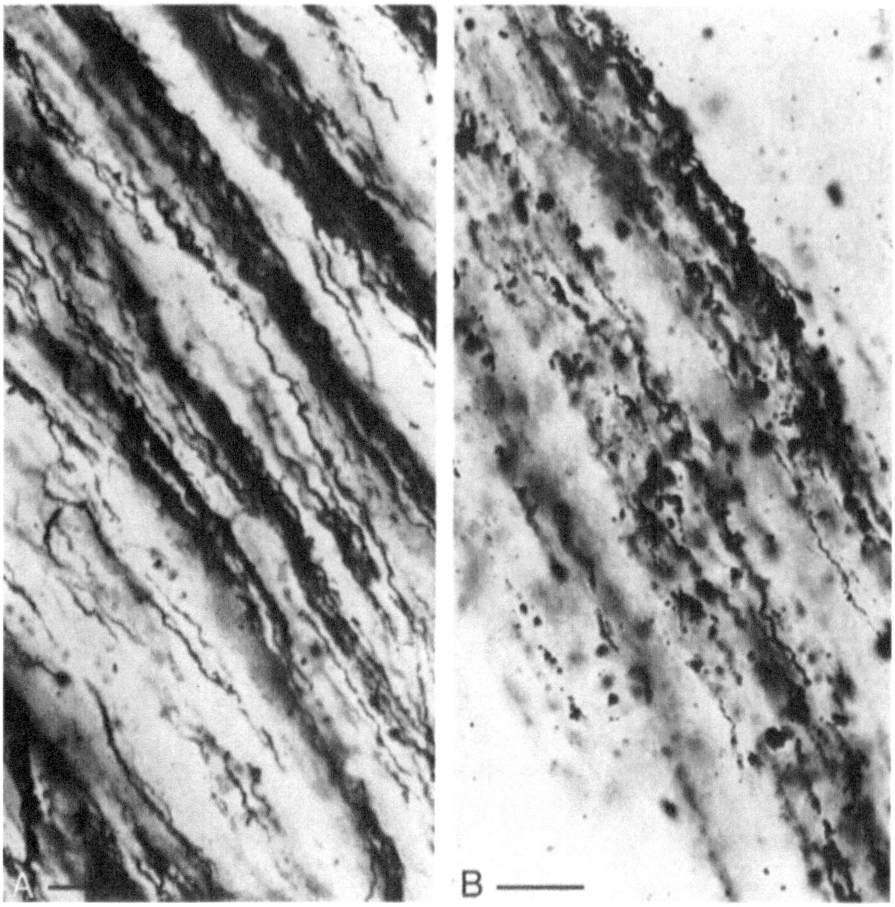

Figure 10. (A) Piriform–amygdaloid fibers in early stage of degeneration impregnated with the cupric silver method. Guinea pig. Survival time, 3 days. Bar, 10 μm. (B) Same as A, but with 5 days survival time. Note the increase in fragmentation of the fibers. Bar, 10 μm.

B. Terminal Degeneration

In typical cases, terminal degeneration can be identified on the basis of a more or less dense accumulation of heavily argyrophilic particles of different size and shape (Figs. 4, 6B, and 9B) and by the fact that the distribution is confined to one or more reasonably well-defined areas (Heimer, 1970a). Furthermore, degenerating axons can typically be traced from the site of the experimental lesion to the areas in question* (Figs. 5, 6A, and 7A).

*Note that there are exceptional cases in which it has proved to be very diffiuclt to trace the degenerating axons in continuity with the field of terminal degeneration. One example is the stria terminalis distribution to the accessory olfactory bulb in the rat (de Olmos and Ingram, 1972).

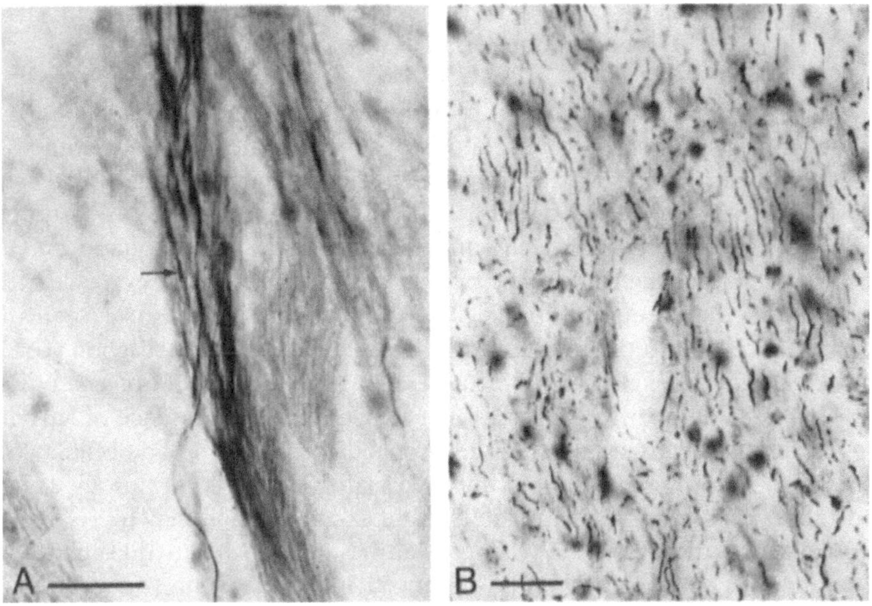

Figure 11. (A) Photomicrograph of a Nauta–Laidlaw-impregnated section from normal material showing fibers with fusiform swellings and myelin cuffs (arrow). Bar, 10 μm. (B) Incomplete impregnation of normal fibers. The large dark profiles represent neuroglial cell bodies. Fink–Heimer, Procedure II. Bar, 20 μm.

In less typical cases, where the projection is limited in extent or diffusely organized, the identification of terminal degeneration can be more problematic, especially if degenerating terminal endstructures are mixed with degenerating fibers of passage. In fact, it may be impossible to interpret the true nature of the degeneration pattern unless some additional criteria can be used. For instance, gradual disappearance of degenerating elements along the course of a pathway may be taken as an indication of a gradual termination along its path.

The terminal degeneration in some corticothalamic fiber systems has a dustlike appearance when impregnated with some of the modern silver techniques (Fig. 8A). As this picture is somewhat atypical, doubts regarding its true nature may arise. However, use of the cupric silver method and a very short survival time (36 hr in the rat and 3 days in the guinea pig) may make a picture more reminiscent of the usual pattern of terminal degeneration become apparent as in the case illustrated in Fig. 8B. The terminal character of the degeneration pattern in this particular case is further emphasized by the fact that there is no evidence of retrograde changes in thalamic neuronal cell bodies at this survival time. For further discussion of this problem, see Section VI.C.

VI. OTHER DEGENERATIVE NEURONAL PHENOMENA

A. Degeneration of Cell Bodies and Dendrites

It has long been known that cellular reactions* may occur in neurons whose axons have been severed, and in recent years it has also become obvious that transneuronal reactions† are considerably more widespread than previously believed (e.g., Cowan, 1970). These changes have been studied both with light microscopic techniques including silver impregnation methods and with the electron microscope. Neurons undergoing severe reactive changes may become argyrophilic, and the argyrophilic response seems to be directly related to the severity and rapidity of the cellular reaction (Grant and Aldskogius, 1967), which, in turn, seem to be dependent on many factors. It has often been suggested, for instance, that the presence of sustaining collaterals greatly diminishes the probability of a retrograde cellular degeneration following axotomy (Ramón y Cajal, 1928; Fry and Cowan, 1972), and the severity of an anterograde transneuronal reaction may be related to the degree of deafferntation (Heimer and Kalil, 1978). Furthermore, degenerative reactions are more pronounced in newborn or young animals than in adult animals, and this fact has been successfully exploited for the purpose of identifying retrogradely degenerating neurons, not only in Nissl-stained preparations (modified Gudden method; Brodal, 1940), but also in silver-impregnated material (Grant, 1975). In adult animals, on the other hand, dense silver staining of degenerating neurons seems to have been described only in a few places, e.g., retrogradely degenerating neurons in the thalamus of the monkey (Tanaka and Chen, 1974; Campos-Ortega *et al.*, 1970) and transneuronally degenerating neurons in olfactory cortex of the rat (Heimer and Kalil, 1978).

Degenerating neurons that have become argyrophilic are usually easy to recognize in silver-impregnated sections (Fig. 12). In cases with short survival times, the cell bodies and dendritic processes are intensely black, imparting a Golgi-like appearance to the neurons (Fig. 12B). At this stage, degenerating neurons may occasionally be difficult to distinguish from Cammermeyer's "dark neurons" (Fig. 13C). Degenerating neurons, however, often exhibit abnormal morphology that becomes more pronounced with longer postoperative survival times. Because of their evolving nature, degenerating neurons can easily be distinguished from "dark neurons" by appropriate gradations of the postoperative survival time. On the other hand, fragmented dendrites of degenerating neurons may, under certain circumstances, be confused with terminal or axonal degeneration. However,

*Grant (1975) distinguishes the term "cellular reaction" from "cellular degeneration." The former term leaves open the possibility that the neuron might recover.

†Transneuronal reactions can be of two types. Retrograde transneuronal reaction may occur in neurons following the death of the cell upon which they project, whereas cells whose afferents have been severed may undergo anterograde transneuronal reaction.

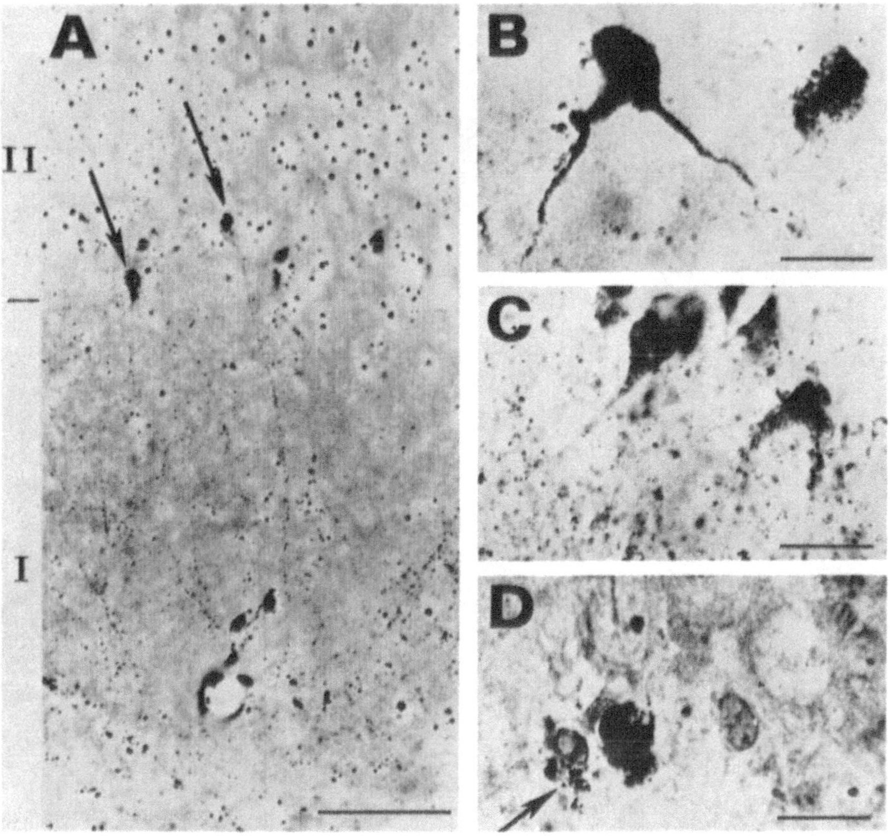

Figure 12. Transneuronally degenerating neurons (arrows) in the olfactory cortex of rats that survived for different times (A and B, 12–14 hr, C, 18 hr, D, 48 hr) following ipsilateral olfactory bulb removal. [From Heimer and Kalil (1978) with permission of the Wistar Institute Press.] Bar in A, 50 μm. Bars in B, C, and D, 10 μm.

the dendritic fragments are usually coarse and are often arranged in a pattern suggesting their dendritic nature.

B. Indirect Wallerian Degeneration

Different cellular reactions may create problems of interpretation when silver methods are used for the mapping of anterograde degeneration. Retrograde neuronal degeneration, for instance, also results in degeneration of the proximal part of the transected axon including its collaterals, i.e., indirect Wallerian degeneration.* Furthermore, axons and their terminal end-

*The term was introduced by van Gehuchten (1903) who pointed out that this type of fiber degeneration occurs secondarily to atrophy and death of the cell body.

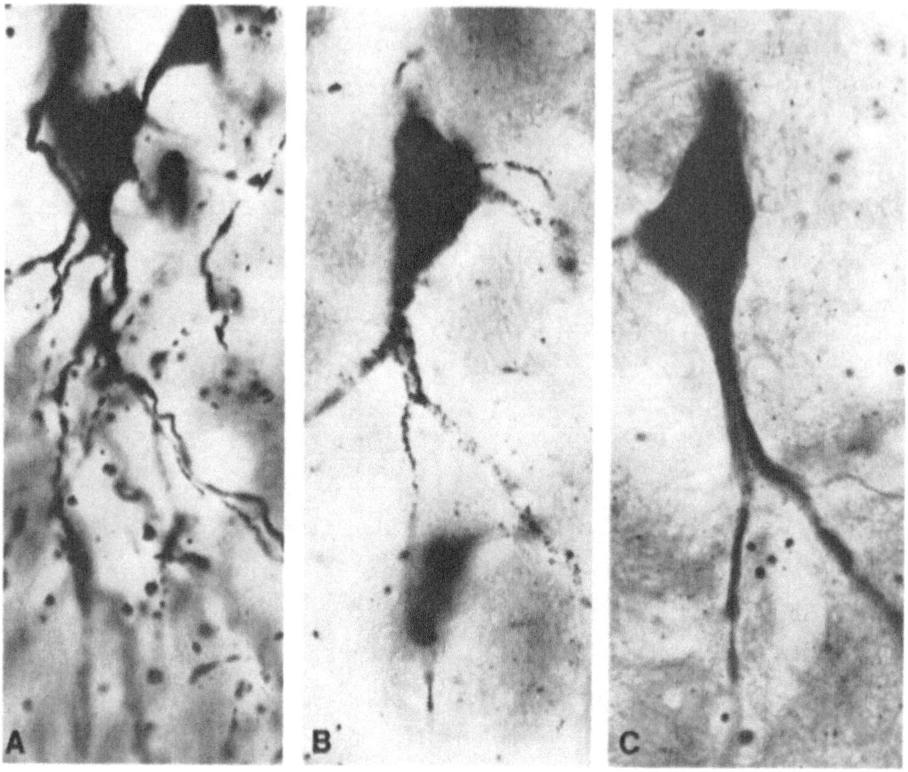

Figure 13. (A) Degenerating neuron in the olfactory cortex of a bulbectomized guinea pig. Cupric silver method. Four days survival time. (B) Granular argyrophilic neuron from the lateral hypothalamic area in the guinea pig. Cupric silver-stained section of normal material. (C) Dark neuron of Cammermeyer. Cupric silver-stained section from reticular formation of a normal guinea pig.

structures belonging to neurons undergoing transneuronal degeneration exhibit morphological and argyrophilic changes similar to those displayed by transected axons (Heimer and Kalil, 1978).

On the other hand, it is also important to realize that both the retrograde and the transneuronal reactions can be used to advantage in the study of central nervous connections. Outstanding examples of this are provided by Grant and Arvidsson (1975) and by Westrum and Canfield (1977) who used the phenomenon of indirect Wallerian degeneration (transganglionic sensory degeneration) to trace the brainstem distribution of the central axons of trigeminal sensory ganglion cells whose peripheral branches were transected. Moreover, indirect Wallerian degeneration secondary to transneuronal cell degeneration in the olfactory cortex was recently used as a means of tracing tertiary olfactory projections (Switzer and Heimer, 1976).

C. "Retrograde Dust" in Thalamus

Following cortical lesions, a uniform dustlike silver deposit appears in the thalamus (Guillery, 1959; Powell and Cowan, 1964). This phenomenon has been studied with different silver techniques, and the picture varies somewhat according to the specific method used (Fig. 8). However, a common feature of the pictures obtained by the different methods is that the "dust" is found almost exclusively in the spaces between the cell bodies.

The most common explanation for the appearance of this deposit has been that it represents a retrograde reaction related to the degeneration of thalamocortical cells. Although this may be partially true in cases where there is a concomitant argyrophilic response in thalamic neurons (Tanaka, 1976), the number of degenerating argyrophilic neurons is usually quite small, and it is unlikely that these neurons, including their dendrites and axon collaterals, are the source of the entire argyrophilic dustlike pattern often observed (cf. Wong-Riley, 1972; Campos-Ortega *et al.*, 1970). Certainly, if there is no argyrophilic perikaryal response as in the case illustrated in Fig. 8, there is little reason to believe that a retrograde component is involved, especially since it has been demonstrated that the argyrophilic response in retrogradely degenerating neurons appears in the cell bodies before it becomes apparent in the axonal processes (Grant and Aldskogius, 1967). Furthermore, EM studies have shown that increased electron density appears simultaneously in both cell bodies and dendrites (Barron *et al.*, 1973). Therefore, there is little reason to believe that an argyrophilic response, usually thought to be related to increased electron density of the cytoplasm, would be confined only to the dendrites as suggested by Leonard (1972). The most reasonable explanation seems to be that the dustlike deposits seen in experiments with short survival times represent terminal degeneration. This view is also supported by the results obtained in a combined Fink–Heimer and autoradiographic study by Kalil (1975) who discusses this problem in more detail.

VII. SOURCES OF ERROR

Although the silver impregnation procedures produce sections in which degenerating axons and terminal degeneration can be relatively easily identified, artifactual deposition of silver in the tissue may cause problems of interpretation. This is especially the case in old animals, where such depositions occur quite frequently. As practically all normal tissue components are potentially argyrophilic, a list of possible sources of error tends to become rather extensive. Fortunately, in a given series of slides there are seldom more than a few types of artifact present, and if one is aware of the problem, it is almost always possible to avoid misinterpretations. As an aid to the beginner, we will discuss the most common artifacts.

Before we turn our attention to the fortuitous staining of normal neuronal and nonneuronal elements, however, it is important to mention the fact that silver may precipitate on the surface of the section (Figs. 3A and B). This is an annoying artifact that occasionally creates difficulties for a beginner. However, it can usually be avoided if a sharp knife is used for sectioning and if special care is taken to use clean dishes and pure solutions in the staining procedure. Should silver precipitate on the surface of the section, its true nature can often be detected by moving the focus through the section. Furthermore, unlike the pattern of degenerating axons and terminals, the specific pattern of a surface precipitate is seldom repeated in adjacent sections.

A. Neuronal Deposits

There are several forms of neuronal silver deposits, some of which may have the appearance of terminal degeneration. This is especially the case in regard to so-called perinuclear deposits and granular argyrophilia. Fusiform axonal swellings, myelin artifacts, and incomplete staining of normal fibers, on the other hand, can sometimes look confusingly similar to degenerating fibers of passage.

1. Perinuclear Deposits

Depending on the age of the animal and the species, a varying number of normal neurons may contain black to red–brown argyrophilic granules of different size (Fig. 14A). These granules are generally found in the perinuclear region and sometimes even in the proximal parts of the dendrites. In general, the phenomenon occurs in neurons containing lipofuchsin ("wear and tear" pigment) or melanin. Although both types of pigment granules occur in neurons of young animals, they do accumulate especially in aging brains. The regional distribution of lipofuchsin seems to vary among species and even among members of the same species (Dayan, 1971), and it is therefore difficult to present a list of structures where the staining of lipofuchsin granules can be confused with terminal degeneration.

Neuromelanin pigments are found in several nuclei including the nucleus coeruleus, nucleus centralis superior, substantia nigra, and dorsal motor nucleus of the vagus in the primate brain. They are also present in the substantia nigra, the locus coeruleus, and certain mesencephalic tegmental nuclei of the cat and dog (Brown, 1943, 1944), but apparently not in other small laboratory animals such as rat, guinea pig, and rabbit (Marsden, 1969). Rather coarse granules of unknown nature are often found in cell bodies in the posterolateral hypothalamus of the guinea pig (unpublished observation).

The perinuclear deposits can be mistaken for degenerating "axoso-

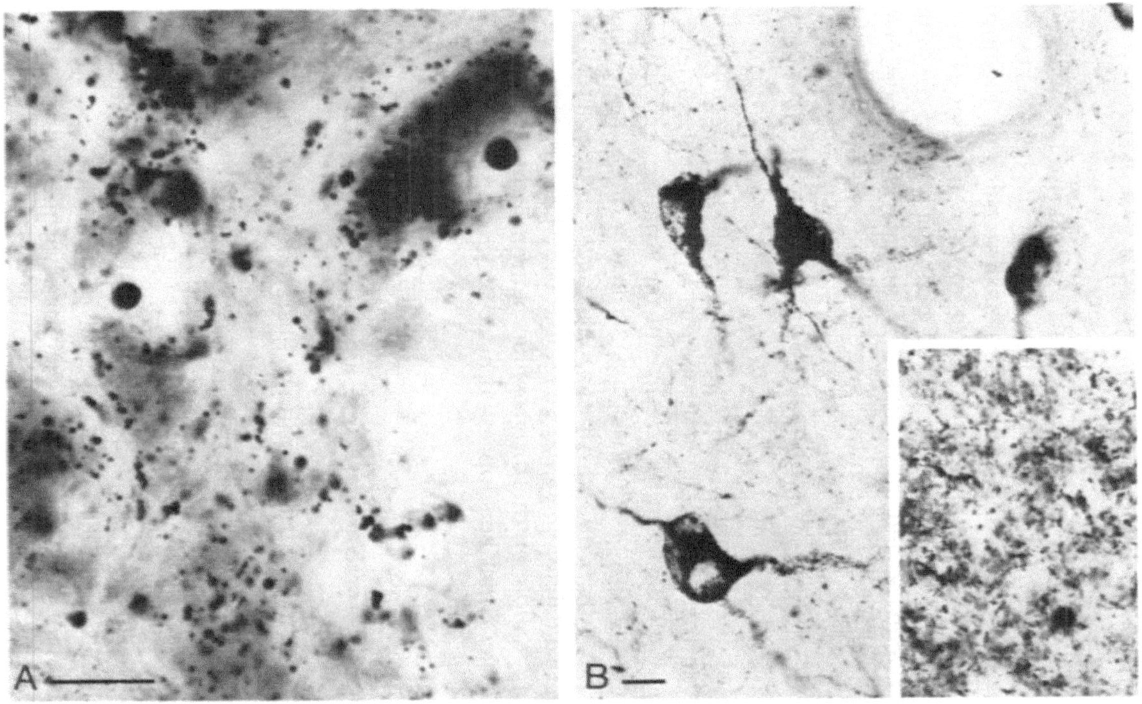

Figure 14. (A) Silver-stained melanin pigments in neurons and dendrites from nucleus centralis superior in the squirrel monkey. Cupric silver method. Bar, 10 μm. (B) Granular argyrophilic neurons in the preoptic periventricular region of the guinea pig. Cupric silver method. The insert demonstrates granular argyrophilic neuropil in the central amygdaloid nucleus. Bar, 10 μm.

matic" terminals, especially since the sizes of the granules are within the range of axon terminals. However, the bilateral occurrence and the absence of granules in the intervals between the cell bodies usually permit the distinction of this type of deposit from true terminal degeneration.

2. Granular Argyrophilic Neurons

This type of deposit, which is often seen in cupric silver preparations, consists of very fine silver granules that completely fill the cell bodies, dendrites, and axons (Figs. 13B and 14B). The neurons exhibiting this granularity have been referred to as granular argyrophilic neurons (Knoche, 1958; Blümcke, 1962), and they are scattered along the walls of the ventral part of the third ventricle and in the dorsal posterolateral and perifornical hypothalamus as well as in the central amygdaloid nucleus and the bed nucleus of the stria terminalis (de Olmos, 1969). Granular argyrophilic neurons have been seen in all mammals in which the cupric silver method has been used.

The axons of the granular argyrophilic neurons form terminal networks, especially at the level of the ventromedial hypothalamic nucleus, the central amygdaloid nucleus, the bed nucleus of the stria terminalis, and the median eminence. The granular argyrophilia in these terminal networks (insert in Fig. 14B) may be confused with terminal degeneration, and anyone using the cupric silver method should be aware of the phenomenon. Fortunately, the impregnation of the argyrophilic neurons and their terminal fields can, to a large extent be suppressed by acetone treatment (see technical note Section X.C.3).

3. Fusiform Swellings of Normal Nerve Fibers

The impregnation of normal fusiform fibers (Figs. 11A and 15C) has been a rather common source of erroneous interpretation, especially since fusiform swellings are a sign of early axonal degeneration (Fig. 10A). Artificial irregularities and varicosities appear more frequently in poorly fixed material. To avoid this type of misinterpretation, it is advisable to choose a postoperative survival time at which most of the degenerating fibers appear fragmented. Another possibility is to use a silver technique that does not usually impregnate normal fibers, e.g., the cupric silver method.

4. Incomplete Staining of Normal Axons

Incomplete staining of normal fibers (Figs. 11B and 15C) is not uncommon, especially in sections impregnated with the Fink–Heimer Procedure II. This type of artifact has a certain resemblance to degenerating fibers,

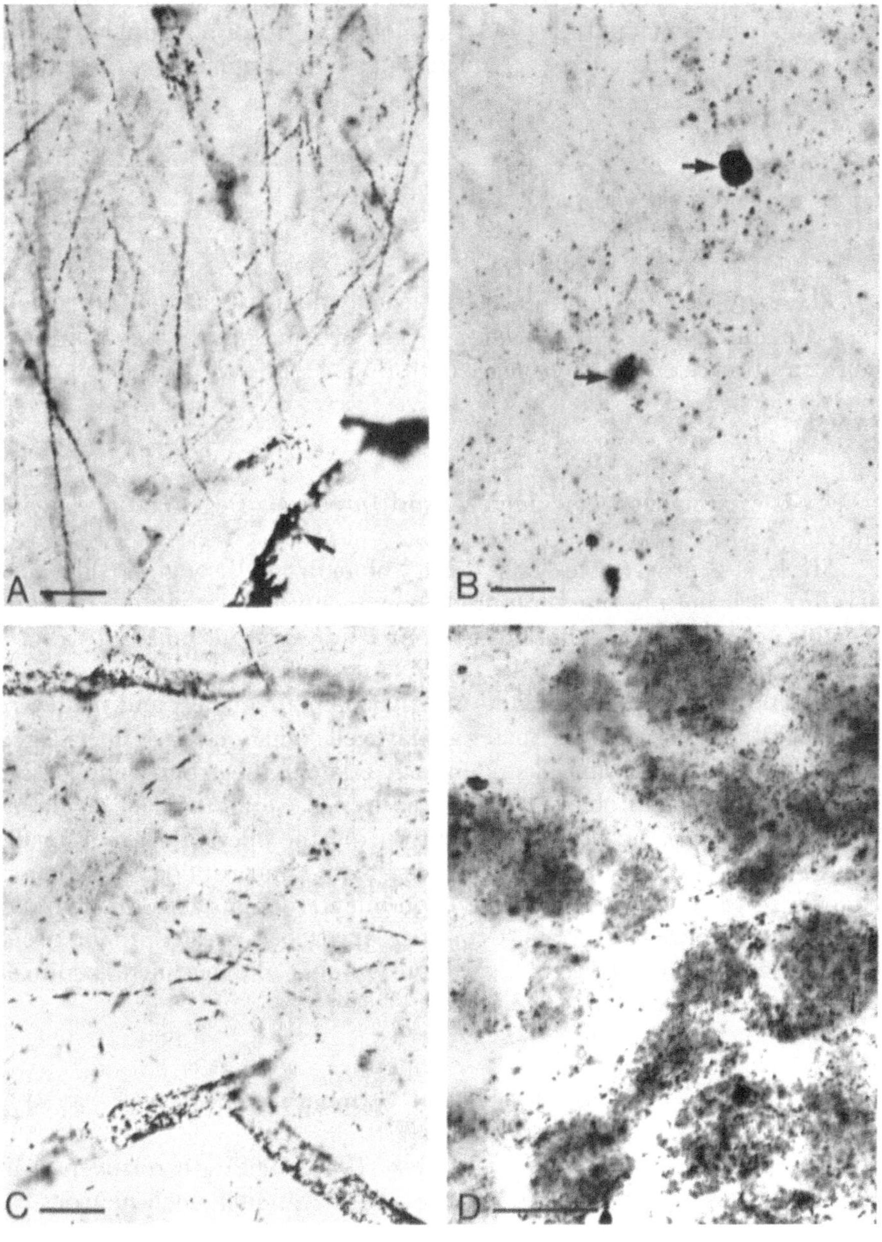

Figure 15. (A) Impregnation of glial processes in the olfactory cortex of the rhesus monkey. The black round profiles represent nucleoli. There is also a partial impregnation of reticular perivascular fibers (arrow). Fink–Heimer, Procedure II. (B) Pericapillary particles probably representing fragments of ruptured blood cells. Cupric silver method. (C) Impregnation of reticular fibers of blood vessels. Note the incomplete impregnation of normal nerve fibers in the intervals between the blood vessels. Fink–Heimer, Procedure II. (D) Fine granular silver deposits in the glomerular layer of the olfactory bulb in the guinea pig. Cupric silver method.

primarily because the irregularly stained normal fibers appear fragmented. The fragments, however, are usually rather uniform and sometimes even rectangular in shape, and they lack the globular, fusiform, droplike, or otherwise highly irregular shapes characteristic of the fragments of degenerating fibers.

5. Myelin Artifacts

These are well known artifacts that usually do not create much problem. They are found especially in Nauta–Gygax sections where they appear in the form of black cuffs of varying length (Fig. 11A).

B. Spontaneous, Accidental, and Infectious Degeneration

There is a progressive "falling out" of neurons throughout life, and, therefore, it is not uncommon to find spontaneously degenerating neuronal elements, including degenerating axons and axon terminals, scattered throughout normal material (Glees and Nauta, 1955).

Infections are not uncommon in experimental animals, and if they result in focal brain lesions, the material may well contain massive fiber degeneration that the inexperienced student may relate to the experimental lesion. A somewhat similar problem is presented by focal brain damage made unintentionally by the surgical instrument during the operation. Occasionally, focal ischemic brain lesions may be caused by the interruption of blood vessels passing through the site of the experimental lesion. Different types of accidental or "irrelevant" brain lesions are invariably revealed by the use of serial sectioning and the use of a sufficient number of experimental animals.

C. Dark Neurons of Cammermeyer

The dark neurons of Cammermeyer (1960, 1962) are relatively common in silver-impregnated sections (Fig. 13C). Although such neurons can be found in well-fixed material, they are more prevalent in material that has been poorly fixed or traumatized. It is usually easy to distinguish dark neurons from degenerating neurons because the former are randomly distributed with no apparent relation to the experimental lesion. Cammermeyer's dark neurons, finally, have a smooth contour and are evenly stained in a nongranular fashion; they do not display the gradual disintegration that characterizes degenerating neurons.

D. Glial Elements and Connective Tissue

1. Glial Processes

Irregular impregnation of glial processes (Fig. 15A) can occasionally be mistaken for degenerating nerve fibers. Although not very common, such accidental glia impregnation occurs particularly near pia mater, close to the ventricles, and in the molecular layer of the cerebellum (Bergmann glia).

2. Pericapillary Particles

This is a characteristic form of artifact that appears as an accumulation of round black corpuscles of different size, usually close to a blood vessel (Fig. 15B). The particles may represent fragments of ruptured blood cells.

3. Reticular Fibers

The impregnation of connective tissue in the vascular adventitia is often, especially in Fink–Heimer-stained sections (Fig. 15C), so massive as to make the blood vessels stand out in sharp contrast to the surrounding tissue. Incomplete impregnation of reticular fibers around blood vessels or in the pia mater sometimes looks confusingly similar to degenerating axons (Fig. 15A).

E. Artifacts in the Olfactory Bulb

Fine, granular silver deposits are relatively common in normal olfactory bulbs of most species. They have a patchy distribution and are often confined to the glomerular layer (Fig. 15D). This type of artifact does occasionally create problems of interpretation, even for experienced neuroanatomists.

VIII. SUMMARY OF ADVANTAGES AND LIMITATIONS

A. Advantages

1. The axonal projections of many fiber systems with widely spread neurons of origin can be more conveniently traced with degeneration methods than with the autoradiographic tracing method. This is especially the case if the pathway is well-confined during part of its course or if the cells of origin are superficially located.

2. One important advantage of the silver methods is their ability to demonstrate the morphological characteristics of degenerating neuronal components, especially those of axons and axon terminals. This feature may allow morphological differentiations of functional significance.
3. With the relatively short survival times used in experimental axonal degeneration studies, the extent of the lesion is relatively easy to determine, at least in sections impregnated with the most recently developed silver techniques. This is because the neuronal tissue affected by the lesion becomes argyrophilic (Fig. 16). To define the effective site of isotope or HRP injections, on the other hand, is considerably more difficult.
4. Although the silver methods are not ideal for the tracing of intrinsic pathways, they have been used with success for the mapping of short neuronal connections. The fact that the experimental lesion can be relatively well-confined is of considerable importance in this regard.
5. In general, the degenerative process remains confined to the neu-

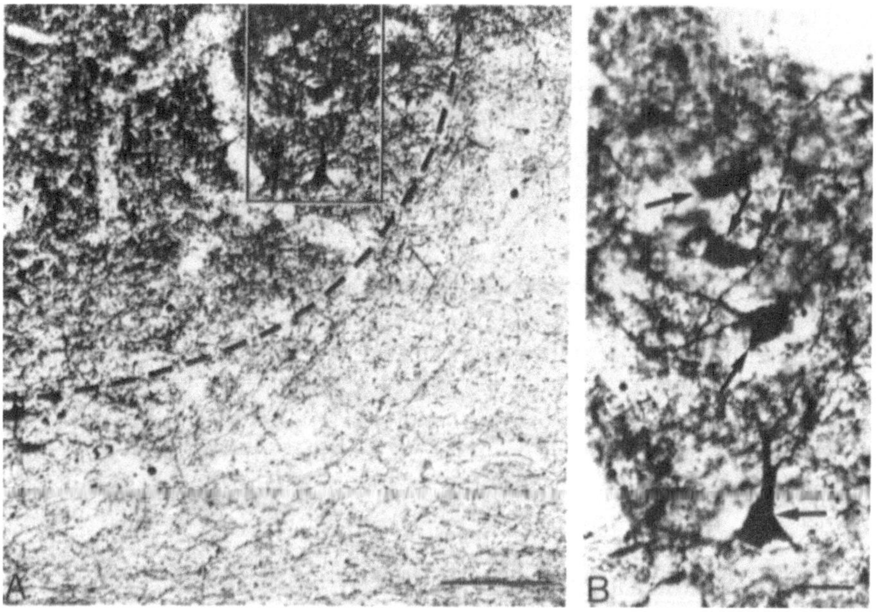

Figure 16. (A) Medium-power micrograph showing a relatively sharp boundary (dashed line) between normal neural tissue and that affected by an electrolytic lesion (upper left corner). The affected tissue is densely stained because of its strong argyrophilia. Cupric silver-stained section from the piriform cortex of a guinea pig. The rectangle outlines the area shown in (B). Bar, 100 μm. (B) High-power micrograph of degenerating neurons (arrows) and neurophil from the area indicated by a rectangle in (A). Although this peripheral zone of the lesioned area appeared largely unaffected in a Nissl-stained preparation, the pronounced argyrophilia seen in a silver-stained preparation indicates that the tissue has been seriously damaged by the electolytic lesion. Postoperative survival time, 4 days. Bar, 20 μm.

ronal system directly damaged by the experimental procedure. Should transneuronal degeneration nevertheless take place, it will reveal itself through the presence of an argyrophilic reaction in the cell bodies and dendrites of the transsynaptically affected neurons. Transsynaptic movement of labeled material, in contrast, cannot be easily excluded.

6. Compared to the autoradiographic technique, which is a lengthy procedure, the results of a silver impregnation experiment are relatively quickly obtained.

B. Limitations

1. As both cells of origin and fibers of passage are destroyed by experimental lesions, questions related to the origin of a specific pathway can usually be more easily solved by the aid of the autoradiographic tracing technique or one of the HRP methods.
2. When an electrolytic lesion is produced by an electrode in the depth of the brain, a disturbing track may appear in the overlying tissue, even if the electrode is properly insulated. The destruction of axons along the electrode track produces degeneration that may interfere with the interpretation of the results. This problem, however, can be greatly reduced by electrophoresis of neurotoxic chemicals through micropipettes (Knook, 1969).
3. The need to determine the appropriate survival time, especially when a less well-known system or animal is studied, may require a large number of experiments.
4. The silver methods are generally more capricious than the other tract-tracing techniques.
5. Although it has generally been maintained that fine unmyelinated fibers are refractory to the silver techniques, it seems that several of the modern procedures do have the capacity to impregnate such fibers. However, the staining of very fine unmyelinated fibers may require a meticulous search for the optimal conditions.

IX. PROSPECTS FOR THE FUTURE

One of the disadvantages of the degeneration studies following experimental lesions is the unavoidable damage of axons passing through the lesioned area. Indeed, this shortcoming has been an important factor in the increasing popularity of the autoradiographic technique. In the last few years, however, several investigators have used a chemical lesion technique with the purpose of producing selective destruction of nerve cell bodies. Silver impregnation of chemically induced neuronal degeneration, for instance, has been used with success both for the tracing of specific axonal projections (e.g., Desclin and Escubi, 1974; Hedreen and Chalmers, 1972) and for the Golgi-like demonstration of neurons (Switzer, 1976). The observation that a localized injection of kainic acid in the brain may cause degeneration of the neurons whose cell bodies are located in the area of injection but not of axons passing through (Coyle *et al.*, 1978) has recently attracted much attention, especially among scientists interested in behavioral effects following destruction of specific cell groups. Figure 17 illustrates a case in which ibotenic acid (Schwarcz *et al.*, 1979) was injected into the striatum of a rat 2 days before sacrifice of the animal. Degenerating axons can be traced from the area of the lesion to their terminations in the globus pallidus and

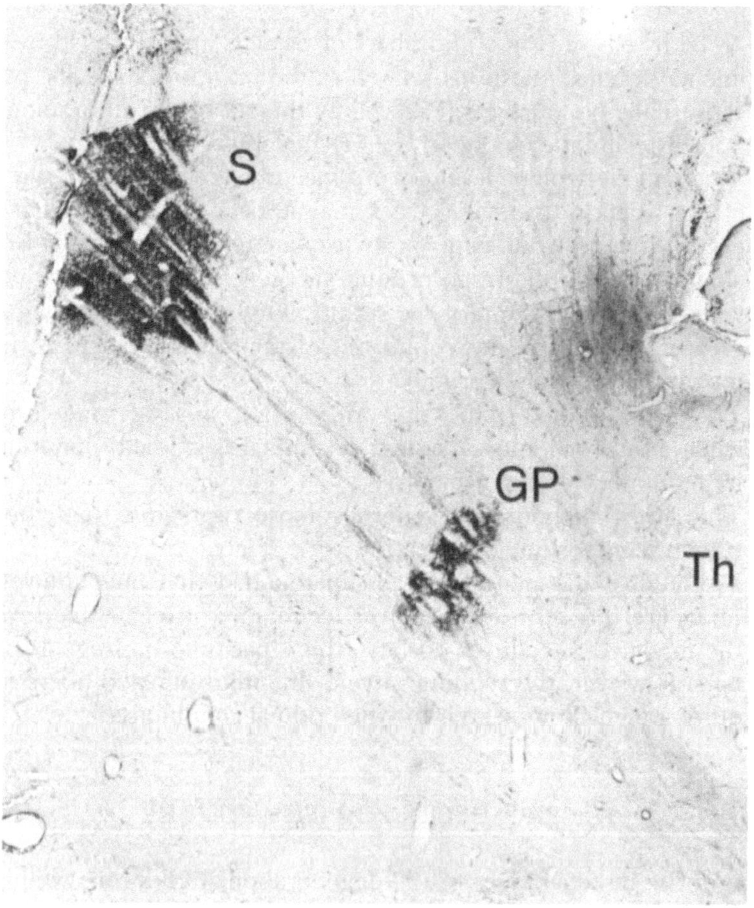

Figure 17. Terminal degeneration in globus pallidus (GP) induced by an injection of 2% ibotenic acid in the striatum (S) of the rat. Degenerating axons can be followed from the injection area to the area of terminal degeneration in GP. The internal capsule fibers passing through the lesion seem to be intact, and there are no significant signs of terminal degeneration in the thalamus (Th). Postoperative survival time 2 days. Cupric silver method.

the substantia nigra. However, there are no significant signs of terminal degeneration in the thalamus, and the internal capsule fibers passing through the lesion seem to be intact. These results indicate that the neuroanatomist might be able to use the silver technique for the mapping of chemically induced axonal degeneration without being hampered by the "fibers of passage" problem. As the injection of the neurotoxic chemical can furthermore be produced by the aid of a micropipette, two of the more serious limitations of the experimental silver technique would be eliminated.

X. APPENDIX

A. The Nauta–Laidlaw Method

1. Preparatory Steps

a. Preliminary Remarks. Utilize analytical grade chemicals and use only chemically clean bottles and glassware. Wash the glassware in nitric acid and rinse thoroughly in tap water followed by double-distilled water. The double-distilled water should be distilled in glass. All stock solutions can be stored at room temperature (18–22°C). During the staining procedure, steps must be taken to insure an adequate and even impregnation of the sections. This is best done by transferring the sections in a basket and by using a mechanical agitator. If sections are carried through the solutions with the aid of a glass rod, the glass rod must be carefully washed at the end of the staining process before it is used again. It should finally be noted that the times suggested in the following staining schedules are estimated for a room temperature of 19–25°C.

b. Fixation. Fixation should be initiated, if possible, by perfusion with 0.9% NaCl solution followed by 10% formalin or formalin–saline. Allow the brain to harden *in situ* for 12–24 hr and then remove it carefully from the skull (Chapter 1, Section III.B.2). Leave the brain in fresh fixative for at least 1–2 weeks.

c. Cryoprotective Treatment. Blocks to be sectioned are soaked in a 30% sucrose solution (made up in 10% formalin) until they have sunk to the bottom of the jar (usually 1–2 days for a rat brain).

d. Sectioning. Blocks are frozen without previous rinsing, and sections of 15–30 μm thickness are collected and stored in dilute (1–2%) formalin either at room temperature or in a refrigerator.

e. Main Stock Solutions. The two principal solutions used in the Nauta–Laidlaw method are prepared as follows:

Laidlaw's Solution. Laidlaw's solution is prepared as follows. (1) Dissolve 12 g silver nitrate in 20 ml distilled water in a 250 ml graduate cylinder. (2) Add 230 ml saturated (1.33% at 9°C) lithium carbonate (C.P.) solution. Shake vigorously and let the silver carbonate precipitate settle to the 70 ml mark. (3) Carefully decant supernatant fluid, refill to 250 ml with distilled water, shake, and let precipipate settle again to 70 ml. Decant as before, and repeat this washing twice more. (4) Again let precipitate settle to the 70 ml mark, decant, and slowly add ammonium hydroxide under constant shaking until solution has become translucent but is still markedly turbid; avoid excess of ammonia. The solution, when complete, should smell only slightly ammoniacal. (5) Dilute with distilled water to a total volume of 120 ml. Filter and store in a chemically clean bottle exposed to daylight for at least 2 weeks before use. Unless an excess of ammonia was added, a heavy silver mirror will form on the glass of the bottle after a few days; this can be disregarded. Filter before use. The solution can be used immediately with almost equally good results if 2 drops of 20% formalin are added to 50 ml of the fluid (shake and filter). Excess ammonia, if present, can be eliminated by adding enough silver carbonate (separately prepared and washed) to turn the fluid turbid, at which stage it is filtered and stored.

Nauta–Gygax Reducer. Nauta–Gygax reducer is prepared by combining the following:

10% Formalin	13.5 ml
1% Citric acid	13.5 ml
95% Alcohol	45 ml
Distilled water	400 ml

2. Staining Procedure

1. Rinse sections briefly in distilled water and soak them in 0.5% phosphomolybdic acid for 15–30 min.
2. Without washing, transfer sections to 0.05% potassium permanganate for 5–15 min (see Technical Note 1, Section X.A.3).
3. Rinse quickly and decolorize in a mixture of equal volumes of 1% hydroquinone and 1% oxalic acid for 1–2 min (see Technical Note 2, Section X.A.3).
4. Wash thoroughly in at least 3 volumes of distilled water and soak in freshly prepared 1.5% silver nitrate for 20–30 min. At this point it is advisable to run trial sections through the following steps (5 and 6).
5. Wash briefly in distilled water and soak in Laidlaw's solution for 45–60 sec. Agitate the sections continuously.
6. Without washing, transfer to a Nauta–Gygax reducer. Prevent the sections from sinking to the bottom of the jar. After about 10 sec, transfer the sections to a second volume of reducer. Leave in reducer for at least 2 min (see Technical Note 3, Section X.A.3).

7. Wash briefly, pass sections through 1% sodium thiosulfate for 10 sec, and wash again in 3 volumes of distilled water.
8. Mount prior to dehydration from a solution composed of equal parts of 0.5% gelatin and 80% alcohol. Mounted sections are dehydrated in 95% and absolute ethanol and cleared in 2–3 volumes of xylene. Sections are enclosed in Permount and covered with a cover slip.

3. Technical Notes

1. Potassium permanganate is a strong oxidizer, and the timing of the potassium permanganate step is highly critical; too short a time will cause impregnation of too many normal fibers; too long a period will suppress the impregnation of degenerating fibers. Optimum duration varies from specimen to specimen and is best established in a trial run. Well-fixed material requires less time. If possible, use an agitator to assure uniform oxidation of the sections.
2. Whereas aqueous oxalic acid solutions are stable, hydroquinone is very unstable and must be prepared fresh. Do not prolong the treatment beyond decoloration.
3. If, at the end of the reduction, the section is too dark, add a few drops of ammonium hydroxide to Laidlaw's solution (e.g., 4 drops to 20 ml of the fluid). An excessively light stain can be counteracted by the addition of a few drops of 2.5% sodium hydroxide to Laidlaw's fluid.

B. The Fink–Heimer Procedures

1. Preparatory Steps

a. Preliminary Remarks. The same precautions as suggested for the Nauta–Laidlaw method should be taken in regard to chemicals, cleaning of glassware, and handling of the sections.

b. Fixation, Cryoprotective Treatment, and Sectioning. These are the same as for the Nauta–Laidlaw method.

c. Main Stock Solutions. The principal solutions are prepared as follows:

Solution A (Procedure I)

Uranyl nitrate	1.0 g
Silver nitrate	5.0 g
Distilled water	1000.0 ml

Solution B (Procedure I)

Uranyl nitrate	1.75 g
Silver nitrate	16.25 g
Distilled water	1000.0 ml

Modified Nauta–Gygax reducer (Procedures I and II)

Distilled water	900.0 ml
95% Alcohol	75.0 ml
10% Formalin	17.5 ml
1% Citric acid	20.0 ml

2. Staining Procedure I

1. Rinse frozen sections in distilled water (three changes, 5 min each).
2. 0.05–0.5% potassium permanganate, 5–15 min (see Technical Note 1, Section X.A.3).
3. Rinse briefly in distilled water.
4. Bleach in a mixture of equal volumes of 1% oxalic acid and 1% hydroquinone (see Technical Note 2, Section X.A.3).
5. Rinse in distilled water (three changes, 5 min each).
6. Solution A, 60–90 min. A considerable lengthening of the impregnation time will occasionally improve the result.
7. Solution B, 30 min. (This step can often be deleted without noticeable impairment of the result.)
8. Rinse in distilled water (three changes, 5 min each).
9. 3–5 min in a freshly prepared ammoniacal silver nitrate solution composed of 30 ml 2.5% silver nitrate, 1.0 ml strong ammonia water, and 2.5% sodium hydroxide (see Technical Note 1, Section X.B.4).
10. Nauta–Gygax reducer. Agitate quickly until tissue turns brown, then transfer to a second volume of reducing solution for 1–2 min (see Technical Note 2, Section X.B.4).
11. Rinse in distilled water.
12. 0.5% sodium thiosulfate, 1 min.
13. Rinse in distilled water (3 changes, 5 min each).
14. Mount from a solution composed of equal parts of 0.5% gelatin and 80% alcohol (see step 8 in the Nauta–Laidlaw procedure, Section X.A.2).

3. Staining Procedure II

1. After rinsing in distilled water, soak sections for 5–10 min in 0.025–0.1% potassium permanganate.

2. Rinse and decolorize sections in an oxalic acid–hydroquinone solution (see Section X.A.2, step 3).
3. Rinse thoroughly and transfer sections for 5–10 min to a 2.5–5% uranyl nitrate solution (see Technical Note 3, Section X.B.4).
4. Rinse thoroughly in three changes of distilled water and transfer sections to a 0.3–0.5% silver nitrate solution for 1–2 hr. The addition of 0.2 ml of pyridine to 10 ml of the silver nitrate solution may improve the results.
5. Without washing, transfer sections for 2–5 min to a freshly prepared ammoniacal silver nitrate solution composed of:

1.5% Silver nitrate	20 ml
95% Ethyl alcohol	12 ml
Strong ammonia water	2 ml
2.5% Sodium hydroxide	1.6–1.8 ml

6. Transfer sections without rinsing to the Nauta–Gygax reducer.
7. Rinse and transfer sections for 1 min to 0.5% sodium thiosulfate solution.
8. If desired, counterstain by immersing the mounted sections briefly in a cresylechtviolet solution (see Technical Note 4, Section X.B.4).
9. Rinse, mount, dehydrate, and enclose.

4. Technical Notes

1. Make a few preliminary trials in order to establish the optimal ammonia–sodium hydroxide ratio. Too high a concentration of sodium hydroxide results in dark sections with increased risk of precipitates, whereas too much ammonia tends to suppress the argyrophilia of degenerating axoplasm. Only freshly prepared ammoniacal silver solutions should be used.
2. If sections become too dark in Procedure I, add a few drops of 1% citric acid to each 50 ml of reducing solution. Keep the citric acid solution in the refrigerator.
3. Since uranyl nitrate, like potassium permanganate, is an oxidant, it may be necessary to reduce or even delete the pretreatment in uranyl nitrate and/or potassium permanganate in order to preserve the argyrophilia of degenerating axons. In other cases a prolonged pretreatment may be needed to suppress the argyrophilia of irrelevant tissue components. *Caution:* Although the radioactive properties of uranyl nitrate do not usually create a hazard, this chemical is very toxic and should be handled with care. Wash hands after handling either the dry salt or solutions.
4. Silver-impregnated sections can be counterstained by immersing them in a 0.5% cresylechtviolet solution for about 3 min, rinsing briefly in distilled water, and dehydrating in 70% and 95% alcohol. Differentiate the sections individually in a mixture of 100 ml 95%

alcohol and 2 ml glacial acetic acid until the desired color is obtained, and continue the dehydration in absolute alcohol. Mount and enclose. An alternative method is to stain sections already mounted on glass slides. To prevent the sections from becoming detached during the staining procedure, it is advisable to mount them on chrome-alum-coated slides.

C. The Cupric Silver Method

1. Preparatory Steps

a. Preliminary Remarks. The preliminary remarks made for the Nauta–Laidlaw method apply here as well (Section X.A.1a).

b. Fixation. Fixation should be initiated by perfusion with a rinsing solution consisting of 0.8% sucrose, 0.4% glucose, and 0.8% NaCl in 0.067 M cacodylate buffer (pH 7.2–7.4). This is followed by 4% formaldehyde (freshly prepared from paraformaldehyde) in 0.067 cacodylate buffer (pH 7.2–7.4) containing 4% sucrose (see Technical Note 1, Section X.C.3).

Allow the brain to harden *in situ* for 12–24 hr and then remove it carefully from the skull (Chapter 1, Section III.B.2). Leave the brain in fresh fixative for 2–8 days, depending on species (e.g., adult rat, 4–6 days; adult cat, 6–8 days) and age (e.g., young rat, 2–4 days). If, of necessity, the postfixation is extended beyond the suggested time, 0.9 ml of 1% cadmium nitrate should be added to every 50 ml of the cupric–silver solution.

c. Cryoprotective Treatment. Block to be sectioned is soaked in a 30% sucrose solution in 0.067 M cacodylate buffer until it has sunk to the bottom of the jar.

d. Sectioning. The block is frozen without previous rinsing, and sections of 25–30 μm thickness are collected and stored in a refrigerator in the cacodylate-buffered fixative for 2–4 days.

2. Staining Procedure

As several of the steps in this procedure require precise timing, it is advisable to transfer the sections through the different solutions with the aid of a basket.

1. Rinse in three changes of distilled water, 3–5 min each.
2. Transfer to an appropriate amount (approximately 50 ml for 60 rat

brain sections) of the following cupric silver solution: 50 ml 2.5% $AgNO_3$, 0.75 ml 0.5% $Cu(NO_3)_2$, 2 ml pyridine, and 4 ml 95% ethyl alcohol (see Technical Note 2, Section X.C.3). The impregnation time varies with species. It is usually 2–4 days for rat, guinea pig, and armadillo (edentate), and 3–6 days for cat.*

3. Transfer sections directly to acetone for 3–5 min. Agitate continuously (see Technical Note 3, Section X.C.3).
4. Transfer the sections to a fresh ammoniacal silver nitrate solution that is prepared by dissolving 6 g of $AgNO_3$ in 20 ml distilled water. Then add distilled water to make the final volume 30 ml. Add 15 ml of 0.4% NaOH followed by 6 ml of concentrated NH_4OH (sp. gr. 0.8990, Fisher Scientific)†. Then add 30 ml distilled water to this concentrated ammoniacal silver nitrate solution (see Technical Note 4, Section X.C.3).
5. Without washing, transfer the sections to a freshly prepared reducing solution consisting of 6.0 ml 10% nonneutralized formalin, 3.5 ml‡ 1% citric acid, and 50.0 ml 100% ethyl alcohol. Add distilled water to make a final reducing solution of 500 ml. Agitate the sections continuously in this solution for 2–3 min or until the sections become dark tobacco in color (see Technical Note 5, Section X.C.3).
6. Rinse in distilled water§ and transfer to a bleaching solution consisting of 0.05% potassium ferricyanide. Stop the process before the bleaching affects the black silver deposits associated with the degenerating fibers and terminals. The optimal time, usually 3–7 min, coincides with the disappearance of the reddish brown impregnation of cell nuclei, most easily observed in the pyramidal cell layer at the hippocampus or the piriform cortex. At this point the sections have obtained a straw yellow color.
7. Rinse thoroughly in distilled water and transfer to 1% sodium thiosulfate for 1 min. Use a fresh solution for every batch.
8. Rinse in three changes of distilled water and dehydrate the sections in 80% ethyl alcohol followed by two changes of a 1:1 mixture of 95% ethyl alcohol and pure acetone.
9. Clear the sections for 12–36 hr in pure creosote.
10. Aided by a fine, soft brush, float each section onto a microscope slide in a clearing mixture made up as follows: dissolve 10 g phenol in 20 ml creosote and add 70 ml xylene while stirring (see Technical Note 6, Section X.C.3).
11. Blot the sections with bibulous paper and coverslip with Permount.

*In a recent modification of the method (Carlsen and de Olmos, 1981) the impregnation time has been considerably reduced by the use of a warm (40°C) cupric silver solution containing allantoin.

†If Merck's concentrated NH_4OH (sp. gr. 0.91) is used, add 7.35 ml.

‡Although this amount works well for rat, guinea pig, and armadillo, 3 ml should be used for cat.

§If needed, the sections can be left overnight in a second bath of distilled water.

3. Technical Notes

1. Although 4% formaldehyde solutions adjusted to pH 7.2–7.4 with cacodylate buffer have given the best results, other buffer systems, e.g., phosphate buffer, can also be used. Unless the pH is adjusted to the value indicated above, the cupric silver procedure will fail. Furthermore, fixatives containing osmic acid or glutaraldehyde should not be used.
2. In the guinea pig, the following formula has given better result for anterograde degeneration: 50 ml 2% $AgNO_3$, 1.0 ml 0.5% $Cu(NO_3)_2$, 2 ml pyridine, and 4 ml 95% ethyl alcohol.
3. Acetone evaporates quickly, and a generous amount should be used. If the sections are not transferred with the aid of a basket, it is convenient to decant off the acetone and immediately pour in the ammoniacal silver nitrate solution. This will insure a uniform treatment of all sections. If too many sections are treated at the same time in the acetone, too much of the chemical will be transferred to the ammoniacal silver nitrate solution. This will decrease the quality of the staining. The treatment in acetone removes unbound silver from the sections and reduces the impregnation of the granular argyrophilic neurons (Section VII.A.2) and other normally argyrophilic structures.
4. If too many normal fibers or cell nucleoli appear in the sections, try an undiluted ammoniacal silver nitrate solution. In order to reduce the impregnation of nucleoli, Dr. Robert Switzer (personal communication) suggests that the ammoniacal silver nitrate solution be prepared by adding each solution in the following order: 28.75 ml 20% $AgNO_3$, 14.65 ml 0.36% NaOH, 6.60 ml concentrated NH_4OH (Fisher Scientific); then dilute 1 : 1 with distilled water.
5. It is useful to control the reduction process by observing the blackening of the lesioned area and of the major degenerating fiber tracts and synaptic fields. The use of a bright spotlight and white background facilitates the control of the reduction process.
6. If creosote is not available, the sections can be mounted in gelatin–alcohol, even if they do lose some of their transparency.

D. The Application of Silver Degeneration Techniques to the Human Brain

Marek-Marsel Mesulam*

1. Introduction

Although the suppressive silver impregnation methods and the intraaxonal transport techniques have introduced great advances in the experimen-

*Bullard and Denny-Brown Laboratories of the Harvard Neurological Unit, Beth Israel Hospital, 330 Brookline Avenue, Boston, Massachusetts 02215 and Aphasia Research Center, Boston University School of Medicine.

tal investigation of neural connections in the brains of nonhuman species, the study of neural connectivity in the human brain has not enjoyed comparable progress. There are several obvious reasons for this. First, many of the manipulations that are required by the more recent neuroanatomical methods cannot be performed in the human brain. For instance, the powerful intraaxonal transport techniques require intracerebral injections of active substances into the living brain. Second, the methodical and meticulous control of lesion size, survival time, and fixation that underlies the successful application of experimental silver impregnation methods is obviously not possible in cases of autopsied human brains that contain naturally occurring lesions. Therefore, questions regarding specific neural connections of the human brain have been answered by extrapolation from information obtained from other species.

Serious objections, however, can be raised to this widespread practice of extrapolating neuroanatomical information from nonhuman animals to man. First, it is quite clear that significant differences in cerebral connectivity do exist among species. For instance, it has been shown that the details of corticospinal, spinothalamic, and hippocampofugal connections do vary from one species to another (Kuypers, 1958; Bowsher, 1957; Mehler, 1957; Valenstein and Nauta, 1959). Hence, it is also reasonable to expect differences in connectivity between the nonhuman and the human brain. Humans possess verbal, cognitive, and social behaviors that are either absent or present only in rudimentary form in other species. Since it is reasonable to assume that the brain is ultimately responsible for these behaviors, it may be predicted that the human brain is characterized by a unique set of neural connections that are absent in other species.

As most experimental procedures used in modern neuroanatomical studies are not applicable to the human brain, it would seem appropriate to encourage the study of the human brain by the aid of the silver techniques.

2. Survival Time

Successful use of the Nauta method in human postmortem material has been reported with survival periods ranging from 8 to 36 days (Bowsher, 1957; Kuypers, 1958), and our own experience confirms the effectiveness of this survival range. As a practical guide it seems reasonable to attempt silver impregnation if the lesion was sustained 1 to 8 weeks before death.*

3. Fixation

Since *in situ* perfusion of the human brain is impossible, the onset of fixation is necessarily delayed. However, human brains remaining unfixed as long as 24 hr after death have, in our experience, yielded excellent results.

*Grafe *et al.* (1978) reported that extensive staining of degenerating elements could still be obtained 2 years after cortical damage.

Our fixation procedure is to immerse the whole brain in unbuffered 10% formalin, suspending it in fixative by a string that is passed under the basilar artery and secured to the rim of the container. This procedure minimizes distortion by compression. After 2 weeks, we change the formalin and continue to fix for another 2 to 6 weeks. In some specimens, it may be helpful to perfuse the excised brain through the carotid and basilar arteries.

4. Embedding and Reconstruction

After fixation, rectangular tissue blocks are prepared from the area of interest. In general, it has been found that limiting them to 7 × 4 cm on the cutting surface facilitates subsequent tissue manipulation. In this context, it should be mentioned that reconstruction of lesion site and plane of sectioning, especially in large brains, should be initiated before embedding. Next, the blocks are immersed in sucrose–formalin for 2–5 weeks and, if necessary, encased in gelatin or an albumin–gelatin matrix (Chapter 1). When the encasing medium has hardened, usually after 1–2 days, the block is immersed in undiluted formalin for another 2–5 weeks before 25–30 μm frozen sections are cut. Before staining, the sections are stored in 10% formalin for 2–20 days. For consistent staining it may be preferable to keep the cut sections in the refrigerator.

5. Staining Procedures

The present author has obtained satisfactory staining of degenerating axonal elements with the help of three different staining methods. *Method 1* is a Nauta procedure with uranyl nitrate as the mordant. This method enables the relative suppression of normal fibers while selectively enhancing the argyrophilia of degenerating axons (Fig. 18). *Method 2* is a Nauta procedure with an extensive pretreatment that improves the suppression of normal fibers and their myelin sheaths. Furthermore, in some overfixed material, the initial calcium chloride–hydroquinone step of Method 2 may restore argyrophilia to tissue that would otherwise be refractory to silver impregnation. *Method 3* is a modified Fink–Heimer procedure. Although this method has been used to visualize terminal boutons in many animals, convincing visualization of degenerating boutons has not been obtained in the human brain. Nevertheless, Method 3 has been very useful in the precise delineation of "terminal fields" within cortical layers (Fig. 19). The physicochemical principles underlying the present techniques are basically the same as for the previously described original Nauta and Fink–Heimer techniques.

In general, Method 1 has given the most consistently good results, even if the staining of some normal fibers may make interpretation somewhat laborious. It is therefore advisable to prepare one series of sections according to Method 1 while additional series may be stained according to Methods

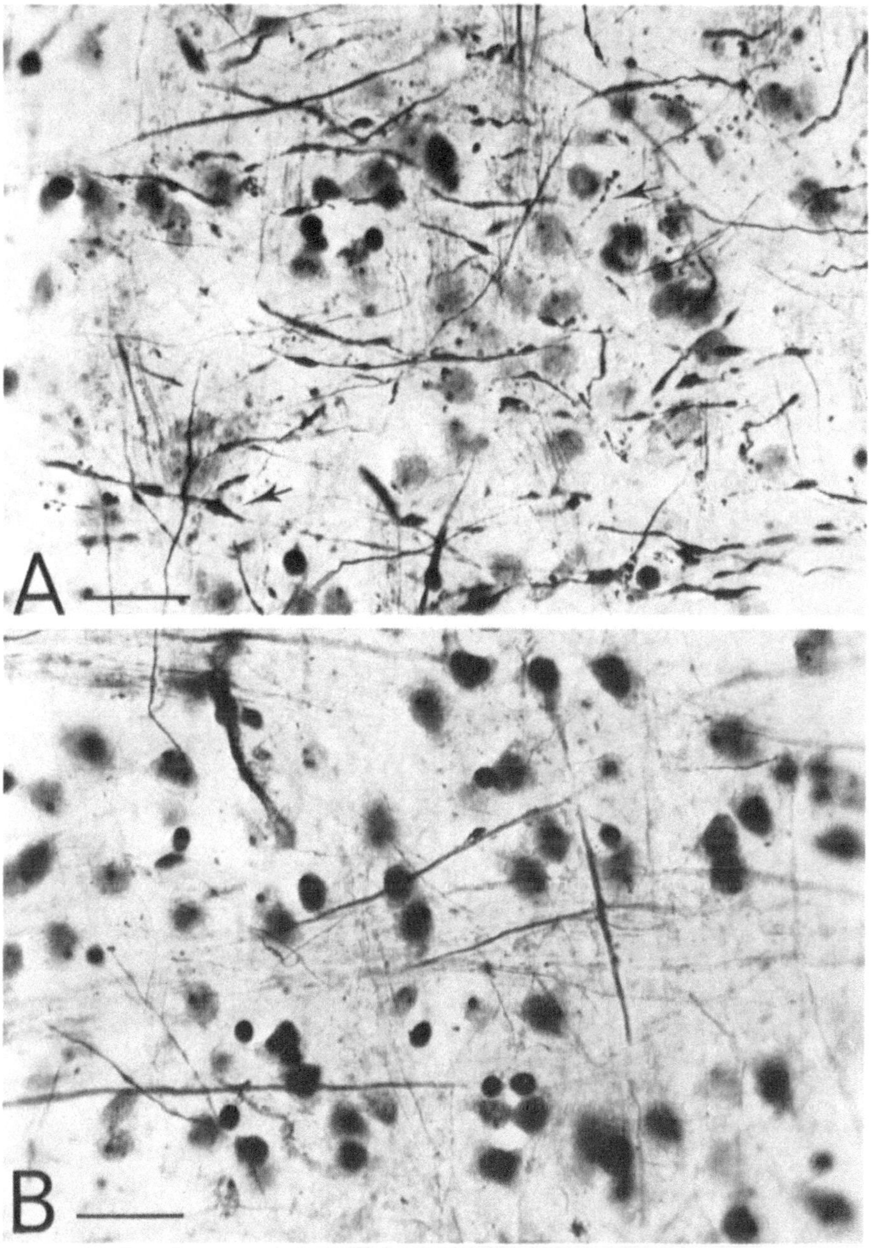

Figure 18. (A) Layer IVc of area 17. This patient had an infarct of the ipsilateral visual radiations 22 days before death. The photomicrograph shows large quantities of degenerating fibers interspersed among the granule cells of layer IVc of area 17. The degeneration in fine-caliber axons that are closely apposed to the nerve cell gives the impression of a terminal pattern (upper right quadrant). The arrowheads point to some of the more obvious degenerated axons. (B) Layer IVc of area 17 from the opposite hemisphere of the same case. This photomicrograph shows some normal fibers but no evidence of degeneration. [From Mesulam (1979).]

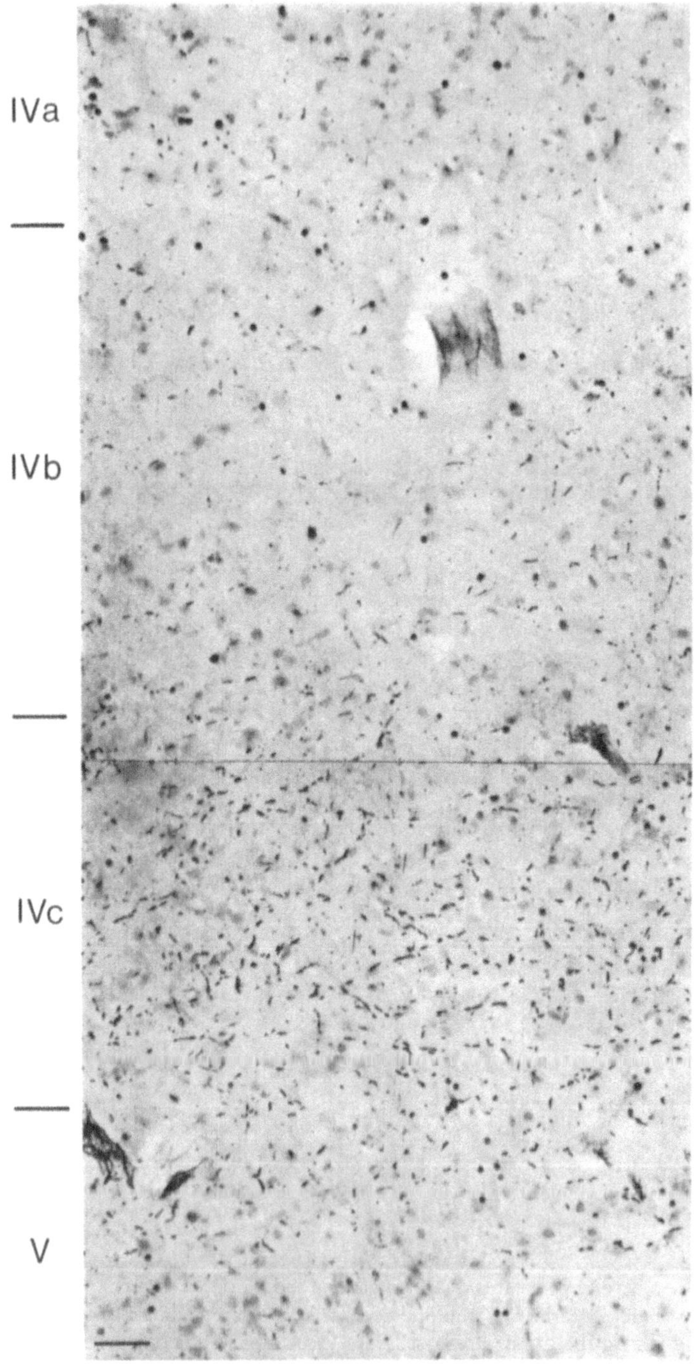

Figure 19. Layers IV and V of area 17 showing the concentration of degeneration in layer IVc. From the same patient as in Fig. 17. The cell-poor region in layer IVb corresponds to the line of Gennari in myelin-stained sections. Method 3. [From Mesulam (1979).]

2 or 3 if greater selectivity is needed. A very rapid method for the simultaneous staining of fibers and cells becomes possible if cresylviolet acetate is used as described in *Method 4*.

In addition to the artifactual silver depositions described in Section VII, human brain tissue is excessively vulnerable to the staining of normal myelin and to the diffuse precipitation of silver grains in the neuropil. It has been possible to minimize these sources of artifactual argyrophilia by varying the pretreatment parameters as described below. In some cases, Method 2 has offered the only means for eliminating excessive argyrophilia. In other cases, the tissue had to be immersed in 5% uranyl nitrate for up to 35 hr in order to prevent this undesirable deposition of reduced silver.

When the necessary blocking and staining have been completed, the rest of the brain should be examined meticulously for evidence of other unsuspected lesions. If such lesions are present, and if the cellular reaction indicates an origin of less than 4 months before death, their possible contribution to the observed pattern of degeneration should be considered. Since it is impractical to stain the whole brain for this purpose, the examination for other unsuspected lesions may be done with reasonable accuracy by cutting the unembedded slabs into 2- to 3-mm thick slices with the help of a commercial meat slicer.

In the methods described below, the staining is done in batches of 8–10 *free floating* sections that are conveniently transferred from one solution to another by using a basket as suggested for the other silver techniques. However, from the Laidlaw bath on, the sections are transferred individually.

a. Method 1. This is a Nauta procedure.

1. Rinse in three changes of distilled water (2 min each).
2. Uranyl nitrate (1% or 5% aqueous solution). The optimal times vary greatly and range from 1–40 hr in a 5% solution.
3. Rinse in three changes of distilled water (5 min each).
4. Potassium permanganate (0.05% aqueous solution). In general 40–60 min yield good results. However, different times should be tried in order to find the optimal time for the individual specimen. If the range of optimal times is too narrow, the 0.05% stock solution may be diluted to 80% or 60% strength in order to expand the range so that a finer adjustment of the timing becomes possible.
5. Rinse in distilled water (2 min).
6. Place sections in a bleaching solution (10 g oxalic acid and 10 g hydroquinone in 2000 ml water) until they are uniformly decolorized, usually after 1–2 min.
7. Rinse in three changes of distilled water (2 min each).
8. Transfer the sections to a 1.5% silver nitrate solution for 35 min.
9. Rinse in two changes of distilled water (2 min each).
10. Transfer to Laidlaw's solution (Section X.A.1e) for 30–60 sec. Agitate gently. The Laidlaw step is best carried out in a crucible containing about 10 ml of solution. Individual sections are transferred

to this crucible from the water rinse in step 9 with the help of an L-shaped glass rod. Regarding the intensity of the staining, see Technical Note 3, Section X.A.3. Discard the solution after 2–3 sections.

11. Transfer to a modified Nauta–Gygax reducer (to 1800 ml water, add 150 ml 95% ethanol, 35 ml 10% formalin, and 40 ml 1% citric acid) for 1 min until a yellowish-brown color appears. If the sections are too dark, diluting the reducer to 80% of its strength may help.
12. Complete the reduction in a second bath for another minute.
13. Rinse in three changes of distilled water (2 min each).
14. Transfer to a 0.5% sodium thiosulphate solution for 10–60 sec.
15. Rinse in three changes of distilled water (2 min each).
16. Mount and enclose (see the Nauta–Laidlaw procedure, step 8, Section X.A.2).

b. Method 2 (Albrecht and Fernstrom, 1959). Following a rinse in two changes of distilled water (2 min each), transfer the sections to the following series of solutions and rinse in two changes of distilled water (2 min each) after each solution except after their treatment in the Laidlaw solution from which the sections go directly to the two reducing steps.

1. 5 ml 10% hydroquinone + 2 ml 10% calcium chloride + 93 ml 10% formalin for 24–48 hr. This solution is used only if the specimen has been overfixed (6 months or longer) and shows a generalized lack of argyrophilia.
2. Oxalic acid–hydroquinone mixture (Method 1, step 6, Section X.D.5a) for 30 min.
3. 0.05% chromic acid for 10 min.
4. 4% hydrobromic acid for 5 min.
5. 1% phosphotungstic acid for 15 min.
6. 0.05% potassium permanganate for 7–9 min (see the Nauta–Laidlaw method, Technical Note 1, Section X.A.3).
7. Oxalic acid–hydroquinone mixture (Method 1, step 6, Section X.D.5a) for 1–2 min.
8. 1.5% silver nitrate for 30 min.
9. Laidlaw's solution for 3 min (see the Nauta–Laidlaw method, Section X.A.1e).
10. Without washing transfer to a modified Nauta–Gygax reducer (800 ml distilled water + 90 ml absolute ethanol + 27 ml 1% citric acid + 27 ml 10% formalin) until the sections are uniformly dark (about 1 min). Dilute the reducer to 80% strength if the sections become too dark.
11. Without washing, transfer to a second reducing bath for 1 min.
12. 0.5% sodium thiosulfate for 10 sec.
13. Mount and coverslip (see the Nauta–Laidlaw procedure, step 8, Section X.A.2).

c. Method 3. This is a modified Fink–Heimer method.

1. Rinse in water (2 min).
2. Transfer to 0.05% potassium permanganate solution for 1–5 min (see Technical Note 1, Section X.A.3).
3. Rinse twice in distilled water (2 min).
4. Bleach as in Method 1 (Section X.D.5a, step 6).
5. Rinse twice in distilled water (2 min each).
6. Transfer to solution A (mix 2 g uranyl nitrate and 10 g silver nitrate in 2000 ml water) for 0–10 min.
7. Rinse in distilled water (2 min).
8. Transfer to solution B (mix 3.5 g uranyl nitrate and 32.5 g silver nitrate in 2000 ml water) for 0–10 min. Note that various combinations of timing must be tried between solutions A and B in order to optimize argyrophilia in degenerating terminals while minimizing the deposition of reduced silver elsewhere. Such optimal combinations may range from a full 10 min in both solution A and solution B to a schedule that completely eliminates either A or B.
9. Rinse in distilled water (2 min).
10. Transfer to solution C. [First prepare a 2.0% solution of silver nitrate; then prepare a base mixture (always freshly prepared) of 2.5% sodium hydroxide with concentrated ammonium hydroxide in a ratio of 9 volumes of the sodium hydroxide to 6 volumes of the ammonia; then to each 100 ml of the silver nitrate solution, add 10 ml of the base mixture and stir until clear.] Sections are placed in this solution for 0.5–2 min. In this procedure, this step is analogous to the Laidlaw step of methods 1 and 2. Note that this is a very crucial step. For optimal staining, it may be necessary to make slight alterations in the 9:6 ratio. Changing the base ratio, even by 0.1 ml, in favor of ammonia may decrease artifactual argyrophilia, but it may also prevent the staining of degeneration. Increasing the ratio in favor of sodium hydroxide may do the opposite. The optimal ratio is crucial for good staining and seems to be very precariously balanced in human tissue.
11. Transfer to reducer (prepared as in Method 1). Leave in this solution until sections have uniformly changed color. This takes about 45–60 sec. If the sections end up being too dark, the reducer may be diluted to 80% strength. A few drops of citric acid may help to eliminate nonspecific precipitation of silver.
12. Repeat step 11.
13. Rinse in distilled water (2 min).
14. Transfer to a 0.5% sodium thiosulfate solution for 1 min.
15. Rinse twice in distilled water (2 min each).
16. Mount and coverslip (see the Nauta–Laidlaw procedure, step 8, Section X.A.2).

d. Method 4. This method is for normal cells and fibers.

1. Rinse in distilled water (5 min).
2. Transfer to a 0.2% aqueous cresyl violet acetate solution for 10 min.
3. Transfer to an acid–alcohol (1 part of 10% acetic acid to 2 parts absolute ethanol) solution and differentiate until the gray matter is pale blue and the white matter is violet.
4. Rinse in distilled water until no more dye comes off.
5. Mount and coverslip (see step 8, Section X.A.2).

ACKNOWLEDGMENTS. This work was supported by USPHS Grant NS 10972 and by the Consejo Nacional de Investigaciones Cientificas Técnicas of Argentina.

Drs. F. Gallyas and J. R. Wolff kindly furnished copies of manuscripts before they were submitted for publication.

REFERENCES

Albrecht, M. H., and Fernstrom, R. C., 1959, A modified Nauta–Gygax method for human brain and spinal cord, *Stain Technol.* **34:**91–94.

Barron, K. D., Means, E. D., and Larsen, E., 1973, Ultrastructure of retrograde degeneration in thalamus of rat. I. Neuronal somata and dendrites, *J. Neuropathol. Exp. Neurol.* **32:**218–244.

Bielschowsky, M., 1904, Die Silberimprägnation der Neurofibrillen, *J. Psychol. Neurol. (Lpz.)* **3:**169–188.

Blümcke, S., 1962, Vergleichende experimentell-morphologische Untersuchungen zur Frage einer retino–hypothalamischen Bahn beim Huhn, Meerschweinchen und Kaze, *Z. Mikrosk Anat. Forsch.* **67:**469–513.

Bowsher, D., 1957, Termination of the central pain pathway in man: The conscious appreciation of pain, *Brain* **80:**606.

Brodal, A., 1940, Modification of Gudden method for study of cerebral localization, *Arch. Neurol.* **43:**46–54.

Brown, J. O., 1943, Pigmentation of substantia nigra and locus coeruleus in certain carnivores, *J. Comp. Neurol.* **79:**393–406.

Brown, J. O., 1944, Pigmentation of certain mesencephalic tegmental nuclei in the dog and cat, *J. Comp. Neurol.* **81:**240–258.

Cammermeyer, J., 1960, The post-mortem origin and mechanism of neuronal hyperchromatosis and nuclear pycnosis, *Exp. Neurol.* **2:**379–405.

Cammermeyer, J., 1962, An evaluation of the significance of the dark neuron, *Ergeb. Anat. Entwickl. Gesch.* **36:**1–61.

Campos-Ortega, J. A., Hayhow, W. R., and Clüver, P. F. de V., 1970, The descending projections from the cortical visual fields of *Macaca mulatta* with particular reference to the question of a cortico–lateral geniculate pathway, *Brain Behav. Evol.* **3:**368.

Carlsen, J., and de Olmos, J. S., 1981, Silver impregnation of degenerating neurons and their processes. A modified cupric silver technique, *Brain Res.,* in press.

Chambers, W. W., Chung-Yu, L., and Chan-Nao, L., 1956, A modification of the Nauta technique for staining of degenerating axons in the central nervous system, *Anat. Rec.* **124:**391–392.

Cook, R. D., Ghetti, B., and Wisniewski, H. M., 1974, The pattern of Wallerian degeneration in the optic nerve of newborn kittens: An ultrastructural study, *Brain Res.* **75:**261–275.

Cowan, W. M., 1970, Anterograde and retrograde transneuronal degeneration in the central and peripheral nervous system, in: *Contemporary Research Methods in Neuroanatomy* (W. J. H. Nauta and S. O. E. Ebbesson, eds.), pp. 217–251, Springer-Verlag, New York.

Coyle, J. T., Molliver, M. E., and Kuhar, M. J., 1978, *In situ* injection of kainic acid: A new method for selectively lesioning neuronal cell bodies while sparing axons of passage, *J. Comp. Neurol.* **180:**301–323.

Dayan, A. D., 1971, Comparative neuropathology of ageing. Studies on the brain of 47 species of vertebrates, *Brain* **94:**31.

Demêmes, D., Fuentes, C., and Marty, R., 1974, Cinétique des processes de dégénérescence axonique dans le système nerveux central: Étude expérimentale à court terme dans le corps calleux chez le rat, *Acta Neuropathol. (Berl.)* **29:**311–323.

de Olmos, J. S., 1966, El sistema granular argirofilico en el cerebro del sapo y de la fortuga, *Acta Physiol. Lat. Am.* **16:**41–42.

de Olmos, J. S., 1969, A cupric–silver method for impregnation of terminal axon degeneration and its further use in staining granular argyrophilic neurons, *Brain Behav. Evol.* **2:**213–237.

de Olmos, J. S., and Ingram, W. R., 1972, An improved cupric–silver method for impregnation of axonal and terminal degeneration, *Brain Res.* **33:**523–529.

Desclin, J. C., and Escubi, J., 1974, Effects of 3-acetylpyridine on the central nervous system of the rat, as demonstrated by silver methods, *Brain Res.* **77:**349–364.

Desclin, J. C., and Escubi, J., 1975, An additional silver impregnation method for demonstration of degenerating nerve cells and processes in the central nervous system, *Brain Res.* **93:**25–39.

Donat, J. R., and Wisniewski, H. M., 1973, The spatio–temporal pattern of Wallerian degneration in mammalian peripheral nerves, *Brain Res.* **53:**41–53.

Eager, R. P., 1970, Selective staining of degenerating axons in the central nervous system by a simplified silver method: Spinal cord projections to external cuneate and inferior olivary nuclei in the cat, *Brain Res.* **22:**137–141.

Ebbesson, S. O. E., 1970, The selective silver impregnation of degenerating axons and their synaptic endings in nonmammalian species, in: *Contemporary Research Methods in Neuroanatomy* (W. J. H. Nauta and S. O. E. Ebbesson, eds.), pp. 132–161, Springer–Verlag, Berlin.

Ebbesson, S. O. E., and Rubinson, K., 1969, A simplified Nauta procedure, *Physiol. Behav.* 4:281–282.

Fink, R. P., and Heimer, L., 1967, Two methods for selective silver impregnation of degenerating axons and their synaptic endings in the central nervous system, *Brain Res.* **4:**369–374.

Freedman, S. L., Voogd, J., and Vielvoye, G. J., 1977, Experimental evidence for climbing fibers in the avian cerebellum, *J. Comp. Neurol.* **175:**243.

Fry, F. J., and Cowan, W. M., 1972, A study of retrograde cell degeneration in the lateral mammillary nucleus of the cat, with special reference to the role of axonal branching in the preservation of the cell, *J. Comp. Neurol.* **144:**1–24.

Gallyas, F., Záborszky, L., and Wolff, J. R., 1980a, Experiments on the mechanism of impregnation methods demonstrating axonal and terminal degeneration, *Stain Technol.* **55:**281–290.

Gallyas, F., Wolff, J. R., Böttcher, H., and Záborszky, L., 1980b, Selective demonstration of axonal degeneration, *Stain Technol.* **55:**291–297.

Gallyas, F., Wolff, J. R., Bottcher, H., and Záborszky, L., 1980c, A reliable and sensitive method to locate terminal degeneration and lysosomes in the CNS, *Stain Technol.* **55:**299–306.

Gamble, H. J., Goldby, F., and Smith, G. M. R., 1957, Effect of temperature on the degeneration of nerve fibres, *Nature* **179:**527.

Glees, P., 1946, Terminal degeneration within the central nervous system as studied by a new silver method, *J. Neuropathol. Exp. Neurol.* **5:**54–59.

Glees, P., and Le Gros Clark, W. E., 1941, Termination of optic fibers in the lateral geniculate body of the monkey, *J. Anat.* **75:**295–308.

Glees, P., and Nauta, W. J. H., 1955, A critical review of studies on axonal and terminal degeneration, *Monatsschr. Psychiatr. Neurol.* **129:**74–91.

Grafe, M. R., Heilman, K. M., Schimpff, R. D., and Leonard, C. M., 1978, Long-lasting degeneration argyrophilia in the human brain, *Anat. Rec.* **190:**405.

Grant, G., 1975, Retrograde neuronal degeration, in: *Golgi Centennial Symposium* (M. Santini, ed.), pp. 195–200, Raven Press, New York.

Grant, G., and Aldskogius, H., 1967, Silver impregnation of degenerating dendrites, cells and axons central to axonal transection. 1. A Nauta study on the hypoglossal nerve of kittens, *Exp. Brain Res.* **3:**150–162.

Grant, G., and Arvidsson, J., 1975, Transganglionic degeneration in trigeminal primary sensory neurons, *Brain Res.* **95:**265–279.

Groenewegen, H. J., and Woogd, J., 1977, The parasagittal zonation within the olivocerebellar projection. I. Climbing fiber distribution in the vermis of cat cerebellum, *J. Comp. Neurol.* **174:**417–488.

Guillery, R. W., 1959, Afferent fibers to the dorso-medial thalamic nucleus in the cat, *J. Anat.* **93:**403–419.

Guillery, R. W., and Oberdorfer, M. D., 1977, A study of fine and coarse retinofugal axons terminating in the geniculate *C* laminae and in the medial interlaminar nucleus of the mink, *J. Comp. Neurol.* **176:**515–526.

Guillery, R. W., and Ralston, H. J., 1964, Nerve fibers and terminals. Electron microscopy after Nauta staining, *Science* **143:**1331–1332.

Guillery, R. W., Shirra, B., and Webster, K. E., 1961, Differential impregnation of degenerating nerve fibers in paraffin-embedded material, *Stain Technol.* **36:**9–13.

Hedreen, J. C., and Chalmers, J. P., 1972, Neuronal degeneration in rat brain induced by 6-hydroxydopamine, a histological and biochemical study, *Brain Res.* **67:**1–36.

Heimer, L., 1970a, Selective silver-impregnation of degenerating axoplasm, in: *Contemporary Research Methods in Neuroanatomy* (W. J. H. Nauta and S. O. E. Ebbesson, eds.), pp. 106–131, Springer-Verlag, New York.

Heimer, L., 1970b, Bridging the gap between light and electron microscopy in the experimental tracing of fiber connections, in: *Contemporary Research Nethods in Neuroanatomy* (W. J. H. Nauta and S. O. E. Ebbesson, ed.), pp. 162–172, Springer-Verlag, New York.

Heimer, L., 1972, The olfactory connections of the diencephalon in the rat, *Brain Behav. Evol.* **6:**484–523.

Heimer, L., and Ekholm, R., 1967, Neuronal argyrophilia in early degenerative states. A light and electron-microscopic study of the Glees and Nauta techniques, *Experientia* **15:**237–239.

Heimer, L., and Kalil, R., 1978, Rapid transneuronal degeneration and death of cortical neurons following removal of the olfactory bulb in adult rats, *J. Comp. Neurol.* **178:**559–610.

Heimer, L., and Peters, A., 1968, An electron microscope study of a silver stain for degenerating boutons, *Brain Res.* **8:**337–346.

Heimer, L., and Wall, P. D., 1968, The dorsal root distribution to the substantia gelatinosa of the rat with a note on the distribution in the cat, *Exp. Brain Res.* **6:**89–99.

Hjorth-Simonsen, A., 1970, Fink–Heimer silver impregnation of degenerating axons and terminals in mounted cryostat sections of fresh and fixed brains, *Stain Technol.* **45:**199.

Hoff, E. C., 1932a, Central nerve terminals in the mammalian spinal cord and their examination by experimental degeneration, *Proc. Roy. Soc. B* **3:**175–188.

Hoff, E. C., 1932b, The distribution of spinal terminals (boutons) of the pyramidal tract, determined by experimental degeneration, *Proc. Roy. Soc. B* **3:**226–237.

Joseph, B. S., 1973, Somatofugal events in Wallerian degeneration. A conceptual review, *Brain Res.* **59:**1–18.

Kalil, K., 1975, A study of so-called "retrograde fine-grain" degneration in the thalamus, *Brain Res.* **93:**189–202.

Knoche, H., 1958, Über die Ausbreitung und Herkunft der nervosen Nodulusfasern in Hypothalamus und Retina, *Z. Zellforsch.* **48:**602–616.

Knook, H. L., 1969, Stereotaxic lesioning with microelectrodes, *Psychiatr. Neurol. Neurochir.* **72:**61–64.

Kuypers, H. G. J. M., 1958, Coritcobulbar connexions to the pons and lower brain-stem in man. An anatomical study, *Brain* **81:**364–388.

Lamothe, C., 1977, Distribution of the tract of Lissauer and dorsal root fibers in primate spinal cord, *J. Comp. Neurol.* **172**:529–562.

Leonard, C. M., 1972, The connections of the dorsomedial nuclei, *Brain Behav. Evol.* **6**:524–541.

Leonard, C. M., 1973, A method for assessing stages of neural maturation, *Brain Res.* **53**:412–416.

Leonard, C. M., 1975, Developmental changes in olfactory bulb projections revealed by degeneration argyrophilia, *J. Comp. Neurol.* **162**:467–486.

Lund, R. D., and Westrum, L. E., 1966, Neurofibrils and the Nauta method, *Science* **151**:1397–1399.

Marchi, V., and Algeri, G., 1885, Sulle degenerazioni discendenti consecutive a lesioni sperimentale in diverse zone della corteccia cerebrale, *Riv. Sper. Freniatr. Med. Leg.* **11**:492–494.

Marsden, C. D., 1969, in: *Pigments in Pathology* (M. Wolman, ed.), pp. 395–420, Academic Press, New York.

Mehler, W. R., 1957, The mammalian "pain tract" in phylogeny, *Anat. Rec.* **127**:332.

Mejia, H., 1976, Rate of climbing fiber degeneration in rabbit cerebellum following paraflocular stalk and medullopontine lesions, *J. Comp. Neurol.* **165**:433–456.

Merzbacher, L., 1903, Untersuchungen an winterschlafenden Fledermausen. II. Die Nervendegeneration während des Winterschlafs. Die Beziehungen zwischen Temperature und Winterschlaf, *Arch. Ges. Physiol. Mensch. Tier.* **100**:562–585.

Mesulam, M.-M., 1979, Tracing neural connections of human brain with selective silver impregnation, *Arch. Neurol.* **36**:814–818.

Monckleberg, G., and Bethe, A., 1899, Die Degeneration der markhaltigen Nervenfasern der Wirbeltiere unter hauptsächlicher Berücksichtigung des Verhaltens der Primitivfibrillen, *Arch. Mikr. Anat.* **54**:135–183.

Morest, D. K., and Baptiste, M. J., 1975, Degeneration and phagocytosis of synaptic endings and axons in the medial trapezoid nucleus of the cat, *J. Comp. Neurol.* **162**:135–156.

Nauta, W. J. H., 1950, Über die sogenannte terminale Degeneration im Zentralnervensystem und ihre Darstellung durch Silberimprägnation, *Arch. Neurol. Psychiatr.* **66**:353–376.

Nauta, W. J. H., 1957, Silver impregnation of degenerating axons, in: *New Research Techniques of Neuroanatomy* (W. F. Windle, ed.), pp. 17–26, Charles C. Thomas, Springfield, Illinois.

Nauta, W. J. H., and Gygax, P. A., 1951, Silver impregnation of degenerating axon terminals in the central nervous system (1) technic (2) chemical notes, *Stain Technol.* **26**:5–11.

Nauta, W. J. H., and Gygax, P. A., 1954, Silver impregnation of degenerating axons in the central nervous system: A modified technique, *Stain Technol.* **29**:91–93.

Nauta, W. J. H., and Ryan, L. F., 1952, Selective silver impregnation of degenerating axons in the central nervous system, *Stain Technol.* **27**:175–179.

Nauta, W. J. H., Butler, A. B., and Jane, J. A., 1973, Some observations on axonal degeneration resulting from superficial lesions of the cerebral cortex, *J. Comp. Neurol.* **150**:349–360.

Powell, T. P. S., and Cowan, W. M., 1964, A note on retrograde fiber degeneration, *J. Anat.* **98**:579–585.

Ramón y Cajal, S., 1904, Quelques méthodes de coloration des cylindres, axes, des neurofibrilles et des nids nerveux, *Trav. Lab. Rech. Biol.* **3**:1–7.

Ramón y Cajal, S., 1928, *Degeneration and Regeneration of the Nervous System* (translated and edited by R. M. May), Hafner, New York.

Saavedra, J. P., Mascitti, T. A., and Lloret, I. L. P., 1973, Increased rate of anterograde degeneration in the visual pathway of kittens, *Brain Res.* **50**:265–274.

Schneider, G. E., 1968, Retinal projections characterized by differential rate of degeneration revealed by silver impregnation, *Anat. Rec.* **160**:423.

Schwarcz, R., Hökfelt, T., Fuxe, K., Jonsson, G., Goldstein, M., and Terenius, L., 1979, Ibotenic acid-induced neuronal degeneration: A morphological and neurochemical study, *Exp. Brain Res.* **37**:199–216.

Switzer, R. C., 1976, Neural argyrophilia induced by puromycin: A directed Golgi-like method, *Neurosci. Lett.* 2:301–305.

Switzer, R. C., and Heimer, L., 1976, A direct olfactory projection to area frontalis in opossum, *Neurosci. Abstr.* **2**:164.

Tanaka, D., Jr., 1976, Thalamic projections of the dorsomedial prefrontal cortex in the rhesus monkey (*Macaca mulatta*), *Brain Res.* **110:**31–38.

Tanaka, D., Jr., and Chen, J. Y. C., 1974, Retrograde thalamic degeneration: Observations using a modification of the Fink–Heimer silver impregnation technique, *Brain Res.* **65:**333–337.

Vaccarezza, O. L., Reader, T. A., Paqualini, E., and Pecci-Saavedra, J., 1970, Temporal course of synaptic degeneration in the lateral geniculate nucleus, *Exp. Neurol.* **28:**277–285.

Valenstein, E. S., and Nauta, W. J. H., 1959, A comparison of the distribution of the fornix system in the rat, guinea pig, cat and monkey, *J. Comp. Neurol.* **113:**337–364.

van Crevel, H., and Verhaart, W. J. C., 1963a. The rate of secondary degeneration in the central nervous system. I. The pyramidal tract of the cat, *J. Anat.* **97:**429–449.

van Crevel, H., and Verhaart, W. J. C., 1963b, The rate of secondary degeneration in the central nervous system. II. The optic nerve of the cat, *J. Anat.* **97:**451–464.

van Gehuchten, A., 1903, La dégénérescence dite rétrograde ou dégénérescence Wallérienne indirecte, *Névraxe* **5:**1–107.

Walberg, F., 1971, Does silver impregnate normal and degenerating boutons? A study based on light and electron microscopical observations of the inferior olive, *Brain Res.* **31:**47–65.

Walberg, F., 1972, Further studies on silver impregnation of normal and degeneration boutons. A light and electron microscopical investigation of a filamentous degenerating system, *Brain Res.* **36:**353–369.

Waller, A. V., 1850, Experiments on the section of the glossopharyngeal and hypoglossal nerves of the frog, and observations of the alterations produced thereby in the structure of their primitive fibers, *Phil. Trans.* **140:**423–469.

Westrum, L. E., and Canfield, R. C., 1977, Light and electron microscopy of degeneration in the brain stem spinal trigeminal nucleus following tooth pulp removal in adult cats, in: *Pain in the Trigeminal Region* (D. J. Anderson and B. Matthews, ed.), pp. 171–179, Elsevier North-Holland, Amsterdam.

Wiitanen, J. T., 1969, Selective silver impregnation of degenerating axons and axon terminals in the central nervous system of the monkey (*Macaca mulatta*), *Brain Res.* **14:**546–548.

Wong-Riley, M. T. T., 1972, Changes in the dorsal lateral geniculate nucleus of the squirrel monkey after unilateral ablation of the visual cortex, *J. Comp. Neurol.* **146:**519–548.

Zaborszky, L., Leranth, C., and Pakovits, M., 1979, Light and electron microscopic identification of monoaminergic terminals in the central nervous system, *Brain Res. Bull.* **4:**99.

The Autoradiographic Tracing of Axonal Connections in the Central Nervous System

5

STEPHEN B. EDWARDS and ANITA HENDRICKSON

I. INTRODUCTION

Since the conception of the autoradiographic tracing method in the late 1960s (Weiss and Holland, 1967; Goldberg and Kotani, 1967; Lasek *et al.*, 1968) and its popularization and first applications in the early 1970s (Cowan *et al.*, 1972, Edwards, 1972), it has rapidly become one of the most widely used techniques for tracing neuroanatomical connections. Based on the process of axoplasmic transport and the technique of autoradiography, the autoradiographic tracing method represents a radical departure from past tracing methods that relied primarily on the visualization of degenerating axons. As such, it offers several new and significant advantages, the most

STEPHEN B. EDWARDS • Department of Anatomy, University of Virginia School of Medicine, Charlottesville, Virginia 22908. ANITA HENDRICKSON • Department of Ophthalmology, University of Washington, Seattle, Washington 98195.

important being the selective demonstration of pathways arising from neurons surrounded by passing fibers. This advantage alone has meant that the connections of whole new brain territories could be established for the first time and the connections of many previously examined areas clarified. But the technique also has definite limitations and certain disadvantages relative to other tracing methods, and, if naively used, it can produce misleading results.

The purposes of this chapter are to discuss the advantages, limitations, and technical pitfalls of the autoradiographic tracing method and, most importantly, to present a workable procedure. It is intended primarily for those beginning to use the method, but it should also be useful to the more experienced. The procedures described here should not be considered exclusive or even the best, since there are various ways to carry out many of the steps involved. We know, however, that these procedures, when carefully followed, have led to many successful experiments.

II. THE PRINCIPLES OF THE METHOD

The rationale for the autoradiographic tracing method dates back more than 40 years to observations based on neurons grown in organ or tissue culture (reviewed by Pomerat *et al.*, 1967). From observations of neurons grown under these specialized conditions, many workers described a considerable traffic of particles or organelles as well as a generalized "bulk flow" of materials away from the cell body along its elongated process and somewhat less material returning along the same process in the retrograde direction. Paul Weiss (1961) termed this movement "axonal flow." Technical limitations prevented detailed studies until the early 1960s when radioactive isotopes, especially tritiated amino acids, were introduced to tag materials transported through the axon. Over the succeeding years, the use of radioactive tracers greatly expanded the understanding of the transport process, and by 1970 it was realized that this technique could also be used to trace neuronal connections.

Today, there is considerable information concerning the mechanism of axoplasmic transport including the types of substances transported as well as the rates at which the different substances are transported. This information has been well reviewed (Barondes and Samson, 1969; Grafstein, 1969, 1975; Droz, 1969; Lasek, 1970; Jeffrey and Austin, 1973; Lubinska, 1975; Ochs, 1972, 1974) and will not be treated in detail. Only a few pertinent generalizations will be made in order to establish the theoretical basis for the autoradiographic tracing method.

1. All neurons synthesize proteins and transport them through their axons. This means that the autoradiographic tracing method can be applied anywhere in the central nervous system.
2. Amino acid incorporation into transportable proteins takes place in

the cell body. If it occurs at all in mammalian axons or synaptic terminals, it is undetectable under the conditions of the autoradiographic tracing method. The selective uptake and transport of amino acids by cell bodies as opposed to axons accounts for the chief advantage of autoradiographic tracing over fiber degeneration methods: the ability to demonstrate selectively projections arising from cells surrounded by axons of passage. The selective uptake by cell bodies also means that retrograde transport from synaptic terminals is generally not a complicating factor in interpreting the results.

3. In mammalian systems, transported protein moves at two general rates: 100–450 mm/day for the fast components and 1–20 mm/day for the slow components. In cold-blooded animals, the rates are about half that in mammals and birds. The fast components, which contain a much smaller proportion of the total transported protein than the slow components, are a complex assortment of glyco- and lipoproteins, nucleic acids, enzymes, membrane components, and transmitters. Many of these fast-moving substances reach the synaptic terminals, but some are incorporated into the axon. The slow-moving proteins are mostly soluble and are less varied in makeup. They remain, for the most part, in the axon, although some reach the synaptic terminals.

It is important to stress that because the biological and technical principles on which the autoradiographic tracing method is based are so well known, the autoradiographic tracing method can be manipulated in a logical manner and with predictable results, and the sources of problems can readily be sought and corrected.

III. METHODOLOGY

Use of the basic principles of axon transport and autoradiography to create an effective neuroanatomical tracing method involves the following steps, each of which will be considered in detail in this section.

1. Selecting a radioactive tracer and injecting it into the central nervous system.
2. Sacrificing and perfusing the animal after a suitable survival time.
3. Sectioning the brain and mounting selected sections on glass slides.
4. Coating the mounted sections with liquid emulsion.
5. Storing the coated sections to allow the radioactivity to expose the overlying emulsion.
6. Developing, fixing, rinsing, and staining the slides.

Before beginning, the novice should consult one of the many monographs on autoradiography such as the one by Rogers (1973). The local radiation safety regulations should also be reviewed.

A. Selection of the Radioactive Tracer

The selection of an effective tracer involves two considerations: the isotope and the compound. With regard to the isotope, there are several choices: ^{3}H, ^{14}C, and ^{35}S. Of these, ^{3}H is by far the most commonly used.

There are several good reasons for the widespread use of tritium as opposed to the other isotopes. The primary reason is that tritium is a relatively low-energy isotope. This means that the particle it emits penetrates only a short distance into the emulsion (2 μm at best), and thus it affects only a small area of emulsion. The developed silver grains observed in the emulsion, therefore, reflect the location of the tritium label to within about 1 μm. This relatively high resolution permits accurate localization of the labeled neuronal structures and provides a relatively sharp image of the neuronal structures when they are heavily labeled (Figs. 1–4). In contrast, structures labeled with ^{14}C or ^{35}S appear to have fuzzy and diffuse outlines. If the budget is a consideration, it should also be pointed out that ^{3}H compounds are less expensive than similar compounds labeled with ^{35}S and much less expensive than ^{14}C compounds. About the only possible advantage of using ^{14}C- or ^{35}S-labeled compounds is that, because of their higher emission energy and because they are available in higher specific activities (a measure of the ratio of labeled *vs.* unlabeled molecules), shorter exposure times can be used.*

The choice of compound is more complicated. Because of the diversity of materials transported through the axon, many different kinds of tracer compounds can be used. Furthermore, many of the tracers become incorporated into substances that are transported in a different manner; thus they can give very different results.

Tritiated amino acids, particularly leucine and proline, are the most commonly used tracer compounds, and there are many reasons for this. *First,* following injection into the brain they are taken up quickly by neuron cell bodies. Thus, they do not diffuse great distances and do not usually enter the vascular system. *Second,* the protein into which these amino acids become incorporated is common to nearly all neuron types, and this enables wide application of the technique. *Third,* since leucine and proline are incorporated into most proteins transported from the cell body, their use as radioactive tracers can result in intense labeling of axons and terminal regions. Despite their general applicability, however, there are also clearcut differences in the type of autoradiographic results produced by leucine and proline.

One major, yet generally unappreciated difference between leucine and proline is evident in their labeling of synaptic terminals as compared to axons. This difference was noted in experiments in which similar solutions of

*The two largest suppliers of labeled compounds in the United States are New England Nuclear, 549 Albany Street, Boston, Massachusetts 02218, and Amersham/Searle, 2636 South Clearbrook Drive, Arlington Heights, Illinois 60005.

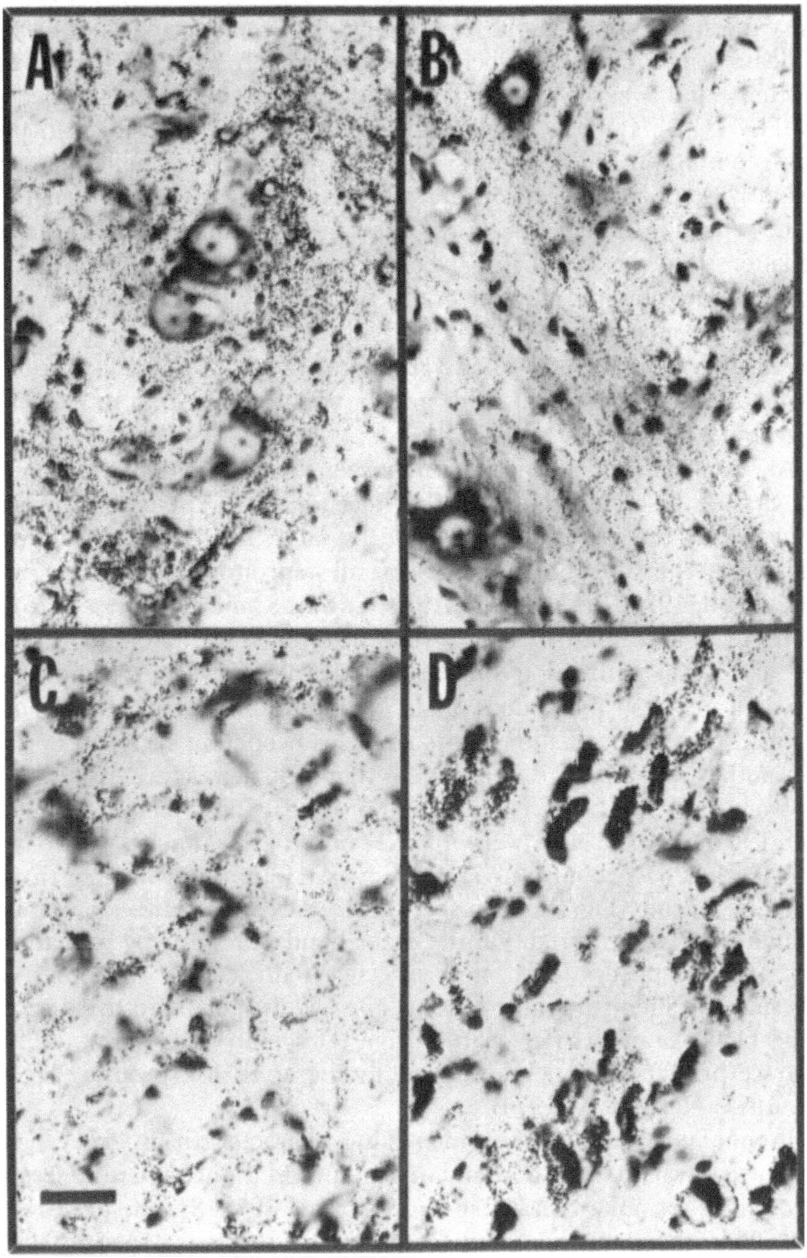

Figure 1. [^{3}H]leucine and [^{3}H]proline, the two most commonly used amino acids for autoradiographic tracing, do not always label axons and terminals with equal effectiveness, as documented in these photographs. After injection of [^{3}H]leucine into one red nucleus and [^{3}H]proline into the opposite red nucleus of a cat, red nucleus terminals in the lateral reticular nucleus are more heavily labeled by [^{3}H]proline (A) than by [^{3}H]leucine (B). However, rubral axons, seen in cross section, are more distinctly labeled by [^{3}H]leucine (D) than by [^{3}H]proline (C). Bar represents 25 μm in A and B and 50 μm in C and D.

leucine and proline were injected, respectively, into the left and right red nuclei of the same animal. The animals were sacrificed after 4–14 days. In these experiments, labeling in terminal fields of the injected nuclei was usually heavier in the proline-labeled system (Fig. 1A,B). Axonal labeling, on the other hand, was always much more distinct in the leucine-labeled pathway (Fig. 1C,D). If these labeling differences between leucine and proline occur generally in the nervous system, the analysis of terminal patterns of previously described projections may be better carried out with proline, and the identification of new pathways more easily accomplished with leucine. The latter is true because lightly labeled fibers can easily be mistaken for the patternless distribution of silver grains that usually indicate a terminal field (Fig. 1). Since leucine labels axons more distinctly than does proline and also labels terminal areas, its use aids greatly in the important distinction that must be made between the axons and terminal areas of an unknown fiber projection.

Another important difference between leucine and proline concerns their uptake by neurons within the injection site. Apparently, proline is taken up less extensively than leucine by large neurons. Leucine, on the other hand, is taken up equally well by all neuron types and sizes (Kunzle and Cuenod, 1973; Berkely *et al.*, 1977; Edwards and de Olmos, 1976). This difference is illustrated in Fig. 2. The difference in uptake of proline by large and small neurons means that if a proline-injected region contains large neurons, their projections might be underestimated or missed entirely. Obviously, careful evaluation of the results is needed in such cases.

Another disadvantage of proline is that it diffuses more easily than leucine from the point of injection. In the previously described experiments in which leucine and proline were injected into the red nuclei, the volume of tissue labeled by proline was always at least four times greater than that labeled by leucine (Fig. 2). Such spread becomes a critical factor when small injection sites are required. On the other hand, for studying retinal projections, proline appears to be more effective than leucine and other amino acids since proline, when injected into the eye, is more effectively excluded by the blood–brain barrier than leucine (Dingman and Sporn, 1959) and therefore produces a much lower autoradiographic background (Crossland *et al.*, 1973; Neale *et al.*, 1974).

Proline possesses another interesting characteristic in that it can be transported transsynaptically and it can proceed through the secondary neuron to its terminals (Grafstein and Laureno, 1973; Specht and Grafstein, 1973). This characteristic, which is shared by fucose—a specific label for

Figure 2. Two major differences between [^{3}H]leucine and [^{3}H]proline can be found at the sites of injection. First, injections of [^{3}H]proline cover a greater volume of tissue than injections of [^{3}H]leucine. This is illustrated in the top photograph which shows a [^{3}H]proline injection site in the right red nucleus and a [^{3}H]leucine injection in the left red nucleus. The injections were nearly identical in concentration, volume, and specific activity, and they were made at the same rate. The second difference is that [^{3}H]leucine is taken up equally well by all types and sizes of neurons (below left), but [^{3}H]proline is taken up preferentially by small neurons (below right).

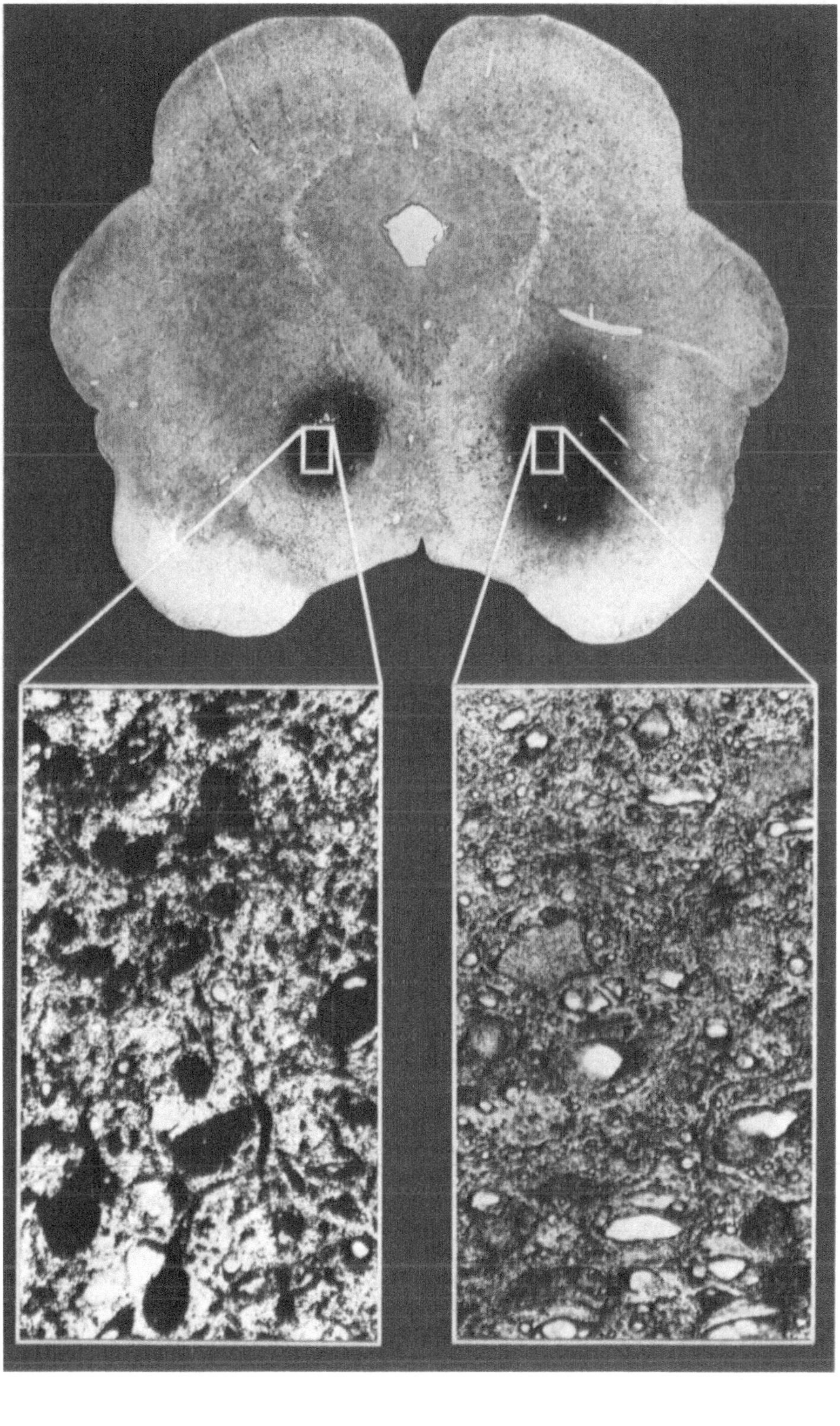

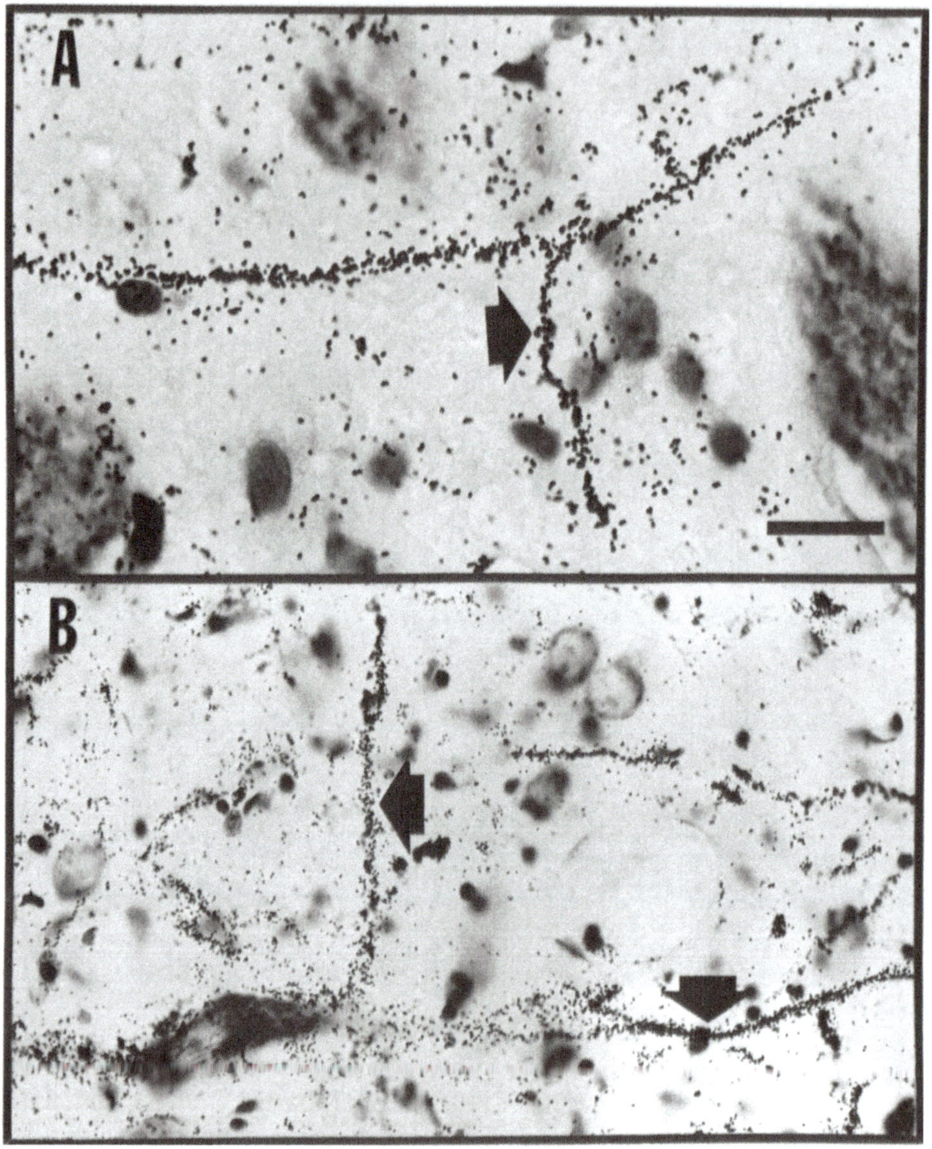

Figure 3. These photographs document a generally unappreciated feature of the autoradiographic tracing method: its ability to demonstrate axon collateralization and preterminal branching patterns. (A) shows a single midbrain reticular axon with a clearly labeled collateral (arrow). (B) shows two midbrain reticular axons (arrows) approaching a spindle-shaped medullary reticular neuron. The juxtaposition of the breakup in labeling of these axons and the faintly stained primary dendrites of the medullary neurons suggests that the axons are climbing along and issuing multiple synapses to the dendrites of the postsynaptic cell. This same terminal arrangement has been observed in Golgi material of the reticular core (Scheibel and Scheibel, 1968). Bar represents 20 μm in A and 30 μm in B.

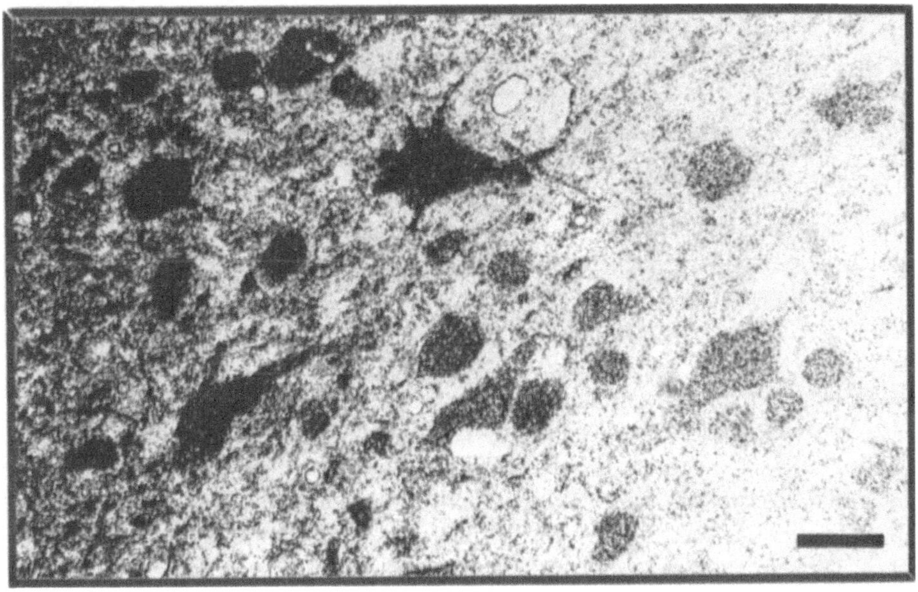

Figure 4. A major problem with the autoradiographic tracing method is defining the injection site or, more accurately, the volume of tissue that contains the neuron cell bodies that have transported the label seen distant from the injection site. This problem is created by the gradual decline in density of perikaryal labeling toward the periphery of the injection site as seen in this photograph of the outer portion of a [^{3}H]leucine injection site in the lateral geniculate nucleus. Bar represents 30 μm.

glycoproteins, can be either troublesome or highly useful, depending on the nature of the study. For example, it is likely to give misleading results if the projections of only the injected region are being studied. On the other hand, if the efferent connections of the structure receiving a projection from the injected area are of interest, then use of proline or fucose can be quite valuable (Wiesel *et al.*, 1974; Dräger, 1974; Kaas *et al.*, 1976). However, to obtain transsynaptic transport in quantities sufficient to study secondary connections, a system must be chosen in which the injected area terminates profusely in the nucleus whose axon distribution is being studied. Also, quantities of 1 mCi or more of proline and fucose are required, and the survival times should be at least 1 month. The large quantities of isotope required make restricted injections into the brain virtually impossible.

The differences in labeling properties between leucine and proline have prompted many to try mixtures of these and other amino acids with the aim of combining the advantages of each. However, such mixtures, or "cocktails," also combine the disadvantages of the amino acids involved and therefore do not always prove more useful than a single amino acid. The most effective amino acid or amino acid combination will depend on the specific fiber system and the requirements of the study, and it is best determined by experimentation.

We have focused attention on the two most commonly used amino acids because they have proved to be effective markers of neuronal pathways and because much is known about their transport. Other amino acids may be equally effective and may also possess unique properties that would provide entirely new ways of studying central nervous connections. Potential tracer compounds other than amino acids are numerous, but little is known about their effectiveness for autoradiographic tracing. These include fucose and glucosamine which label glycoproteins; cholesterol, choline, and glycerol which label lipids; nucleosides to tag nucleic acids; and norepinephrine and 5-hydroxytryptamine which label transmitters. The transmitter compounds are taken up only by neurons utilizing these transmitters. Lipid precursors are not often used for the autoradiographic tracing method because tissue processing for histology produces a severe loss of many labeled lipids and also because lipid precursors freely exchange with preformed *in situ* lipids (Droz, 1975). Nucleosides, such as [^{3}H]adenosine and [^{3}H]uridine, appear to be transported either as the free nucleoside or within nucleic acids. There is disagreement as to whether the transport is orthograde, retrograde, or in both directions and whether these substances move transsynaptically (Schubert and Kreutzberg, 1974, 1975; Wise and Jones, 1976; Hunt and Kunzle, 1976; Kruger and Saporta, 1977; Wise *et al.*, 1978).

B. Injection of the Tracer into the Brain

This step involves concentration of the isotope as it is shipped from the manufacturer and delivery of the isotope into the brain. The discussion will be limited to the [^{3}H]amino acids, since these are the most commonly used tracers, but most comments apply to other compounds as well.

1. Concentration of the Isotope

Tritium-labeled amino acids are usually supplied at a concentration of 1 mCi/ml,* although higher concentrations can be obtained by special order. Amino acids are relatively stable molecules, but they are attacked by the free radicals formed by radioactive decay; the higher the specific activity of the amino acid, the greater this autolysis will be. Most [^{3}H]amino acids are supplied in 0.01 N HCl or 2% ethanol to quench free radical formation. Unless specifically instructed otherwise, store them at 4–6°C. They should not be frozen, since free radical formation is greater in ice than in water. At 4°C in 0.01 N HCl, the rate of decay is about 1%/month for a concentration of 1 μCi/μl. If relatively large volumes can be injected (e.g., 500 μl intravitreal injections), sufficient labeling is obtained using the standard 1 μCi/μl con-

*C or Ci is the abbreviation for curie, a standard measure of radioactivity. A curie is that quantity of any isotope undergoing 3.7×10^{10} disintegrations per second.

centration that comes from the supplier. However, injections made in the CNS require smaller volumes (0.1 to 1.0 μl) to avoid tissue damage and to produce restricted injection sites. To achieve sufficient neuronal label and to avoid excessively long exposures, one must concentrate [^{3}H]amino acid to 10–100 μCi/μl (a concentration of 25 μCi/μl is satisfactory in most situations). If the labeled amino acids are stored at 4–6°C, the loss in concentration by autolysis is about 10%/month (New England Nuclear, personal communication). Since the quenching agent is removed during drying, very concentrated (50–100 ×) solutions should be used within 2 weeks.

Concentration of the amino acids can be done in a variety of ways. The simplest is by heat evaporation in a 55°C oven. Alternatively, the amino acid can be dried at room temperature in a vacuum desiccator using NaOH pellets as dessicant. Generally this is a faster method, but it can easily cause the formation of bubbles that may "boil" the entire amount of amino acid out of the drying tube. The liquid can also be evaporated under a gentle stream of nitrogen with warming of the isotope to speed evaporation. A rapid but more expensive method is to dry the isotope in a vacuum centrifuge equipped to handle small volumes.* Whatever method is used, the isotope, when dry, is redissolved to the desired concentration in sterile water, saline, or buffer. It is important that this be done as soon as the isotope reaches the crystal state in order to avoid excessive degradation of the amino acid. Also, it is advisable to maintain sterile conditions during the reconstituting procedure to avoid carrying infection to the brain during the injection.

2. Injection of the Isotope

The procedures for injection of radioactive tracers are discussed in Chapter 3. Perhaps the easiest method is to use a Hamilton fixed-needle syringe (Chapter 3, Section II.A).

The size of the actual injection site depends on many factors including the diameter of the needle, the volume injected, the rate of injection, and the composition of the tissue into which the injection is made. Reducing the volume and rate of injection will produce smaller injection sites and will also minimize the tissue damage produced by hydrostatic pressure. The denser the cell bodies are at the injection site, the more restricted will be the spread of label. The spread of label during the injection also tends to be restricted by compact fiber systems. In general, 1 μl of [^{3}H]leucine at 25 μCi/μl injected with a 26-gauge needle at a rate of 0.20 μl/min produces a sphere of labeled tissue 1.5–2.0 mm in diameter, whereas 0.4 μl of the same solution injected with a 30-gauge needle at a rate of 0.03 μl/min produces an injection site less than 1 mm in diameter.

A cautionary note is appropriate for those wishing to inject less than about 0.3 μl of solution for the purpose of producing even smaller injection

*SVC 100 Speed Vac, Savant Instrument, Incorporated, Hicksville, New York 11802.

sites. Unless the concentration of the isotope is increased, so little label will be available for uptake by the neurons that it may be difficult to trace the labeled projections, particularly if they are diffuse.

Very small areas or even a single neuron can be labeled by iontophoretic delivery of the isotope. The advantage of iontophoresis is that only the charged tracer molecules are deposited in the tissue. Since no bulk flow occurs, there are insignificant tissue damage and little diffusion of the isotope away from the point of deposit. Iontophoresis, however, requires more equipment than the pressure injection method, and it is prone to unexplained failures in which no isotope is deposited. The iontophoretic technique is described in detail in Chapter 3.

C. Survival Time

Because the rates and destinations of axoplasmic transport are well known, postinjection survival time can be determined in a rational manner. The appropriate survival time will depend mainly on the species used, the length of the fiber system being studied, and the type of data desired. If, for example, identification of terminal fields is of primary concern, a relatively short survival time can be used, one just long enough to allow the fast component of transport to proceed. Since the fast component travels about 250 mm/day in mammals, a suitable survival time can be obtained by dividing this figure by the estimated length of the fiber system in millimeters. Though electron microscopic evidence suggests that most of the fast component of transport ends up in the terminals (Schonbach *et al.*, 1971, 1973; Hendrickson, 1972), it is clear that this component also labels axons. The axons, furthermore, are often labeled indistinctly (cf. Figs. 1C and 5B) and therefore can easily be mistaken for terminal fields. Care must be taken in interpreting the labeling patterns in experiments with short survival times, especially when new projections are being defined.

Perhaps the best way to distinguish terminal fields from axons and to get the fullest appreciation for the trajectory and structural details of the fiber system being studied is to use a survival time long enough to allow both the fast and slow components of transport to proceed through the axons. The slow component travels at 1–20 mm/day in mammals, and in calculating the survival time, it is probably wise to use a rate of 1 mm/day. The voluminous slow component of transport distributes heavily to the axon and also delivers additional material to the terminals. The results are heavily labeled terminal fields plus more clearly labeled axons. In such material, fine collaterals and thin branching points can sometimes be identified and the mode of termination of labeled afferent fibers on the postsynaptic cell indicated (Fig. 3). The clarity of axonal labeling is maximized when [^{3}H]leucine is used together with long survival times.

In studies using fish, reptiles, or amphibians, survival times should be based on a rate of 100 mm/day for the fast component and 0.5 mm/day for the slow component of transport.

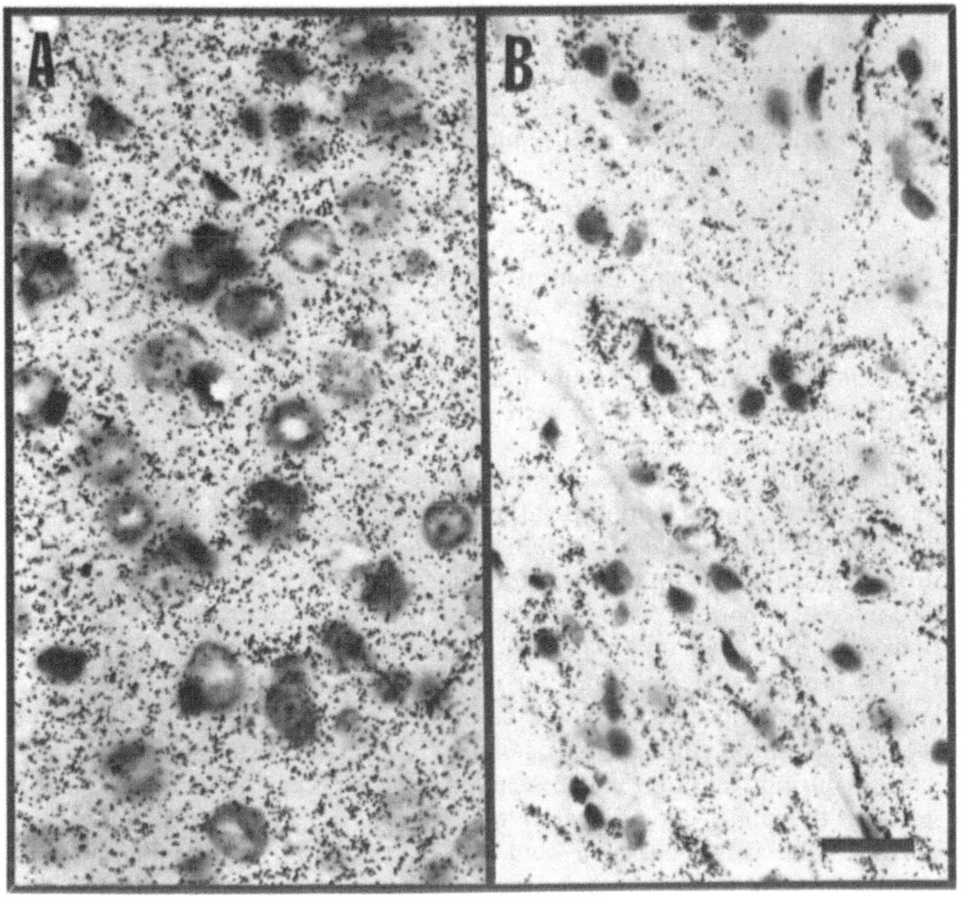

Figure 5. These photographs illustrate the remarkable similarity in grain patterns that can occur between a labeled terminal field and indistinctly labeled fibers. The silver grains in (A) represent labeled geniculate terminals and preterminal arbor in layer 4 of cortical area 17, whereas the grain pattern in (B) represents labeled fibers in the subcortical white matter of the same experiment. When such indistinct axonal labeling occurs in a neuronal region, it can easily be mistaken for a terminal field or give the impression that a terminal field is denser than it really is. Bar represents 30μm.

D. Perfusion and Fixation

At the end of the survival period, the animal is sacrificed and perfused with saline followed by a fixative.

Proteins containing an isotope are reliably fixed by a variety of aldehyde fixatives, including formalin, paraformaldehyde, and glutaraldehyde. These aldehyde fixatives will retain 92–98% of the labeled protein (Droz *et al.*, 1963). Glutaraldehyde will also bind free amino acids in tissue that is fixed minutes after very high doses of radioactive amino acids (Peters and Ashley, 1967), but since the half-life of free amino acids in brain is 1 hr or less (Lajtha, 1975) and survival times for autoradiographic tracing are usually much longer than 1 hr, the use of glutaraldehyde-containing fixatives generally does not present any problems. Fixatives containing mercury or lead salts should be avoided, since they cause chemical reduction of the emulsion (Kopriwa and LeBlond, 1962). Tissue should be left in fixative at least 3 days before cutting. Since the half-lives of tritium and carbon-14 are so long, the tissue can be stored for a much longer period without risk of diminished autoradiographic reaction.

E. Cutting and Mounting the Sections on Glass Slides

Either frozen or paraffin sections can be used for autoradiography. Frozen sections are easier to prepare if the pieces of tissue are large, and they must be used if the tissue is to be reacted for horseradish peroxidase combined with autoradiography. Paraffin-embedded material, on the other hand, is better suited for sections containing separate pieces of tissue and for serial sectioning. The thin paraffin sections, furthermore, often give better cytological detail than do frozen sections. Finally, because of the tissue shrinkage that occurs during paraffin embedding, it is possible that paraffin sections give a more intense autoradiographic reaction than frozen sections. Both types of sections take about the same time to prepare.

To prepare frozen sections for autoradiography, the tissue is fixed at least overnight and then kept in a solution of 20–30% sucrose in fixative at room temperature until the tissue sinks. The sucrose prevents ice crystal damage during freezing. Frozen sections are cut serially at 10–50 μm and stored in the bins of a plastic tackle box* in 2% formalin. The sections can be stored for many months at room temperature. However, if some of the sections are to be processed for a HRP technique, this should be done within the first few days following sectioning, and the sections must then be stored in buffer.

Frozen sections can be mounted either from alcoholic gelatin or onto chrome alum–gelatin-subbed slides (Hendrickson *et al.*, 1972). Frosted-end "write-on" slides are strongly recommended because in the darkroom they permit easy identification of the side containing the tissue. One should be sure to use a No. 4 pencil to write on the frosted end; the loose graphite particles of softer pencils can contaminate the sections. If frozen sections are defatted before coating, they give more reliable staining, less background, and a more even emulsion coat (Hendrickson *et al.*, 1972). To defat, dry the sections thoroughly after mounting and dehydrate them by passing them through 70%, 95%, and 100% ethanol (two 4-min changes in each). Then place them in xylene for 30 min. Take the sections back through descending (100%, 95%, 70%, 30%) concentrations of ethanol, wash them for 10 min in running water, and allow them to dry.

For paraffin embedding, the brain is cut into small blocks no greater

*Vichel Plastics, Middlefield, Ohio 44062.

than 4 cm^3. These are then dehydrated slowly in a series of increasing concentrations of ethanol; cleared in benzene, xylene, or other clearing agent; infiltrated with warm paraffin; and finally cooled to form a solid paraffin block. A sample embedding procedure useful for cat brain is provided in the Appendix (Section VII.A). The best procedure will vary depending on the species or the brain area to be sectioned. Paraffin sections are cut on a rotary microtome at 5–20 μm. Thirteen-micrometer sections have been found optimal for cat brain.

Paraffin sections can be mounted on slides from a 46°C water bath containing 0.2% gelatin and 0.06% chromium potassium sulfate as an adherent. After drying overnight in a 37°C oven, the mounted paraffin sections are dewaxed by passing the slides through four changes of xylene followed by two changes of absolute ethanol. The conventional procedure then calls for taking the sections through a series of decreasing concentrations of al-

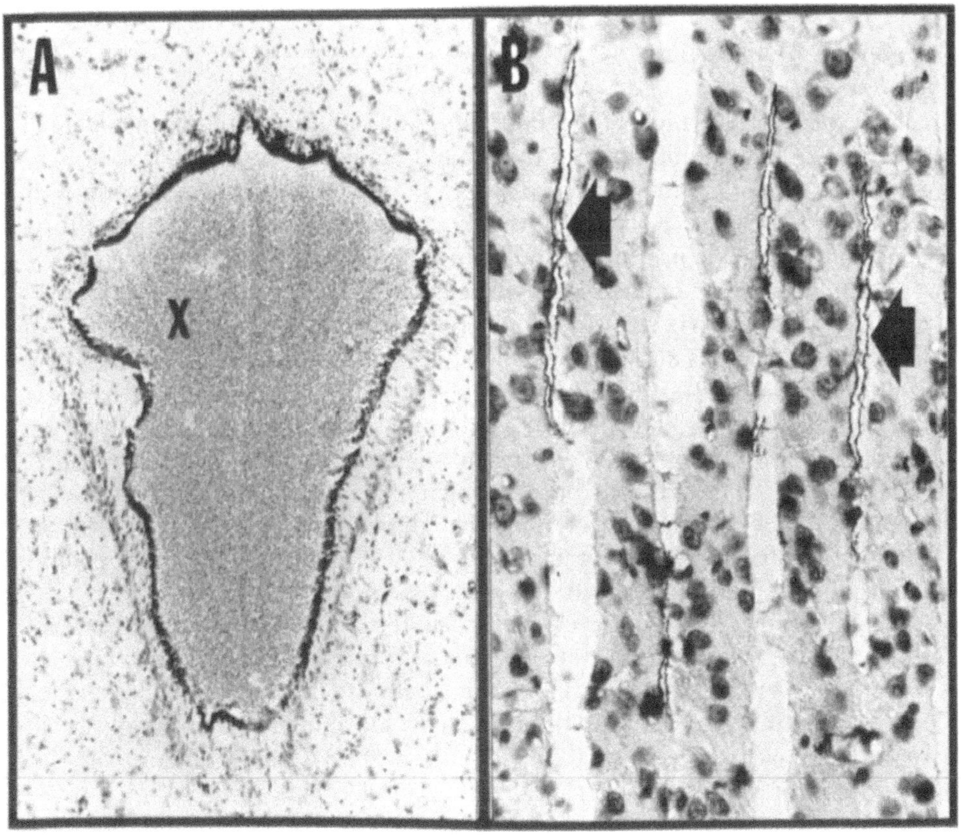

Figure 6. One of the most common artifacts in autoradiographic tracing occurs as dense grain distributions over discontinuities in the section. Although it does not pose any interpretive problems in places like the cerebral aqueduct (X and A) it can look strikingly like labeled fibers over the valleys of knife chatter (arrows in B). This artifact is caused by mechanical stress of the emulsion as it dries.

cohol to water. However, if the sections have not been fixed well or embedded properly, they may split and detach from the slides in this procedure. This can be prevented by removing the sections from absolute alcohol and placing them in a large, covered staining dish prewarmed to about 70°C. In the bottom of the dish there should be a small amount of water absorbed in a paper towel or cloth. By covering the dish and placing it in a 70°C oven for about 30 min, the alcohol-saturated sections are gently hydrated by high humidity. Allow the sections to cool and dry before dipping them.

Whatever technique is used, it is crucial that sections free of holes and tears be prepared because even minute disruptions in the tissue can produce labeling artifact (see Section IV.B.2; Fig. 6). Critical factors are thorough fixation of the brain, complete sucrose infiltration, a sharp microtome knife, and gentle hydration of paraffin sections during dewaxing.

F. Coating the Mounted Sections

1. Choice of Emulsion

Kodak NTB-2 and NTB-3 bulk emulsions are generally used for light microscopic autoradiography in the United States, whereas Ilford K-2, K-5, and G-5 emulsions are more common in Europe.* This discussion will mainly concern the Kodak emulsions. An excellent discussion of Ilford emulsions and their use can be found in Rogers' textbook (1973).

A photographic emulsion is a suspension of sensitized silver halide crystals in a gelatin matrix. All emulsions are unstable, and their shelf life is 1–2 months at 4°C. They should be stored at 4°C and never frozen. When a new emulsion is received, it should be tested by melting a small amount, coating a test slide containing a tissue section, drying, and developing after a 24-hr exposure. Silver grains that appear in an unexposed emulsion are called "background." They result from stress, gamma rays, or spontaneous events. Acceptable background varies somewhat depending on laboratory criteria but will probably be 1–3 grains/1000 μm^2. Excessively high background emulsions should be discarded.

Although either NTB-2 or NTB-3 emulsions will give satisfactory results, NTB-3 has certain advantages for autoradiographic tracing. First, it has a larger grain size which makes the labeled structures more impressive. Second, it is more sensitive, and this means a more intense autoradiographic reaction for a given exposure time. Because NTB-3 is more sensitive, one would expect a higher background with extended exposure times. In practice, however, there is no significant difference in background between NTB-3 and NTB-2 with the 2- and 6-week exposure times commonly used for autoradiographic tracing.

*Kodak emulsions can be obtained from Special Products Division, Eastman Kodak, Rochester, New York 14650. Ilford emulsions are available in the United States through Polysciences Incorporated, Warrington, Pennsylvania 18976.

2. Darkroom Requirements

The darkroom should meet certain minimum requirements. *First,* the room should be clean and dust-free. Preferably, the room should be used for autoradiography only, because inadequately covered photographic fluids can contaminate the emulsion during dipping and drying. *Second,* the room should have a double-door system so that a person can go in and out without exposing the darkroom to light. *Third,* the room must be perfectly light-tight. This feature should be checked from all angles, including the floor, by a fully dark-adapted observer. All electrical equipment connections and switches should be checked for sparking because the autoradiographic emulsion is very sensitive to light flashes. *Fourth,* the darkroom should be slightly cool (17–20°C) and humid (60–80% relative humidity). Humidity can be increased with an inexpensive "cool mist" type humidifier sold for home use.

Recommended safelight conditions for Kodak NTB emulsion vary in published reports from total darkness to use of a Wratten 1A filter over a 15-watt bulb. The latter gives off so little light as to be almost useless. However, we have found the Thomas Duplex-Super Safelight®* extremely valuable. The sodium light emitted by this safelight permits easy visualization of all autoradiographic procedures but does not affect the emulsion. It is used with the red-taped glass filters in the vanes and the yellow-taped glass filters inside the lamp. If the lamp is not fully warmed up, the sodium tube can emit wavelengths that will expose the emulsion; therefore, the safelight should be turned on for 30 min before the emulsion is exposed to it.

Additional equipment needed in the darkroom is listed in the Appendix (Section VII.B).

3. Coating Procedure

A working solution of Kodak emulsion consists of a 1 : 1 dilution with distilled water and is made up as follows. Fill the dipping chamber halfway with distilled water and cover. Under safelighting, place the chamber, the bulk emulsion, and the capped jar of distilled water in the water bath pre heated to 45°C. If necessary, begin to bring the temperature and humidity of the darkroom to proper levels. It takes 30–45 min for the emulsion to melt. Fill the dipping jar to the top with melted emulsion and stir gently with a clean glass rod to mix the emulsion with the distilled water. Return the stock emulsion to the water bath. Allow about 15 min for the bubbles that are inevitably produced by this mixing to rise to the surface. Remove the bubbles with a Q-tip® and by dipping several clean blank slides into the emulsion. Do not begin dipping the experimental slides until dipped blank slides appear bubble-free when examined under the safelight.

*Thomas Instrument Company, 1313 Belleview, Charlottesville, Virginia 22901.

Arrange or prearrange the experimental slides in pairs, back-to-back. The fronts of the slides can easily be identified in the darkroom by the presence of the frosted end. Dip the slides, back-to-back, into the emulsion and withdraw them at a moderate and steady rate. The faster the slides are withdrawn, the thicker the emulsion coat will be. Separate the slides and place them briefly on end on an absorbent towel to drain excess emulsion. Wipe the backs of the slides on a towel and place them vertically in a drying rack. Make sure the slides are not touching each other when placed in the rack because the drying emulsion may stick them together. Do not allow anything to touch the emulsion on the front of the slides after the dipping.

At intervals, check the dipped slides under the safelight to determine if the dipping chamber contains enough emulsion to coat the tissue on the slides. If the emulsion in the chamber is too low, replenish it with one part bulk emulsion to one part distilled water, the latter taken from the capped jar in the water bath. Mix gently and check again for bubbles.

Allow the slides to dry about 3 hr. It is important that the drying not be hurried, since this stresses the emulsion and produces artifacts. Controlled drying is achieved by the cool and humid darkroom conditions.

Place the dried slides in black plastic boxes with a packet of desiccant and seal the boxes with black plastic tape. Return the unused bulk emulsion to a 4°C refrigerator for storage and discard the remaining diluted emulsion. Acid clean the dipping chamber and keep it tightly covered when not in use. Bulk Kodak emulsions can be remelted for several coating sessions without adverse effects.

This procedure is useful when one is dipping large quantities of slides that require frequent refilling of the dipping chamber. If less than about 30 slides are to be dipped, an appropriate amount of unmelted bulk emulsion can be scooped from the shipping jar into the dipping chamber and diluted 1:1 with water. This avoids having to melt the entire bottle of emulsion for each coating session.

G. Exposure of the Emulsion

Autoradiography provides a cumulative record in the emulsion of the radioactive decay within the tissue over the period of time that the emulsion is in contact with the section. During this exposure time, the β particles given off by the isotopically labeled molecules in the tissue interact with the silver halide crystals in the emulsion to form an unstable condition called the latent image. A latent image is converted to a real image in the form of silver grains when it is exposed to the developer. Exposure conditions must be suitable for the creation of latent images, the preservation of these latent images, and the suppression of latent image formation not caused by radioactive decay in the tissue, i.e., the "background."

Since the work of Kopriwa and LeBlond (1962), most exposures have been carried out at 4–6°C. Recently Boren *et al.* (1975) have reexamined the question of temperature during exposure over the range −192° to 37°C. They found that below −20°C there was a severe loss of emulsion sensitivity, and very few grains were produced. Between −20°C and 24°C there was an almost linear increase in the number of grains, while above 24°C the number dropped off sharply, presumably because of a combination of high humidity and melting of the emulsion. If Kodak emulsions are exposed under high humidity, regardless of temperature, severe latent image fading occurs; thus there is need for fresh dessicant during exposure. A good compromise exposure temperature is 4°C. At that temperature the emulsion is reasonably sensitive, and the humidity can easily be kept at low levels by including a desiccant in the slide boxes. However, if slides are exposed at room temperature in a desiccator or with large amounts of desiccant in the boxes, Boren *et al.*, (1975) found that exposure at 24°C will produce 50% more grains in the same period of time than exposure at 4°C with no increase in background. We have repeated this experiment with uniformly ^{3}H-labeled liver tissue embedded in epon and found that there was a 19% difference between the grain counts of tissue exposed at 8°C (refrigerator) and at −15°C (freezer). This significant difference with no difference in background supports Boren *et al.*'s (1975) suggestion that exposures should be done at the warmest possible temperature.

In addition to heat and humidity, some chemical substances can either produce latent images (positive chemography) or cause fading of formed latent images (negative chemography). When autoradiography is first done or when some major aspect of an established technique is changed, it is wise to include both a light-flashed slide containing a sample section and a coated section of similarly processed but nonradioactive tissue to check for chemography during exposure.

Determination of the optimal exposure time is best accomplished by running a series of test sections taken from areas suspected to contain labeled axons or terminals. These sections should be developed at regular intervals (e.g., 2, 4, 6, and 8 weeks) after dipping. When the terminal areas and axons in these sections are distinctly labeled and the background is low (<5 grains/1000 μm^2), the experimental slides can be developed. For most experiments this occurs between 3 and 6 weeks. The main dangers with a too-short exposure time are that areas of light termination might be confused with background and lightly labeled axons might be mistaken for terminal areas. Since, during exposure, background increases relatively little compared to the isotopically labeled areas, extension of the exposure time usually produces more readily interpretable results.

H. Development and Fixation of the Emulsion

Development of the latent images in the emulsion is a critical step, since it determines the final appearance of the latent images that have formed in the emulsion. There are several important variables in the development process, each of which can influence the developed image. These are the type of developer, its concentration, its temperature, the length of development, and the degree of agitation.

We have experimented with several developers, including D-19®, D-72®, Microdol®, D-8®, and HC-110® at different concentrations, temperatures, and times. The results showed that fresh concentrated D-19® at 12–15°C gave the best results in the sense that the grains were fully developed and there was sharp contrast with low background. Development at temperatures higher than 15°C noticeably increased the background. Other developers produced either too fine grains (Microdol®) or high background (D-8®). The optimum development time (without agitation) was 3.5 min. Shorter times produced underdevelopment, and longer times produced larger grains with an increase in background.

Whatever development procedure is used, it is important to keep it constant to insure comparable results among slides of the same or different experiments. This requires attention not only to the time and temperature of development but also to keeping the developer and the fixer fresh. To insure the freshness of these solutions, one should change them after every 75 slides.

After development, transfer the slides immediately to a stop-bath solution of distilled water with a few drops of glacial acetic acid. Rinse the slides briefly by agitation and place them in a fixer for 3 or 4 min at 12–15°C. Various types of photographic fixers are suitable for autoradiography, e.g., 30% sodium thiosulfate (hypo), undiluted Kodak hypo, and undiluted Kodak Rapid Fix®. A fixer that contains a hardener for the gelatin in the emulsion (Kodak hypo, Rapid Fix® with hardener added) will provide greater protection against slippage or scratching of the emulsion but can also make subsequent staining of the tissue more difficult. The hardener can be left out without any adverse effects.

Thorough rinsing after fixing is absolutely essential. Thiosulfate in fixer becomes adsorbed to the surface of the grains and, unless removed, will gradually and finally destroy the grains. Traces of fixer also cause major staining irregularities.

Rinse the slides in running tap water at least 8 hr and then in several changes of distilled water. If the tissue comes loose from the slides with this procedure, as is likely with frozen sections, the slides can be treated with Kodak Hypo Clearing Agent® for 5 min, rinsed in several changes of distilled water, and allowed to soak in distilled water for about 1 hr.

I. Staining of the Tissue

A variety of stains are useful for the autoradiographic tracing method, and they have been discussed in several papers (Thurston and Joftes, 1963; Sidman, 1970; Hendrickson *et al.*, 1972). Staining can be done before coating, after developing, or both. The most widely used stains are those that stain cell bodies, such as cresyl violet, toluidine blue, and thionin. These are applied after development and rinsing. A cresyl-violet-staining protocol suitable for cat or rat paraffin sections is presented in the Appendix (Section VII.C).

IV. ANALYSIS OF THE DATA

A. Definition of a Labeled Pathway

The value of the autoradiographic tracing method is its ability to display accurately the origin, course, and termination of a fiber system. This information is displayed in the form of developed grain patterns in the emulsion overlying the tissue. The grain patterns can differ depending on the labeled structure that produced them, or they can be similar regardless of the type of labeled structure. To be able to extract accurate neuroanatomical data from the grain patterns, it is imperative that they be interpreted carefully.

1. Defining the Origin of a Pathway

That cell bodies and not axons take up injected amino acids means that the origin of the labeled fibers and terminals in an experiment can be assigned accurately to the labeled cells within the injection site. It further means that possible topographical arrangements in the projections can be explored by making small injections in various parts of a structure, provided it is large enough. In these ways, the autoradiographic tracing method is more useful than fiber degeneration techniques or even the orthograde use of horseradish peroxidase in establishing the origin of a fiber pathway. With these latter methods, the lesion produced or the enzyme injected can affect both cell bodies and axons of passage. The retrograde use of horseradish peroxidase, on the other hand, has the advantage of being able to demonstrate the cells of origin that project to a specific locus.

Although the selective uptake of amino acids by cell bodies is the basis for one of the major advantages of the autoradiographic tracing method, it is sometimes a disadvantage. For instance, the autoradiographic tracing method is impractical for studying pathways of wide origin such as corticofugal systems or ascending spinal pathways because of the large amount of tissue that would have to be injected. The course and distribution of such fiber systems, however, can be studied by the aid of degeneration methods following transection of the axons where they form a compact bundle or by taking advantage of the orthograde transport of HRP following an HRP injection into the bundle.

Although the labeled fibers and terminals can be attributed to labeled cells within the injection site, it is difficult to define the boundaries of the injection site. More precisely, it is not known how heavily labeled a cell body must appear in order to be sure that it has transported a detectable amount of label through its axon. Within the center of the injection site neurons appear heavily labeled, and it is reasonable to assume that they have contributed to the transported label seen distant to the injection site. Proceeding outward from the center of the injection site, however, the neurons show a gradual decline in labeling density until background levels are reached (Fig. 4). To complicate matters further, the apparent size of the injection sites will increase dramatically with exposure time. When exposure times are increased, the lightly labeled periphery of the injection site appears more densely labeled, and it is probably fair to assume that the axons and terminals of these cells are now visibly labeled. However, since it is not definitively known how exposure time affects the labeled appearance of cells within the injection site as well as their axons and terminals, it is important to specify the length of exposure used, particularly in the legends to drawings and photographs of injection sites. It is possible that survival time also affects the size of the injection site. With survival times of 2 weeks or more, the injection site seems faded or retracted compared to those of experiments with comparable injections but shorter survival times. This apparent diminution of the injection site could be the result of continued transport of labeled material from the injected cells.

2. Tracing the Course of a Pathway

A complete picture of a neuronal connection requires not only a description of the origin of the pathway but also the course of the axons and the identification of their terminal fields. It is not generally appreciated that the autoradiographic tracing method can give considerable information in this regard. It can demonstrate axon trajectories (Fig. 7) and in some cases give information regarding collaterals, relative fiber size, and preterminal branching patterns (Fig. 3). Also, it should be repeated that it is not only possible to label axons distinctly, but it is important to do so to avoid confusing them with terminal fields. This is especially true when tracing new projections. When axons are lightly labeled or when they are thin, highly branching, or turning in various directions to the plane of section, they can give grain patterns that are indistinguishable from those seen over labeled terminal fields (Fig. 5).

As suggested in Section III, axonal labeling can be maximized by using

Figure 7. These photographs document the differences between dark- and bright-field illumination of autoradiograms. In the dark-field view (A) labeled axons and the labeled terminal area over the oculomotor complex (III) are clearly visible. In the bright-field picture of the same region (B), these labeled axons and terminals are not apparent, but cytological details can now be distinguished. Such pairing of dark- and bright-field photomicrographs is an effective way to document autoradiographic results. int. mlf, Interstitial nucleus of the medial longitudinal fasciculus.

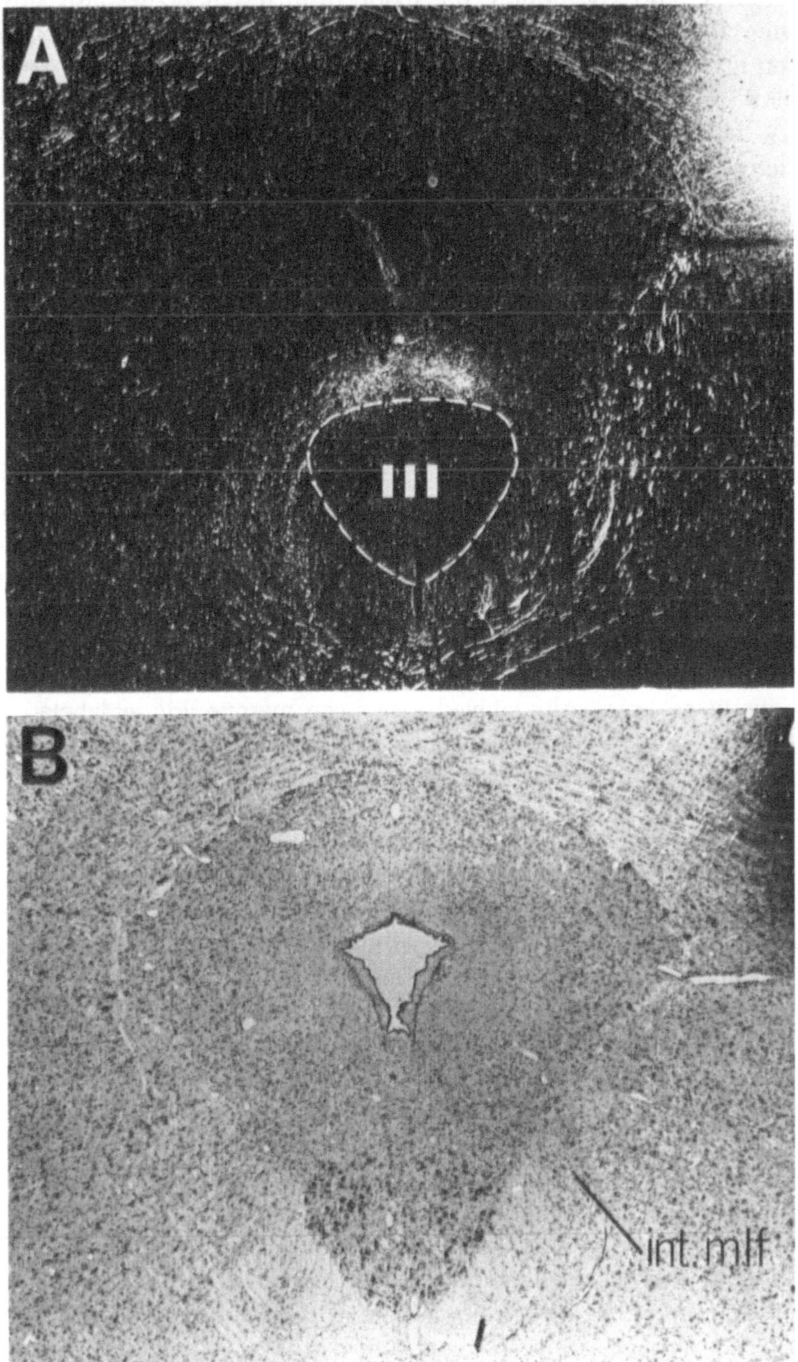
A
III
B
int.mlf

[^{3}H]leucine, survival times long enough to permit both fast and slow components of axonal transport to proceed through the axons, and exposure times of 4 weeks or more. Sectioning the tissue in different planes also aids in identifying and tracing axons. What looks like a terminal field in one plane of section may turn out to be labeled axons cut in cross-section.

The dispersion of β particles in the emulsion makes the grain pattern slightly wider than the labeled fiber it represents. Therefore, it is not possible to determine the actual size of the labeled fiber by measuring the grain pattern in the emulsion. However, if axons are distinctly labeled, one can easily get an impression of their relative widths.

3. Identifying the Terminal Fields of a Pathway

Identifying the terminal fields of a labeled pathway can be a difficult task. The general characteristics of grain patterns produced by labeled terminal fields can usually be appreciated if one examines the grain distribution over known sites of termination (Fig. 5). Terminal areas are always marked by a concentration of single grains, sometimes together with fused clusters of grains. Naturally, the exact character of the grain pattern (size, shape, density, etc.) will be determined by the anatomy of the labeled pathway.

The major problem in identifying labeled terminal regions is to distinguish them from similar grain patterns produced in other ways. We have already discussed how easily labeled axons can masquerade as labeled terminals. But it is also possible to confuse artifacts for labeled terminal fields. For example, it may be difficult to distinguish between background and diffuse terminal labeling, particularly if the former is dense and the latter is light. In this situation, the observer must take into account the overall background of the section and then decide whether there are enough additional grains in a specific region to justify calling it a terminal area. Several factors enter into this decision. For instance, terminal fields are usually not evenly spread throughout a region; there is almost always a focal grain concentration or a grain cluster that is distinguishable by its pattern from the even and random spread of background grains. Furthermore, the same grain distribution should be present in more than one section of the same experiment. A focus of labeling found in a single section could be the result of a localized artifact. Another important feature of a labeled terminal area is that it is often approached by labeled fibers that do not proceed beyond the area in question. This is another good reason for manipulating the technique to label fibers distinctly. Finally, one should keep in mind that a new finding should always be confirmed in additional experiments.

A possible aid in the identification of labeled terminal areas is the counting of silver grains. This involves counting the number of grains in a unit area of the suspected labeled region at 1000 × magnification and then counting the grains in the same unit area over tissue that is outside the suspected labeled region. This background count is then substracted from the count over the labeled region. How much denser a region should be labeled to qualify as a terminal area is arbitrary, but levels of at least four times the background density are usually needed to be convincing. Grain counting is tedious and time-consuming and often of little help in deciding the existence of a terminal area. For purposes of graphic presentation, however, grain count distributions offer an excellent documentation of the extent and intensity of the label. In cases of diffuse, light label, this may be the only way to document the data convincingly. Wider availability of automatic grain-counting devices (Rogers, 1972; Wann *et al.*, 1974; Goldstein and Williams, 1974; Price and Wann, 1975; Woolsey and Dierker, 1978) should make the quantitative method more feasible as a routine procedure.

Whatever method is used, it must be stressed that the mere presence of grain densities higher than background, even when documented with grain counts, is not sufficient evidence for the existence of a terminal area. Consideration must be given to what labeled structures or artifacts produced the grains in question, particularly if the axons in the experiments were indistinctly labeled.

4. Dark- and Bright-Field Illumination

The appearance of isotopically labeled material differs markedly between bright- and dark-field viewing. Dark-field illumination enables easy visualization of labeled pathways and terminal areas at low magnifications (Fig. 7). It is therefore useful for rapid scanning of sections and for photographic documentation of large portions of labeled pathways. Dark-field illumination also makes the injection site appear much larger than it looks in a bright-field view. Bright-field illumination, on the other hand, provides details of the grain pattern together with full appreciation of the cytology of the labeled region (Figs. 2–6). These factors are often obscured under dark-field illumination.

B. Common Artifacts

Although some of the aggravating artifacts listed below have already been mentioned, they will now be discussed in a more detailed and systematic fashion. Most of them can easily be identified and prevented.

1. High General Background

Many silver grains evenly distributed over both section and slide probably constitute the most common and also the most easily recognized artifact. The causes are diverse: old emulsion, a light leak in the darkroom, incorrect safelight, overheating of emulsion during melting, chemical traces in the tissue, and drying of the dipped slides at too-low humidity or too-high temperature. It is because of this capricious artifact that it is unwise to coat all the sections of an experiment. Naturally, there will always be some general background, but if it is kept below 5 grains/1000 μm^2, it should pose no problem.

2. Local Grain Aggregations

These artifacts fall into two categories. The first and more common consists of dense clusters of grains along the edge of the section and over holes and tears in the section (Fig. 6A). This artifact is caused by mechanical stress that occurs within the emulsion as it dries and stretches over discontinuities in the section. When severe, it is found around lysed and contracted cells, over prominent fiber tracts, and along the valleys of knife chatter (Fig. 6B). Around disrupted cells, this artifact can give the appearance of a terminal field. Over knife chatter or long thin tears, it can look strikingly like labeled fibers (Fig. 6B). Stress artifact is prevented by preparing well-fixed, intact, and evenly cut sections and, furthermore, making sure the sections dry slowly at reduced temperatures and increased humidity. If these steps fail, add a few drops of glycerine to the diluted emulsion used for coating (enough to make a 1% glycerine solution). This has proved very effective in reducing stress developed in the emulsion during drying.

The second type of local grain aggregation appears in the form of localized deposits of silver grains related to a specific region or type of cell. This is usually caused by chemical reduction of the emulsion by a substance within the section (positive chemography) and is more commonly seen in the unfixed frozen sections that are used for histochemistry. However, it has also been described in autoradiographic material (Hendrickson *et al.*, 1972; Hackman and Vappatalo, 1972; Daniels and Gilmore, 1975). Chemography can also be generalized throughout the entire section, but it can be distinguished from the other sources of general background by the absence of high background in the emulsion adjacent to the section.

3. Latent Image Fade

This artifact results in the disappearance of the formed latent image during exposure so that a few or no silver grains appear over regions known to be radioactive. It can be caused by a chemical in the section suppressing latent image formation (negative chemography) or by high humidity during exposure. This artifact can be checked by exposing a coated radioactive section that has been light-flashed so that the maximum number of latent images are present from the beginning. If there is a general or localized reduction of grains over such slides, negative chemography is taking place. If experiments routinely produce lightly labeled slides even though reasonable amounts of radioactivity are being injected, this artifact could be the cause.

4. Stain Artifacts

One type of stain artifact involves an interaction of the stain with the emulsion either to cause or to suppress latent image formation. It is very important to use stains in their proper sequence for autoradiography (Thurston and Joftes, 1963; Hendrickson *et al.*, 1972). Since most of the basophilic stains cause chemical reduction of the emulsion, clean alcohols should be used when defatting the sections before coating so that the sections, are not contaminated by stains in these solutions. Another stain artifact is precipitation of the stain itself, causing deposits that resemble silver grains. This seems to be especially prominent with cresyl violet which can form red–brown particles similar to silver grains in size and density. This artifact can be eliminated by using cresylecht violet instead of cresyl violet acetate. Finally, if fixer is not thoroughly rinsed from the slides, it can produce major staining irregularities. A preliminary rinse before staining in buffer of the same pH as the stain and use of freshly filtered staining solutions will help prevent staining problems.

V. ELECTRON MICROSCOPIC AUTORADIOGRAPHY

One of the advantages of the autoradiographic tracing method is that it can be extended to the EM level without significant changes in the technique. Electron microscopic autoradiographic tracing has two main advantages over tracing pathways at the EM level using degeneration following a lesion. Since synaptic terminals remain labeled for long periods after an injection (Cuénod *et al.*, 1972; Hendrickson, 1972), survival times have a relatively wide latitude. Degeneration of synaptic endstructures, on the others hand, is often characterized by a very rapid or erratic course requiring very precise survival times. This may explain why in certain cases EM autoradiography gives a more accurate count of the number of synaptic endings than the analysis of degenerating axon terminals. LeVay and Gilbert (1976) found 28% of the terminals in layer IV of striate cortex were labeled after [^{3}H]proline injection into the dorsal lateral geniculate nucleus instead of the 5–10% found degenerating after dorsal lateral geniculate lesions. Similarly, Dekker and Kuypers (1976) found that [^{3}H]leucine labeled 14 times the number of terminals originating from the mammillary body than did degeneration after lesions. The second advantage is that normal morphology is retained in Em autoradiography, facilitating identification of a specific type of bouton (Fig. 8). Unless degenerating terminals are caught at an early stage, they may be very dense or even engulfed by glial cells; this usually prevents an examination of their postsynaptic relationships. The long exposure time needed for EM, on the other hand, is a disadvantage. In order to keep the exposure time within reasonable limits, the injected amino acid should be as concentrated as possible, preferably 100–500 μCi/μl.

Although there are many excellent reviews dealing with the basic methodology of EM autoradiography (Rogers, 1971; Fischer and Werner, 1971;

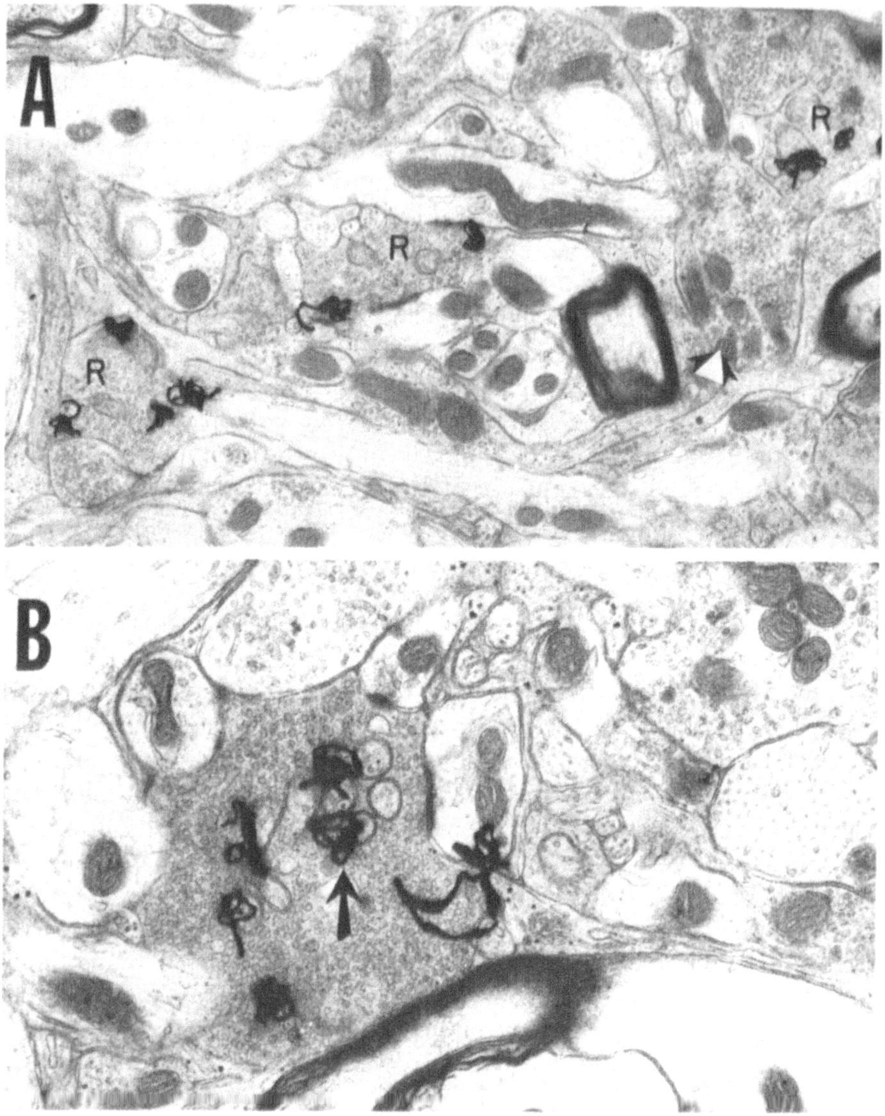

Figure 8. (A) An EM autoradiograph of the rabbit superior colliculus one day after an intraocular injection of ^{3}H-leucine. The silver grains are over retinal terminals (R) which contain pale mitochondria and round vesicles. Another terminal type, which has dark mitochondria, is not labeled (arrow), 14,000×. (B) A retinal terminal labeled with silver grains (arrow) showing the localization of silver grains and the good morphological detail available in EM autoradiography. Rabbit superior colliculus, 32,000×.

Salpeter and Bachmann, 1972; Kopriwa, 1973; Hendrickson, 1975; Droz, 1975), some additional comments are needed in regard to EM autoradiographic tracing of central nervous connections. As in any other type of experimental electron microscopic tracing technique, it is important to realize that a proper selection of tissue samples for electron microscopy is usually impossible without the assistance of light microscopic techniques. Several techniques can be used for accurate sampling of the labeled terminal field. A simple and practical method is to cut alternate 30-μm and 200-μm serial sections, and to use the 30-μm sections for light microscopic autoradiography. Following the light microscopic identification of the exact focus of activity, the focus can be cut out from a nearby 200-μm section and processed for EM autoradiography (Hendrickson, 1975).

Two basic methods are used to coat the sections. The flat substrate method requires putting the thin section on a plastic-coated slide that is then dipped in dilute emulsion. After developing, the emulsion–section–plastic sandwich is floated onto water and picked up onto a grid. This method gives a uniform emulsion coat, and, since section thickness can be measured directly in an interferometer, the method is suitable for quantitative studies (e.g., Salpeter and Eldefrawi, 1973). The major disadvantage is that the emulsion–section–plastic layers have a tendency to stick to the glass slide, and part or all of an experiment can be lost. Salpeter and Bachmann (1972) mention ways to cope with this problem.

The second EM coating method involves coating the section on the grid. Thin sections are cut as usual and picked up on 200- or 300-mesh grids. Two grids are attached at the very edge of a tiny piece of double-sided cellophane tape that is placed near the end of a glass slide. The grids on the slide are coated with a "soap bubble" of emulsion using a wire loop (Caro and van Rubergen, 1962; Hendrickson, 1975). After exposure, the slides are developed, and the grids are removed for staining. This method has the advantage of a >95% recovery rate, but the emulsion coat is often not as regular as with the dipping method. This variation in emulsion thickness induces some uncertainty about resolution parameters, thus making the loop method unsatisfactory for rigid quantitative studies. However, the method is reliable and much easier than the flat substrate method for pathway identification.

In general, the analysis of EM autoradiographs can be straightforward (Hendrickson, 1972). However, if needed, it can also be taken to much more sophisticated quantitative levels (Salpeter and Bachmann, 1972; Salpeter and McHenry, 1973; Parry and Blackett, 1976).

VI. SUMMARY OF ADVANTAGES AND LIMITATIONS

A. Advantages

1. The unique and most important advantage of the autoradiographic tracing method is its ability to demonstrate selectively projections arising from neuron cell bodies surrounded by axons of passage. This advantage is based on the fact that only neuron cell bodies take up and transport injected amino acid. The selective nature of the

autoradiographic tracing method also makes it useful for analyzing the topographic organization of the origins of fiber systems.

2. When the isotope is carefully injected or when iontophoresis is used, there is no spread of the isotope along the needle track. Thus, there is no equivalent of the tissue damage produced by lesioning electrodes used in fiber degeneration studies or the spread of horseradish peroxidase that often appears along the track of the needle used to deposit the enzyme.
3. Since all neurons synthesize and transport protein, the autoraographic tracing method can be used to demonstrate all types of fiber connections including the very fine and unmyelinated or immature systems often hard to visualize with fiber degeneration methods.
4. The easy use of cell stains allows ready identification of the labeled nuclei, nuclear subdivisions, and major fiber pathways.
5. The autoradiographic tracing method is based on well-known biological and technical principles. It can therefore be manipulated in a rational manner and with predictable results. This contrasts with the fiber degeneration methods and the HRP techniques in which many of the steps must be determined empirically.

B. Limitations

1. Because of the difficulty of injecting isotope into broad areas of brain tissue, the autoradiographic tracing method is impractical for tracing pathways of wide origin, such as the corticofugal projections. Such systems are more easily studied with fiber degeneration methods by severing the axons of the system where they are compact.
2. The difficulty in defining the boundaries of the injection site introduces constant uncertainty about the extent of injected tissue that has contributed to the labeled projections observed.
3. The autoradiographic tracing method is not useful for demonstrating short axonal connections, that is, local connection at the periphery of the injection site. This is because the presence of such connections is usually obscured by the halo of diffuse labeling that surrounds all injection sites.
4. Compared to other tracing techniques, the autoradiographic tracing method is lengthy. This means that mistakes take a relatively long time to discover and correct.
5. Because of the unusual techniques required to produce it, transneuronal transport of labeled material is not a common limitation. Nevertheless, its known presence in certain fiber systems demands cautious interpretation of labeling in all cases to insure that secondary projections are not misinterpreted as primary projections.

VII. APPENDIX

A. Paraffin-Embedding Schedule for Cat Brain

40% EtOH	4 hr
70% EtOH	4 hr
95% EtOH	16 hr (overnight)
100% EtOH	2 hr
100% EtOH	2 hr
Benzene	until tissue clears
Benzene	(usually 3–5 hr total in both benzenes)
1:10 Benzene–Paraplast®	16 hr in 60°C oven
Paraplast®	2 hr in 60°C oven
Paraplast®	2 hr in 60°C oven
Vacuum (15 mm Hg)	20 min or until no bubbles drawn from tissue

Place the tissue in a clean paraffin mold.

B. Darkroom Equipment Needed for Emulsion Coating

1. Temperature-controlled water bath set at 45°C.
2. A 30–50 cc dipping chamber made of glass or plexiglass and just large enough to accommodate 1 × 3 inch slides.
3. A small capped bottle containing distilled water.
4. Black plastic slide boxes.
5. A glass stirring rod.
6. 17–20 g of desiccant such as silica gel or Drierite® wrapped in a Kimwipe® or similar material.
7. Black plastic tape.
8. Clean blank slides.
9. Absorbent paper towels.
10. A rack to support slides in a vertical position during drying.

All items should be dust-free, and the dipping chamber and stirring rod should be acid-cleaned and thoroughly rinsed in distilled water.

C. Cresyl Violet Staining for Cat and Rat Paraffin Sections

Cut 5–15 μm paraffin sections.

Stain	5 min at room temperature
D-$H_2$0	rinse, change often
70% EtOH	3 min
95% EtOH	1 min

95% EtOH	1 min
Abs. EtOH	1 min
1 : 1 Xylene : Abs. EtOH	5 min (this step often increases differentiation)
Abs. EtOH	1 min
95% EtOH	1 min
95% EtOH + 0.5% HoAc	differentiate to desired extent
95% EtOH	1 min
Abs. EtOH	1 min
Abs. EtOH	½ min
Xylene	5 min
Xylene	2 min
Xylene	2 min (change often)
Coverslip	

Stain with 0.5% Cresylechtviolet (2.5 g) buffered to pH 3.9 and filtered. (Do not use cresyl violet acetate because it reacts adversely with the emulsion.)

ACKNOWLEDGMENTS. This work was supported by Grant NS11254 to Stephen B. Edwards and Grant EY01208 to Anita Hendrickson from the National Institutes of Health.

REFERENCES

Barondes, S. H., and Samson, F. E., Jr., 1969, Axoplasmic transport, *Neurosci. Res. Prog. Bull.* **3:**307–419.

Berkely, D. J., Graham, J., and Jones, E. G., 1977, Differential incorporation of tritiated proline and leucine by neurons of the dorsal column nuclei in the cat, *Brain Res.* **132:**485–505.

Boren, H. G., Wright, E. C., and Harris, C., 1975, Quantitative light microscopic autoradiography emulsion sensitivity and latent image fading, *J. Histochem. Cytochem.* **23:**901–909.

Caro, L. G., and Van Rubergen, R. P., 1962, High resolution autoradiography, I. Methods, *J. Cell Biol.* **15:**173–188.

Cowan, W. M., Gottlieb, D. I., Hendrickson, A. E., Price, J. L., and Woolsey, T. A., 1972, The autoradiographic demonstration of axonal connections in the central nervous system, *Brain Res.* **37:**21–51.

Crossland, W. J., Cowan, W. M., and Kelly, J. P., 1973, Observation on the transport of radioactively labeled proteins in the visual system of the chick, *Brain Res.* **56:**77–105.

Cuénod, B., Boesch, J., Marko, P., Perisic, M., Sandri C., and Schonbach, J., 1972, Contributions of axoplasmic transport to synaptic structure and functions, *Int. J. Neurosci.* **4:**77–87.

Daniels, J. S., and Gilmore, S. A., 1975, Chemography associated with specific anatomic areas in autoradiographs of brain stems from adult rats, *Brain Res.* **98:**343–347.

Dekker, J. J., and Kuypers, H. G. J. M., 1976, Quantitative E. M. study of projection terminals in the rat's AV thalamic nucleus. Autoradiographic and degeneration techniques compared, *Brain Res.* **117:**399–422.

Dingman, W., and Sporn, M. B., 1959, The penetration of proline and proline derivatives into the brain, *J. Neurochem.* **4:**148–153.

Dräger, U. C., 1974, Autoradiography of tritiated proline and fucose transported transneuronally from the eye to the visual cortex in pigmented and albino mice, *Brain Res.* **82:**284–292.

Droz, B., 1969, Protein metabolism in nerve cells, *Int. Rev. Cytol.* **25:**363–390.

Droz, B., 1975, Autoradiography as a tool for visualizing neurons and neuronal processes, in: *The Use of Axonal Transport for Studies of Neuronal Connectivity* (W. M. Cowan and M. Cuenod, eds.), pp. 127–154, Elsevier, Amsterdam.

Droz, B., Warshawsky, B., and Warshawsky, H., 1963, Reliability of the radioautographic technique for the detection of newly synthesized protein, *J. Histochem. Cytochem.* **11**:426–435.

Edwards, S. B., 1972, The ascending and descending projections of the red nucleus in the cat: An experimental study using an autoradiographic tracing method, *Brain Res.* **48**:45–63.

Edwards, S. B., and De Olmos, J. S., 1976, Autoradiographic studies of the midbrain reticular formation; ascending projections of nucleus cuneiformis, *J. Comp. Neurol.* **165**:417–432.

Fischer, H. A., and Werner, G., 1971, A contribution to electron microscopic autoradiographic technique, *Histochemie* **29**:44–53.

Goldberg, S., and Kotani, M., 1967, The projection of the optic fibers of *Rana catesbiana* as demonstrated by radioautography, *Anat. Rec.* **154**:325–332.

Goldstein, D. J., and Williams, M. S., 1974, Quantitative assessment by photometric reflectance microscopy. An improved method using polarized light, *Histochem. J.* **6**:223–230.

Grafstein, B., 1969, Axonal transport: Communication between soma and synapse, in: *Advances in Biochemical Psychopharmacology, Vol. 1* (E. Costa and P. Greengard, eds.), pp. 11–25, Raven Press, New York.

Grafstein, B., 1975, Principles of anterograde axonal transport in relation to studies of neuronal connectivity, in: *The Use of Axonal Transport for Studies of Neuronal Connectivity* (W. M. Cowan and M. Cuenod, eds.), pp. 47–68, Elsevier, Amsterdam.

Grafstein, B., and Laureno, R., 1973, Transport of radioactivity from eye to visual cortex in the mouse, *Esp. Neurol.* **39**:44–57.

Hackman, C. R., and Vappatalo, H., 1972, Reproducible chemography in autoradiographs of the rat brain, *Experientia* **28**:492–493.

Hendrickson, A., 1972, Electron microscopic distribution of axoplasmic transport, *J. Comp. Neurol.* **144**:381–397.

Hendrickson, A., 1975, Technical modifications to facilitate tracing synapses by electron microscopic autoradiography, *Brain Res.* **85**:241–247.

Hendrickson, A., Moe, L., and Nobel, B., 1972, Staining for autoradiography of the central nervous system, *Stain Technol.* **47**:283–290.

Hunt, S. P., and Kunzle, H., 1976, Bidirectional movement of label and transneuronal transport phenomena after injection of [H^3]adenosine into the central nervous system, *Brain Res.* **112**:127–132.

Jeffrey, P. L., and Austin, L., 1973, Axoplasmic transport, *Prog. Neurobiol.* **2**:207–255.

Kaas, J. H., Lin, C. S., and Casagrande, V. A., 1976, The relay of ipsilateral and contralateral retinal input from the lateral geniculate nucleus to striate cortex in the owl monkey: A transneuronal transport study, *Brain Res.* **105**:371–378.

Kopriwa, B. M., 1973, A reliable standardized method for ultrastructural electron microscopic radioautography, *Histochemie* **37**:1–18.

Kopriwa, B. M., and LeBlond, C. P., 1962, Improvements in the coating technique of radioautography, *J. Histochem. Cytochem.* **10**:269–284.

Kruger, L., and Saporta, S., 1977, Axonal transport of [^3H]adenosine in visual and somatosensory pathways, *Brain Res.* **122**:132–136.

Kunzle, H., and Cuénod, M., 1973, Differential uptake of [H^3]proline and [H^3]leucine by neurons: Its importance for the autoradiographic tracing of pathways, *Brain Res.* **63**:213–217.

Lajtha, A., 1975, Transport and incorporation of amino acids in relation to measurement of axonal flow, in: *The Use of Axonal Transport for Studies of Neuronal Connectivity* (W. M. Cowan and M. Cuénod, eds.), pp. 25–46, Elsevier, Amsterdam.

Lasek, R. J., 1970, Protein transport in neurons, *Int. Rev. Neurobiol.* **13**:289–324.

Lasek, R. J., Joseph, B. S., and Whitlock, D. G., 1968, Evaluation of a radioautographic neuroanatomical tracing method, *Brain Res.* **8**:319–336.

Levay, S., and Gilbert, C. D., 1976, Laminar patterns of geniculocortical projections in the cat, *Brain Res.* **113**:1–19.

Lubinska, L., 1975, On axoplasmic flow, *Int. Rev. Neurobiol.* **17**:241–296.

Neale, J. H., Elam, J. S., Neale, E. A., and Agranoff, B. W., 1974, Axonal transport and turnover of proline- and leucine-labeled protein in the goldfish visual system, *J. Neurochem.* **23:**1045–1055.

Ochs, S., 1972, Fast transport of materials in mammalian nerve fibers, *Science* **176:**252–260.

Ochs, S., 1974, Fast axoplasmic transport of materials in mammalian nerve and its integrative role, *Ann. N.Y. Acad. Sci.* **193:**43–58.

Parry, D. M., and Blackett, N. M., 1976, Analysis of electron microscope autoradiographs using the hypothetical grain analysis method, *J. Microsc. (Oxf.)* **106:**117–124.

Peters, T., Jr., and Ashley, C. A., 1967, An artifact in radioautography due to binding free amino acids to tissues by fixatives, *J. Cell. Biol.* **33:**53–60.

Pomerat, C. M., Hendleman, W. J., Raiborn, C. W., Jr., and Massey, J. F., 1967, Dynamic activities of nervous tissue *in vitro,* in: *The Neuron* (H. Hyden, ed.), pp. 119–178, Elsevier, Amsterdam.

Price, D. L., and Wann, D. F., 1975, The use of quantitative autoradiography for axonal tracing experiments and an automated system for grain counting, in: *The Use of Axonal Transport for the Studies of Neuronal Connectivity* (W. M. Cowan and M. Cuénod, eds.), pp. 155–172, Elsevier, Amsterdam.

Rogers, A. W., 1971, Recent developments in the use of autoradiographic tedhniques with electron microscopy, *Phil. Trans. R. Soc. Lond.* [*Biol.*] **261:**159–171.

Rogers, A. W., 1972, Photometric measurements of grain density in autoradiographs, *J. Microsc. (Oxf.)* **96:**141–153.

Rogers, A. W., 1973, *Techniques of Autoradiography, 2nd Edition,* Elsevier, Amsterdam.

Salpeter, M. M., and Bachmann, L., 1972, Electron microscope autoradiography, in: *Principles and Techniques of Electron Microscopy, Biological Applications, Vol. II,*(M. A. Hayat, ed.), pp. 221–278, Van Nostrand–Reinhold, New York.

Salpeter, M. M., and Eldefrawi, M. E., 1973, Sizes of end plate compartments, densities of acetylcholine receptors and other quantitative aspects of neuromuscular transmission, *J. Histochem. Cytochem.* **21:**769–778.

Salpeter, M. M., and McHenry, F. A., 1973, Electron microscope autoradiography, analysis of autoradiograms, in: *Advanced Techniques of Biological Electron Microscopy* (J. K. Koehler, ed.), pp. 114–151, Springer-Verlag, New York.

Scheibel, M. E., and Scheibel, A. B., 1968, The brain stem reticular core—An integrative matrix, in: *Systems Theory and Biology* (M. Mesarovic, ed.), pp. 261–285, Springer-Verlag, New York.

Schonbach, J., Schonbach, C., and Cuénod, M., 1971, Rapid phase of axoplasmic flow and synaptic proteins: An electron microscopical autoradiographic study, *J. Comp. Neurol.* **141:**485–498.

Schonbach, J., Schonbach, C., and Cuénod, M., 1973, Distribution of transported proteins in the slow phase of axoplasmic flow. An electron micrscopical autoradiographic study, *J. Comp. Neurol.* **152:**1–16.

Schubert, P., and Kreutzberg, G. W., 1974, Axonal transport of adenosine and uridine derivatives and transfer to postsynaptic neurons, *Brain Res.* **76:**526–530.

Schubert, P., and Kruetzberg, G. W., 1975, [^{3}H] Adenosine, a tracer for neuronal connectivity, *Brain Res.* **85:**317–320.

Sidman, R. L., 1970, Autoradiographic methods and principles for study of the nervous system with thymidine-^{3}H, in: *Contemporary Research Methods in Neuroanatomy* (W. J. H. Nauta and S. O. E. Ebbeson, eds.), pp. 252–274, Springer-Verlag, New York.

Specht, S., and Grafstein, B., 1973, Accumulation of radioactive protein in mouse cerebral cortex after injection of ^{3}H-fucose into the eye, *Exp. Neurol.* **41:**705–722.

Thurston, J. M., and Joftes, D. L., 1963, Stain compatible with dipping autoradiography, *Stain Technol.* **38:**231–235.

Wann, P. F., Price, J. L., Cowan, W. M., and Agulnek, M. A., 1974, An automated system for counting silver grains in autoradiographs, *Brain Res.* **81:**31–58.

Weiss, P., 1961, The concept of perpetual neuronal growth and proximo-distal substance convection, in *Regional Neurochemistry* (S. S. Kety and J. Elkes, eds.), pp. 220–242, Pergamon Press, Oxford.

Weiss, P., and Holland, Y., 1967, Nuronal dynamics and axonal flow: II. The olfactory nerve as model test object, *Proc. Natl. Acad. Sci. USA* **57:**258–264.

Wiesel, T. N., Hubel, D. H., and Lam, D. M. K., 1974, Autoradiographic demonstration of ocular dominance columns in the monkey striate cortex by means of transneuronal transport, *Brain Res.* **79:**273–279.

Wise, S. P., and Jones, E. G., 1976, Transneuronal or retrograde transport of [^{3}H]adenosine in rat somatic sensory system, *Brain Res.* **107:**127–131.

Wise, S. P., Jones, E. G., and Berman, N., 1978, Direction and specificity of the axonal and transcellular transport of nucleosides, *Brain Res.* **139:**197–217.

Woolsey, T. A., and Dierker, M. L., 1978, Computer assisted recording of neuroanatomical data, in: *Neuroanatomical Research Techniques* (R. T. Robertson, ed.), Academic Press, New York.

Horseradish Peroxidase 6

The Basic Procedure

W. BRUCE WARR, JOSÉ S. de OLMOS, and LENNART HEIMER

I. INTRODUCTION

Since its introduction in 1971 (Kristensson *et al.*, 1971, LaVail and LaVail, 1972), the use of retrograde axonal transport of macromolecular tracers, particularly horseradish peroxidase (HRP), has proven to be both a reliable and sensitive method for the positive identification of cell bodies giving origin to fiber projections to a given site within the central or peripheral nervous system (for reviews, see LaVail, 1978, and Jones and Hartmann, 1978). The subsequent demonstration that HRP is transported by axons in the anterograde direction (Hansson, 1973; Lynch *et al.*, 1974; Scalia ad Colman, 1974; LaVail and LaVail, 1974; Sotelo and Riche, 1974; Repérant, 1975; Walberg *et al.*, 1976; de Olmos and Heimer, 1977; Robertson, 1977; Streit and Reubi, 1977; Mesulam and Mufson, 1980) and can be readily detected if histochemical processing is optimal extended the utility of this technique even further. Indeed, innovations and improvements in technique as well as

W. BRUCE WARR • The Boys Town Institute for Communication Disorders in Children, Omaha, Nebraska 68131. JOSÉ S. de OLMOS • Instituto de Investigación Medica Mercedes y Martin Ferreyra, Cordoba, Argentina. LENNART HEIMER • Department of Neurology, University of Virginia School of Medicine, Charlottesville, Virginia 22908.

the discovery of new tracer molecules (Kuypers *et al.*, 1977; Schwab and Thoenen, 1977) continue to expand the potential applications of macromolecular tracers in unraveling the intricate circuitry of the nervous system.

The purpose of this chapter is to describe the principles of incorporation and transport of HRP, the basic neuroanatomic applications, and the different methodological steps involved in an HRP experiment. Special attention will also be given to problems related to the analysis of the data. An appendix, finally, contains detailed descriptions of several of the most generally useful histochemical procedures for the demonstration of HRP.

II. BASIC APPLICATIONS

The HRP method has three basic applications.

1. *Identification of sources of synaptic input.* Delivery of HRP to a region containing axon terminals results in the uptake and retrograde axonal transport of the tracer to the neuronal cell bodies that provide synaptic input to the region injected. The use of highly sensitive histochemical methods (e.g., de Olmos *et al.*, 1978) may often reveal the entire extent of a labeled neuron, including its dendritic arbor, axon, and axon collaterals. Delivery of HRP may be usefully combined with an independent injection of a second label, either a radioactively labeled or fluorescent compound, in order to determine the presence of collateral projections by double retrograde labeling (Chapter 8).
2. *Determination of efferent projections.* Injections of HRP into gray matter may result in the substantial uptake of the tracer by neuronal cell bodies in the injection site. These neurons will transport the material orthogradely through the axon to its synaptic terminals. However, such anterogradely transported HRP can be visualized only with histochemical methods of optimal sensitivity. Retrograde transport, with axonal and perikaryal labeling, is also demonstrated in these preparations.
3. *Golgi-like staining of neurons.* Injections of HRP into gray matter may produce, within a few minutes to a few hours, solidly filled, Golgi-like impregnations of neuronal cell bodies, dendrites, and their local axonal ramifications (Hedreen and McGrath, 1977; Keefer, 1978; Vanegas *et al.*, 1978). Dense filling of cell bodies and axons, either by retrograde or anterograde transport, may also result from the direct application of HRP to severed axons within grey matter, tracts, and peripheral nerves (DeVito *et al.*, 1974; Scalia and Colman, 1974; Furstman *et al.*, 1975; Adams and Warr, 1976). As an ultimate refinement of this application, HRP can be iontophoresed intracellularly (Chapter 7).

III. INCORPORATION AND TRANSPORT OF HRP

A. Characteristics of HRP

Horseradish peroxidase is a glycoprotein containing a heme group that is essential for its histochemical activity (Straus, 1974). In the presence of its substrate, hydrogen peroxide, HRP is able to catalyze the oxidation of certain compounds (chromogens), thus yielding a visible reaction product.

The molecular mass of HRP is 40,000 daltons, and its molecular radius is 3.0 nm (Rennke and Venkatachalam, 1979). Some commercial sources of HRP contain variable mixtures of several closely related enzymes or isozymes that differ in the extent to which they are retrogradely transported (Bunt and Haschke, 1978). Unfortunately, most commercial sources of HRP are not identified as to their isozyme content. Nevertheless, empirical studies demonstrate that Worthington's Type HPOFF and Miles and Serva brands of the enzymes result in greater retrograde axonal transport than does Sigma Type VI (Adams, 1977; Hedreen and McGrath, 1977; Keefer, 1978). Boehringer's purest grade HRP is said to be equivalent to Worthington's HPOFF.

B. Diffusion of HRP

The HRP molecule is sufficiently small that, even if it is introduced into the brain nontraumatically by way of certain permeable capillaries or by intraventricular injection, it diffuses through the extracellular space at a minimum rate of approximately 0.5 mm/hr. Clearance from these spaces, which is generally complete by 24 hr, occurs by means of bulk flow of the cerebrospinal fluid, uptake by capillaries (Van Deurs, 1977), uptake and degradation by phagocytic elements such as pericytes (Cancilla *et al.*, 1972) and microglia, uptake and degradation by neuronal perikarya (Broadwell and Brightman, 1976, 1979), and transport from the area following uptake by axon terminals or preterminal axons.

Direct injection of HRP into the brain results in spread of the enzyme that can vary from several millimeters to even centimeters from its site of release, depending on the amount injected, the rate of injection, and the extent of mechanical damage and resulting brain edema (Kim and Strick, 1976; Hedreen and McGrath, 1977; Vanegas *et al.*, 1978). The zone of visible diffusion expands for several hours following an injection, begins to shrink within approximately 18–24 hr, and disappears after 5–7 days. Phillipson (1979) found that following iontophoretic delivery, the HRP diffusion spot reached a maximum size at 4 hr and contracted to approximately one-third this diameter after 24 hr.

A typical injection site consists of a central region in which all tissue elements appear to be filled with HRP surrounded by a larger halo zone in which labeling becomes progressively weaker with increasing distance from the center (Figs. 1 and 2). As will be discussed in Section VI.A.2, it is usually not possible to delineate precisely the portion of the diffusion zone that is the source of observed retrograde labeling.

C. Incorporation of HRP by Neurons

1. Uptake by Terminals

When injections of HRP are delivered nontraumatically, the marker is taken up principally by synaptic terminals (Turner and Harris, 1974; Broadwell and Brightman, 1979) including peripheral endings of sensory ganglion cells (Arvidsson, 1975; Furstman *et al.*, 1975; Broadwell and Brightman, 1976; Persson and Kristensson, 1979). Electron microscopic studies have shown that at various times after injection, HRP appears successively in pinocytotic vesicles, membraneous sacs and tubules, and multivesicular and dense bodies (Fig. 3A). With longer intervals after injection, the HRP disappears from these elements in this same sequence, thus suggesting that the stages of incorporation, packaging, and lysosomal degradation of the tracer involve those ultrastructural organelles.

The uptake of HRP by terminals is coupled directly to their synaptic activity (Heuser and Reese, 1975; Holtzman, 1977). For example, HRP uptake is reduced markedly by inhibition of synaptic activity, either by systemic pentobarbital (Turner, 1977) or by removal of excitatory input from the neurons (Singer *et al.*, 1977), and is increased by stimulation. Cells in a growth phase—whether in a juvenile animal or in one experiencing reinnervation by regenerating (Kristensson and Sjöstrand, 1972) or sprouting axons—take up HRP most readily (Kristensson *et al.*, 1971; LaVail and LaVail, 1972; Walberg *et al.*, 1976).

2. Uptake by Axons

Studies with the electron microscope suggest that pinocytosis of HRP by axons, particularly myelinated ones, is an exceedingly rare phenomenon (LaVail and LaVail, 1974, 1975; Turner and Harris, 1974). Nevertheless, studies with the light microscope suggest that the tracer may enter transiently traumatized axons, such as those in injection sites, from which it then travels both proximally and distally. Remarkably, the HRP taken up by these injured axons appears to diffuse out, since labeling is rarely observed in them, in their parent cell bodies, or in their terminal arborizations 24 hr after the injection (Hedreen and McGrath, 1977; Oldfield and McLachlan, 1977; Phillipson, 1979). Axons subjected to severe trauma, such as crushing or cutting, take up HRP in large amounts (Kristensson and Olsson, 1974, 1975, 1976; Adams and Warr, 1976). The HRP may virtually fill the axon

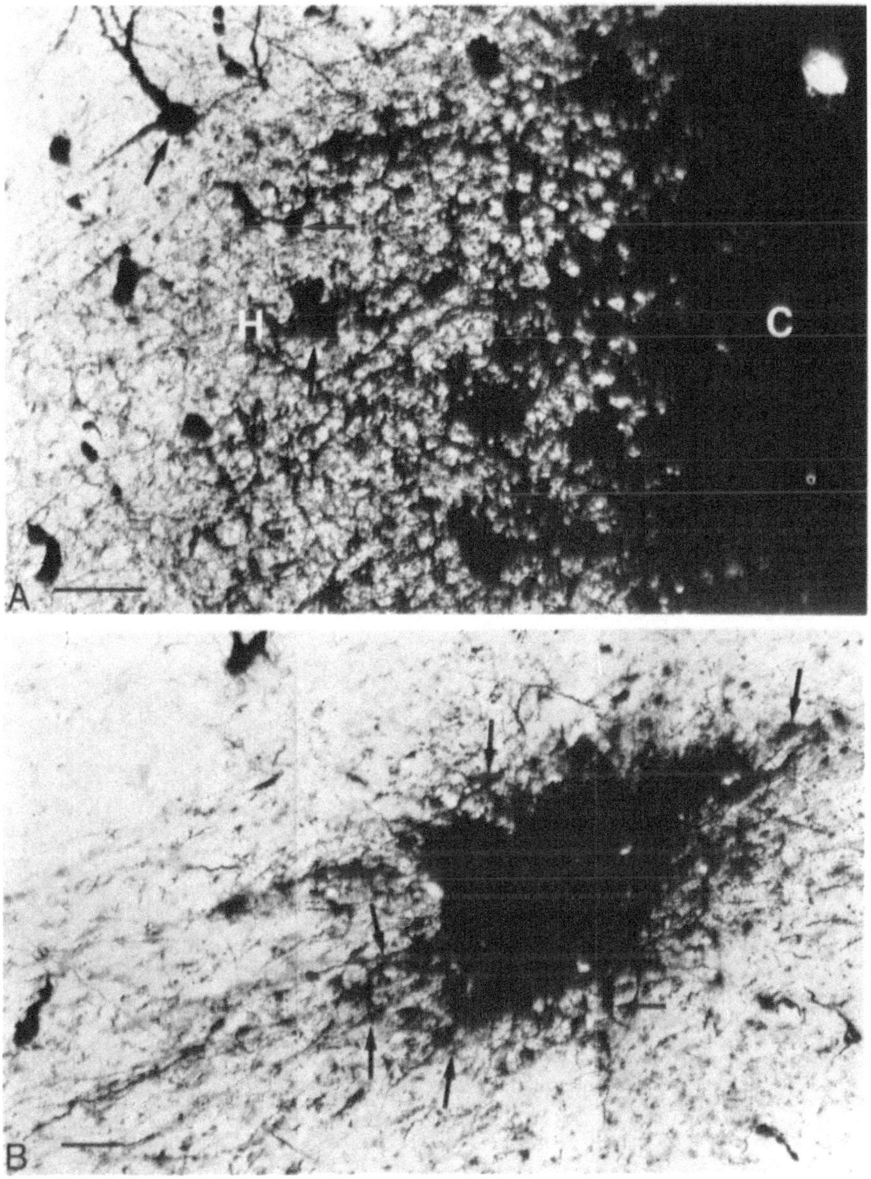

Figure 1. (A) A micrograph of a pressure injection site in the cerebral cortex of a young rat. The core (C) is uniformly dense, and no structural details can be seen. In the halo (H) that surrounds the core, the apparent density of staining decreases with distance from the core. A few heavily labeled neurons (arrows) are present in the periphery of the halo. Postoperative survival time, 18 hr. De Olmos BDHC technique. Bar, 100 μm. (B) An iontophoretic injection site in the rat medial hypothalamus. Many labeled cell bodies (arrows) and dendrites are visible in the halo, which is smaller and less densely stained than in (A.) The injection was made with 1 μA current for a total "on" time of 10 min. Postoperative survival time, 18 hr. De Olmos BDHC method. Bar, 100 μm.

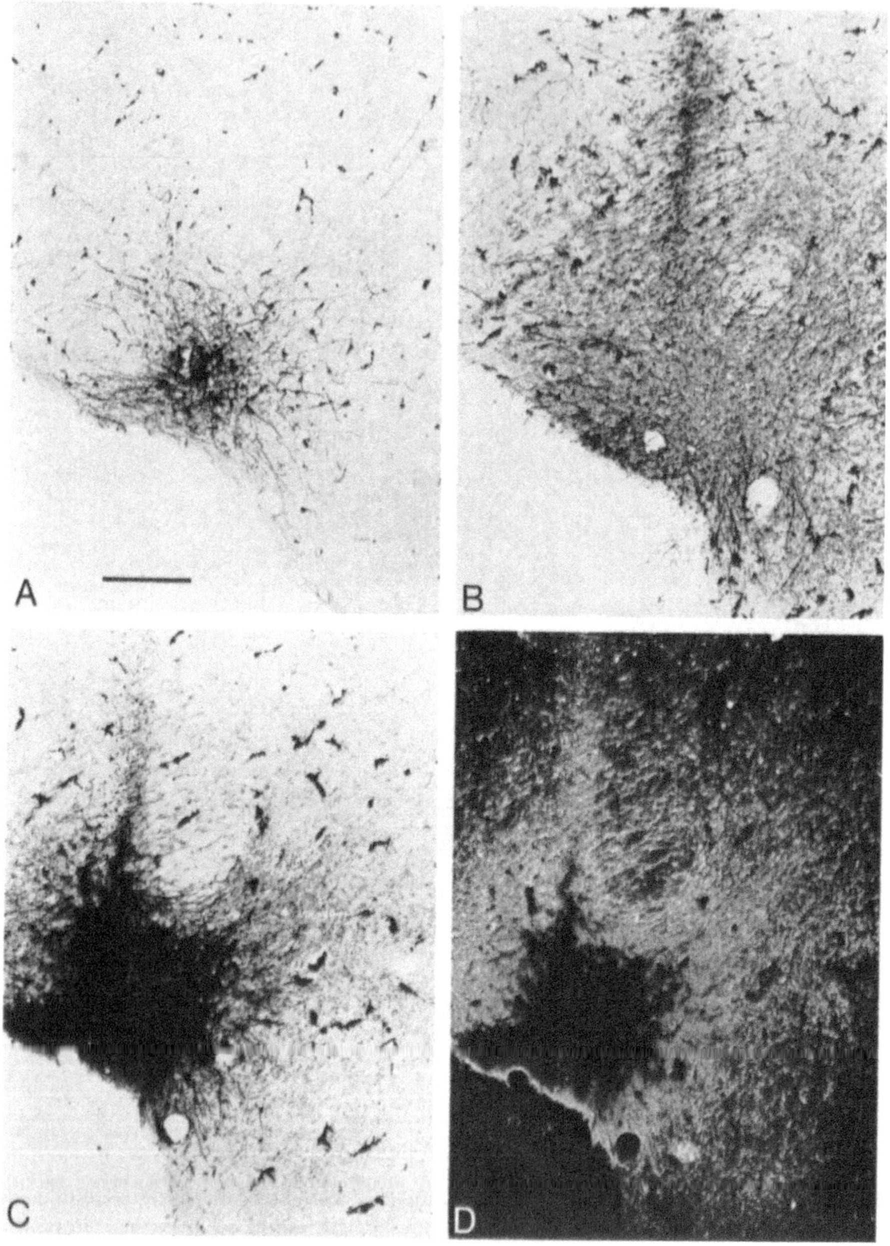

Figure 2. Photographs of an iontophoretic injection site stained by the DAB (A), TMB (B) and BDHC (C and D) methods. Comparison of the sections demonstrates that the choice of HRP technique greatly affects the apparent size of the injection site. The bright-field and dark-field photographs of same section (C and D) reveal that the method of illumination also has an effect on apparent size of injection site. Note that the extent of the injection site in (D) is comparable to that in (B).

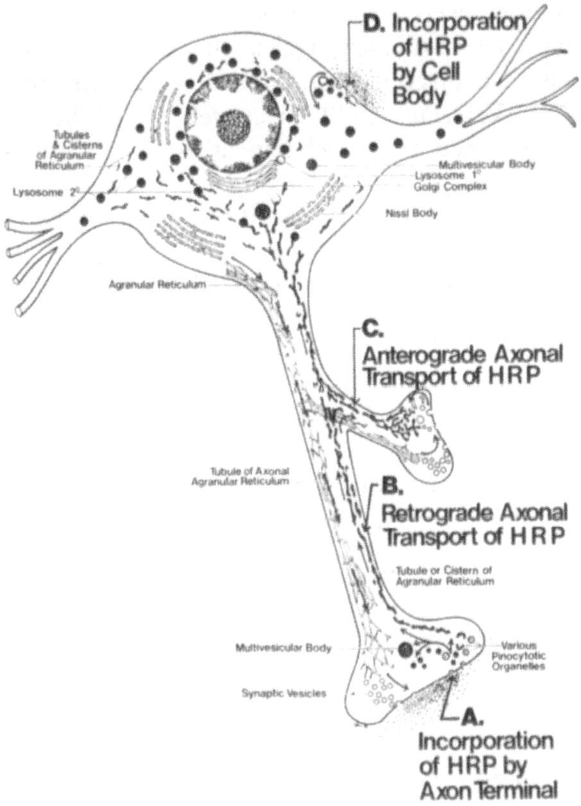

Figure 3. Schematic summary of incorporation, retrograde axonal transport and lysosomal accumulation of HRP (illustrated in black) in neurons. Anterograde axonal transport of HRP from the perikaryon is not illustrated in the drawing. See text for explanation.

for ten or more millimeters proximal and distal to the injury. Much of the label in the proximal stump of the axon is ultimately packaged as previously described and transported to the perikaryon for lysosomal degradation. HRP in the distal stump is transported to the synaptic terminals and is ultimately phagocytized when the axon inevitably undergoes Wallerian degeneration. Anterograde and retrograde filling of traumatized axons can be exploited experimentally to determine the origins and terminations of tracts and peripheral nerves. However, since transport of HRP within severed axons of passage can result in labeling of neuronal perikarya and of axon terminals, these phenomena must be taken into account whenever the HRP is injected in the nervous system for the purpose of tracing connections.

3. Retrograde Axonal Transport

The organelles within which HRP is conveyed to the soma (Fig. 3B) consist primarily of membrane-bound oval or tubular elements and a smaller number of multivesicular bodies (LaVail *et al.,* 1980). Although these structures have sometimes been identified as belonging to the smooth endoplasmic reticulum, recent evidence suggests that they comprise a distinct system of organelles involved in degradative, i.e., lysosomal, functions (Holtzman, 1977; LaVail *et al.*, 1980; Broadwell *et al.*, 1980).

The retrograde transport of HRP appears to depend on the integrity of microtubules, since microtubule-blocking agents, such as colchicine and vinblastine, block their movement to the perikaryon (LaVail and LaVail, 1974; Bunt *et al.*, 1974).

The rate of retrograde transport of HRP in axons of warm-blooded animals, as determined by the relatively insensitive 3,3′-diaminobenzidine (DAB) method, is generally in the range of 70 to 120 mm/day (Hansson, 1973; LaVail and LaVail, 1974), whereas its movement appears to be considerably slower in cold-blooded animals (Wilczynski and Northcutt, 1977). It is likely that the application of more sensitive histochemical procedures to this question would result in rate estimates considerably in excess of those cited above.

4. Fate of HRP Transported to the Cell Body

The HRP-labeled tubules, sacs, and vesicles are transported to the perikaryon where they accumulate in the perinuclear region in close association with the Golgi complex (Fig. 3). They appear to fuse with one another, forming larger structures usually identified as vacuoles, multivesicular bodies, and dense bodies (Sotelo and Riche, 1974; LaVail and LaVail, 1975), the latter of which are in reality lysosomes (Broadwell and Brightman, 1979). The entire soma, including the dendrites, may ultimately fill with HRP-containing lysosomes.

As early as 3 days after its initial arrival in the soma, there is a noticeable reduction in the amount of HRP in the lysosomes as the enzyme undergoes degradation; and between 4–8 days, most lysosomes are depleted of HRP (Turner and Harris, 1974; LaVail and LaVail, 1975; Price and Fisher, 1978). Thus, in terms of practical experimentation, allowing too long a postinjection survival time may produce a spurious absence of HRP labeling in the perikaryon.

There seems to be a consensus among experimental neuroanatomists that there is no significant transport of HRP from a labeled cell body to surrounding neurons.

5. Uptake by Cell Bodies and Anterograde Axonal Transport

Direct uptake of HRP by the neuronal soma occurs almost exclusively at or near injection sites or in regions where the neurons are exposed to high concentrations of the tracer. As illustrated in Fig. 3D, the pinocytotic organelles involved in perikaryal incorporation of HRP are similar to those found in the axon terminal. Moreover, the fate of the HRP is the same: it is delivered to lysosomes where it is gradually degraded. If a massive influx of HRP into the soma occurs, as can be produced with dimethyl sulfoxide (Keefer, 1978) or poly-L-ornithine (Hadley and Trachtenberg, 1978; Itaya *et al.*, 1978), pronounced anterograde axonal transport of HRP follows. It should be noted that anterograde transport is detectable even in cases where cellular damage cannot be demonstrated, as in the retinal ganglion cells after intravitreal injections (Hansson, 1973; Repérant, 1975; Colman *et al.,* 1976; Mesulam and Mufson, 1980). In such instances, the HRP is apparently transported in the same sacs and tubules that are involved in its retrograde transport (Sotelo and Riche, 1974; Broadwell *et al.*, 1980).

The terminal arborizations of the labeled axons are often readily seen with the light microscope, but their parent axons are frequently difficult to discern unless histochemical processing is optimal. Electron microscopic examination of terminals labeled by anterograde axonal transport shows them to contain labeled organelles resembling smooth endoplasmic reticulum, but these are probably derived from the lysosomal system in the cell soma. Synaptic and pinocytotic vesicles, however, do not contain HRP (Fig. 3C). In agreement with this, there are no confirmed instances of exocytosis of HRP at synapses following its anterograde transport, and evidence of such a phenomenon with resultant labeling of postsynaptic neurons, i.e., transneuronal transfer, has so far not been detected at the light microscopic level.

IV. METHODOLOGY

A. Choice of Anesthetic

The influence of systemic anesthetic agents on the uptake and axonal transport of HRP has only begun to be explored; however, preliminary evidence suggests that at sufficient dose levels, both of these phenomena are suppressed by pentobarbital (Turner, 1977; Mesulam and Mufson, 1980; Rogers *et al.*, 1980). In the most pertinent study yet reported, Rogers *et al.* (1980) found that pentobarbital at the relatively high dosage of 50 mg/kg in rats reduced both retrograde cell labeling and anterograde axonal labeling after HRP application to the severed vagus nerve, as compared with animals anesthetized with urethane (1.5 g/kg).

B. Methods of Extracellular Delivery

The method of delivering HRP to the nervous system depends on the purpose of the study. HRP can be applied to the surface of the brain, either in dry form or in solution, or it can be applied directly to the cut ends of nerve fibers. The most common method is to inject the enzyme with the aid of pressure or electrophoresis. Since the techniques of injecting tracers into the nervous system have a wide application in tract-tracing experiments, the equipment and procedures for different types of injections are discussed in detail in Chapter 3.

1. Pressure Injections

When this type of procedure is used, the HRP is usually dissolved in distilled water, sterile saline, or phophate-buffered saline at a concentration of 4–50%. Figure 4 presents a series of functions useful in preparing HRP solutions of various concentrations. The extent of diffusion increases both with the concentration and the volume of injected HRP solution. In regard to the volume, it is illuminating to examine the graph in Fig. 5 which shows the diameter of a sphere containing volumes of the magnitude normally injected. Most authors use a volume of 0.05–0.5 μl, which generally produces diffusion spots with diameters varying between one to several millimeters (LaVail and LaVail, 1974; Beitz and King, 1976; Vanegas *et al.*, 1978)

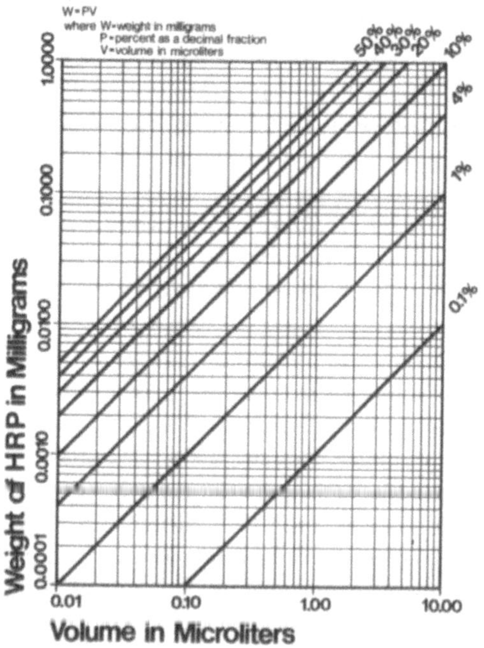

Figure 4. Functions relating weight and volume at various concentration of HRP over the range commonly used in experimentation. These functions are useful in preparing working solutions of HRP and in determining the dry weight of HRP actually injected in a given case when the volume and concentration are known. For example, in order to determine the amount of HRP needed to prepare 50 μl of 20% HRP, find the intersection of the 20% line with the 1-μl line. Read weight to the left (0.2 mg) and multiply by 50 (10 mg). To determine how much HRP, in milligrams, is actually delivered if 0.1 μl is injected, find the intersection of the 20% line with the 0.1-μl line and read answer to the left (0.02 mg).

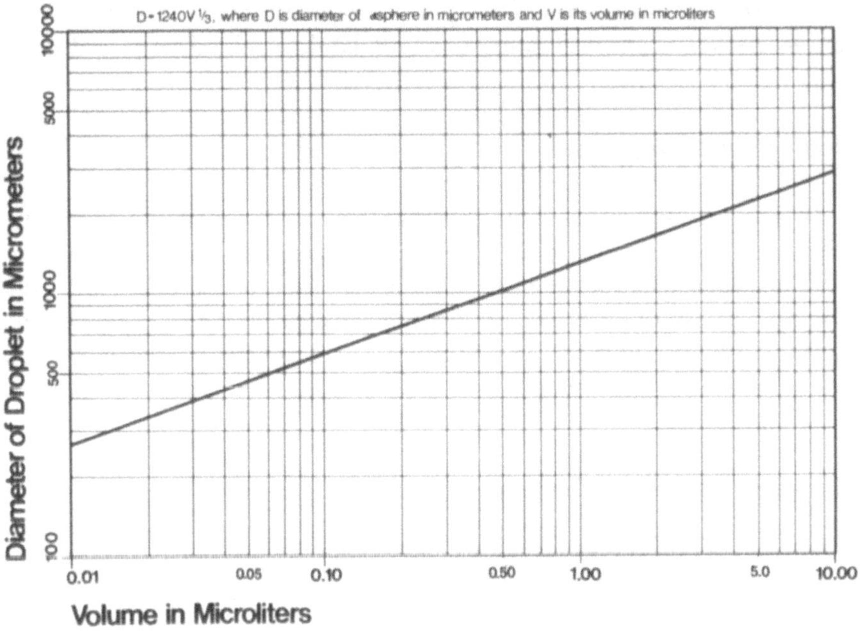

Figure 5. The diameter of a spherical droplet as a function of volume, plotted for volumes readily delivered with microsyringes. Note that the smallest volume plotted, 0.01 μl, which corresponds to the smallest graduation mark on a Hamilton 1-μl syringe, has a diameter of 264 μm. The actual diameter of the diffusion spot resulting from an intracerebral injection of this amount depends on factors discussed in the text but will generally exceed the diameter of the spherical volume delivered.

in material stained with the Graham and Karnovsky (1966) method. To obtain diffusion spots with a diameter of 0.2–0.5 mm in DAB-stained material, the volumes must be considerably smaller, i.e., 0.002–0.01 μl (Price *et al.*, 1977).

The relation between the concentration of the HRP solution and the size of the injection has been studied by Kim and Strick (1976) who used concentrations between 1.5% and 30% and found, not surprisingly, that the strongest solution resulted in the largest diffusion spots. The strength of the solution may, at least to a certain extent, also determine the intensity of the labeling of individual cells (LaVail, 1978).

The size of the diffusion spot is related not only to the concentration and the volume of the injected solution but also to the tissue characteristics at the site of the injection (LaVail and LaVail, 1974; Walberg *et al.*, 1976; Beitz and King, 1976). For instance, a given amount generally involves a larger region in subcortical gray matter than in cortical tissue. Another significant variable is the survival time of the animal (e.g., Vanegas *et al.*, 1978; Hedreen and McGrath, 1977; Haberly and Price, 1978) as is discussed in more detail below (Section IV.B). It is also important to recognize that the size of the "apparent" diffusion spot will vary with the histochemical method used to demonstrate the HRP (Fig. 2). Considering the many variables that influence the size of the diffusion spot, it becomes apparent that the choice of parameters in a given situation can only be determined empirically.

2. Iontophoretic Injections

To achieve smaller injection sites and to avoid the tissue damage caused by the injected volume of HRP solution, Graybiel and Devor (1974) developed a microelectrophoretic procedure for the delivery of HRP. Difficulties in the application of that specific method have prompted the development of various modifications of the iontophoresis parameters. The practical application of the iontophoretic injection procedure is discussed in Chapter 3.

3. Labeling after Axon Trauma

Labeling can be accomplished by exposing the cut ends of nerve fibers to HRP, either in solution or in dry form. This procedure produces anterograde and retrograde filling of axons and cell bodies and has been used for the mapping of origin and termination of both central tracts and peripheral nerve fibers (e.g., Scalia and Colman, 1974; Oppenheim and Heaton, 1975; Adams and Warr, 1976; Grant *et al.*, 1979).

C. Survival Time

1. Labeling by Retrograde Transport of HRP

The optimal postinjection survival time depends on the rate of axonal transport of HRP, the length of the fiber system being studied, and the rate of lysosomal degradation of the tracer once it reaches the cell soma. Since studies using the DAB method have yielded estimates of the rate of retrograde movement of HRP in the range of 50–120 mm/day (LaVail and LaVail, 1972, 1974; Hansson, 1973; Kristensson, 1975), the minimum survival time in a given experimental situation can be calculated if the maximum distance over which the tracer must be transported is known. It must be realized, however, that because of the relative insensitivity of the DAB method, the actual rate of retrograde axonal transport of HRP may well be considerably greater than is widely supposed. But variables other than transport rate and fiber length affect the detectability of accumulated HRP in the cell bodies, and, thus, these variables also have a direct bearing on the choice of survival time. For example, the density of the terminal arborization in the region of the injection (Jones, 1975) and probably the diameter of the axons transporting the HRP (Nauta *et al.*, 1975; Jones *et al.*, 1977) are important. In addition, the fact that axon collaterals of retrogradely labeled cells can be anterogradely filled with HRP (de Olmos and Heimer, 1977) indicates that the extent of collateralization can also have an influence on the amount of HRP accumulating in the cell body. The functional state of the system, too, governs the rate of uptake by the axon terminals (e.g., Heuser and Reese, 1975; Singer *et al.*, 1977), and electrical or natural stimulation of a pathway under study may enhance the incorporation and transport of HRP.

The time needed for maximal cell labeling is dependent not only on the rate of accumulation of HRP but also on its removal by lysosomal inactivation,* which may be very rapid. For instance, Bunt and her colleagues (1974) and LaVail and LaVail (1974) reported that HRP granules are almost completely cleared from retinal ganglion cells in the rat and the chick 3–4 days following tectal injections. Considering the many factors mentioned above, it is not surprising to find that different neuron populations that share the same projection field sometimes display a pronounced asynchrony in their optimal accumulation of HRP (Leger *et al.*, 1975; Hedreen and McGrath, 1977).

As it is difficult to predict the relative effect of many of the variables that are involved in the accumulation and removal of HRP in a given population of cells, it is almost impossible to give precise information about the optimal survival time for study of a particular fiber system. Therefore, a series of different survival times should be tried in order to establish the optimal timing in each of the several sources of afferents to the target injected. A cursory survey of the literature on mammalian species indicates that most investigators use 24–48 hr survival intervals, while studies of cold-blooded animals require survival times that are several days longer.

In this connection, it should be noted that the HRP method works best in young or immature animals (e.g., Kristensson *et al.*, 1971; LaVail and LaVail, 1972; Jacobson and Trojanowski, 1975; Colman *et al.*, 1976; Walberg *et al.*, 1976), perhaps because of pronounced pinocytotic activity at the tips of growing axons. As a result, the optimal survival times for newborn and young animals may well be different from—probably somewhat shorter than—those for mature animals (e.g., Clarke and Cowan, 1976; Dennis and Kerr, 1976). On the other hand, if it is important to achieve a retrograde marking not only of cell bodies but also of the dendritic tree, a slight extension of the survival time may be necessary (Jacobson and Trojanowski, 1975; Wiksten, 1979).

2. Labeling by Anterograde Transport of HRP

Since HRP has not been used extensively as an anterograde tracer, the information regarding the effect of different survival times on the labeling of synaptic fields is somewhat limited. The consensus seems to be that the

*It is interesting to note that the rate of enzymatic degradation seems to vary among different commercial brands of HRP (Keefer, 1978).

anterograde transport of HRP is related to the fast axoplasmic flow (Hansson, 1973; Scalia and Colman, 1974; Sotelo and Riche, 1974; Colman *et al.*, 1976; Brownson *et al.*, 1977; Spencer *et al.*, 1978). In substantial support of this, Mesulam and Mufson (1980) determined that in axons of the rat's retinal ganglion cells, the rate of anterograde transport of HRP is at least 300 mm per day. These authors used the sensitive tetramethylbenzidine (TMB) method which is likely to give a higher and more accurate rate of transport than the DAB method. Many investigators who have used HRP for both retrograde and anterograde tracing experiments have chosen the same survival time for both types of experiments (Sotelo and Riche, 1974; Sherlock and Raisman, 1975; Halpern *et al.*, 1976; Hedreen and McGrath, 1977; Frankfurter *et al.*, 1978; de Olmos *et al.*, 1978; Kicliter, 1979), and this would appear to be a reasonable practice until the rate of retrograde transport has also been determined in experiments aided by the TMB method.

In general, it is reasonable to assume that many of the variables that affect retrograde labeling will also affect anterograde labeling. The species and age of the animal, the rate of uptake and removal of HRP, and the degree of collateralization are all of special significance in anterograde tracing experiments.

D. Fixation and Sectioning

1. Fixation

Fixation is undoubtedly one of the most crucial steps in preparing the tissue for HRP histochemistry. It is well-known that fixation reduces the enzymatic activity, thereby decreasing the likelihood of detecting small amounts of HRP. For that reason, Courville and Saint-Cyr (1978) proposed that the HRP method be carried out without previous fixation.* Although these authors claim to have improved the sensitivity of the HRP method significantly, problems of enzyme diffusion and absorption (Straus, 1964) as well as the presence of endogenous HRP activity may offset the advantage of working with unfixed tissue. As a consequence, it seems advisable to look for a compromise between the conditions that provide maximal enzyme activity and those that reduce diffusion, absorption, and endogenous HRP activity.

To achieve the abovementioned goal, two parameters must be given special attention: choice of fixative and duration of fixation. There does not seem to be any consensus regarding the best fixative. Most of the fixatives that have been proposed contain varying amounts of both paraformaldehyde and glutaraldehyde. Some investigators (e.g., Bunt *et al.*, 1975; Lund

*The animals are perfused with 6% dextran and the brains immediately removed and immersed in a beaker containing freon or 2-urethyl butane. Rapid freezing of the tissue is obtained by placing the beaker in liquid nitrogen. Following sectioning on a cryostat and mounting of the sections on slides, the material can be stained with one of the HRP procedures.

et al., 1975) have used formaldehyde alone, whereas Adams (1977) and Malmgren and Olsson (1978) have used only glutaraldehyde. Indeed, Malmgren and Olsson, using the DAB method, made a careful study of the effects of various fixatives, and they found that the number of labeled cells was greatly reduced if the fixative contained paraformaldehyde. However, Rosene and Mesulam (1978) and de Olmos (see Appendix, Sections VIII.B–D), who have developed methods using benzidine dihydrochloride (BDHC) and TMB as chromogens, have obtained very good results with a double-aldehyde fixative. Furthermore, the incorporation of tannic acid (Andersson *et al.*, 1975) does seem to optimize the demonstration of retrogradely labeled cells, at least in BDHC-stained material.

In order to wash out the blood as completely as possible, the perfusion of the animal starts with a washing solution, usually consisting of an isotonic sodium chloride or buffer solution that sometimes includes different amounts of sucrose, glucose, dextran, etc. (see the different HRP procedures in the Appendix, Sections VIII.A–D). The washing is important because blood cells have endogenous peroxidase activity that may cause artifacts. Following the washing and the perfusion of fixative, which usually is prolonged for 15–30 min, the brain is either removed immediately from the skull or perfused further with a sucrose solution (Rosene and Mesulam, 1978) in order to wash excess unbound fixative from the tissue. In our experience, however, this latter procedure has not worked well for the BDHC procedure described in the Appendix (Section VIII.B). Instead, we prefer to remove the brain from the skull and either transfer it directly to a cold (4°C) 30% phosphate-buffered sucrose solution or postfix it for 4 hr (Malmgren and Olsson, 1978) in order to assure the inactivation of endogenous peroxidase. Following the postfixation, the brain is placed in the sucrose solution and kept at 4°C until it sinks to the bottom of the jar. The treatment in sucrose has three objectives: to remove the excess of fixative, to serve as a cryoprotective agent, and to reduce the formation of crystalline artifacts in the subsequent staining. To increase the penetration of the sucrose, thereby shortening the time in this solution, Adams (1977) recommends that 1% dimethylsulfoxide (DMSO), a detergent, be added to the solution.

2. Sectioning

If necessary, the tissue can be encased in gelatin or albumin–gelatin before sectioning on a freezing microtome or in a cryostat. Although it is customary to cut the sections somewhere between 20 and 50 μm, it should be emphasized that the penetration capacities of the various chromogens differ. BDHC, for instance, seems to penetrate poorly through the tissue. Therefore, in order to appreciate the fine details of HRP transport in BDHC-stained material, the thickness of the sections should not exceed 30–35 μm. Material stained with the TMB method, on the other hand, can be cut as thick as 50 μm. It goes without saying that a sharp knife should be used for the cutting. This is especially important for material intended for the BDHC or TMB procedures, which are prone to produce chromogen crystals on rough surfaces created by a dull knife.

E. Potentiation of Uptake and Transport of HRP

Several procedures have been developed to increase the effectiveness of the HRP method. The simplest is to add to the HRP solution any of several substances that may effect both somatofugal and somatopetal transport. Another method, which is specifically designed for retrograde tracing experiments, is based on the elimination of other afferent inputs to the area of injections.

1. Additives to Potentiate Uptake and Transport

Although high concentrations of DMSO reportedly have a blocking effect on the fast axoplasmic transport (Donoso *et al.*, 1977), several authors have found that the addition of 2% DMSO to the HRP solution increases the amount of intracellular HRP available for intraaxonal transport (Keefer *et al.*, 1976; West and Black, 1979), probably by increasing the membrane permeability for HRP.

The observation that the presence of large basic macromolecules facilitates the uptake of proteins by cells (Ryser and Hancock, 1965) prompted the use of poly-L-ornithine (ORN) as an additive (Hadley and Trachtenberg, 1978; Itaya *et al.*, 1978). According to Hadley and Trachtenberg, who used ORN in concentrations of 0.1–0.5%, the HRP uptake is enhanced fourfold. A recent communication by Spencer *et al.* (1978), furthermore, suggests that the addition of 0.05% (w/v) of adenosine may improve the uptake of HRP.

Other alternatives are provided by conjugation of HRP with other substances such as poly-L-lysine (Shen and Ryser, 1978) or wheat germ agglutinin (Gonatas *et al.*, 1979). The conjugation of HRP with the latter substance, which is a highly sensitive retrograde intraaxonal tracer (Schwab *et al.*, 1978), has reportedly improved the sensitivity of the HRP technique 40 times.

2. Expansion of the Terminal Field

This procedure was first used by Siegel *et al.* (1977) who enhanced the labeling of amygdaloid projections to the mediodorsal thalamic nucleus by making prefrontal cortex lesions 1 week before injecting HRP. Similar results have also been obtained for entorhinal cortex projections to the dentate gyrus (Steward and Vinsant, 1978; Goldschmidt and Steward, 1980). The mechanism of enhancement of HRP uptake in a particular afferent pathway following the elimination of other sources of afferent supply to a certain region is poorly understood. The two most plausible explanations for the enhancement are that it occurs either because the terminal field of undamaged axons expands as a result of sprouting of axon terminals in the denervated zone (Goldschmidt and Steward, 1980) or because competing terminals for HRP uptake have been eliminated, thereby increasing the amount of available HRP for uptake by undamaged terminals (Siegel *et al.*, 1977). A combination of both mechanisms cannot be ruled out.

V. GENERAL CHARACTERISTICS OF THE DIFFERENT HRP METHODS

There are, in principle, three different techniques available for the study of transported HRP. One possibility is to label the HRP or its apoenzyme with a radioactive isotope, e.g., tritium, and study its localization in the neuron with the autoradiographic technique (Chapter 5). Another alternative is to localize the intraneuronal HRP by the use of immunologic techniques. Although the immunohistochemical technique was claimed to be very sensitive and specific (Vacca *et al.*, 1975), the results obtained so far have not met expectations.* The third and most popular procedure is based on a histochemical reaction. To produce this reaction, the tissue is incubated in a medium containing hydrogen peroxide and chromogen, usually a benzidine derivative, that is oxidized to a colored product by the peroxide–peroxidase system. The four most commonly used chromogens are DAB (3,3′-diaminobenzidine tetrahydrochloride), BDHC (benzidine dihydrochloride), TMB (3,3′,5,5′tetramethylbenzidine), and *o*-dianisidine (3,3′-dimethoxybenzidine dihydrochloride). Unfortunately, benzidine and its derivatives are generally carcinogenic and should be handled with care. Use gloves, lab coat, and face mask when handling the substrate, and set aside a safe hood where the substance can be weighed. Although tetramethylbenzidine (TMB) has been claimed to be noncarcinogenic (Holland *et al.*, 1974), it would also seem advisable to follow precautionary measures when dealing with this substance. Recently a noncarcinogenic substrate, a combination of *p*-phenylenediamine dihydrochloride and pyrocatechol (Hanker–Yates product), has been introduced as a chromogen (Hanker *et al.*, 1977). However, the Hanker–Yates product does not seem to be as sensitive as the other chromogens (Mesulam and Rosene, 1979). Homovanilic acid has also been proposed as a chromogen (Papadimitrou *et al.*, 1976), but it has not been widely used.

*Whereas the more recently developed histochemical techniques are sensitive enough to display both retrogradely and anterogradely transported HRP, Vacca and his collaborators failed to provide evidence for orthograde enzyme transport with the immunohistochemical technique.

A. The DAB Method

Since the introduction of the HRP method for the demonstration of neuronal retrograde transport (LaVail and LaVail, 1972), DAB has been the most commonly used chromogen. The popularity of the DAB method is based on several factors. The technique is simple, and the results obtained are easily reproducible. The brown reaction product, furthermore, is osmiophilic, thereby making DAB-stained material suited for electron microscopic studies. The reaction product is also highly refractive and therefore easily visible in dark-field illumination (bright metallic color with a green or golden tinge). An additional advantage is that the DAB reaction, when used to display diffusely filled neurons, generally gives a good definition of the morphology of the neuron (Figs. 6D and 7). However, when DAB is used to reveal cell bodies that have been labeled by axoplasmic transport, the brown reaction product is not easily visible in bright-field microscopy, even if the use of special filters improves the image considerably (Haberly and Price, 1978). Therefore, dark-field microscopy is often used.

Although the original DAB method by Graham and Karnovsky (1966) and especially its modification by LaVail (1975) still seem to be commonly used HRP procedures, it is evident that these methods do not always produce optimal visualization of HRP-labeled profiles in the light microscope. In the last few years, however, several improved versions of the DAB method have been published (Adams, 1977; Streit and Reubi, 1977; Malmgren and Olsson, 1978).

Adams introduced a pretreatment with cobalt chloride and claims that "the darker reaction product gives the added advantage of extending the sensitivity of the HRP method, and it obviates the need for darkfield illumination." According to a comparative study by Mesulam and Rosene (1979), however, the modification by Adams does not appreciably improve the sensitivity.

Streit and Reubi added *p*-cresol to the incubation medium and prescribed a pH of 5–6 for the incubation. The *p*-cresol probably acts as an oxidogenic donor that stimulates the peroxidatic reaction with other donors (Nicholls, 1962). By introducing a buffered *p*-cresol-DAB-H_2O_2 solution, the authors claim to have achieved a much higher sensitivity than that obtained by the original DAB method. This view is supported by experiments in our own laboratory (de Olmos *et al.*, 1978) as well as by the comparative study by Mesulam and Rosene (1979).

Malmgren and Olsson (1978), finally, introduced a concentrated cacodylate buffer in order to adjust the pH of the incubation solution to 5.1, at which point, they report, the labeling of neurons reaches its peak. They also claim that a higher DAB concentration (0.2%) improves the result. If a neutral pH is needed, as in the preparation of material for electron microscopy, Malmgren and Olsson suggest adding imidazole (0.1 M) to the incubation medium. The imidazole broadens the optimal pH range for the catalytic activity of the HRP.

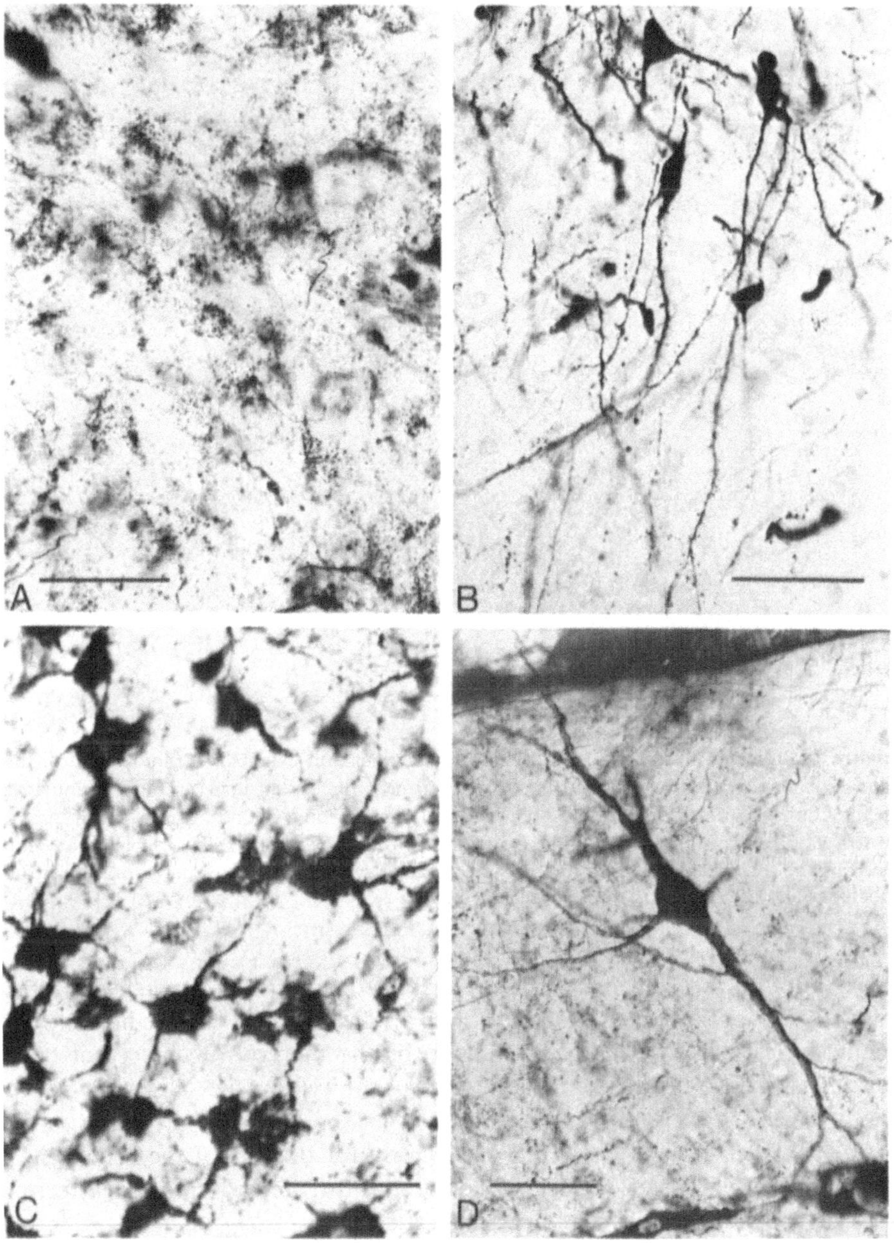

Figure 6. Different types of labeling of cell bodies. Bar is 50 μm in all figures. (A) Granular staining of cell bodies in the halo zone of an iontophoretic injection site. DAB stain. (B) Dense, Golgi-like staining of cell bodies, dendrites, and dendritic spines in neurons in the halo zone of an iontophoretic injection site. DAB stain. (C) Heavy granular staining of cell bodies and proximal dendrites that have been retrogradely labeled with HRP and stained with TMB. (D) Diffusely labeled neuron located near an injection site. The processes have a translucent appearance. DAB stain.

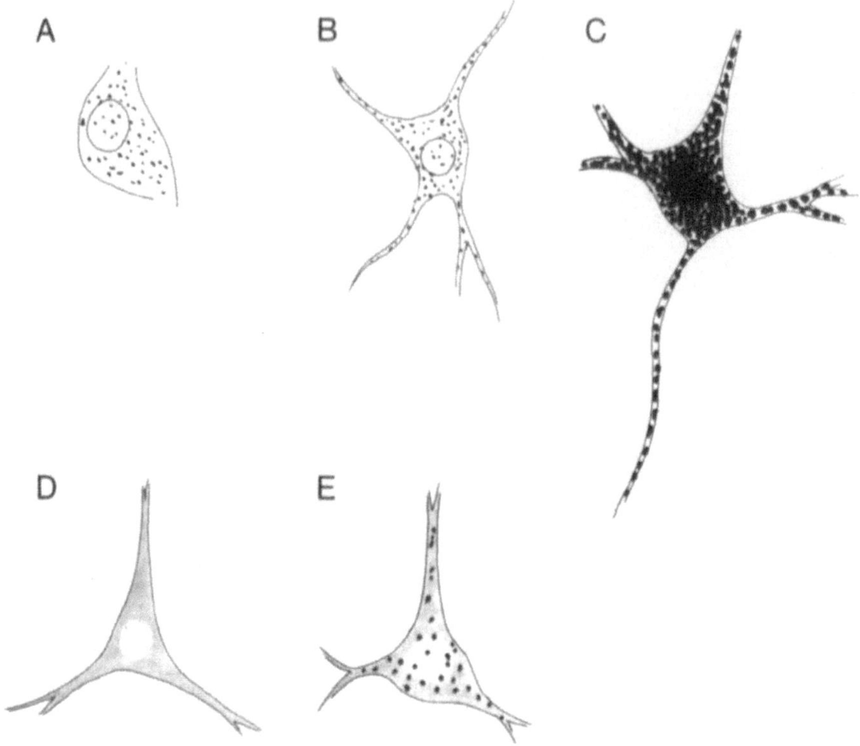

Figure 7. Drawings illustrating the various combinations and extents of granular and diffuse labeling commonly encountered in neuronal perikarya. Bar is 200 μm. (A) Light granular labeling confined to cell soma. (B) Moderate granular labeling extending into dendrites. (C) Heavy granular labeling involving dendrites and proximal axon. (D) Diffuse labeling of entire perikaryon. (E) Combination of diffuse and granular labeling. In extreme cases, the diffuse label may almost completely obscure the granular.

Although the sensitivity of the DAB method does not match the sensitivity of either the BDHC method or the TMB method on the light microscopic level, the DAB method seems to be the method of choice for those interested in the ultrastructural localization of HRP, especially since recent modifications have significantly improved its rendition on the electron microscopic level (Itoh *et al.*, 1979). The reason for this is that the brown DAB reaction product is very electron-opaque and, furthermore, is characterized by a high degree of localization (Goldfischer, 1969).

B. The *o*-Dianisidine Method

Although *o*-dianisidine has been used much less frequently than DAB as a chromogen, the *o*-dianisidine reaction product is more easily visible in bright-field microscopy than the DAB product. Unfortunately, however, the reaction product obtained with *o*-dianisidine has a tendency to fade rather easily.

There are several versions of the *o*-dianisidine method. Graham and Karnovsky (1966), who studied the absorption of horseradish peroxidase in the kidney, used an incubation solution with pH 7.6. Warr (1975), who apparently was the first to use *o*-dianisidine for the demonstration of HRP enzymatic activity in tract-tracing experiments, followed the technical directions provided by Graham and Karnovsky. Colman *et al.* (1976), on the other hand, used an unbuffered *o*-dianisidine medium (pH 4.5), which results in a green reaction product instead of the less visible brown color reaction obtained by the original Graham–Karnovsky procedure. A buffered (pH 4.5) *o*-dianisidine medium was used by de Olmos (1977) who in addition took advantage of cryoprotective agents, thereby allowing the enzymatic reaction to take place at sub-zero temperatures. Under these conditions, the enzymatic reaction could apparently be better controlled, thereby allowing the reaction to be "pushed" to its maximum without excessive crystallization.

C. The BDHC Method

In 1974, Lynch *et al.* introduced a tract-tracing HRP method in which BDHC was used as a chromogen. This procedure was based on a benzidine method developed by Straus (1964). In recent years, several BDHC modifications have been published, and excellent results have been obtained with the aid of these new protocols. Indeed, the quantitative study by Mesulam and Rosene (1977) has convincingly shown that the blue BDHC reaction product is superior to the brown DAB product for the detection of HRP-labeled neurons in bright-field microscopy. These BDHC modifications, which have also been increasingly more successful in demonstrating axonal projections and collateral pathways (Lynch *et al.*, 1974; Mesulam, 1976; de Olmos and Heimer, 1977), are the product of a careful analysis of different preparatory steps and incubation parameters. The BDHC protocol presented in the Appendix (Section VIII.B) takes advantage of the experience amassed by several laboratories (Sasaki, 1952; Straus, 1967; Novikoff and Goldfischer, 1969; Pearse, 1972; Adams, 1977; de Olmos, 1977; de Olmos and Heimer, 1977).

With respect to the staining characteristics of the BDHC reaction product, it should be noted that reaction products of differing color can be obtained depending on the pH of the BDHC–peroxide medium. The blue pigment that is obtained in an acidic medium is much easier to visualize than the brown pigment produced in a basic medium. The difference in visualization is even more apparent in dark-field microscopy where the reaction product obtained in an acidic medium appears reddish-silver in color (Fig. 8A).

Finally, in assessing the value of the different techniques, it should be noted that the BDHC reaction product, like the DAB reaction product, does not seem to fade as easily as either the *o*-dianisidine or the TMB reaction product. However, BDHC, like DAB and *o*-dianisidine, is a suspected carcinogenic agent, and the substance must therefore be handled with great caution.

D. The TMB Method

In 1974, Holland *et al.* suggested that TMB could be used as a substitute for benzidine in the detection of blood. Although the main reason for proposing TMB was its apparent noncarcinogenic qualities, the authors also found TMB to be considerably more sensitive than benzidine. This combination of high sensitivity and apparent noncarcinogenicity has rapidly made the TMB-based methods the procedures of choice for the light microscopic mapping of HRP-labeled pathways (Hardy and Heimer, 1977; Mesulam and Rosene, 1977; Mesulam, 1978; de Olmos *et al.*, 1978). The TMB reaction product is easily visible in the light microscope (Fig. 9), where it appears as grayish blue granules in bright-field and as golden granules in dark-field.

Although TMB appeared to be a very effective chromogen, it was soon realized that the practical application of the TMB method to the nervous system was hampered by the tendency for excessive crystallization of the product and the instability of the reaction over time, with its eventual fading and disappearance. The TMB protocol published by Mesulam (Appendix VIII.D) represents one attempt to solve this problem. Another TMB protocol developed by de Olmos (Appendix VIII.C) has, at least in our hands, provided the most successful solution to the problem of crystallization. This has been accomplished by adding gelatin to the incubation medium. In addition, tracers of metals are added, increasing both the sensitivity and the stability of the reaction product.

To date, TMB has not proven useful for electron microscopic studies. Two reasons are apparent. First, it is not osmiophilic as is DAB. Second, the TMB reaction product is highly soluble in the organic solvents essential for preparing tissue for electron microscopic study.

Figure 8. (A) Sagittal section through the basal forebrain of the rat following an HRP injection in the ipsilateral olfactory bulb. The heavily labeled cell group in the right half of the picture is the nucleus of the horizontal limb of the diagonal band. Retrogradely HRP-labeled axons (white arrow) emanating from cells in this nucleus proceed rostrally deep to the olfactory tubercle. The HRP-labeled terminals of the lateral olfactory tract form a dense bright lamina at the base of the brain. De Olmos BDHC procedure. Survival time: 18 hr. (B) Large cochlear nerve terminals in the anteroventral cochlear nucleus (solid arrows) labeled with HRP after applying the enzyme to severed axons in the internal auditory meatus in the adult cat. A perivascular macrophage (open arrow) is also labeled. Graham and Karnovsky DAB procedure, pH 7.6.

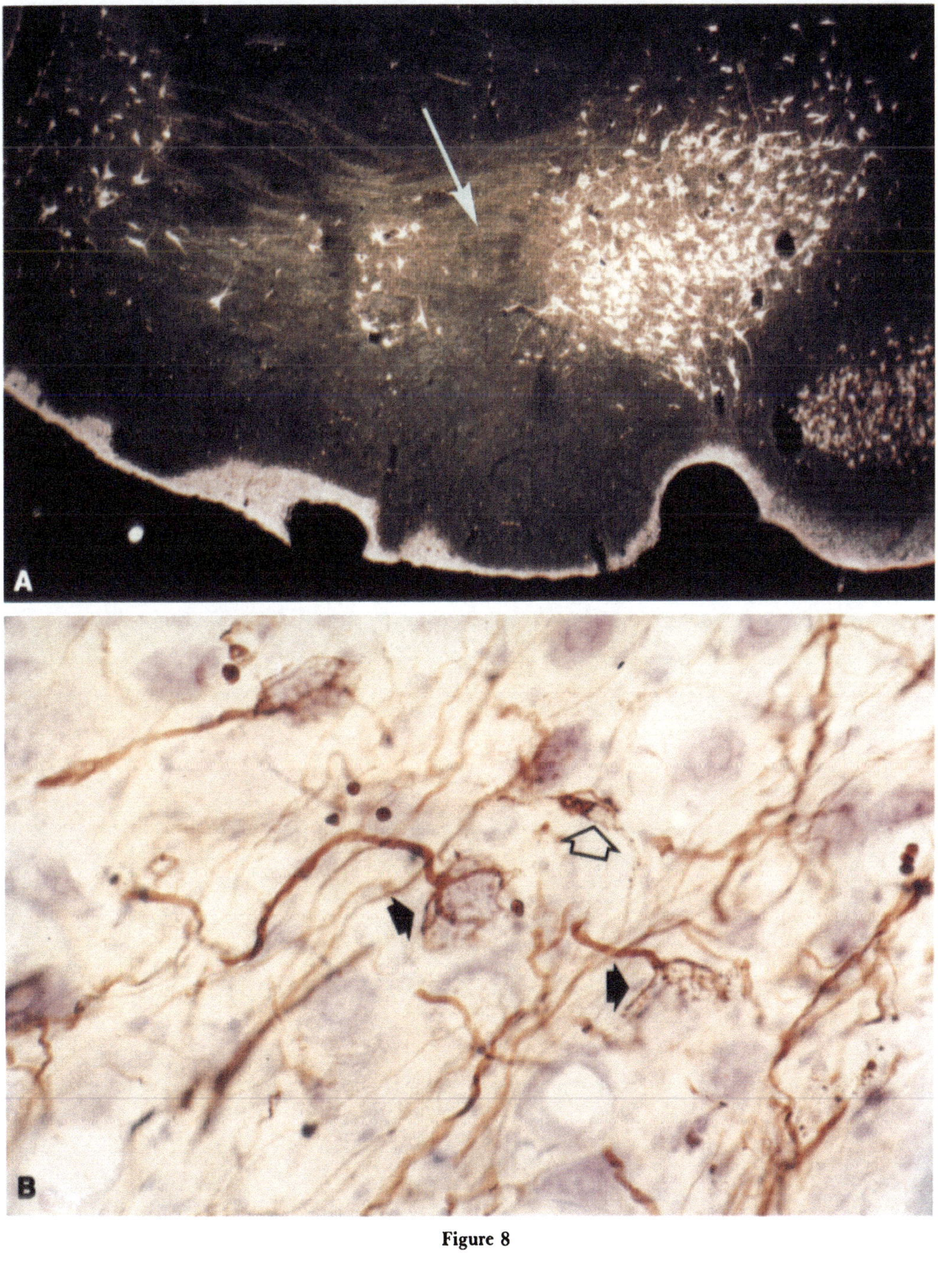

Figure 8

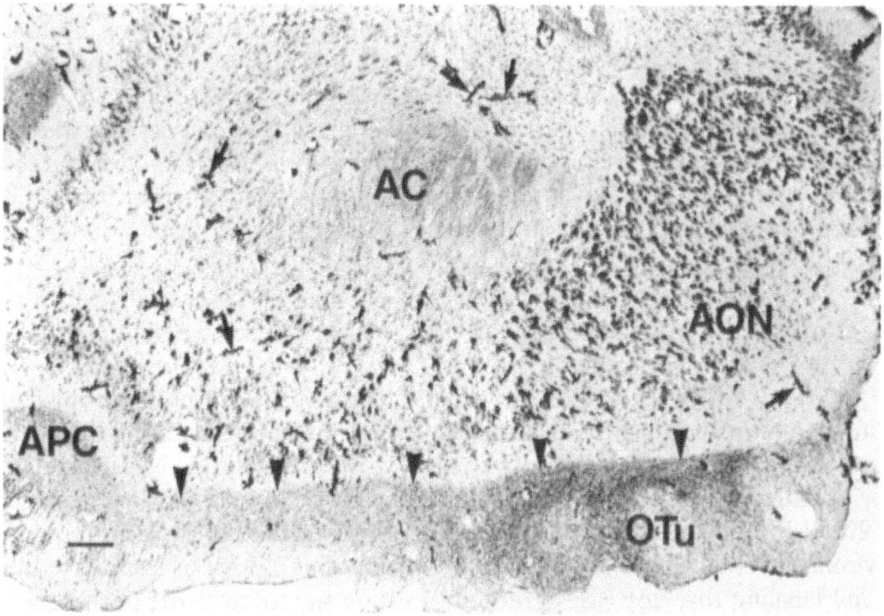

Figure 9. Bright-field photomicrograph illustrating retrogradely labeled neuron cell bodies in the anterior olfactory nucleus (AON) and anterior piriform cortex (APC) following an HRP injection into the main olfactory bulb of the rat. Also visible is an area containing anterogradely labeled lateral olfactory tract terminals (arrow heads) in APC and olfactory tubercle (OTu). The pale staining of the anterior commissure (AC) is caused by labeled axons.

VI. RESULTS AND INTERPRETATIONS

In examining the material, one has to remember that many variables besides the histochemical procedure itself have a significant effect on the result. Some of these variables, such as species, age, and survival time of the animal, as well as fixation, have already been mentioned. The type and quality of the microscope equipment is also important (e.g., Illing and Wässle, 1979; Haberly and Price, 1978). While these variables have a general bearing on the overall evaluation of the material, other factors may have a more specific relation to certain aspects of the results. For example, the chosen method of delivery of the HRP has a clear effect on the appearance of the injection site.

A. The Site of Injection

We must now address one of the most difficult issues of interpretation in the HRP tract-tracing method: evaluation of the injection site. Visually, HRP injections almost always appear as a dense core surrounded by a paler halo. There has been a natural tendency to correlate the core with the transport zone and the halo with the nontransporting diffusion area. Unfortunately, the functional transport zone may not always correspond to the dense center of an injection site visible under the microscope. To further complicate matters, the apparent size of the injection site varies considerably depending on such variables as survival time, choice of chromogen, method of injection, and type of ilumination of the specimen during microscopic examination. We will discuss these issues at length here.

1. Morphologic Characteristics

To introduce the subject, we shall illustrate a typical pressure injection site as seen in bright-field microscopy (Fig. 1A). Generally, a medium-sized pressure injection with 15–18 hr postoperative survival time is characterized by a central core of damage and dense staining (Zone I of Vanegas *et al.*, 1978) where details cannot be seen. Surrounding this core is a more or less symmetric halo (Zones II and III of Vanegas *et al.*, 1978) of less dense neuropil labeling that decreases gradually in intensity towards the periphery. A few densely labeled cell bodies appear in the halo, especially at the periphery where the background staining is lighter. Although similar HRP diffusion zones appear in iontophoretic injections, the halo has a somewhat different appearance in the sense that background labeling is less pronounced. Instead, the halo is characterized by the predominance of dense, Golgi-like perikaryal labeling (Figs. 1B and 2).

As both the core and the halo seem to play a significant role in the transport of HRP (see Section VI.A.2), they should be carefully charted. Whereas the core has a rather stable appearance, at least during the first few days following the injection, the halo undergoes significant changes in size and appearance with time (Hedreen and McGrath, 1977; Haberly and Price, 1978; Hadley and Trachtenberg, 1978; Vanegas *et al.*, 1978). Apparently, the size of the halo, which constitutes a major part of the area of diffusion of the enzyme, reaches its maximum a few hours following the injection. After 4–8 hr, the halo begins to diminish, and it is appreciably reduced after 18 hr. After 24 hr, the reduction is even more pronounced, and, particularly after a small injection, the halo may have disappeared completely after 48 hr. Considering the change in apparent injection site there may be a hazard of underestimating the transport zone in material from animals that have survived postoperatively for a long period and of overestimating the transport zone in short-survival cases.

Although the three HRP methods referred to previously produce injection sites with a dense core and a surrounding halo, a comparison of matching sections in bright- or dark-field illumination reveals that the size of the injection site appears very different with the three methods (Mesulam, 1978). The injection site appears largest in material stained with the TMB procedure, smallest with DAB, and intermediate with BDHC (Fig. 2).

Another factor of potential importance is the illumination. An injection site usually appears smaller in bright-field than in dark-field microscopy (Fig. 2C and D).

2. The Effective Injection Site

Considerable effort has been devoted to the problem of defining the so-called "effective injection site." Since survival time, fixation, tissue damage, and histochemical processing are known to affect the apparent size of the HRP diffusion zone and the extent of retrograde cell labeling, it is not surprising that different experiments have arrived at widely differing estimates of the extent and morphology of the zone from which retrograde cell labeling actually occurs.

The method of choice for determining the effective injection site entails making a series of closely spaced injections in a pathway whose connections are already independently known. Using this approach, Bunt *et al.* (1976) found that in most cases, labeling of retinal ganglion cells occurred in isolated clusters corresponding to the centers of several injection sites in the geniculate nucleus. This occurred in spite of the fact that the HRP from the multiple injections coalesced to fill the entire nucleus. However, in one case, diffusion of HRP into the lateral geniculate nucleus from an injection centered some distance away in the pulvinar resulted in extensive retinal labeling. Other authors provide evidence that, indeed, retrograde labeling can sometimes occur from the margins of the zone of diffusion (Hedreen and McGrath, 1977; Dürsteler *et al.*, 1977). Thus, the findings of Jones and Leavitt (1974) and Vanegas *et al.* (1978) that the effective injection site corresponds approximately to a central dense zone cannot safely be generalized for all retrograde tracing experiments.

Anterograde transport can also occur from neurons located in the halo zone of the injection. In a recent study of hypothalamic connections (de Olmos and Carrer, 1978), a significant labeling of terminals appeared in both the anterior thalamic nucleus and the ventral and dorsal tegmental nuclei, although the core of the injection was restricted to the ventromedial nucleus. In addition, a faint axonal labeling of the mammillothalamic and mammillotegmental tracts was seen. In all of these cases the injection halo had invaded the mammillary body, the site of origin for the well-known mammillary projections to the anterior thalamus and the tegmental nuclei. Since these projections were seen only in cases where the halo did involve the mammillary body, it seems reasonable to conclude that mammillary cells situated in the halo zone of diffusion were responsible for the anterograde transport of HRP. In light of these results, it seems advisable to interpret unexpected findings from HRP studies only after thoroughly exploring the possibility that the halo zone may be originating the unusual projections. Thus, the injection site should be evaluated using the most sensitive reaction available, and at least some short (12 hr) survival cases should be included in order to assess the maximum size of the injection (de Olmos *et al.*, 1978).

Thus, conservatism recommends that part or all of the halo should be included in the presumed transport zone until adequate control injections are completed. Beyond this, we must conclude that establishing a rule for the visual identification of the actual transport site within an apparent injection area is, for all practical purposes, impossible. This limitation is by no means unique to the HRP method. For example, the precise delineation of the effective extent of isotope injections used in tract tracing is not really possible either, but a number of well-established strategies are available that largely circumvent this shortcoming. Briefly stated, these strategies are comparison of results in several cases with slightly differing injection sites, refinement and standardization of injection and processing procedures in order to make relatively restricted injections and to produce truly comparable cases in material reacted with the most sensitive chromogen, and use of other methods, such as an anterograde transport technique, to verify conclusions about the source of afferents to the HRP injection site.

B. Labeling of Cell Bodies

1. Granular and Diffuse Labeling

The appearance of labeled cells in DAB-reacted material generally ranges from a light granularity of the perikaryon to a dense Golgi-like "staining" of the entire neuron (Figs. 6 and 7). All types can, however, probably be explained on the basis of two reaction phenomena of HRP—a diffuse and a granular component—that may be present in varying proportions in given perikarya.

By far, the granular type of labeling is the most commonly observed in tract-tracing experiments. Although neurons with granular labeling typically appear in regions that project to the area of the injection site, they may also be located in the halo zone surrounding the dense central core of the injection. The granules in DAB-stained cells are usually confined to the perikaryal cytoplasm where the nucleus is negatively outlined by the absence of granules. The more sensitive methods, such as the BDHC, the *o*-dianisidine, and the TMB methods, usually result in more heavily labeled cells in which cell bodies and proximal dendrites are often seen clearly (Figs. 7B and 10). Furthermore, whereas the DAB granules are fine and dustlike in appearance, the granules produced by the aid of the other methods are often quite large, and they may be packed so densely that part of the neuron appears solidly filled. This feature also facilitates the identification of labeled neurons with bright-field microscopy (Fig. 9).

Diffusely labeled neurons are often located in the halo zone of the injection site (e.g., LaVail and LaVail, 1974; Keefer, 1978; Vanegas *et al.*, 1978), but they may also appear in distant areas projecting to the injection site. They appear typically in DAB-reacted material in which they usually have a light or dark brown color (Figs. 6 and 7D). Several of them contain

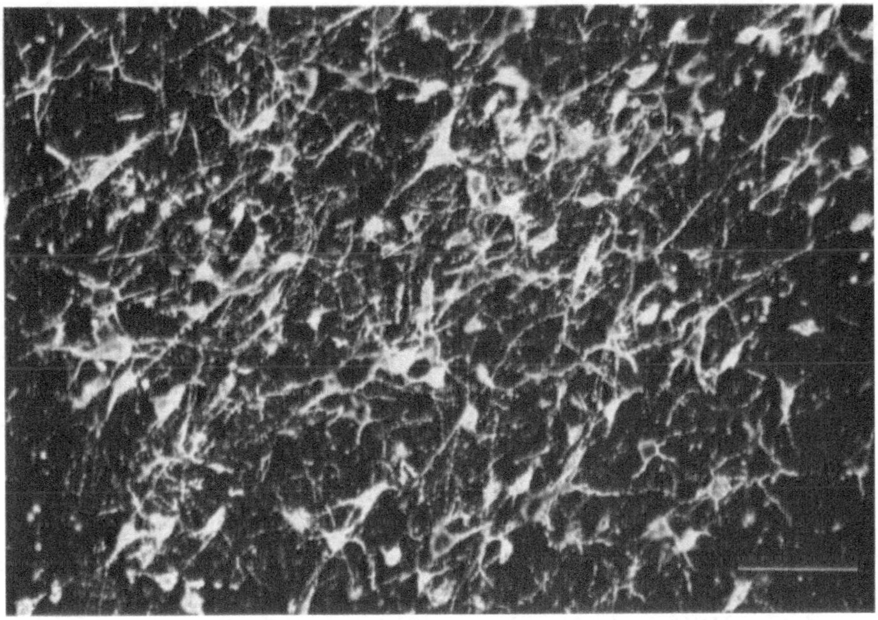

Figure 10. Dark-field photograph illustrating a mixture of labeled perikarya and axons in the basal forebrain of a rat. Labeled axons of the medial forebrain bundle that penetrate this nucleus cannot be distinguished from either the small dendritic branches or the axonal systems of the HRP-filled neurons in the nucleus. De Olmos BDHC stain. Bar, 100 μm.

dark granules as well (Fig. 7E), in which case they are referred to as "diffuse-granular" HRP neurons by Keefer. Golgi-like "staining" of neurons (Fig. 6B) in experiments with extracellular HRP injections has been observed in DAB-stained material both at the site of the injection (Vanegas *et al.*, 1978) and in distant regions projecting to the injection site (Keefer, 1978). The structural details provided by this type of HRP labeling apparently equal that produced by the Golgi method. Although Golgi-like neurons can also be obtained in BDHC- and TMB-stained material, especially in the region of the injection site (Lynch *et al.*, 1974; Spencer *et al.*, 1978), it should be pointed out that these two chromogens do not seem to produce the same type of diffuse labeling as the DAB method. Instead, the staining has a coarse granular or "frosty" appearance, and such material is not as suitable for the appreciation of fine dendritic details as DAB-stained material. Similar observations have also been reported on the basis of intracellular injections of HRP (Deschênes *et al.*, 1979).

2. Assessment of Labeled Cell Bodies

An intracerebral injection of HRP usually produces a complex array of labeled neuronal elements both within and distant from the injection site. To experienced users of the HRP method, it is obvious that the extent to which these elements can be interpreted in connectional terms depends largely on such factors as the complexity of the system under study, the amount of connectional and cytological data already available on the system, the level of refinement with which the experiment was carried out, the thoroughness of the observer in reading and charting the findings, and the availability of adequate control cases. The purpose of what follows is to discuss some of the more important obstacles and pitfalls commonly encountered in analyzing the data from experiments using the HRP method.

a. Within the Zone of Diffusion. Labeled cell bodies located within the zone of diffusion of an HRP injection generally represent an uncertain mixture of both direct perikaryal uptake and retrograde axonal transport of HRP. Although these labeled elements sometimes are sufficiently filled with HRP reaction product to provide excellent opportunities for cytological study, as with the Golgi method, the connectional utility of such labeling is limited and will not be considered further.

b. Outside the Zone of Diffusion. Labeled cell bodies located at a distance from the injection site are, of course, the principal focus in many studies with the HRP method, and they offer less difficulty in interpretation. Such labeling is virtually always via retrograde axonal transport, and it is taken to signify that the neurons have axons that either terminate in or pass through the injection site. Axons of passage damaged during the injection procedure produce retrograde labeling that, although often greater in degree, is usually indistinguishable from that seen following uptake by axon terminals (e.g., Aghajanian and Wang, 1977; Herkenham and Nauta, 1977). Labeling of cell bodies belonging to damaged fibers may create interpretative problems that often can be solved only by making control HRP injections or by studying the fiber system with the aid of other tract-tracing methods. As pointed out already, severely damaged axons incorporate HRP readily. Consequently, intentionally damaging fibers whose projections are to be studied is an effective approach for tracing cells of origin of well-defined pathways (e.g., Adams and Warr, 1976) or for studying cytological details of Golgi-like HRP-labeled neurons (Keefer, 1978).

Although several studies have shown that HRP can also be taken up by undamaged axons, there is mounting evidence that retrograde or anterograde transport will not occur in such axons except to a minor extent (e.g., Krishnan and Singer, 1973; Dürsteler *et al.*, 1977; Phillipson, 1979).

c. Degree of Labeling. The degree of labeling of neuronal cell bodies remote from the injection site often ranges from solid filling of the soma and dendrites to faint labeling of the soma alone (Fig. 11). Although it might be argued that this variation is a natural consequence of the HRP concentration gradient created at a typical injection site, little connectional significance should be attached to this phenomenon because a variety of factors contribute to the apparent "density" of perikaryal labeling. These include axon

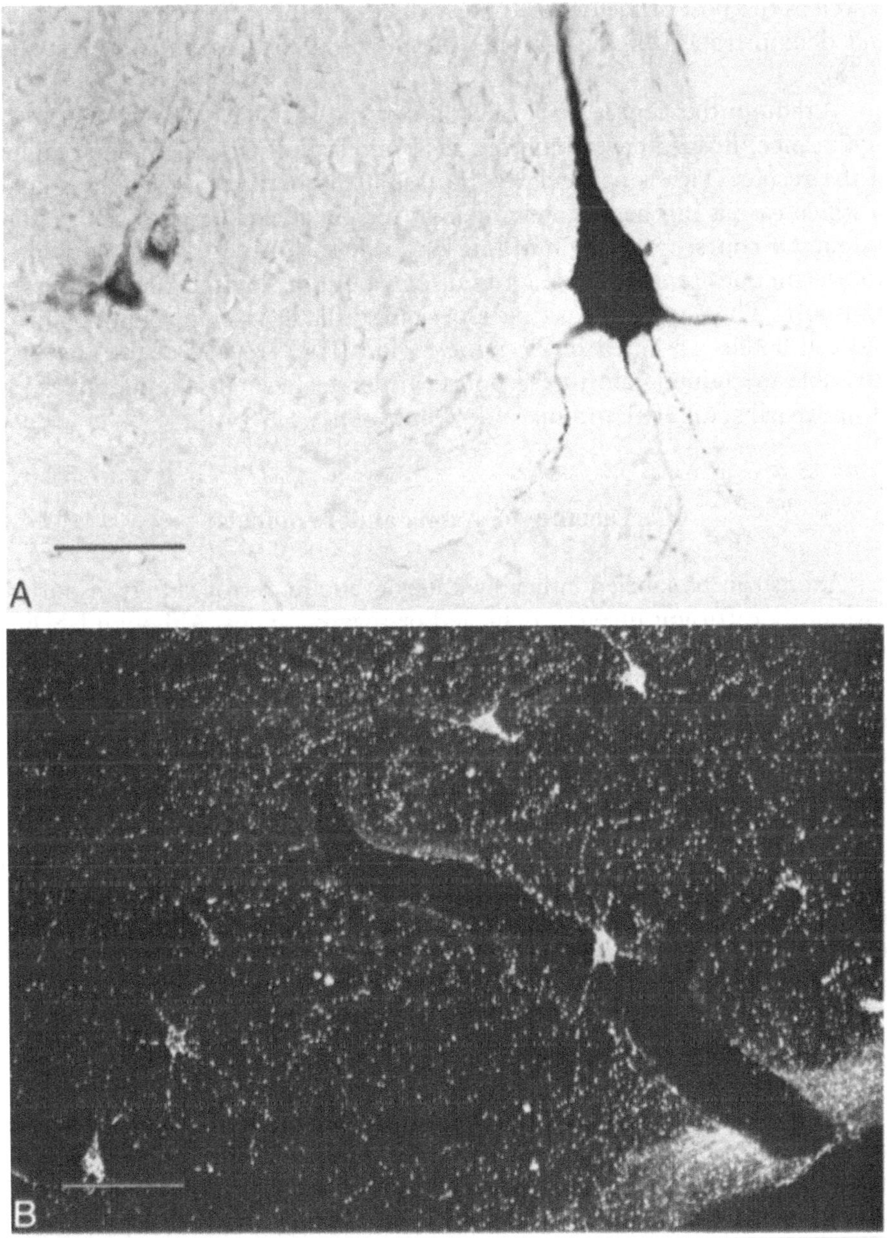

Figure 11. Variations in the degree of labeling with HRP. (A) Retrogradely labeled neurons in the cerebral cortex following HRP injection in the superior colliculus. The two cell bodies to the left are relatively lightly labeled in comparison with the neuron to the right. Bar, 50 μm. (B) Dark-field micrograph of retrogradely labeled neurons. Several of the neuronal cell bodies have accumulated enough HRP to be readily identified. Others, however, do not contain as much label and can be misinterpreted as labeled axons and terminals, particularly in dark-field and in material not counterstained. Bar, 100 μm.

damage, extent of terminal arborization having access to HRP, small perikaryal size with an attendant spatial concentration of the reaction product, as well as the host of factors previously discussed that affect the preservation and demonstration of the enzyme (Mesulam, 1976; Rosene and Mesulam, 1978).

Although the degree of perikaryal labeling may have little connectional significance, it does have some practical implications for the interpretation of the results. This is particularly the case if the perikaryal labeling is light, in which case it can be misinterpreted as terminal labeling in a diffuse fiber system. Of course, a problem of this kind will only occur in material stained with techniques sensitive enough to display both orthograde and retrograde transport. The situation may be especially difficult if both axon terminals and cell bodies are filled in the same region (Fig. 11B). In such cases, it is advisable to counterstain the section in order to facilitate the identification of intraperikaryal HRP granules (de Olmos *et al.*, 1978).

C. Labeling of Axons and Terminals

Axons can be labeled either by anterograde or retrograde transport of HRP or by diffusion through damaged or severed axons. Although labeling by diffusion can be displayed by any of the current HRP techniques, axons labeled by axoplasmic transport of HRP can usually be detected on the light microscopic level, and with varying degree of success only by the more sensitive techniques such as the BDHC, *o*-dianisidine, TMB, and *p*-cresol-DAB techniques (e.g., Lynch *et al.*, 1974; Scalia and Colman, 1974; Colman *et al.*, 1976; Streit and Reubi, 1977; de Olmos, 1977; de Olmos *et al.*, 1978; Mesulam, 1978). Although axons labeled by axoplasmic transport have been displayed by the aid of the original DAB technique under optimal conditions in dark-field microscopy (Haberly and Price, 1978), this seems to be the exception rather than the rule.

1. Axons Labeled by Diffusion

Since damaged axons and retrograde labeling of their parent cell bodies are unavoidable concomitants of the HRP method, it is important for the sake of accurate interpretation that the signs of such damage be recognized and charted in the analysis of the data. The labeling of a severely damaged axon by diffusion (Fig. 8B) is, in general, characterized by a dense, uniform distribution of the HRP reaction product in DAB-stained material (Adams and Warr, 1976). Such axons may appear solidly filled for ten or more millimeters from the injection site. Beyond this distance, however, the HRP is apparently packaged, and the labeling becomes granular. A solid appearance of damaged axons is usually also obtained in material stained with the BDHC, *o*-dianisidine, and TMB methods. However, when such material is observed with high-power objectives, it becomes obvious that the reaction product is deposited in the form of granules that are so densely packed that the axons appear homogeneously labeled, thereby making them easily visible in bright-field microscopy. In material where a fiber system is labeled by diffusion, it is usually possible to display not only the axons but also the preterminal and terminal portions with great clarity (e.g., Scalia and Colman, 1974; Adams and Warr, 1976; Light and Perl, 1977; Kusuma and Donkelaar, 1979). Diffusely labeled preterminal and terminal projections provide excellent opportunities for ultrastructural studies (Beatti *et al.*, 1978; Gobel and Falls, 1979).

A distinction between retrograde and anterograde diffusion in solidly filled axons can sometimes be made by tracing them to their somatic origins or synaptic terminations. It must be admitted, however, that the utility and feasibility of this task are limited by the complexity of the system under study and by the extent of the injection. Following application of HRP to the cut peripheral processes of primary sensory neurons, it is possible to label the central projections including the terminal fields within the central nervous system (e.g., Grant *et al.*, 1979; Berger, 1979). This type of transganglionic diffusion seems to result in a granular rather than a diffuse homogeneous filling of the central processes.

2. Axons Labeled by Axoplasmic Transport

When axons are labeled by axoplasmic transport, it is usually much more difficult to recognize individual fibers, especially if the labeling has occurred through anterograde transport.* Such a pathway is characterized by a diffuse accumulation of small granules (Fig. 12). It may, in some cases at least, be easier to identify individual axons labeled by retrograde transport, probably because the density of intraaxonal HRP is higher. The granules are also bigger, and they may occur in such density and with typical alignment that there is little doubt about the presence of a continuous fiber, at least in dark-field microscopy (Fig. 8A). Nevertheless, an accurate identification of somatopetally labeled axons can be made only in those cases in which the axons can be traced to their parent cell bodies.

3. Labeling of Axon Terminals by Axoplasmic Transport

The labeling of axon terminals by somatofugal transport is characterized by certain features that usually make a correct interpretation possible.

*Although anterograde transport of HRP has been questioned (Broadwell and Brightman, 1979), there seems to be a considerable amount of evidence that such transport occurs and can be detected with modern sensitive HRP procedures (e.g., Colman et al., 1976; Sotelo and Riche, 1974). Especially convincing is the recent work by Mesulam and Mufson (1979) who have demonstrated that colchicine, which inhibits axoplasmic transport, temporarily blocked the labeling of terminals in the visual pathways following intraocular injection of HRP.

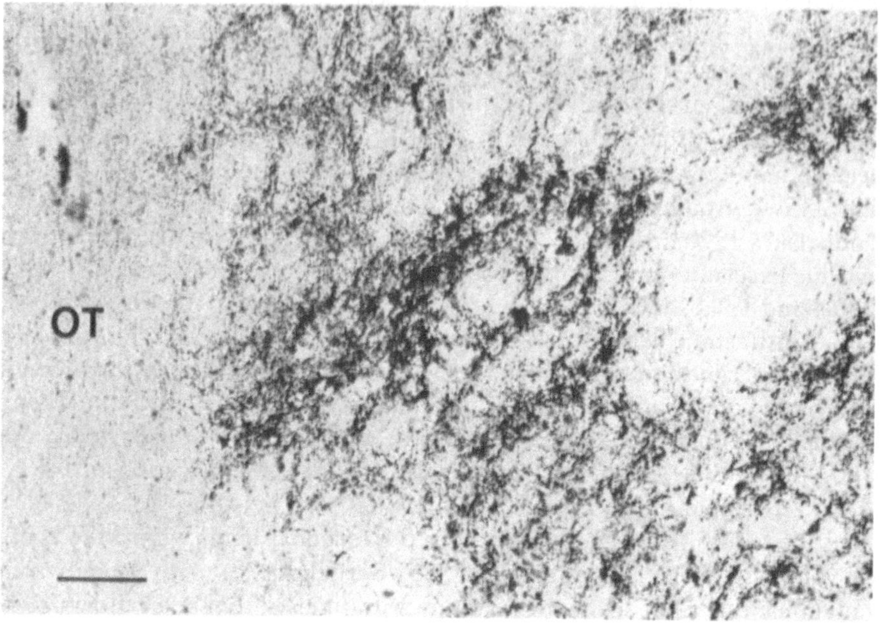

Figure 12. Low-power micrograph through the lateral geniculate body of a monkey. Compare the faintly stained optic tract (OT) axons with their densely labeled terminals which surround unstained neuron cell bodies. The staining pattern seen in the optic tract is typical of uninjured axons. In this experiment, HRP was injected into the anterior chamber of the eye. Survival time, 4 days. De Olmos TMB method. Bar, 100 μm.

First, as with autoradiographic studies, the terminals become heavily labeled before the parent axons show an appreciable amount of labeling. *Second*, a terminal field is usually recognized by a relatively high density of granules, and the granules are generally larger than the ones that occur in labeled axons (Fig. 12).

The identification of terminal labeling, however, is not always easy, and there are two types of situations that may create special problems. As in autoradiographic tracing of axonal projections, it is very difficult to separate labeled axons from labeled terminals if the pathway is loosely arranged and has a diffuse distribution, especially if the sections are cut in a plane more or less transverse to the course of the pathway (Fig. 10). The other situation arises in cases where reciprocal connections are suspected, and one sees a dustlike field of HRP granules among densely labeled cell bodies. In such cases it may be difficult to distinguish labeled terminals of the afferent pathway from terminals of intrinsic collaterals or from labeling in dendritic branches (Robertson, 1977). It is usually possible to exclude a significant dendritic component for the simple reason that even the most sensitive HRP techniques do not reveal labeling in peripheral dendritic ramifications. However, it may be difficult to rule out an exclusive contribution from an intrinsic collateral system, in which case one may have to resort to control experiments using other techniques.

4. Labeling of Collateral Projections

The use of sensitive HRP techniques makes it possible to trace the primary or secondary branches of dichotomizing axons (de Olmos and Heimer, 1977; Frankfurter *et al.*, 1978). Such collateral pathways do not differ in their labeling characteristics from other axons labeled by anterograde HRP transport. Although this phenomenon can be used to guide and complement studies with double-labeling techniques (Chapter 8), it also creates problems of interpretation, since there is no way by which a collateral system of retrogradely labeled cells can be distinguished from an axonal pathway that has the cells of origin in the area of injection. However, an experimental neuroanatomist with some knowledge of the pathway being studied, will usually be able to clarify the situation with the aid of other tract-tracing procedures.

D. Sources of Error

1. Cell Bodies

The neurons of young animals exhibit little endogenous peroxidase activity if the pH of the incubation medium is maintained at 6.8 or below, as is recommended in most HRP protocols. However, it may be prudent to process sections from uninjected control animals for HRP demonstration in order to rule out endogenous, heme-containing enzymes (Wong-Riley, 1976) and to examine unincubated sections as well for the presence of endogenous pigments (Fig. 13A) such as lipofuchsin and the melanin granules in neurons of the locus coeruleus and substantia nigra. These latter may be particularly deceptive under dark-field illumination.

In contrast to neurons, Keefer and Christ (1976) show that in the normal rat, certain areas contain many small cells, presumed to be glial elements, which exhibit endogenous staining with diamino-benzidine at pH 5.0–8.5. Their paper should be consulted for details of the distribution and appearance of these reactive cells.

Another possible source of misinterpretation arises from the fact that granules of HRP reaction product in perivascular macrophages (Fig. 13B) and phagocytic glia (Fig. 13C) can resemble neuronal labeling in some planes of section. Red blood cells and the cytoplasmic granules of white blood cells, particularly polymorphonuclear elements, exhibit endogenous peroxidase activity that can occasionally be confusing, but, because of their small size and intravascular location, these elements can usually be recognized.

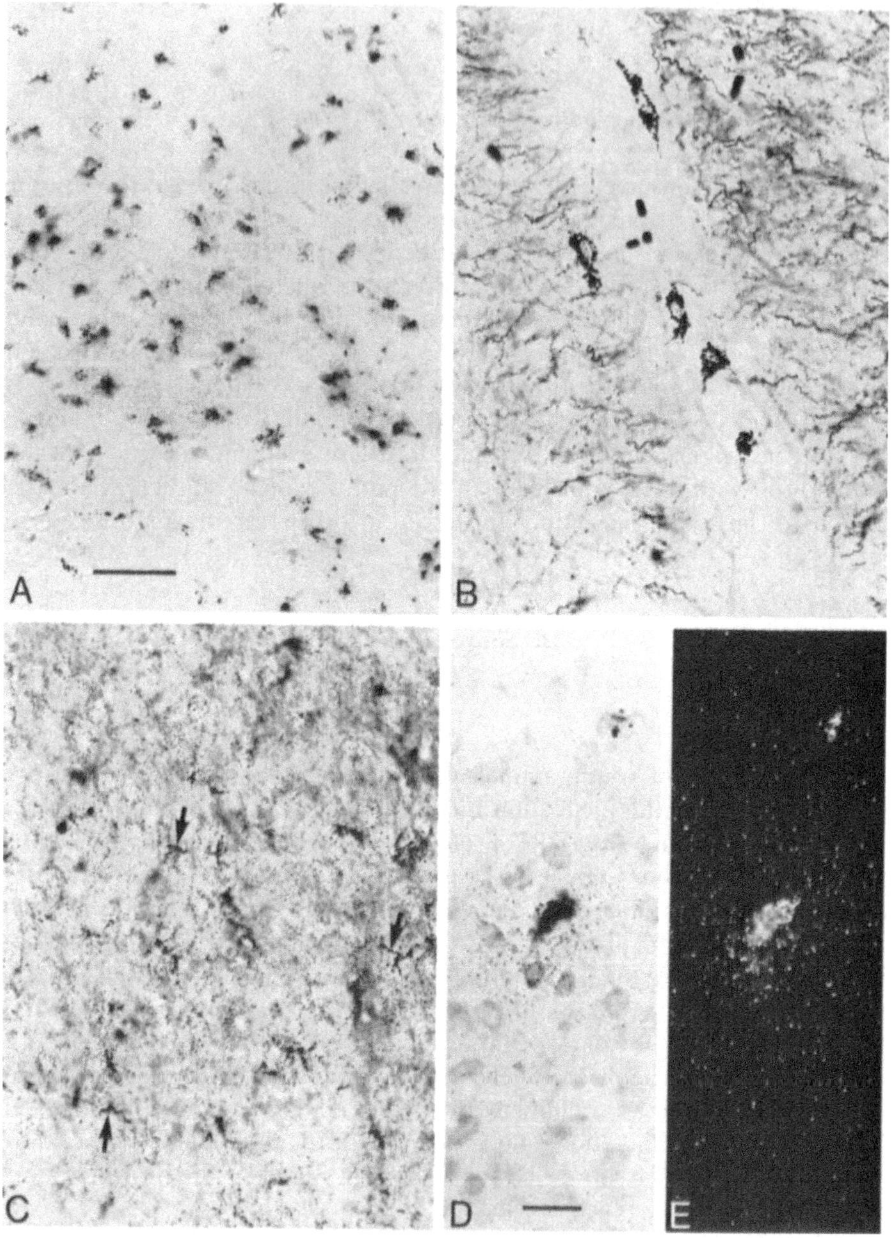

Figure 13. Sources of errors of interpretation. (A) An example of endogenous pigment in the rat hypothalamic arcuate nucleus. DAB stain. Bar, 50 μm. (B) Labeled pericytes in association with a large blood vessel that crosses through a zone of HRP-filled axons. The perivascular location of these cellular elements aids in their identification. Note the intravascular red blood cells which exhibit HRP activity because of their hemoglobin content. (C) Dense accumulations of stain in phagocytic glia (arrows). The fine, dispersed granules represent intraneuronal HRP label. (D and E) Bright- and dark-field micrographs of the same field showing a damaged pericyte, probably ruptured during sectioning. In the dark-field picture (E), the artifact is especially reminiscent of an HRP-labeled terminal field. Bar, 20 μm.

2. Terminal Fields

In addition to their occasionally being mistaken for labeled neuronal cell bodies, pericytes may occasionally rupture, giving rise to artifacts reminiscent of small HRP-labeled terminal fields (Fig. 13D and E). The most likely explanation for this is that an excessive hydrostatic pressure during perfusion causes rupture of capillaries and destruction of the pericytes with subsequent dispersion of their phagocytized HRP granules. Destroyed blood cells may also be the source of this type of artifact. The perivascular location and random distribution of these "exploded" cells usually permit their identification.

Deposition of chromogen crystals on the surface of the section is another artifact that is frequently observed in poorly fixed BDHC- and TMB-stained material. The occurrence of this artifact is especially frequent in material cut with a dull knife, since rough surfaces tend to accumulate crystalline artifacts. Areas of poor fixation are sometimes confined to particular cell groups, in which case the crystalline precipitate may easily be mistaken for HRP-labeled terminal fields, since this artifact has a tendency to occur consistently through a series of sections. Naturally, this problem can be solved easily by the use of a sufficient number of experimental animals.

VII. SUMMARY OF ADVANTAGES AND LIMITATIONS

A. Advantages

1. One of the unique advantages of the HRP method is its ability to localize specific cells of origin in a mixed population of neurons.
2. The recently developed HRP techniques may be as sensitive as the autoradiographic method for the tracing of axonal projections and terminal areas, but with the HRP method, these projections may be studied within a few days of the experiment rather than after the several weeks required for autoradiographic exposure and processing. Furthermore, the morphology of axons and axon terminals can be displayed in great detail (Fig. 8B) if one takes advantage of the diffusion process by applying HRP to the cut end of nerve fibers.
3. If DAB is used as the chromogen, the HRP reaction product obtained is osmiophilic and therefore suited for electron microscopic studies.
4. The HRP technique can be used for the tracing of collateral projections, either alone or in combination with other techniques (Chapter 8). It can also be combined successfully with various histochemical procedures capable of demonstrating transmitter-related enzymes.
5. A Golgi-like labeling of cell bodies and dendrites can be obtained, not only following intracellular injections of HRP (Chapter 7) but also in material with extracellular injections. This is an advantage that does not seem to have been exploited to its fullest extent.

B. Limitations

1. As with the autoradiographic technique, it is often very difficult to determine the effective injection site.
2. Several of the chromogens in regular use have undisputed carcinogenic properties.
3. The labeling of collateral pathways that is likely to occur when sensitive HRP techniques are used may create problems of interpretation.
4. In contrast to the autoradiographic tract-tracing procedure, which does not seem to affect fibers of passage, HRP can be transported through damaged axons, creating problems of interpretation.
5. The uptake of HRP by neurons is dependent on the concentration of free HRP and on the functional state of the neuron. As neurons with depressed activity have a tendency to take up less HRP, this may be a potential limitation. The uptake of HRP, however, can be made independent of these variables by conjugation with other substances such as wheat germ agglutinin (WGA).

VIII. APPENDIX

A. The 3,3′-Diaminobenzidine (DAB) Method

Based on the method of LaVail (1975)

1. Preparatory Steps

a. Fixation. Perfuse the animal with 100 ml (rat) rinsing solution consisting of 0.8% sucrose, 0.4% glucose, and 0.8% sodium chloride in 0.067 M phosphate buffer, pH 7.2–7.4, followed by 400 ml (rat) aldehyde mixture containing 1.5% glutaraldehyde, 1% paraformaldehyde, 4% sucrose, and 0.02% calcium chloride in 0.067 M phosphate or cacodylate buffer, pH 7.2–7.4 at room temperature. Following removal of the brain from the skull, place the brain in a jar with the same fixative at 4°C for 4 hr.

b. Cryoprotective Treatment. Transfer the brain to a cold (4°C) solution containing 30% sucrose in 0.067 M phosphate buffer. Leave the brain in the cold sucrose solution until it has sunk to the bottom of the jar, usually after 24 hr, at which time it is ready for sectioning.

c. Sectioning. Collect frozen sections (30–50 μm thick) serially in a compartmentalized plastic tray filled with a solution of chilled 5% sucrose in 0.1 M phosphate buffer, pH 7.2–7.4. The sections should be incubated for HRP demonstration as soon as possible because the enzymatic activity of

HRP diminishes with time, and the reaction product may be noticeably less intense in sections processed after 1 week of storage in buffer at 4°C.

d. Solutions and Reagents for Incubation. Prepare the following solutions for incubation:

0.1 M Phosphate Buffer, pH 5.5

The buffer is prepared by combining, in a proportion of 47 parts to 3 parts, 0.1 M solutions of $NaH_2PO_4 \cdot H_2O$ and Na_2HPO_4, respectively. Measure the pH of the resulting solution to insure that it is within ±0.1 of pH 5.5. The pH can be raised or lowered by adding, in increments, small amounts of the 0.1 M disodium or monosodium phosphate solutions, respectively. One liter of this buffer is required for each group of sections being incubated.

Chromogen Solution

Prepare a solution (approximately 0.5%) of 3,3′-diaminobenzidine (or of 3,3′-diaminobenzidine tetrahydrochloride) by adding 50 mg of the chromogen to 100 ml of distilled water. Agitate vigorously on a magnetic stirrer for 10 min. Filter through Whatman No. 1 paper.

Substrate-Working Solution (1% H_2O_2)

For convenience, 10 ml of a 1% working solution of hydrogen peroxide may be prepared from the commercially supplied stock (generally 30%) shortly before the HRP incubation is to be carried out. Dispense 0.33 ml of 30% H_2O_2 into 10 ml of distilled water in a brown, ground-glass-stoppered bottle.

Deactivator

A solution of approximately 5% sodium hypochlorite (Clorox® or equivalent household bleach).

2. Incubation Procedure

Transfer as many as 40 sections from a rat brain to a plastic sieve constructed, for example, from a polypropylene beaker, the bottom of which has been cut off and replaced by fine nylon mesh. All steps in the following procedure may be carried out at room temperature.

1. Wash sections briefly in three changes of the pH 5.5 buffer.
2. Incubate the sections for 20 min in a freshly prepared (less than 1-hr-old) medium consisting of chromogen solution, 100 ml, plus substrate-working solution, 1 ml. Gentle agitation at 5-min intervals is advisable.
3. Wash sections in three 5 min changes of the pH 5.5 buffer.

3. Mounting

Since the reaction product is extremely stable and virtually insoluble in alcohol, the sections may be safely stored for a day or two in buffer at 4°C or may be promptly mounted on slides from a gelatin–alcohol mixture prepared by combining equal parts of 1% gelatin solution and 80% ethyl alcohol. The mounted sections are allowed to air dry in a dust-free environment and may be stored indefinitely in an incubator at 37°C or may be dehydrated in alcohol, starting with 95%, cleared in xylene, and coverslipped.

4. Counterstaining

It is usually advisable to counterstain lightly at least some of the sections with cresyl violet or any one of a number of other Nissl stains.

5. Staining Characteristics

a. Reaction Product. Brown.

b. Labeled Neurons. Perikaryal lysosomes densely brown. Reaction product faithfully reflects, subject to the limitations of light microscopy, the characteristic localization of HRP in various membrane-bound organelles as seen ultrastructurally and often delineates with exquisite precision such intricate features of cellular morphology as dendritic spines and axonal end arborizations and their terminals.

c. Red Cells, White Cells, and Phagocytes. Positive reaction.

d. Melanin. No reaction.

e. Precipitate. Little if any.

6. Technical Notes

The most salient virtues of the DAB method described here are the high degree of morphological fidelity achieved in localizing HRP, the stability and resistance to fading of the reaction product, and the relative simplicity of the procedure. None of these advantages, however, compensates entirely for the now well-recognized shortcoming of the method, its relative insensitivity. It is therefore advisable to use one or more of the HRP methods described next.

B. Benzidine Dihydrochloride (BDHC) Method

J. S. de Olmos

1. Preparatory Steps

a. Fixation. Perfuse the animal with 100 ml (rat) solution consisting of 0.8% sucrose, 0.4% glucose, and 0.8% sodium chloride in 0.067 M phosphate buffer, pH 7.2–7.4, followed by 400 ml (rat) aldehyde mixture containing 1.5% glutaraldehyde, 1% paraformaldehyde, 1% tannic acid,* 4% sucrose, and 0.02% calcium chloride in 0.067 M phosphate buffer,† pH 7.2–7.4, at room temperature. After 15 min, the fixation continues for 30 more minutes with a solution of the same kind but without tannic acid (see Technical Note 1, Section VIII.B.6). Following removal of the brain from the skull, place the brain in a jar with the same fixative at 4°C for 4 hr.

b. Cryoprotective Treatment. Place the brain in a jar containing cold (4°C) 30% sucrose in 0.067 M phosphate buffer. Leave the brain in the sucrose solution until it has sunk to the bottom of the jar, usually after 24 hr, at which time it is ready for sectioning.

c. Sectioning. Collect 30–35 μm‡ thick frozen sections serially in a compartmentalized plastic tray filled with a solution containing 30% sucrose and 1% polyvinylpyrrolidone (PVP-40, Sigma) and 30% ethylene glycol buffered with 0.067 M sodium phosphate. Keep the solution chilled in an ice bath while collecting the sections. Store the tray in the freezer (−10 to −15°C) for at least 12 hr before staining. If needed, the sections can be kept in the freezer for at least 3 weeks without losing their enzymatic activity.

d. Stock Solutions. Prepare the following solutions for use in the staining procedure:

Gelatin–Methanol–Ethylene Glycol Solution (GME Solution)

0.2 M Acetic acid–sodium acetate buffer (pH 4.3–4.5)	60 ml
0.125% Aqueous gelatin§	240 ml
Absolute methanol	200 ml

*Tannic acid reacts slowly with the aldehydes, and a precipitate is eventually produced. Therefore, it is very important that the fixative solution be prepared immediately before fixation. The best results have been obtained with tannic acid ("analytical reagent") sold by Mallinckrodt, Inc., Paris, Kentucky 40361.

†If some of the material is being prepared for EM, cacodylate buffer should not be used in the preparation of fixative containing tannic acid (Anderson *et al.*, 1975).

‡For the study of anterogradely transported HRP, it is advisable not to use sections thicker than 35 μm.

§Gelatin 270 Bloom, Fisher Scientific Corporation, has been used regularly in our laboratory. Any gelatin of similar purity will probably be equally suitable. When dissolving the gelatin, the temperature of the water should not exceed 70°C.

Ethylene glycol	100 ml
Distilled water	420 ml

Keep the GME solution in the freezer at −10° to −20°C.

Heavy Metal Solutions

Prepare and keep separate the following aqueous metal solutions: 0.005 M manganese chloride and 0.005 M nickel ammonium sulfate.

Chromogen Solution

Sonicate 224 mg BDHC in 40 ml 50% methanol until the chromogen is completely dissolved. Then add 20 ml of ethyl ether and use the solution the same day. If benzidine free base (powerful carcinogen) is used instead of BDHC, 160 mg of the free base is dissolved directly in 60 ml ethyl ether. For preparation of the free base, see Straus (1967).

Stabilizer

0.2 M sodium nitroprusside (anhydrous). Keep at 4°C in the dark and use within a day or two.

Ozonic Ether (Pearse, 1972)

Ethyl ether (anhydrous, reagent grade)	50 ml
Absolute methanol	100 ml
3% Hydrogen peroxide	50 ml

The solution can be kept in the freezer (−10 to −20°C) for several weeks. Use a tightly stoppered bottle.

2. Staining Procedure

Select the sections from one of the tray compartments and place them in a plastic cup in which the bottom has been replaced by a fine mesh net. Make sure that the cup as well as the net are resistant to organic solvents. The sections in the plastic cup can then be carried with ease from solution to solution.

1. Wash briefly in 3 changes of the chilled GME solution.
2. Move the sections to the following incubating bath kept at −15° to −20°C.

GME solution	85.00 ml
0.005 M Manganese chloride	0.10 ml
0.005 M Nickel ammonium sulfate*	0.20 ml
Chromogen solution	15.00 ml

*0.005 M cobalt chloride can also be used.

This amount is adequate for 60–80 transversely cut rat brain sections.

3. After 10 min in the incubating bath, add 1 ml of the stabilizer. Mix well by shaking the jar and add 0.35 ml of the ozonic ether. Mix well again.
4. Additional amounts of 0.20 ml ozonic ether should be added after 180, 60, 60, and 30 min (see Technical Note 2, Section VIII.B.6).
5. Remove the sections from the freezer and wash them quickly in three successive baths of methanol (−20°C) followed by two baths of cold distilled water (kept at 4°C in an ice bath).

3. Mounting

Mount the sections on gelatin-coated slides direct from the second water bath. Blot the sections in a gentle manner with bibulous paper and dry them in a stream of dry cold air (see Technical Note 3, Section VIII.B.6).

Complete the dehydration by carrying the slides through successive jars of absolute methanol–xylene, 1:1; methanol–xylene, 1:6; and two jars of pure xylene. Pass quickly through the first two baths to avoid extraction of the stain (see Technical Note 4, Section VIII.B.6).

Blot the sections quickly with bibulous paper to extract excess xylene and coverslip with permount.

4. Counterstaining

Light counterstaining seems to render the blue reaction product more stable, and it is therefore advisable to prepare a certain number of counterstained sections. This should be done in the following way.

After mounting and drying, transfer sections directly to a solution containing 0.01% toluidine blue for about 30 sec or until they show a light bluish color. Wash in distilled water. Dehydrate quickly and cover.

5. Staining Characteristics

a. Reaction Product. Very fine blue, predominantly spherical granules in bright-field microscopy. Reddish color in dark-field.

b. Labeled Cells. More or less pronounced cytoplasmic deposition of granular reaction product in perikarya and major dendritic branches.

c. Axons. Rows of more or less densely packed granules.

d. Axon Terminals. *Active transport*: More or less random distribution of granular particles in terminal fields. *Diffusion*: Granular particles arranged in patterns closely indicating the underlying morphology.

e. Red Cells and Pericytes. Positive reaction.

f. Melanin. Negative reaction.

g. Precipitate. Small round or oval crystals, sometimes forming small star-shaped aggregates.

6. Technical Notes

1. Although other fixatives such as those proposed by Rosene and Mesulam (1978) or Malmgren and Olsson (1978) can be used for the BDHC method, the fixative containing tannic acid has given more consistent results in our laboratory.
2. It is advisable to run a series of test sections with varying incubation times. In doing so, it is convenient to separate sections containing the injection site from those that do not. A clear background is often obtained at the expense of sensitivity, and it is therefore advisable to carry the reaction beyond the stage where crystal artifacts start to appear. However, if the sections are incubated for too long, heavy crystallization will eventually make evaluation of the sections almost impossible.
3. The drying of the sections in warm and humid air should be avoided. Otherwise, a pinkish staining of the background will develop. The sections should be mounted or counterstained immediately following the drying. If for any reason the sections cannot be mounted immediately after the staining, they can be left in 35% ethylene glycol in the freezer until a more convenient time.
4. If the sections are carried too quickly through the alcohol-containing baths, incomplete dehydration with subsequent loss of transparency may occur. A more prolonged dehydration process can apparently be achieved by replacing alcohol with acetone, which seems to preserve the blue reaction product better (Dr. Anthony Frankfurter, unpublished observation).

C. Tetramethylbenzidine (TMB) Method

J. S. de Olmos

1. Preparatory Steps

a. Fixation. Perfuse the animal with 100 ml (rat) rinsing solution consisting of 0.8% sucrose, 0.4% glucose, and 0.8% sodium chloride in 0.067 M phosphate buffer, pH 7.2–7.4, followed by 400 ml (rat) aldehyde mixture containing 1.5% glutaraldehyde, 1% paraformaldehyde, 4% sucrose, and 0.02% calcium chloride in 0.067 M phosphate or cacodylate buffer, pH 7.2–7.4, at room temperature. Following removal of the brain from the skull, place the brain in a jar with the same fixative at 4°C for 4 hr.

b. Cryoprotective Treatment. Transfer the brain to a cold (4°C) solution containing 30% sucrose in 0.067 M phosphate buffer. Leave the brain in the cold sucrose solution until it has sunk to the bottom of the jar, usually after 24 hr, at which time it is ready for sectioning.

c. Sectioning. Collect frozen sections (30–50 μm thick) serially in a compartmentalized plastic tray filled with a solution containing 30% sucrose and 1% polyvinylpyrrolidone (PVP-40, Sigma) and 30% ethylene glycol buffered with 0.067 M sodium phosphate. Keep the solution chilled in an ice bath while collecting the sections. Store the tray in the freezer (– 10 to – 15°C) for at least 12 hr before staining. If needed, the sections can be stored in the freezer for at least 3 weeks without losing their enzymatic activity.

d. Stock Solutions. Prepare the following solutions for use in the staining procedure.

Ethylene Glycol–Buffer Solution (EGB Solution)

0.05 M Acetate buffer	20 ml
Ethylene glycol	100 ml
Distilled water	880 ml

Gelatin–Ethanol–Dimethyl Sulfoxide (DMSO) Solution

0.05 M Acetate buffer (pH 3.5)	20 ml
1% Aqueous gelatin*	400 ml
Absolute ethanol	100 ml
DMSO	10 ml
Distilled water	470 ml

Keep the solution at room temperature.

Heavy Metal Solutions

Prepare and keep separate the following aqueous metal solutions: 0.005 M manganese chloride and 0.005 M cadmium sulfate.

Chromogen Solution

Sonicate 14 mg TMB in 10 ml ethyl ether. Prepare fresh.

*Gelatin 270 Bloom, Fisher Scientific Corporation, has been used regularly in our laboratory. Any gelatin of similar purity will probably be equally suitable. When dissolving the gelatin, the temperature of the water should not exceed 70°C.

Stabilizer

0.2 M Sodium nitroprusside. Prepare fresh.

Ozonic Ether (Pearse, 1972)

Ethyl ether (anhydrous, reagent grade)	50 ml
Absolute methanol	100 ml
3% Hydrogen peroxide	50 ml

The solution can be kept in the freezer (−10 to −20°C) for several weeks. Use a tightly stoppered bottle.

2. Staining Procedure

Select the sections from one of the tray compartments and place them in a plastic cup in which the bottom has been replaced by a fine mesh net. Make sure that the cap as well as the net are resistant to organic solvents. The sections in the plastic cup can then be carried with ease from solution to solution.

1. Wash briefly in three changes of the chilled (0°C) EGB solution.
2. Move the sections to the following incubating bath kept at 4°C by the aid of an ice bath.

Gelatin–ethanol–DMSO solution	190.00 ml
0.005 M Manganese chloride	0.25 ml
0.005 M Cadmium sulfate*	0.50 ml
Chromogen solution	10.00 ml

   This amount is adequate for 60–80 transversely cut rat brain sections.
3. After 10–20 min in the incubating bath, add 2.4 ml of the stabilizer. Mix well by shaking the jar and add 0.7 ml of the ozonic ether. Mix well again.
4. Additional amounts of 0.4 ml ozonic ether should be added after 20, 40 and 60 min (see Technical Note 1, Section VIII.C.6).
5. Wash the sections quickly in three successive baths of chilled distilled water.

3. Mounting

Mount the sections on gelatin-coated slides direct from the third water bath. Blot the sections in a gentle manner with bibulous paper and dry them in a stream of dry cold air (see Technical Note 2, Section VIII.C.6).

Complete dehydration by carrying the slides through successive jars of

*0.005 M nickel amonium sulfate or 0.005 M cobalt chloride can also be used.

absolute ethanol–xylene, 1:1; ethanol–xylene, 1:6; and two jars of pure xylene. Pass quickly through the first two baths to avoid extraction of the stain (see Technical Note 3, Section VIII.C.6).

Blot the sections quickly with bibulous paper to extract excess xylene and coverslip with permount.

4. Counterstaining

Light counterstaining seems to render the grayish blue reaction product more stable, and it is therefore advisable to prepare a certain number of counterstained sections. This should be done in the following way.

After being mounted and dried, sections are transfered directly to a solution containing 0.01% toluidine blue for about 30 sec or until they show a light bluish color. Wash in distilled water. Dehydrate quickly and cover.

5. Staining Characteristics

a. Reaction Product. Grayish blue fine spherical to oval granules in bright-field microscopy. Golden granules in dark-field.

b. Labeled Cells. More or less pronounced cytoplasmic deposition of granular reaction product in perikarya and a major dendritic branches. Irregular contours of labeled cell bodies and dendrites.

c. Axons. Rows of more or less densely packed granular particles.

d. Axon Terminals. *Active transport*: More or less random distribution of granular particles in terminal fields. *Diffusion*: Granular particles arranged in patterns closely indicating the underlying morphology.

e. Red Cells and Pericytes. Positive reaction.

f. Melanin. Negative reaction.

g. Precipitate. Rodlike crystals occasionally forming star-shaped aggregates.

6. Technical Notes

1. It is advisable to run a series of test sections with varying incubation times. In doing so, it is convenient to separate sections containing the injection site from those which do not. A clear background is often obtained at the expense of sensitivity, and it is therefore advisable to

carry the reaction beyond the stage where crystal artifacts start to appear. However, if the sections are incubated for too long, heavy crystallization will eventually make evaluation of the sections almost impossible.

2. The drying of the sections in warm and humid air should be avoided. Otherwise, fading may occur. If needed, dried sections can be stored for several days in refrigerator.
3. If the sections are carried too quickly through the alcohol-containing baths, incomplete dehydration with subsequent loss of transparency may occur. A more prolonged dehydration process can apparently be achieved by replacing alcohol with acetone, which seems to preserve the blue reaction product better (Dr. Anthony Frankfurter, unpublished observation).

D. Tetramethylbenzidine (TMB) Method

M.-M. Mesulam

1. Preparatory Steps

a. Fixation. The perfusion is initiated by a rapid bolus of 50 (rat)–200 (monkey) ml of physiological saline or Ringer's solution at room temperature. This is followed by 500 (rat)–2000 (monkey) ml of fixative at room temperature. The fixative contains 1.25% glutaraldehyde and 1% paraformaldehyde in an 0.1 M phosphate buffer at pH 7.4. The flow rate is adjusted so that half the volume of fixative flows as rapidly as possible while the other half is administered more slowly in order to keep the total time for the perfusion with fixative at 30 min. Then the animal is perfused with 500 (rat)–2000 (monkey) ml of the 0.1 M, pH 7.4 phosphate buffer which has a sucrose concentration of 10%. This sucrose–buffer solution should have a temperature of 4°C.

b. Cryoprotective Treatment. The brain is then removed and stored in the same sucrose–buffer solution at 4°C for 2–24 hr.

c. Sectioning. This is performed on a freezing microtome. Although satisfactory results can be obtained with sections varying in thickness from 30 to 80 μm, 40-μm-thick sections are preferred. The sections are collected at room temperature in a 0.1 M phosphate buffer at pH 7.4 (no sucrose added). Special precautions must be taken if sections are stored before staining (see Technical Note 1, Section VIII.D.6).

d. Stock Solutions. The following solutions are prepared for use in the staining procedure.

Buffer at pH 3.3

Add 200 ml of 1.0 M sodium acetate to 200 ml distilled water; add 190 ml of 1.0 M HCl; make up the volume to 1000 ml with distilled water. Check the final pH and titrate with concentrated acetic acid or sodium hydroxide in order to bring the final pH to 3.3.

Solution A

Mix 92.5 ml of distilled water, 5 ml of the pH 3.3 buffer, and 100 mg of sodium nitroferricyanide (Sigma). Multiples of these quantities may be used for larger volumes.

Solution B

Add 5 mg of 3,3′,5,5′-tetramethylbenzidine (TMB, Sigma) to each 2.5 ml of absolute ethanol. This may be heated to 37–40°C to dissolve the TMB.

Incubation Solution

This is made by adding 2.5 ml of solution B to each 97.5 ml of solution A. Neither solution A nor solution B should be older than 2 hr. The actual mixing of A to B should take place within the incubation vessel and only seconds before the sections are introduced.

Peroxide Solution

This should have a concentration of 0.3% of H_2O_2 in water.

Neutral Red Solution

Add 40 ml of pH 4.8 acetate buffer to each liter of a 1% aqueous neutral red solution. To prepare the buffer, add 500 ml of 0.1 N acetic acid to 750 ml of 0.1 M sodium acetate. Filter before using.

Postreaction Storage and Rinsing Solution

This is prepared by adding 5 ml of the pH 3.3 buffer to each 95 ml of distilled water.

e. Glassware. This should be free from oxidizing agents. Specifically, bleach (Clorox®) should not be used, and the glassware should be cleaned with soap and water.

f. Chemicals. For most of the chemicals, the exact supplier is not critical. The one exception is TMB. We have had consistently excellent results with TMB purchased from Sigma, whereas TMB purchased from two other suppliers has given strikingly inferior results.

2. Staining Procedure

1. Rinse in three changes of distilled water (1 min each).
2. Prereaction soak for 20 min at room temperature (10–23°C). Sections are immersed in the "incubation solution" on a rocker table. The "incubation solution" is prepared by mixing solutions A and B as described in Section VIII.D.1d. The actual mixing should be done

within the reaction vessel and only seconds before the sections are introduced. This solution should maintain a clear amber emerald color during the 20 min; otherwise, chemical contamination must have occurred.

3. Enzymatic reaction for 20 min at room temperature (19–23°C). This is initiated by adding 2.5–4.5 ml of the 0.3% H_2O_2 solution to each 100 ml of the "incubation solution." The sections are removed, the H_2O_2 is rapidly stirred in, and the sections are reintroduced. This reaction is carried out on a rocker table with gentle agitation. The optimal quantity of H_2O_2 may vary from one case to another. This should be determined by performing five "trial" sets on a limited number of representative sections by adding 2.5, 3, 3.5, 4, or 4.5 ml of the 0.3% H_2O_2 per 100 ml of the "incubation solution." The optimal concentration of H_2O_2 may be determined by examining labeled perikarya and fields of anterograde transport in the trial sections. The objective is to find an H_2O_2 concentration that demonstrates the most transport but that does not produce excessive levels of artifact.
4. Rinse three times, 1 min at a time, in the "postreaction storage and rinsing solution" prepared as described in Section VIII.D.1d. The sections are left in this solution until they are mounted. They can be left here for up to 4 hr at 0–4°C without appreciable loss of reaction product (see Technical Note 2, Section VIII.D.6).

3. Mounting

The sections are mounted at room temperature from the "postreaction storage and rinsing solution" onto glass slides subbed with chrome alum. These are left to air dry for 5–48 hr at room temperature.

4. Counterstaining, Dehydration, Clearing, and Coverslipping

It is most convenient to counterstain half the sections in each series, while the alternate sections may be dehydrated (2 changes, 15 sec each, in 100% ethanol), cleared (2 changes, 1–15 min each, in xylene), and coverslipped (Permount) without counterstaining.

The sections to be counterstained are immersed in the neutral red solution for 3 min. Then they are dehydrated and cleared according to the following schedule: 15 sec in water, 15 sec in 70% ethanol, 15 sec in 95% ethanol, 15 sec in 100% ethanol, 15 sec in a second bath of 100% ethanol, 1 min in xylene, and then 1–30 min in a second bath of xylene. Coverslipping is satisfactory with Permount.

As described in Technical Note 3, Section VIII.D.6, counterstaining offers advantages as well as disadvantages. Avoid prolonged treatment in ethanol during the dehydration (see Technical Note 4, Section VIII.D.6).

5. Staining Characteristics

See previous TMB methods, Section VIII.C.5.

6. Technical Notes

1. Once the tissue has been cut, the HRP activity in the tissue is rapidly depressed unless proper precautions are taken. Cell counts from matching tissue sections of the same case have shown that the number of labeled perikarya that can be demonstrated does not diminish if the tissue is stored at 4°C in 1.0 M pH 7.4 phosphate buffer for up to 96 hr before it is reacted (Mesulam *et al.*, 1980). We have obtained similar results when the tissue is stored at − 10°C in phosphate buffer that contains 30% sucrose and 30% ethylene glycol as suggested by de Olmos *et al.* (1978).
2. It is important that the sections that have already undergone the TMB reaction be rinsed and stored only in this "postreaction storage and rinsing solution." The use of either distilled water or phosphate buffer at this stage results in significant loss of the TMB reaction product.
3. An obvious advantage of counterstaining is that it facilitates the identification of cytoarchitectonic and cytological detail. Another less well-known advantage of counterstaining is related to the stability of the TMB reaction product. Preliminary observations suggest that the TMB reaction product occasionally tends to show gradual fading at room temperature in coverslipped sections. This fading appears to occur only in sections that are not counterstained. Counterstaining with neutral red seems in most cases to eliminate the fading. On the other hand, the process of counterstaining may itself depress the visibility of the TMB reaction product and may interfere with the clarity of detail during microscopic examination or photography.

   Counterstained sections may be kept at room temperature for us to 3 years (and probably for much longer) without any appreciable loss of the TMB reaction product.

   The neutral red stain may itself fade in time, especially when exposed to bright light. However, the mere precaution of keeping the slides in a covered box when not in use is sufficient for maintaining an adequate counterstain for 2–3 years.
4. The TMB reaction product is vulnerable to ethanol. In fact, a 30-min exposure to 95 or 100% ethanol is sufficient to result in the complete loss of the TMB reaction product from the tissue. Therefore, it is important to limit the exposure to alcohols as suggested by the dehydration protocol described in Section VIII.D.4. A more prolonged dehydration process can apparently be achieved by replacing

alcohol with acetone which seems to preserve the blue reaction product better (Dr. Anthony Frankfurter, unpublished observation).

ACKNOWLEDGMENTS. This work was supported by USPHS NS 10972 and NS 14832 and by Consejo Nacional de Investigationes Cientificas Tecnicas of Argentina.

REFERENCES

Adams, J. C., 1977, Technical considerations on the use of the HRP as a neuronal marker, *Neuroscience* **2:**141–145.

Adams, J. C., and Warr, W. B., 1976, Origins of axons in the cat's acoustic striae determined by injection of horseradish peroxidase into severed tracts, *J. Comp. Neurol.* **170:**107–122.

Aghajanian, G. K., and Wang, R. Y., 1977, Habenular and other midbrain raphe afferents demonstrated by a modified retrograde tracing technique, *Brain Res.* **122:**229–242.

Andersson, W. A., Trantalis, J., and Kany, Y.-H., 1975, Ultrastructural localization of endogenous mammary gland peroxidase during lactogenesis in the rat, *J. Histochem. Cytochem.* **23:**295–302.

Arvidsson, J., 1975, Location of cat trigeminal ganglion cells innervating dental pulp of upper and lower canines studied by retrograde transport of horseradish peroxidase, *Brain Res.* **99:**135–139.

Beatti, M. S., Bresnahan, J. C., and King, J. S., 1978, Ultrastructural identification of dorsal root primary afferent terminals after anterograde filling with horseradish peroxidase, *Brain Res.* **153:**127–134.

Beitz, A. J., and King, G. W., 1976, An improved technique for the microinjection of horseradish peroxidase, *Brain Res.* **108:**175–179.

Berger, A. J., 1979, Distribution of carotid sinus nerve afferent fibers to solitary tract nuclei of the cat using transganglionic transport of horseradish peroxidase, *Neurosci. Lett.* **14:**153–158.

Broadwell, R. D., and Brightman, M. W., 1976, Entry of peroxidase into neurons of the central and peripheral nervous system from extracerebral and cerebral blood, *J. Comp. Neurol.* **166:**257–284.

Broadwell, R. D., and Brightman, M. W., 1979, Cytochemistry of undamaged neurons transporting exogenous protein in vivo, *J. Comp. Neurol.* **185:**31–74.

Broadwell, R. D., Oliver, C., and Brightman, M. W., 1980, Neuronal transport of acid hydrolases and peroxidase within the lysosomal system of organelles: Possible involvement of a specialized compartment of the agranular reticulum, *J. Comp. Neurol.* **190:**519–530.

Brownson, R. H., Uusitalo, R., and Palkama, A., 1977, Intraaxonal transport of horseradish peroxidase in the sympathetic nervous system, *Brain Res.* **120:**407–422.

Bunt, A. H., and Haschke, R. H., 1978, Features of foreign proteins affecting their retrograde transport in axons of the visual system, *J. Neurocytol.* **7:**665–678.

Bunt, A. H., Lund, R. D., and Lund, J. S., 1974, Retrograde axonal transport of horseradish peroxidase by ganglion cells of the albino rat retina, *Brain Res.* **73:**215–228.

Bunt, A. H., Hendrickson, A. E., Lund, J. S., Lund, R. D., and Fuchs, A. F., 1975, Monkey retinal ganglion cells: morphometric analysis and tracing of axonal projections, with a consideration of the peroxidase technique, *J. Comp. Neurol.* **164:**265–285.

Bunt, A. H., Haschke, R. H., Lund, R.. D., and Calkins, D. F., 1976, Factors affecting retrograde axonal transport of horseradish peroxidase in the visual system, *Brain Res.* **102:**152–155.

Cancilla, P. A., Baker, R. N., Pollack, P. S., and Frommes, S. P., 1972, Reaction of pericytes of the central nervous system to exogenous protein, *Lab. Invest.* **26**(4)**:**376–383.

Clarke, P. G. H., and Cowan, W. M., 1976, The development of the isthmo–optic tract in the chick, with special reference to the occurrence and correction of developmental errors in the location and connections of isthmo–optic neurons, *J. Comp. Neurol.* **167:**143–163.

Colman, D. R., Scalia, F., and Cabrales, E., 1976, Light and electron microscopic observations on the anterograde transport of horseradish peroxidase in the optic pathway in the mouse and rat, *Brain Res.* **102:**156–163.

Courville, J., and Saint-Cyr, J. A., 1978, Modification of the horseradish peroxidase method avoiding fixation, *Brain Res.* **142:**551–558.

Dennis, B. J., and Kerr, D. I. B., 1976, Origins of olfactory bulb centrifugal fibres in the cat, *Brain Res.* **110:**593–600.

de Olmos, J. S., 1977, An improved HRP method for the study of central nervous connections, *Exp. Brain Res.* **29:**541–551.

de Olmos, J. S., and Carrer, H., 1978, A horseradish peroxidase study of the afferent connections to the medial basal hypothalamus in the rat, *Anat. Rec.* **190:**380.

de Olmos, J. S., and Heimer, L., 1977, Mapping of collateral projection with the HRP method, *Neurosci. Lett.* **6:**107–114.

de Olmos, J. S., Hardy, H., and Heimer, L., 1978, The afferent connections of the main and accessory olfactory bulb formations in the rat. An experimental HRP study, *J. Comp. Neurol.* **181:**213–244.

Deschênes, M., Landry, P., and Labelle, A., 1979, The comparative effectiveness of the "brown and blue reactions" for tracing neuronal processes of cells injected intracellularly with horseradish peroxidase, *Neurosci. Lett.* **12:**9 15.

DeVito, J. L., Clausing, K. W., and Smith, O. A., 1974, Uptake and transport of horseradish peroxidase by cut end of the vagus nerve, *Brain Res.* **82:**269–271.

Donoso, J. A., Illanes, J.-P., and Samson, F., 1977, Dimethylsulfoxide action on fast axoplasmic transport and ultrastructure of vagal axons, *Brain Res.* **120:**287–301.

Dürsteler, M. R., Blakemore, C., and Garey, L. J., 1977, Uptake of horseradish peroxidase by geniculo–cortical axons in the golden hamster: Analysis by computer reconstruction, *Exp. Brain Res.* **29:**487–500.

Frankfurter, A., Persing, J., and Steward, O., 1978, Collateral afferents to the inferior olive and superior colliculus originating in the spinal trigeminal nucleus, *Neurosci. Abstr.* **4:**551.

Furstman, L., Saporta, S., and Krüger, L., 1975, Retrograde axonal transport of HRP in sensory nerves and ganglion cells of the rat, *Brain Res.* **84:**300–304.

Gobel, S., and Falls, W. M., 1979, Anatomical observations of horseradish peroxidase-filled terminal primary axonal arborizations in layer II of the substantia gelatinosa of Rolando, *Brain Res.* **175:**335–340.

Goldfischer, S., 1969, Further observations on the peroxidatic activities of microbodies (peroxisomes), *J. Histochem. Cytochem.* **17:**61–65.

Goldschmidt, R., and Steward, O., 1980, Time course of increases in retrograde labeling and increases in cell size in entorhinal cortex neurons sprouting in response to unilateral entorhinal lesions, *J. Comp. Neurol.* **189:**359–380.

Gonatas, N. K., Harper, C., Mizutani, T., and Gonatas, J. O., 1979, Superior sensitivity of conjugates of horseradish peroxidase with wheat germ agglutinin for studies of retrograde axonal transport, *J. Histochem. Cytochem.* **27:**728–734.

Graham, R. C., and Karnovsky, M. J., 1966, The early stages of absorption of injected horseradish peroxidase in the proximal tubules of mouse kidney: Ultrastructural cytochemistry by a new technique, *J. Histochem. Cytochem.* **14:**291–302.

Grant, G., Arvidsson, J., Robertson, B., and Ygge, J., 1979, Transganglionic transport of horseradish peroxidase in primary sensory neurons, *Neurosci. Lett.* **12:**23–28.

Graybiel, A. M., and Devor, M., 1974, A microelectrophoretic delivery technique for use with horseradish peroxidase, *Brain Res.* **68:**167–173.

Haberly, L. B., and Price, J. L., 1978, Association and commissural fiber systems of the olfactory cortex of the rat. I. Systems originating in the piriform cortex and adjacent areas, *J. Comp. Neurol.* **178:**711–740.

Hadley, R. T., and Trachtenberg, M. C., 1978, Poly-L-ornithine enhances the uptake of horseradish peroxidase, *Brain Res.* **158:**1–14.

Halpern, M., Want, R. T., and Colman, D. R., 1976, Centrifugal fibers to the eye in a nonavian vertebrate: Source revealed by horseradish peroxidase studies, *Science* **194:**1185–1187.

Hanker, J. S., Yates, P. E., Metz, C. B., Carson, K. A., Light, A., and Rustioni, A., 1977, A new specific, sensitive and noncarcinogenic reagent for the demonstration of horseradish peroxidase (HRP), *Neurosci. Abstr.* **3:**30.

Hansson, H.-A., 1973, Uptake and intracellular bidirectional transport of horseradish peroxidase in retinal ganglion cell, *Exp. Eye Res.* **16:**377–388.

Hardy, H., and Heimer, L., 1977, A safer and more sensitive substitute for diaminobenzidine in the light microscopic demonstration of retrograde and anterograde axonal transport of HRP, *Neurosci. Lett.* **5:**235–240.

Hedreen, J. C., and McGrath, S., 1977, Observations on labeling of neuronal cell bodies, axons, and terminals after injection of horseradish peroxidase into rat brain, *J. Comp. Neurol.* **176:**225–246.

Herkenham, M., and Nauta, W. J. H., 1977, Afferent connections of the habenular nuclei in the rat. A horseradish peroxidase study, with a note on the fiber-of-passage problem, *J. Comp. Neurol.* **173:**123–146.

Heuser, J. E., and Reese, T. S., 1975, Evidence for recycling of synaptic vesicle membrane during transmitter release at the frog neuromuscular junction, *J. Cell Biol.* **57:**315–344.

Holland, V. R., Saunders, B. C., Rose, F. L., and Walpole, A. L., 1974, A safer substitute for benzidine in the detection of blood, *Tetrahedron* **30:**3299–3302.

Holtzman, E., 1977, The origin and fate of secretory packages, especially synaptic vesicles, *Neuroscience* **2:**327–355.

Illing, R.-B., and Wässle, H., 1979, Visualization of the HRP reaction product using the polarization microscope, *Neurosci. Lett.* **13:**7–11.

Itaya, S. K., Williams, T. H., and Engel, E. L., 1978, Anterograde transport of horseradish peroxidase enhanced by poly-L-ornithine, *Brain Res.* **150:**170–176.

Itoh, K., Konishi, A., Nomura, S., Mizuno, N., Nakamura, Y., and Sugimoto, T., 1979, Application of coupled oxidation reaction to electron microscopic demonstration of horseradish peroxidase: cobalt–glucose oxidase method, *Brain Res.* **175:**341–346.

Jacobson, S., and Trojanowski, J. Q., 1975, The appearance of dendrites of callosal and corticothalamic neurons in somatosensory cortex of immature rats demonstrated by horseradish peroxidase, *Adv. Neurol.* **12:**319–333.

Jones, E. G., 1975, Possible determinants of the degree of retrograde neuronal labeling with horseradish peroxidase, *Brain Res.* **85:**249–253.

Jones, E. G., and Hartman, B. K., 1978, Recent advances in neuroanatomical methodology, *Annu. Rev. Neurosci.* **1:**215–296.

Jones, E. G., and Leavitt, R. Y., 1974, Retrograde axonal transport and the demonstration of non-specific projections to the cerebral cortex and striatum from thalamic intralaminar nuclei in the rat, cat and monkey, *J. Comp. Neurol.* **154:**349–378.

Jones, E. G., Coulter, J. D., Burton, H., and Porter, R., 1977, Cells of origin and terminal distribution of corticostriatal fibers arising in the sensory motor cortex of monkeys, *J. Comp. Neurol.* **173:**53–80.

Keefer, D. A., 1978, Horseradish peroxidase as a retrogradely-transported, detailed dendritic marker, *Brain Res.* **140:**15–32.

Keefer, D. A., and Christ, J. F., 1976, Distribution of endogenous diaminobenzidine-staining cells in the normal rat brain, *Brain Res.* **11:**312–316.

Keefer, D. A., Spatz, W. B., and Misgeld, U., 1976, Golgi-like staining of neocortical neurons using retrogradely transported horseradish peroxidase, *Neurosci. Lett.* **3:**233–237.

Kicliter, E., 1979, Some telencephalic connections in the frog, *Rana pipiens*, *J. Comp. Neurol.* **185:**75–86.

Kim, C. C., and Strick, P. L., 1976, Critical factors involved in the demonstration of horseradish peroxidase retrograde transport, *Brain Res.* **103:**356–361.

Krishnan, N., and Singer, M., 1973, Penetration of peroxidase into peripheral nerve fibers, *Am. J. Anat.* **136:**1–14.

Kristensson, K., 1975, Retrograde axonal transport of protein tracers, in: *The Use of Axonal*

Transport for Studies of Neuronal Connectivity (W. M. Cowan and M. Cuénod, eds.), pp. 69–82, Elsevier, Amsterdam.

Kristensson, K., and Olsson, Y., 1974, Retrograde transport of horseradish peroxidase in transected axons. 1. Time relationships between transport and induction of chromatolysis, *Brain Res.* **79:**101–109.

Kristensson, K., and Olsson, Y., 1975, Retrograde transport of horseradish peroxidase in transected axons. 2. Relations between rate of transfer from the site of injury to the perikaryon and onset of chromatolysis, *J. Neurocytol.* **4:**653–661.

Kristensson, K., and Olsson, Y., 1976, Retrograde transport of horseradish peroxidase in transected axons. 3. Entry into injured axons and subsequent localization in perikaryon, *Brain Res.* **115:**201–213.

Kristensson, K., and Sjöstrand, J., 1972, Retrograde transport of protein tracer in the rabbit hypoglossal nerve during regeneration, *Brain Res.* **45:**175–181.

Kristensson, K., Olsson, Y., and Sjöstrand, J., 1971, Axonal uptake and retrograde transport of exogenous proteins in the hypoglossal nerve, *Brain Res.* **32:**399–406.

Kusuma, A., and Donkelaar, H. J. T., 1979, Staining of the dorsal root primary afferent fibers by anterograde movement of horseradish peroxidase and retrograde labelling of motoneurons and preganglionic autonomic cells in the turtle spinal cord, *Neurosci. Lett.* **14:**141–146.

Kuypers, H. G. J. M., Catsman-Berrevoets, C. E., and Padt, R. E., 1977, Retrograde axonal transport of fluorescent substances in the rat's forebrain, *Neurosci. Lett.* **6:**127–135.

LaVail, J. H., 1975, The retrograde transport method, *Fed. Proc.* **34:**1618–1624.

LaVail, J. H., 1978, A review of the retrograde transport technique, in: *Neuroanatomical Research Techniques* (R. T. Robertson, ed.), pp. 355–384, Academic Press, New York.

LaVail, J. H., and LaVail, M. M., 1972, Retrograde axonal transport in the central nervous system, *Science* **176:**1416–1417.

LaVail, J. H., and LaVail, M. M., 1974, The retrograde intraaxonal transport of horseradish peroxidase in the chick visual system: A light and electron microscopic study, *J. Comp. Neurol.* **157:**303–358.

LaVail, M. M., and LaVail, J. H., 1975, Retrograde intraaxonal transport of horseradish peroxidase in retinal ganglion cells of the chick, *Brain Res.* **85:**273–289.

LaVail, J. H., Rapisardi, S. and Sugino, I. K., 1980, Evidence against the smooth endoplasmic reticulum as a continuous channel for the retrograde axonal transport of horseradish peroxidase, *Brain Res.* **191:**3–20.

Leger, L., Sakai, K., Salvert, D., Touret, M., and Jouvet, M., 1975, Delineation of dorsal lateral geniculate afferents from the cat brain stem as visualized by the horseradish peroxidase technique, *Brain Res.* **93:**490–496.

Light, A. R., and Perl, E. R., 1977, Differential termination of large-diameter and small-diameter primary afferent fibers in the spinal dorsal gray matter as indicated by labeling with horseradish peroxidase, *Neurosci. Lett.* **6:**59–63.

Lund, J. S., Lund, R. D., Hendrickson, A. E., Bunt, A. H., and Fuchs, A. F., 1975, The origin of efferent pathways from the primary visual cortex, area 17, of the macaque monkey as shown by retrograde transport of horseradish peroxidase, *J. Comp. Neurol.* **164:**287–304.

Lynch, G., Gall, C., Mensah, P., and Cotman, C. W., 1974, Horseradish peroxidase histochemistry: A new method for tracing efferent projections in the central nervous system, *Brain Res.* **65:**373–380.

Malmgren, L., and Olsson, Y., 1978, A sensitive method for histochemical demonstration of horseradish peroxidase in neurons following retrograde axonal transport, *Brain Res.* **148:**279–291.

Mesulam, M.-M., 1976, The blue reaction product in horseradish peroxidase neurohistochemistry: Incubation parameters and visibility, *J. Histochem. Cytochem.* **24:**1273–1280.

Mesulam, M.-M., 1978, Tetramethyl benzidine for horseradish peroxidase neurohistochemistry. A non-carcinogenic blue reaction-product with superior sensitivity for visualizing neural afferents and efferents, *J. Histochem. Cytochem.* **26:**106–117.

Mesulam, M.-M., and Mufson, E. J., 1980, The rapid anterograde transport of horseradish peroxidase, *Neuroscience* **5:**1277–1286.

Mesulam, M.-M., and Rosene, D. L., 1977, Differential sensitivity between blue and brown reaction procedures for HRP neurohistochemistry, *Neurosci. Lett.* **5**:7–14.

Mesulam, M.-M., and Rosene, D. L., 1979, Sensitivity in horseradish peroxidase neurohistochemistry: A comparative and quantitative study of nine methods, *J. Histochem. Cytochem.* **27**:763–773.

Mesulam, M.-M., Hegarty, E., Barbas, H., Carson, K. A., Gower, E. C., Knapp, A. G., Moss, M. B., and Mufson, E. J., 1980, Additional factors influencing sensitivity in the tetramethyl benzidine method for HRP neurohistochemistry, *J. Histochem. Cytochem.* **28**:1255–1259.

Nauta, H. J. W., Kaiserman-Abramof, I. R., and Lasek, R. J., 1975, Electron microscopic observations of horseradish peroxidase transported from the caudoputamen to the substantia nigra in the rat: Possible involvement of the agranular reticulum, *Brain Res.* **85**:373–384.

Nicholls, P., 1962, The reduction of catalase by azide and peroxides, *Biochim. Biophys. Acta* **58**:386–388.

Novikoff, A. B., and Goldfischer, S., 1969, Further observations on the peroxidatic activities of microbodies (peroxisomes), *J. Histochem. Cytochem.* **17**:55–60.

Oldfield, B. J., and McLachlan, E. M., 1977, Uptake and retrograde transport or HRP by axons of intact and damaged peripheral nerve trunks, *Neurosci. Lett.* **6**:135–141.

Oppenheim, R. W., and Heaton, M. B., 1975, The retrograde transport of horseradish peroxidase from the developing hind limb of the chick embryo, *Brain Res.* **98**:291–302.

Papadimitrou, J. M., van Duijn, P., Brederoo, P., and Streefkerk, J. G., 1976, A new method for the cytochemical demonstration of peroxidase for light, fluorescence and electron microscopy, *J. Histochem. Cytochem.* **24**:82–90.

Pearse, A. G. E., 1972, *Histochemistry, Theoretical and Applied, Vol. 2,* Churchill Livingstone, Edinburgh and London, p. 1387.

Persson, L. A., and Kristensson, K., 1979, Uptake of horseradish peroxidase in sensory nerve terminals of mouse trigeminal nerve. *Acta Neuropathol.* **46**:191–196.

Phillipson, O. T., 1979, Afferent projections to the ventral tegmental area of Tsai and interfascicular nucleus: A horseradish peroxidase study in the rat, *J. Comp. Neurol.* **187**:117–144.

Price, P., and Fisher, A. W. F., 1978, Electron microscopical study of retrograde axonal transport of horseradish peroxidase in the supraoptico–hypophyseal tract in the rat, *J. Anat.* **125**:137–147.

Price, P., Fisher, A. W. F., and Redstone, P., 1977, A simple apparatus for injection of small (nanolitre) volumes of horseradish peroxidase, *Neurosci. Lett.* **6**:21–25.

Rennke, H. G., and Venkatachalam, M. A., 1979, Chemical modification of horseradish peroxidase, *J. Histochem. Cytochem.* **27**:1352–1353.

Repérant, J., 1975, The orthograde transport of horseradish peroxidase in the visual system, *Brain Res.* **85**:307–312.

Robertson, R. T., 1977, Bidirectional movement of horseradish peroxidase and the demonstration of reciprocal thalamocortical connections, *Brain Res.* **130**:538–544.

Rogers, R. C., Butcher, L. L., and Novin, D. 1980, Effects of urethane and pentobarbital anesthesia on the demonstration of retrograde and anterograde transport of horseradish peroxidase, *Brain Res.* **187**:197–200.

Rosene, D. L., and Mesulam, M.-M., 1978, Fixation variables in horseradish peroxidase neurohistochemistry. I. The effect of fixation time and perfusion procedures upon enzyme activity, *J. Histochem. Cytochem.* **26**:28–39.

Ryser, H.-J.-P., and Hancock, R., 1965, Histones and basic polyamine acids stimulate the uptake of albumin by tumor cells in culture, *Science* **150**:501–503.

Sasaki, M., 1952, On the various alums in the peroxidase stain of blood cells, *Okajimas Folia Anat. Jpn.* **24**:81–87.

Scalia, F., and Colman, D. R., 1974, Aspects of the central projection of the optic nerve in the frog as revealed by anterograde migration of horseradish peroxidase, *Brain Res.* **79**:496–504.

Schwab, M. E., and Thoenen, H., 1977, Retrograde axonal and transsynaptic transport of macromolecules: Physiological and pathophysiological importance, *Agents Actions* **7**:361–368.

Schwab, M. E., Javoy-Agid, F., and Agid, Y., 1978, Labeled wheat germ agglutinin (WGA) as a new, highly sensitive retrograde tracer in the rat brain hippocampal system, *Brain Res.* **152:**145–150.

Shen, W.-C. and Ryser, H.-J.-P., 1978, Conjugation of poly-L-lysine to albumin and horseradish peroxidase: A novel method of enhancing the cellular uptake of proteins, *Proc. Natl. Acad. Sci. USA* **75:**1872–1876.

Sherlock, D. A., and Raisman, G., 1975, A comparison of anterograde and retrograde axonal transport of horseradish peroxidase in the connections of the mammillary nuclei in the rat, *Brain Res.* **85:**321–324.

Siegel, A., Fukushima, T., Meibach, R., Burke, L., Edinger, H., and Weimer, S., 1977, The origin of afferent supply to the mediodorsal thalamic nucleus: Enhancement of HRP transport by selective lesions, *Brain Res.* **135:**11–23.

Singer, W., Holländer, H., and Vanegas, H., 1977, Decreased peroxidase labeling of lateral geniculate neurons following deafferentation, *Brain Res.* **120:**133–137.

Sotelo, C., and Riche, D., 1974, The smooth endoplasmic reticulum and the retrograde and fast orthograde transport of horseradish peroxidase in the nigro–striatal loop, *Anat. Embryol.* **146:**209–218.

Spencer, H. J., Lynch, G., and Jones, R. K., 1978, The use of somatofugal transport of horseradish peroxidase for tract tracing and cell labeling, in: *Neuroanatomical Research Techniques* (R. Robertson, ed.), pp. 291–315, Academic Press, New York.

Steward, O., and Vinsant, S., 1978, Identification of the cells of origin of a central pathway which sprouts following lesions in mature rats, *Brain Res.* **146:**223–243.

Straus, W., 1964, Factors affecting the state of injected horseradish peroxidase in animal tissues and procedures for the study of phagosomes and phago–lysosomes, *J. Histochem. Cytochem.* **12:**470–480.

Straus, W., 1967, Methods for the study of small phagosomes and their relationship to lysosomes with horseradish peroxidase as a "marker protein," *J. Histochem. Cytochem.* **15:**375–380.

Straus, W., 1974, Cleavage of heme from horseradish peroxidase by methanol with inhibition of enzymic activity, *J. Histochem. Cytochem.* **22:**908–915.

Streit, P., and Reubi, J. C., 1977, A new and sensitive staining method for axonally transported horseradish peroxidase (HRP) in the pigeon visual system, *Brain Res.* **126:**530–537.

Turner, P. T., 1977, Effect of pentobarbital on uptake of horseradish peroxidase by rabbit cortical synapses, *Exp. Neurol.* **54:**24–32.

Turner, P. T., and Harris, A. B., 1974, Ultrastructure of exogenous peroxidase in cerebral cortex, *Brain Res.* **74:**305–326.

Vacca, L. L., Rosario, S. L., Zimmerman, E. A., Tomaskofsky, P., Ng, P. Y., and Hsu, K. C., 1975, Application of immunoperoxidase techniques to localize horseradish peroxidase-tracer in the central nervous system, *J. Histochem. Cytochem.* **23:**208–215.

Van Deurs, B., 1977, Vesicular transport of horseradish peroxidase from brain to blood in segments of the cerebral microvasculature in adult mice, *Brain Res.* **124:**1–8.

Vanegas, H., Holländer, H., and Distel, H., 1978, Early stages of uptake and transport of horseradish peroxidase by cortical structures, and its use for the study of local neurons and their processes, *J. Comp. Neurol.* **177:**193–212.

Walberg, F., Brodal, A., and Hoddevik, G. H., 1976, A note on the method of retrograde transport of horseradish peroxidase as a tool in studies of afferent cerebellar connections, particularly those from the inferior olive; with comments on the orthograde transport in Purkinje cell axons, *Exp. Brain Res.* **24:**483–501.

Warr, W. B., 1975, Olivocochlear and vestibular efferent neurons of the feline brain stem: Their location, morphology and number determined by retrograde axonal transport and acetylcholinesterase histochemistry, *J. Comp. Neurol.* **161:**159–182.

West, J. R., and Black, A. C., Jr., 1979, Enhancing the anterograde movement of HRP to label sparse neuronal projections, *Neurosci. Lett.* **12:**35–40.

Wiksten, B., 1979, The central cervical nucleus in the cat. II. The cerebellar connections studied with retrograde transport of horseradish peroxidase, *Exp. Brain Res.* **36:**155–173.

Wilczynski, W., and Northcutt, R. G., 1977, Afferents to the optic tectum of the leopard frog: An HRP study, *J. Comp. Neurol.* **173**:219–230.

Wong-Riley, M. T. T., 1976, Endogenous peroxidatic activity in brain stem neurons as demonstrated by their staining with diaminobenzidine in normal squirrel monkeys, *Brain Res.* **108**:257–278.

Horseradish Peroxidase 7

Intracellular Staining of Neurons

STEPHEN T. KITAI and GEORGIA A. BISHOP

I. INTRODUCTION

A new application of the horseradish peroxidase (HRP) technique has recently been developed in which the enzyme is injected, either by pressure (Muller and McMahan, 1976; Sakai *et al.*, 1978) or by electrophoresis (Cullheim and Kellerth, 1976; Jankowska *et al.*, 1976; Kitai *et al.*, 1976; Light and Durkovic, 1976; Snow *et al.*, 1976), into a single neuron through a recording micropipette. Following appropriate histochemical procedures, the HRP reaction product completely fills the neuron—soma, dendrites, and axon—thus allowing the examination of the morphological characteristics of the cell. There are several advantages to this technique. *First,* its histological resolution is fine enough to visualize minute details such as dendritic spines, beaded endings of axons, or fine axon collaterals (Bishop *et al.*, 1979; Burke *et al.*, 1979; Christensen and Ebner, 1978; Cullheim and Kellerth, 1976, 1978; Gilbert and Wiesel, 1978; Friedlander *et al.*, 1978; Jankowska *et al.*, 1976; Kitai *et al.*, 1976; Light and Durkovic, 1976; McCrea *et al.*, 1976; Muller and McMahan, 1976; Neale *et al.*, 1978; Preston *et al.*, 1980; Sakai *et al.*, 1978; Snow *et al.*, 1976). Clearly, this represents an advantage over fluores-

STEPHEN T. KITAI and GEORGIA A. BISHOP • Department of Anatomy, Michigan State University, East Lansing, Michigan 48824.

cent dyes, e.g., procion yellow, that are also used as intracellular stains (Nicholson and Kater, 1973). *Second,* since the injected neuron is the only stained element in the tissue, it can be reconstructed serially without the confusion that inevitably accompanies the reconstruction of Golgi-impregnated neurons. Neuronal processes such as axons can therefore be followed for long distances (Burke *et al.*, 1979; Preston *et al.*, 1980). *Third,* since the HRP reaction product is electron-dense, the ultrastructural features of a neuron can be determined in the same tissue examined with the light microscope (Christensen and Ebner, 1978; Cullheim and Kellerth, 1976, 1978; Jankowska *et al.*, 1976; Neale *et al.*, 1978; Wilson and Groves, 1979).

Finally, the most significant advantage of the HRP intracellular staining technique is its application in studies involving a physiologically and morphologically heterogeneous population of neurons. In these areas, the investigator can attribute the particular physiological response recorded prior to injection of the enzyme to a specific neuron type, thus correlating the morphological and physiological characteristics of a neuron.

In comparison with HRP, intracellular cobalt represents an inferior single-cell marker. *First,* cobalt is toxic to impaled neurons, thus altering their internal structure unpredictably. *Second,* the use of cobaltous electrolyte solutions increases the probability of pipette clogging—a devastating problem if electrophysiological recording is essential (Tweedle, 1978). In addition, the long time between the beginning of the injection and the histological processing frequently results in the diffusion of the cobalt out of the previously filled neuron (Tweedle, 1978). Minimizing this interval increases the probability of success but also makes it difficult to stain more than one neuron per experiment.

II. METHODS

A. Preparation Procedures

For preparation of the different solutions, see Appendix, Section V.A.

1. HRP Preparation

The enzyme is prepared in a tris buffer–KCl solution as follows. Prepare tris buffer at pH 7.6. Add KCl to tris buffer to make a 0.5 M KCl solution. Filter tris–KCl solution through fine-pore filter paper (e.g., Whatman No. 50). Make a 5–10% solution of HRP in filtered tris–KCl (w/v).

It is advisable to keep the HRP solution in a slightly humid environment. This can be achieved by placing the vial with the HRP in a petri dish filled with water and covered with a glass beaker.

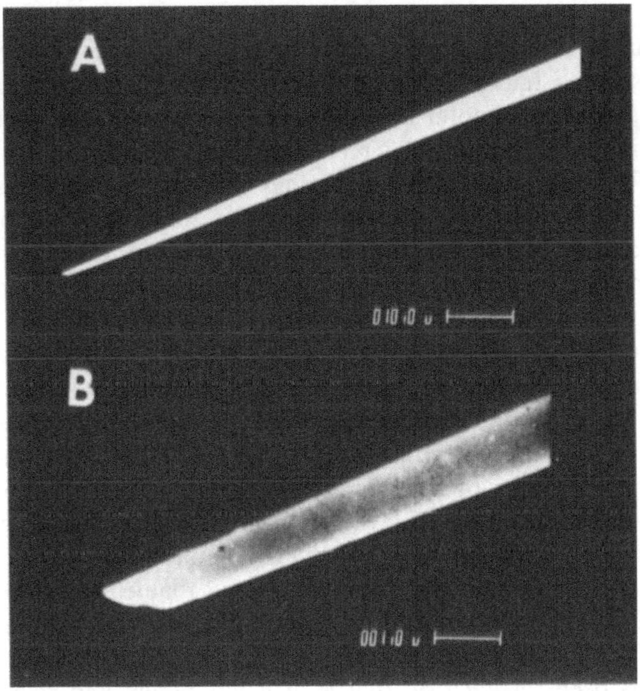

Figure 1. (A) Scanning electron micrograph of a recording microelectrode beveled by a mechanical bumping method. (B) Higher-power electron micrograph of the electrode shown in (A).

2. Electrode Preparation

The preparation of microelectrodes is described in detail in the Appendix of Chapter 3. Briefly, glass tubing with an outside diameter of 1–3 mm is pulled to a fine tip using a standard pipette puller. Tubing with an internal filament to facilitate capillary and, thus, electrode filling is preferred. The tip is then beveled using either a commercial beveler (Narishige, Kopf) or a mechanical bumping method. In the latter technique, the electrode tip is touched against some solid object under microscopic observation until appropriate tip diameter and bevel are obtained (Fig. 1). Electrodes optimal for intracellular penetration, with tip diameters of 0.5–1.0 μm and D.C. resistance values ranging between 30 and 60 MΩ, can be obtained in this manner. The beveled electrodes are then filled with the HRP solution and immersed in tris buffer to prevent clogging of the tip with the HRP.

3. Animal Preparation

The animal is anesthetized, placed in the stereotaxic head holder, and surgically prepared. Holes are made in the skull for placement of stimulating and recording electrodes.

B. Recording and Injection

HRP can be injected intracellularly by pressure or iontophoresis as discussed in detail in Chapter 3. We have primarily used iontophoresis, and the reader is referred to Chapter 3, Section III.C for a detailed description of the requirements for intracellular recording and injection.

The HRP-filled micropipette is visually or stereotaxically positioned in the area to be studied. The optimal electrode impedance in the tissue is between 35 and 60 MΩ. Upon penetration of a neuron, physiological events such as the resting membrane potential, excitatory or inhibitory postsynaptic potentials, or action potentials are recorded. When sufficient data have been obtained, the HRP is injected, providing the impaled neuron appears to be stable; specifically, if the resting membrane potential is stable, the spike mechanism is still intact, and injury discharge has ceased. Direct current pulses ranging in amplitude from 3–15 nA with 100–200 msec duration and delivered at frequencies of 3–5 Hz are applied for 30 sec to 20 min. The specific parameters will vary with different areas of the nervous systems. In general, however, lower currents (5–15 nA) applied for a short time (2–5 min) are sufficient to stain neurons with soma diameters in the range of 10–13 μm. Upon completion of the injection, the electrode is removed from the brain. The same electrode may be used for subsequent probes if its performance has been satisfactory; that is, if it has penetrated cells easily and cleanly with little apparent injury to the cells. Subsequent probes should be well separated to avoid damage to the area in which the previous injection was made and to avoid confusion when analyzing data at a later time. (See Table I for variations in injection procedures.)

C. Animal Perfusion

1. Light Microscopy

The survival time of the animal will depend on the goal of the project. If the purpose is to identify the perikaryal and dendritic features of a neuron, survival times as short as 30 min are adequate. If, on the other hand, the purpose is to trace axons over long distances, survival times of 12–24 hr may be better, although we have no conclusive data on this latter point. At the time of sacrifice, the animal is perfused with 0.9% saline followed by 1–2% paraformaldehyde and 1–1.25% glutaraldehyde in phosphate buffer (pH 7.3). The general procedures for the perfusion of animals are discussed in Chapter 10. The volume of the perfusate will depend on the species used. For a rat, 500 ml is sufficient, while larger animals, such as rabbits or cats, require about 1000 ml. Following the perfusion, the brain is removed from the skull and placed in cold (4°C) phosphate buffer to which sucrose (30% w/v) has been added. The brain is kept in a refrigerator at 4°C until it sinks to the bottom of the jar, usually after 1–3 days.

Table I. Experimental Protocols for Electrophoretic and Pressure Application of HRP through Intracellular Electrodes

Electrophoretic application

	HRP in KCl solution		Tris buffer			Injection current				
	Concentration (%)	Molarity (M)	Molarity (M)	pH	Electrode resistance (M Ω)	Intensity (nA)	Duration (msec)	Frequency (Hz)	Time (min)	Injection site
Jankowska *et al.*[a] (1976)	4	0.2	0.05	8.5	25–35	5–10	Continuous	1.7	5–25	Spinocervical neurons
Kitai *et al.* (1976)	4	0.2	0.05	8.6	70–100	10–20	200	5	10–20	Caudate neurons
Light and Durkovic (1976)	4	0.5	0.05	7.6	50–80	40–200	—	200	0.5–2.5	Spinal motoneuron
McCrea *et al.* (1976)	4	0.2	—	8.6	50–100	15–40	200	3.3	12–45	Purkinje cell
Snow *et al.* (1976)	4	0.2	0.05	8.6	5–20	20–40	450	1.7	—	Spinocervical neurons
Christensen and Ebner (1978)[a]	4	0.5	0.05	7.6	—	20	—	—	<5	Cortical neurons
Cullheim and Kellerth (1976)[a]	25	0.1	(NaOH solution)	—	30–90	20–30	Continuous	—	3–15	Spinal motoneuron
Friedlander *et al.* (1978)	3	0.2	—	8.6	100–175	5–10	—	200	5–10	Lateral geniculate neurons
Neale *et al.* (1978)	4	0.2	0.05	8.6	50–150	5–20	500	1	5–30	Cultured spinal neuron
Burke *et al.* (1979)	4	0.2	0.05	8.6	—	25	25	20	30–40	Ia afferent fiber
	4	0.2	0.05	8.6	—	15–20	25	20	10–15	spinal motoneurons
Preston *et al.* (1980)	4	0.5	0.05	7.6	35–60	3–15	100	3–5	0.5–10	Caudate neurons

Pressure application

	HRP in KCl solution		Tris buffer			Injection		
	Concentration (%)	Molarity (M)	Molarity (M)	pH	Electrode resistance (M Ω)	Pressure	Time (sec)	Injection site
Muller and McMahan (1976)	2	0.1	—	—	100	30–100 k Pa (5–15 lb/m²)	5–30	Leach ganglion cell
Sakai *et al.* (1978)	4	0.6–1.0	—	8.6	—	60–90 lb/m²	1–10	Cortical neuron

[a]EM analysis in addition to LM

2. Electron Microscopy

If the brain is to be prepared for electron microscopy, the animal is perfused with 3% paraformaldehyde and 1% glutaraldehyde in 0.1 M phosphate buffer (pH 7.3) to which dextrose is added to increase buffer osmolarity to 360 mOsm. (Christensen and Ebner, 1978). The final osmolarity of the perfusion solution is 1410–1430 mOsm. Following perfusion, the animal's head is removed and placed in a jar containing the perfusate. After 8–12 hr the brain is removed from the skull and transferred to cold phosphate buffer.

D. Histological Processing

For preparation of the different solutions, see Appendix, Section V.A.

1. Light Microscopy

Frozen sections are cut 100–125μm thick, collected over ice in phosphate buffer (pH 7.3), and processed as follows.

1. Place sections in a 5% aqueous solution of cobaltous chloride for 10 min to intensify the reaction product (Adams, 1977).
2. Rinse sections thoroughly in three changes of phosphate buffer (5–10 min each).
3. Transfer sections to a solution of 0.05% diaminobenzidine tetrahydrochloride (DAB) and incubate at 38°C for 30 min. Since DAB may be carcinogenic, appropriate precautions should be taken when using it (e.g., gloves, surgical mask, lab coat, fume hood).
4. Add 0.06% H_2O_2 (4 ml/100 ml DAB) to the DAB solution. Agitate for 30 min at room temperature.
5. Rinse sections thoroughly in three changes of phosphate buffer (5–10 min each).
6. Final 10-min rinse in distilled water.
7. Mount sections on glass slides, counterstain with cresyl violet, and coverslip.

2. Electron Microscopy

The brain is blocked appropriately, and 100–125 μm sections are cut on a tissue chopper, a Vibratome, or by hand. The tissue is then processed as described in step 5 (Section II.D.1). Following the final buffer rinse, sections are examined in buffer for light microscopic identification of injected neurons. When the light microscopic analysis is completed, the coverslip is

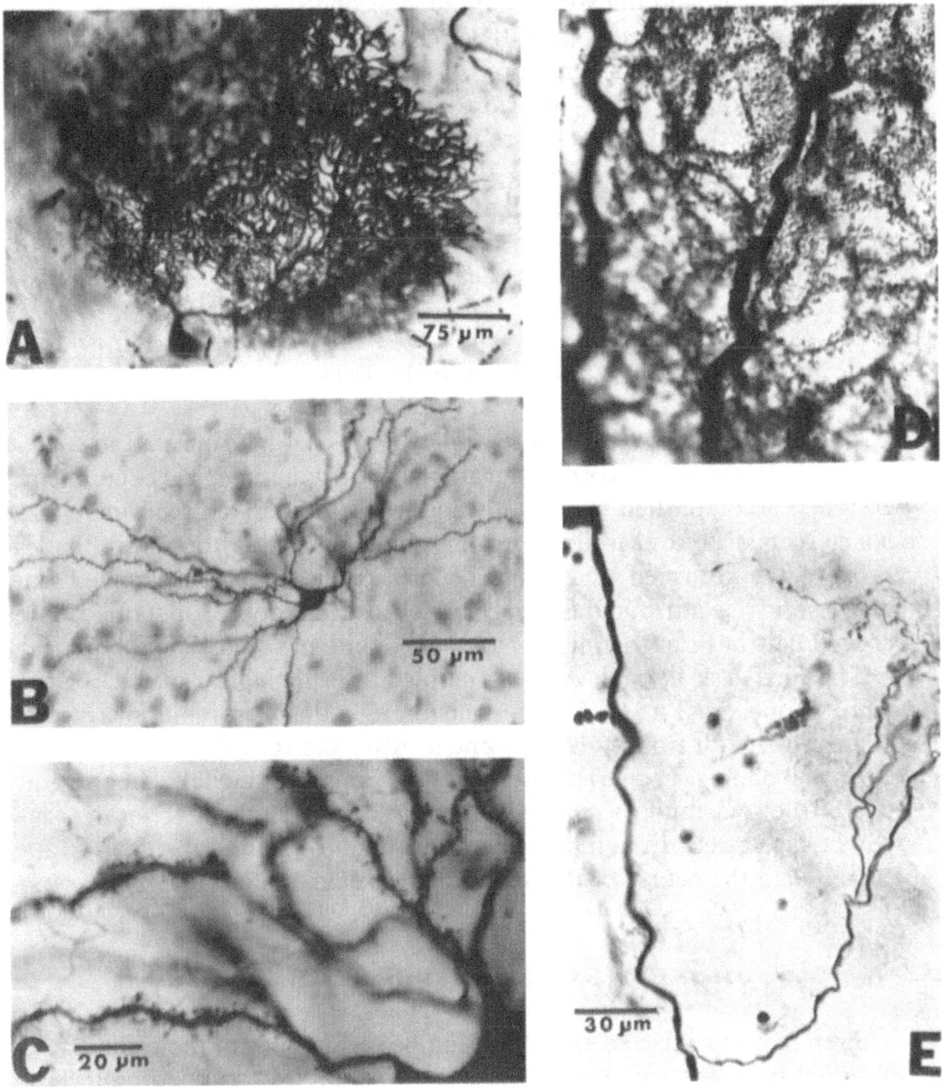

Figure 2. Photomicrographs of neurons intracellularly labeled with HRP. (A) Cat cerebellar Purkinje cell. (B) Rat striatal medium spiny neuron. (C) Portion of the soma, spine-free primary and spine-laden secondary dendrites of a striatal medium spiny neuron. (D) Spine-laden distal dendrites of a Purkinje cell. (E) Axon and single recurrent collateral from a Purkinje cell. A portion of the soma is seen in the upper right-hand corner. Calibration in (C) also applies to (D).

gently removed, and the section containing the stained neuron is prepared for electron microscopy as follows:

1. Rinse in phosphate buffer (three changes, 5 min each).
2. Place in 2% osmium tetroxide for 1–2 hr.

3. Rinse in phosphate buffer (three changes, 5 min each), place in covered vials, and dehydrate in graded alcohols.
4. Rinse in propylene oxide (three changes, 10 min each).
5. Place in a (1:1) mixture of propylene oxide and epon (or epon–araldite) for 8–10 hr. Leave vial uncovered.
6. Change the mixture to pure plastics and place on rotator for 8 hr.
7. Embed sections by placing them in a mold, pour fresh plastic over them, and cure them in an oven at 60°C for 12 hr or more.
8. Cut ultrathin sections for electron microscopy and stain them with lead citrate and uranyl acetate.

E. Analysis of the Data

The first step is to locate all of the HRP-injected neurons and to correlate them with the sites recorded during the electrophysiological experiment. This is best accomplished by carefully noting the location of recording tracks from stereotaxic and microdrive readings and correlating these with the locations of the recorded neurons. The physiological records obtained can then be positively attributed to a specific cell as each probe is identified. The recorded neurons can then be photographed and analyzed (Fig. 2).

The next task is to develop a complete morphological picture of the neuron by means of a serial reconstruction. For this purpose a microscope equipped with a drawing tube is essential. After the reconstruction is made, various morphological characteristics become apparent (Fig. 3). The size and shape of the soma and the shape and extent of the dendritic tree can accurately be illustrated. In addition, the distribution of spines or dendritic appendages and the course of the axon will be visible.

III. APPLICATION OF THE TECHNIQUE

Since neurons injected with HRP can be reconstructed, the technique is well suited for examining the morphology of all parts of the neuron including its axon, which often can be traced over long distances. Important information can be obtained with regard to specific features of axonal trajectory and site of termination. This type of analysis, in which local circuit neurons have been distinguished from projection neurons in a nucleus, is illustrated in the recent publication by Preston *et al.* (1980).

The technique is also very useful in examining the ultrastructural characteristics of a neuron (Christensen and Ebner, 1978; Cullheim and Kellerth, 1976; Janokowska *et al.*, 1976; Neale *et al.*, 1978). Since the HRP reaction product is electron-dense, injected neurons and their processes are clearly-differentiated from unstained neural elements in electron micrographs (Fig. 4).

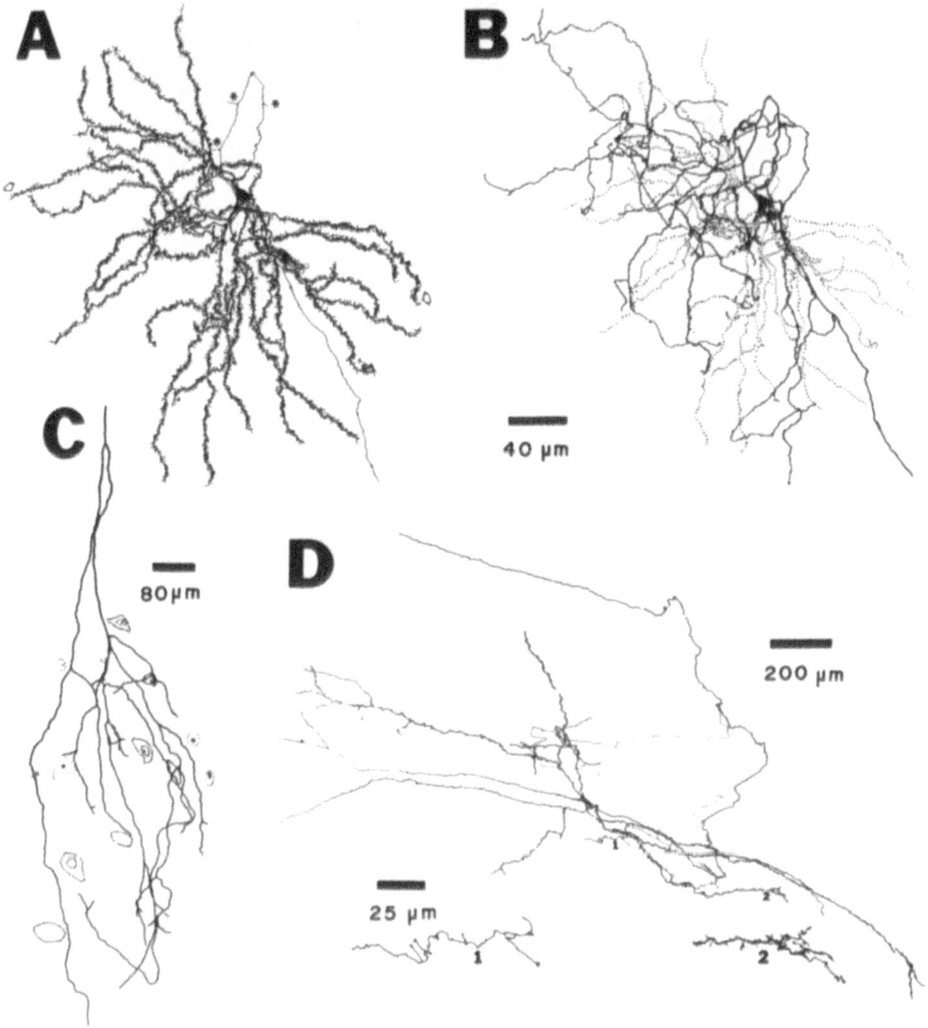

Figure 3. Reconstruction of HRP-labeled neurons from serial sections. (A) Somadendritic morphology and partial illustration of the axon of a rat striatal medium spiny neuron. Asterisks indicate points of collateral origin from the parent axon. Arrows indicate tapering of the distal extreme dendrites to fine spine-studded filament. (B) The axon collateral plexus of the same cell superimposed on the cell's dendritic field (stippled). (C) Terminal arborization of the axon of a Purkinje cell in the nucleus interpositus of the cat cerebellum. Nuclear cells are drawn with stippled lines. (D) Somadendritic morphology and partial illustration of the axon system of a cat substantia nigra neuron. (D_1) Enlargement of a portion of the axon collateral containing beaded processes suggestive of boutons termineaux. (D_2) Enlargement of distal portion of dendrites illustrating varicosities and dendritic appendages.

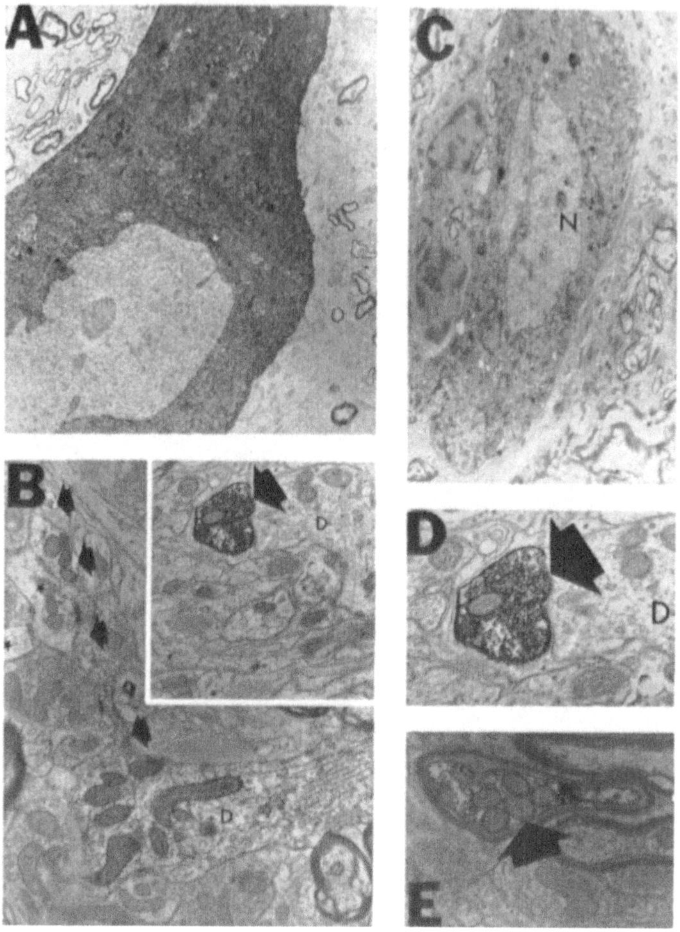

Figure 4. Electron micrographs of neurons intracellularly labeled with HRP. (A) Soma of a Purkinje cell of the cat cerebellum (×6250). (B–E) Neuron of the dorsal accessory nucleus of the cat inferior olive complex. (B) A labeled dendritic shaft (D) gives rise to a spiny appendage (block arrows) that, together with other unlabeled appendages (asterisks), makes up the central core of a synaptic cluster (×37,500). The insert in the upper right hand corner shows a labeled axon terminal (block arrow) located adjacent to an unlabeled dendrite (D) (×37,500). An enlargement of this insert is shown in (D). (C) Soma of the same neuron. (E) The myelinated axon is illustrated by the block arrow (×37,500). Reproduced at 45%. [Electron micrographs were made in collaboration with Dr. James King of the Department of Anatomy, School of Medicine, Ohio State University, Columbus, Ohio.]

Detailed studies of synaptic interrelations are possible on the electron microscopic level. For instance, discrete lesions can be made in areas that are presumed to provide afferents to another area. After several days' survival time, neurons in the target nucleus can be penetrated and recorded from; any changes in their physiological response can be attributed to the lesion. Subsequently, injections of HRP can be made and the tissue prepared for electron microscopy. On examination of the HRP-injected neuron in the electron micrographs, one may succeed in demonstrating degenerating terminals in contact with the HRP-stained neuron.

Several successful attempts have been made to render Golgi-impregnated neurons acceptable for electron microscopic analysis (Chapter 12). However, there are some difficulties associated with the Golgi–EM technique. For instance, the Golgi technique usually stains many neurons in an area, making it difficult to trace the processes of any single cell without confusion from intervening processes of other cells. Intracellular injection of HRP obviates this problem.

Intracellular injections of HRP and correlated light and electron microscopic analysis were recently used to study synaptic interactions in dissociated cell cultures of spinal cord neurons (Neale *et al.*, 1978). With this technique, complex interneuronal relationships can be visualized, and it is possible to identify morphological substrates—i.e., the number, location, and ultrastructure of synaptic contacts—for electrophysiological events such as EPSPs or IPSPs.

IV. SUMMARY OF ADVANTAGES AND LIMITATIONS

A. Advantages

1. The most important advantage of the present technique is its ability to display in great detail the morphology of a neuron that has been characterized in physiological terms.
2. As the injected neuron is the only stained element in the tissue, a long process can be traced by the use of serial sections, provided it is well stained (see Limitations, Section IV.B, 3).
3. Since the DAB reaction product is osmiophilic, the ultrastructural features of the labeled neuron and its relation to other tissue components can be conveniently studied in the electron microscope.

B. Limitations

1. The sampling of different types of neurons and the number of samples are dependent on the investigator's ability to penetrate different types of cells with a microelectrode. There may be a bias toward injecting larger cells in a heterogenous population of neurons, since it is generally more difficult to penetrate small cells.
2. As a result of injury during penetration or injection, the cell may display various signs of degeneration ranging from total destruction to swollen or fragmented dendrites, in which case the cell is not suitable for analysis.
3. When tracing an axonal projection, it is not uncommon to find that

the axon gradually becomes faint and finally fades entirely in a somatofugal direction. The use of an intensification procedure, such as pretreatment in cobaltous chloride, greatly improves the chance of staining the entire axon. Improved axonal filling can sometimes be achieved simply by increasing the surivial time. Another possiblity is to use an animal, such as the mouse, in which central pathways are of short length.

V. APPENDIX

A. Chemicals for HRP Histological Processing (Intracellular Staining)

Phosphate Buffer (pH 7.3)

NaH_2PO_4	2.2075 g
Na_2HPO_4	11.9275 g
Dextrose	25.0 g

Bring to 100 ml volume with distilled H_2O.

Tris Buffer (pH 7.6)

Tris base	0.139 g
Tris HCl	0.606 g

Bring to 100 ml volume with distilled H_2O.

Diaminobenzidine Tetrahydrochloride (DAB)

DAB	50.0 mg
Phosphate buffer	100.0 ml

Filter before use.

Cobalt Chloride ($CoCl_2$)

Prepare 5% solution of $CoCl_2$ in phospate buffer. Filter before use.

Hydrogen Peroxide (0.06%)

H_2O_2	1 part
Distilled water	49 parts

Horseradish Peroxidase

Sigma Type VI or Worthington HPOFF.

Fixative (1.5% Paraformaldehyde–1% Glutaraldehyde): Light Microscopy

Add paraformaldehyde (15.0 g) to 500 ml distilled H_2O. Add 8 drops 1N NaOH. Heat at 60°C until paraformaldehyde dissolves. Cool. Add 40 ml 25% glutaraldehyde and 460 ml phosphate buffer (pH 7.3).

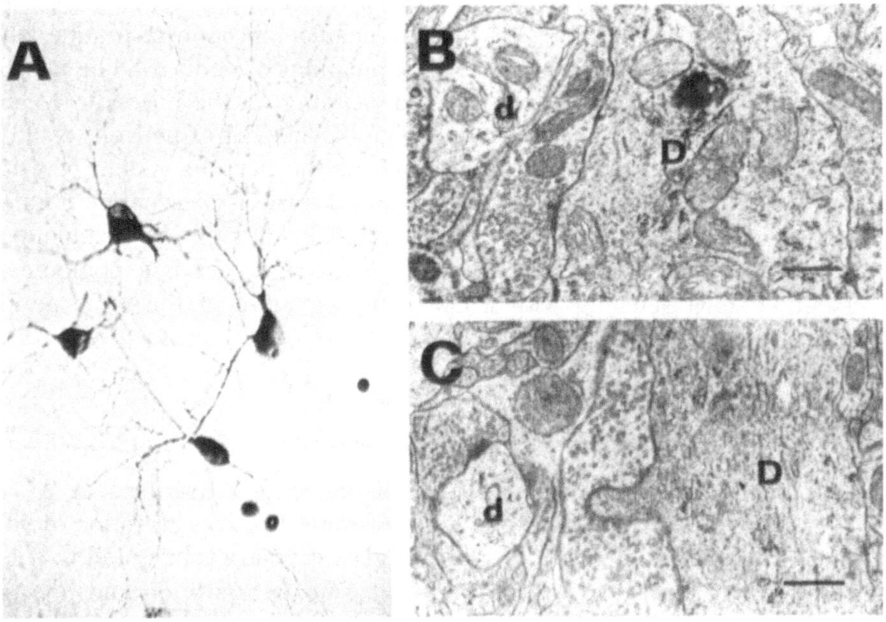

Figure 5. Examples of solidly labeled rabbit thalamic neurons 48 hr after striatal injection of HRP in DMSO. (A) A cluster of four solidly labeled thalamic neurons (×280). (B, C) Solidly labeled (D) and unlabeled (d) postsynaptic dendrites. Note the overall increased electron density and dark granules in the labeled dendrites. Bar represents 0.5 μm. Reproduced at 70%. [Figs. B and C are from Keefer *et al.* (1980).]

Fixative (3% Paraformaldehyde–1% Glutaraldehyde): Electron Microscopy

Dissolve paraformaldehyde (60.0 g) in phosphate buffer (580.0 ml) at 60°C. Cool to 30°C. Add 25% glutaraldehyde, 80.0 ml, dextrose, 50.0 g, and 10% calcium chloride, 1–3 ml (until cloudiness goes away with stirring). Add enough distilled H_2O to bring volume to 2000 ml.

B. An Alternative Approach Using Retrograde Golgi-like Labeling of Neuronal Populations

Donald Keefer*

It has recently become possible to label select populations of neurons in a Golgi-like fashion (Fig. 5A) by injecting HRP at sites that intercept their efferent projections (Keefer *et al.*, 1976; Hedreen and McGrath, 1977). This technique shares many of the advantages that characterize the intracellular HRP-labeling technique described in this chapter. In contrast to the intracellular injection techniques, however, the cells are characterized by their

*Department of Anatomy, University of Virginia, Charlottesville, Virginia 22901.

efferent projections rather than by their physiological properties. Another main difference is that whereas the intracellular injection technique labels one neuron only, this technique labels populations of neurons. The neurons appear either in clusters or scattered, depending on the injection (Keefer, 1978). If the method for retrograde Golgi-like labeling of neurons could be applied more generally to different parts of the nervous system, it would offer the same possibilities for ultrastructural analysis of synaptic relations (Keefer *et al.*, 1980; Fig. 5B,C) as the intracellular injection technique. A promising way to improve the efficiency of this approach is to combine the retrograde HRP labeling with a Golgi impregnation of the same neuron (Somogyi and Smith, 1979).

1. Procedure for Retrograde Golgi-like Labeling

1. Horseradish peroxidase (Miles, Sigma, etc.) is dissolved in 2% dimethylsulfoxide (DMSO) in distilled water to a concentration of 30%.
2. The cannula is made of a drawn glass capillary tube glued to a 1-μl syringe and filled with paraffin oil. During aspiration and injection for the HRP–DMSO, allow an air bubble to separate the injection solution from the paraffin oil. A 100-μm-diameter cannula tip has allowed good selectivity of labeling for small populations of neurons.
3. Survival times of 6 hr to 5 days have been used to visualize Golgi-like labeling in neurons afferent to and distant from the injection site.
4. Fixation and staining for light or electron microscopy is similar to that described in Section II.D. As for the chromogen, DAB is preferable to TMB or BDHC since it produces an amorphous rather than a crystalline reaction product. The decreased sensitivity of the DAB is more than offset by the greater morphological detail seen in fine dendrites and spines. However, the TMB staining methods have been used successfully to visualize the Golgi-like staining of retrogradely labeled neurons (West and Black, 1979).

ACKNOWLEDGMENT. This work was supported by USPHS Grants NS14866 and NIH BRSG, Michigan State University College of Osteopathic Medicine.

REFERENCES

Adams, J. C., 1977, Technical considerations on the use of HRP as a neuronal marker, *Neuroscience* **2:**141–146.

Bishop, G. A., McCrea, R. A., Lighthall, J. W., and Kitai, S. T., 1979, An HRP and autoradiographic study of the projection from the cerebellar cortex to the nucleus interpositus anterior and nucleus interpositus posterior of the cat, *J. Comp. Neurol.* **185:**735–756.

Burke, R. E., Walinsley, B., and Hodgson, J. A., 1979, HRP anatomy of group Ia afferent contacts on alpha motoneurons, *Brain Res.* **160:**347–352.

Christensen, B. N., and Ebner, F. F., 1978, the synaptic architecture of neurons in opossum somatic sensory-motor cortex: A combined anatomical and physiological study, *J. Neurocytol.* **7**:39–60.

Cullheim, S., and Kellerth, J. O., 1976, Combined light and electron microscopic tracing of neurons, including axons and synaptic terminals after intracellular injection of horseradish peroxidase, *Neurosci. Lett.* **2**:307–313.

Cullheim, S., and Kellerth, J. O., 1978, A morphological study of the axons and recurrent axon collaterals of cat sciatic α-motoneurons after intracellular straining with horseradish peroxidase, *J. Comp. Neurol.* **178**:537–558.

Friedlander, M. J., Lin, C. S., and Sherman, S. M., 1978, Structure and functional correlation for single neurons in cat dorsal lateral geniculate nucleus, *Soc. Neurosci. Abstr.* **4**:388.

Gilbert, C. D., and Wiesel, T. N., 1978, Intracellular staining of physiologically identified cells and afferents in cat visual cortex, *Soc. Neurosci. Abstr.* **4**:389.

Hedreen, J. C., and McGrath, S., 1977, Observations on labeling of neuronal cell bodies, axons, and terminals after injection of horseradish peroxidase into rat brain, *J. Comp. Neurol.* **176**:225–246.

Jankowska, E., Rostad, J., and Westman, J., 1976, Intracellular application of horseradish peroxidase and its light and electron microscopial appearance in spinocervical tract cells, *Brain Res.* **105**:557–562.

Keefer, D. A., 1978, Horseradish peroxidase as a retrogradely-transported, detailed dendritic marker, *Brain Res.* **140**:15–32.

Keefer, D. A., Spatz, W., and Misgeld, V., 1976, "Golgi like" staining of neocortical pyramidal neurons with horseradish peroxidase, *Neurosci. Lett.* **3**:233–237.

Keefer, D. A., Chung, J. W., and Heimer, L., 1980, Morphology and ultrastructure of anterior thalamic neurons solidly labeled with horseradish peroxidase, *Brain Res.* **191**:301–311.

Kitai, S. T., Kocsis, J. D., Preston, R. J., and Sugimori, M., 1976, Monosynaptic inputs in caudate neurons identified by intracellular injection of HRP, *Brain Res.* **109**:601–606.

Light, A. R., and Durkovic, R. G., 1976, Horseradish peroxidase: An improvement in intracellular staining of single, electrophysiologically characterized neurons, *Exp. Neurol.* **53**:847–853.

McCrea, R. A., Bishop, G. A., and Kitai, S. T., 1976, Intracellular straining of Purkinje cells and their axons with HRP, **118**:132–136.

Muller, K. J., and McMahan, U. J., 1976, The shapes of sensory and motor neurons and the distributions of their synapses in ganglia of the leech: A study using intracellular injection of horseradish peroxidase, *Proc. R. Soc. Lond.* [*Biol.*] **194**:481–499.

Neale, E. A., McDonald, R. L., and Nelson, P. G., 1978, Intracellular horseradish peroxidase injection for correlation of light and electron microscopic anatomy with synaptic physiology of cultured mouse spinal cord neurons, *Brain Res.* **152**:265–282.

Nicholson, C., and Kater, S. B., 1973, The development of intracellular staining, in: *Intracellular Staining in Neurobiology* (S. B. Kater and C. Nicholson, eds.), pp. 1–19, Springer-Verlag, Berlin and New York.

Preston, R. J., Bishop, G. A., and Kitai, S. T., 1980, Medium spiny neuron projection from the rat striatum. An intracellular horseradish peroxidase study, *Brain Res.* **183**:253–263.

Sakai, M., Sakai, H., and Woody, C., 1978, Intracellular staining of cortical neurons by pressure injection of horseradish peroxidase and recovery by core biopsy, *Exp. Neurol.* **58**:138–144.

Snow, P. J., Rose, P. K., and Brown, A. G., 1976, Tracing axons and axon collaterals of spinal neurons using intracellular injection of horseradish peroxidase, *Science* **191**:310–311.

Somogyi, P., and Smith, A. D., 1979, Projection of neostriatal spiny neurons to the substantia nigra. Application of combined Golgi staining and horseradish peroxidase transport procedure at both light and electron microscopic level, *Brain Res.* **178**:3–15.

Tweedle, C. D., 1978, Single-cell staining techniques, in: *Neuroanatomical Research Techniques* (R. T. Robertson, ed.), pp. 141–174, Academic Press, New York.

West, J. R., and Black, A. C., Jr., 1979, Enhancing the anterograde movement of HRP to label sparse neuronal projections, *Neurosci. Lett.* **12**:35–40.

Wilson, C. J., and Groves, P. M., 1979, A simple and rapid section embedding technique for sequential light and electron microscopic examination of individually-stained central neurons. *J. Neurosci. Methods* **1**:383 – 391.

Horseradish Peroxidase and Fluorescent Substances and Their Combination with Other Techniques

8

OSWALD STEWARD

I. INTRODUCTION

The various horseradish peroxidase (HRP) methods for labeling of neuronal pathways have become well-characterized tools in the armamentarium of the neuroscientist despite the relatively short period during which the methods have evolved. Another stage of methodological approach is now emerging in which the basic HRP method is being combined with other techniques for special applications. To be sure, such procedures are not restricted to combinations in which the HRP method is a necessary component. The Golgi method can be combined with the analysis of experimentally induced degeneration for the study of interneuronal connections (Chapter 12), and double labeling of neuronal cell bodies, using immunofluorescence for neurotransmitter analyses and a fluorescent retrograde tracer for the study of the neu-

OSWALD STEWARD • Departments of Neurological Surgery and Physiology, University of Virginia School of Medicine, Charlottesville, Virginia 22908.

rons' axonal projections, is described in the Appendix (Section VII.C) of this chapter.

The present chapter will consider some of these combined techniques. Procedural matters as well as advantages and limitations of the different methods will be discussed, and several recipes will be included in the Appendix (Section VII). When two methods are combined, the following general issues must be considered.

What neuronal properties are being tapped by the various methods? All neuroanatomical methods take advantage of some specific property of neurons. For example, the retrograde labeling methods take advantage of the capacity of the terminal processes of neurons to take up some exogenous proteins and transport these proteins in a retrograde direction back to the cell body. When combining two neuroanatomical methods, it is important to realize that manipulations required for one method may influence the *neuronal properties* on which the other method is based. For example, transport of one foreign protein may alter significantly the capacity to transport other proteins (see Section II).

What methodological compromises must be made to combine techniques? Most of the neuroanatomical techniques have particular procedural requirements that are more or less unique. For some procedures, animals are perfused with a fixation solution, whereas for other methods, fixatives are detrimental. For many basic histological procedures, embedding in a supporting medium is either desirable or essential, whereas for histochemical procedures that measure enzyme activity, embedding is usually impossible, at least prior to the histochemical procedures. It is thus rare that two techniques are immediately compatible, and a compromise must usually be made in order to combine two methods.

To what extent are the markers for the various methods mutually distinguishable? Histological staining techniques result in the deposition of visible substances in different tissue components, and many of these substances are quite difficult to distinguish from each other. An unusually difficult example is the problem of visualizing a histochemical reaction product in cells stained by the Golgi method. In attempting to develop combined procedures, special modifications in one or another procedure are sometimes required to insure that the markers involved can be visualized and distinguished.

In Section II of this chapter, we will consider techniques used in the study of collateral projections. The majority of these techniques are double-labeling procedures in which either the HRP method is combined with another retrograde method or two retrograde fluorescent tracers are used. Subsequent sections will deal with other double-labeling procedures that take advantage of neuronal properties unrelated to retrograde transport.

II. THE TRACING OF COLLATERAL PROJECTIONS

A feature of neurons recognized in early Golgi studies is that many axons bifurcate and project to separate target areas. Such collateralization may have important functional implications. If the same cells in a neuronal pool project to two different regions, it can be concluded that the two regions obtain the same information from the communal pool. However, if the two regions receive their projections from different cells within the pool, the two regions may well receive different kinds of messages.

Until recently it has been impossible to determine the extent of collateralization with experimental anatomical methods. Although both the reduced silver stains for anterograde degeneration and the autoradiographic method can show that a given area projects to multiple sites, they do not disclose whether the different projections originate in different cell populations or from a single neuronal population with collateral projections. The HRP retrograde cell-labeling method can identify the cells that give rise to the projection pathways, but even if a single cell type is labeled following injections into two different terminal fields, it is still impossible to determine whether single neurons project to both sites via collaterals or whether a heterogeneous population of neurons, similar in appearance, project selectively to each terminal field.

Recent methodological advances have provided techniques that permit a direct experimental analysis of collateralization. There are two principal methods available (Fig. 1). One is a retrograde double-labeling procedure in which different markers are injected in the two synaptic fields. The other method takes advantage of the fact that HRP may be transported through-

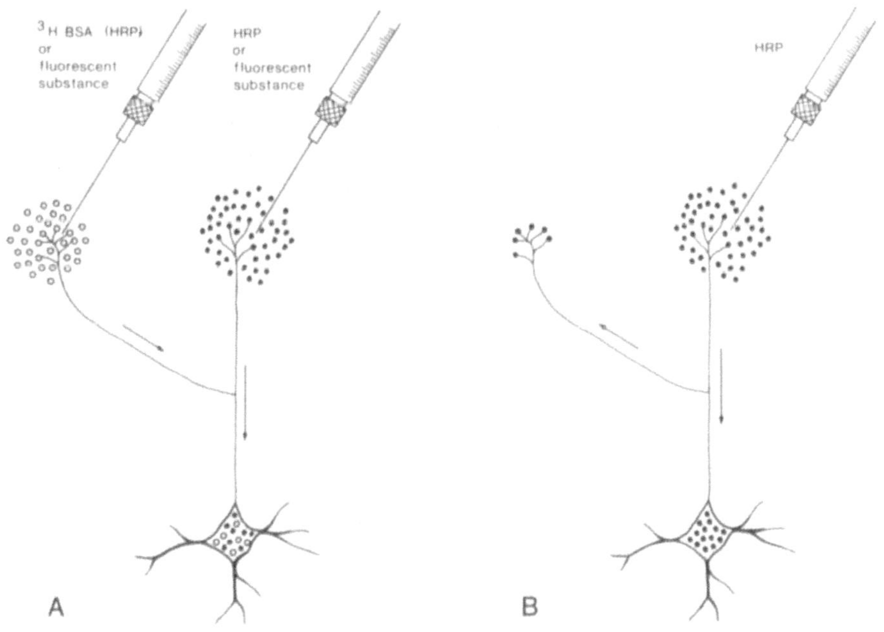

Figure 1. Two methods for demonstrating collateral projections. (A) Double retrograde labeling with two tracers (for example, HRP and [^{3}H]-BSA). (B) Labeling of collateral terminal fields as a result of the transport of the enzyme throughout the processes of a labeled cell.

out the processes of cells whose terminal arborizations take up the protein, resulting in labeling both of neuronal cell bodies and of the terminal fields of existing axon collaterals.

A. Retrograde Double-Labeling Procedures Using HRP in Different Combinations

1. Double Labeling with HRP and [^{3}H]-HRP

The first successful attempt to develop a double retrograde labeling method utilized tritiated HRP ([^{3}H]-HRP) in combination with HRP histochemistry to study collateral projections of thalamic neurons (Geisert, 1976). The labeled HRP was injected into one terminal field of geniculate neurons in area 17 of the visual cortex, and nonradioactive HRP was injected into area 18 which also receives geniculocortical afferents. Since the [^{3}H]-HRP could be visualized autoradiographically in lower concentrations than those producing histochemical reactivity, and because the labeling procedure at least partially inactivated the enzyme, residual enzymatic activity of the radiolabeled molecule was an insignificant problem. After such double injections, the tissue was processed for both HRP histochemistry and autoradiography. Cells in the lateral geniculate nucleus, which were labeled with the HRP reaction product and with silver grains overlying their cytoplasm, were thus interpreted as giving rise to collateral projections to areas 17 and 18.

Although this method can demonstrate collateral projections, it could conceivably produce false positive results. The labeled HRP molecule must be injected in concentrations low enough not to be visible after the histochemical reaction, since enzymatically active [^{3}H]-HRP could in itself result in double labeling even if the labeled neurons do not project to both terminal fields. Unfortunately, various batches prepared in the same way and with comparable specific activity may differ in enzymatic activity, making the use of [^{3}H]-HRP unpredictable. At a minimum, the problem demands that each batch of [^{3}H]-HRP be analyzed for enzyme activity. To circumvent this difficulty, attempts have been made to inactivate the enzymatic moiety of the HRP molecule without changing its transport capacities.

2. The Use of Enzymatically Inactive HRP

New England Nuclear* has recently made available a preparation of [^{3}H]-HRP that has been enzymatically inactivated but which does not seem to have any unusual transport properties. The substance, [^{3}H]-apo-HRP, has recently been used in combination with HRP for the analysis of collateral pathways in corticothalamic systems as well as in the primary sensory projec-

*Boston, Massachusetts.

tions from cells in the spinal ganglia (Hayes and Rustioni, 1979; Spreafico *et al.*, 1979).

The procedure for double-labeling with HRP and [^{3}H]-apo-HRP is described in detail in Appendix, Section VII.B. Animals are perfused as usual for HRP histochemistry. Following removal of the brain from the skull, tissue blocks including the relevant areas are prepared and reacted in a histochemical medium containing paraphenylene diamine pyrocatechol (the Hanker–Yates reagent). The blocks are then embedded in paraffin, sectioned at 15 μm, and prepared for light microscopic autoradiography. The results obtained from such preparations are illustrated in Fig. 2A. Both the brown HRP reaction product and the autoradiographic silver grains that indicate the presence of [^{3}H]-apo-HRP can be found over the cytoplasm of double-labeled cells.

Although the use of enzymatically inactive HRP is promising, it should not be assumed that the enzymatically inactive HRP behaves exactly like the enzymatically active molecule. For example, since the ^{3}H label is bound to amino acids that compose the HRP enzyme, those amino acids may be utilized either by the labeled cell or by its neighbors when the protein is degraded. Unlike ordinary HRP, which is inactivated by cellular proteases, the labeled amino acids may be transported in proteins assembled from the breakdown products of the ^{3}H-labeled HRP. Thus, silver grains in the autoradiogram may reflect not only the location of tritiated HRP but also of tritiated amino acids secondarily derived from the injected protein. This potentially confounding factor is shared by any radiolabeled protein including [^{3}H]-bovine serum albumin (see Section II.A.3).

3. Double Labeling with HRP and Bovine Serum Albumin (BSA)

A second approach to developing double retrograde labeling methods is to make use of proteins other than HRP. A number of different proteins have been reported to be transported in a retrograde fashion by neurons. These proteins include bovine serum albumin (Kristensson, 1970; Steward and Scoville, 1976), tetanus toxin, cholera toxin, nerve growth factor, and wheat germ agglutinin (Stoeckel *et al.*, 1977), as well as dopamine-β-hydroxylase antibodies (Fillenz *et al.*, 1976). Some of the substances are transported more effectively by some neurons than by others. This may be the case particularly for substances that relate to the presumed transmitter mechanisms of the neurons in question. The choice of protein, therefore, may depend on the system under investigation, and preliminary experiments with each tracer are necessary to insure that the substances are being transported.

In our laboratory, tritiated bovine serum albumin ([^{3}H]-BSA) in combination with HRP has found practical application (Steward *et al.*, 1977). Unilateral lesions of the entorhinal cortex of the rat induce sprouting of the projections from the intact contralateral entorhinal area to the denervated dentate gyrus ipsilateral to the lesion (Steward *et al.*, 1976). To determine whether the reinnervating fibers originate as collaterals of the normal ipsilateral projection system, HRP was injected into the reinnervated dentate gyrus, and [^{3}H]-BSA was injected into the dentate gyrus ipsilateral to the intact entorhinal area. Animals were perfused with aldehydes, and the brains were processed for HRP histochemistry, embedded in plastic, and processed for autoradiography as described in the Appendix (Section VII.A.1). Neurons containing both tracers were found in the surviving entorhinal area (Fig. 2B), indicating that the sprouting fibers arise as collaterals of the normal ipsilateral projection system (Steward and Vinsant, 1978b).

Double-labeled cells with large concentrations of the HRP reaction product are usually not as heavily labeled with [^{3}H]-BSA as are neighboring cells that are labeled only with BSA. A similar phenomenon has also been noted with other double retrograde labeling methods (A. Rustioni, unpublished observations; Van der Kooy *et al.*, 1978; see also Figs. 3A–D). The difference is most evident when HRP appears in large accumulations in the cell cytoplasm (Steward and Vinsant, 1978b). Although the reason for this phenomenon is not clear, it is possible that cells with major collateral projections to two or more sites give rise to less extensive arborizations in each of their terminal fields than do other neighboring cells that project predominantly or exclusively to a single site. This difference could explain the difference in labeling intensity if the less extensive arborizations permit the cell less of an opportunity to incorporate and transport the tracers. Alternatively, it is possible that either the retrograde transport mechanism or the storage sites for exogenous protein become saturated by one of the protein tracers, resulting in less efficient transport of the other tracer. Such "overloading" may well exist, especially if both protein molecules are sequestered within similar cytoplasmic inclusions (see Electron Microscopy section). Finally, it is possible that high concentrations of the exogenous protein are toxic. This could reduce the viability of the cell, thereby disabling its transport or storage capacity. If "overloading" or toxicity can influence transport, caution must be exercised in interpreting the results of double-labeling experiments in which only single-labeled cells appear in a collection of cells known to project to two sites.

a. Electron Microscopy of Double-Labeled Cells. The autoradiographic and histochemical procedures utilizing DAB or the Hanker–Yates reagent are readily adaptable to electron microscopic studies, particularly if plastic embedding is used for the preparation of the light microscopic material. As illustrated in Fig. 4, both the HRP reaction product and the silver

Figure 2. (A and B) A neuron that is double-labeled with HRP and [^{3}H]-apo-HRP is illustrated at two focal planes to illustrate both the HRP reaction product (A) and the silver grains (B). [Courtesy of A. Rustioni.] (C) The neuron on the right is double-labeled with HRP and [^{3}H]-BSA, whereas the neuron on the left is labeled only with [^{3}H]-BSA. (D) Neuron showing AChE reaction product (black) and HRP reaction product (brown). [Courtesy of L. Heimer.] (E) This neuron also exhibits AChE and HRP reaction products, but in this case the AChE reaction product is brown, and the HRP reaction product dark blue.

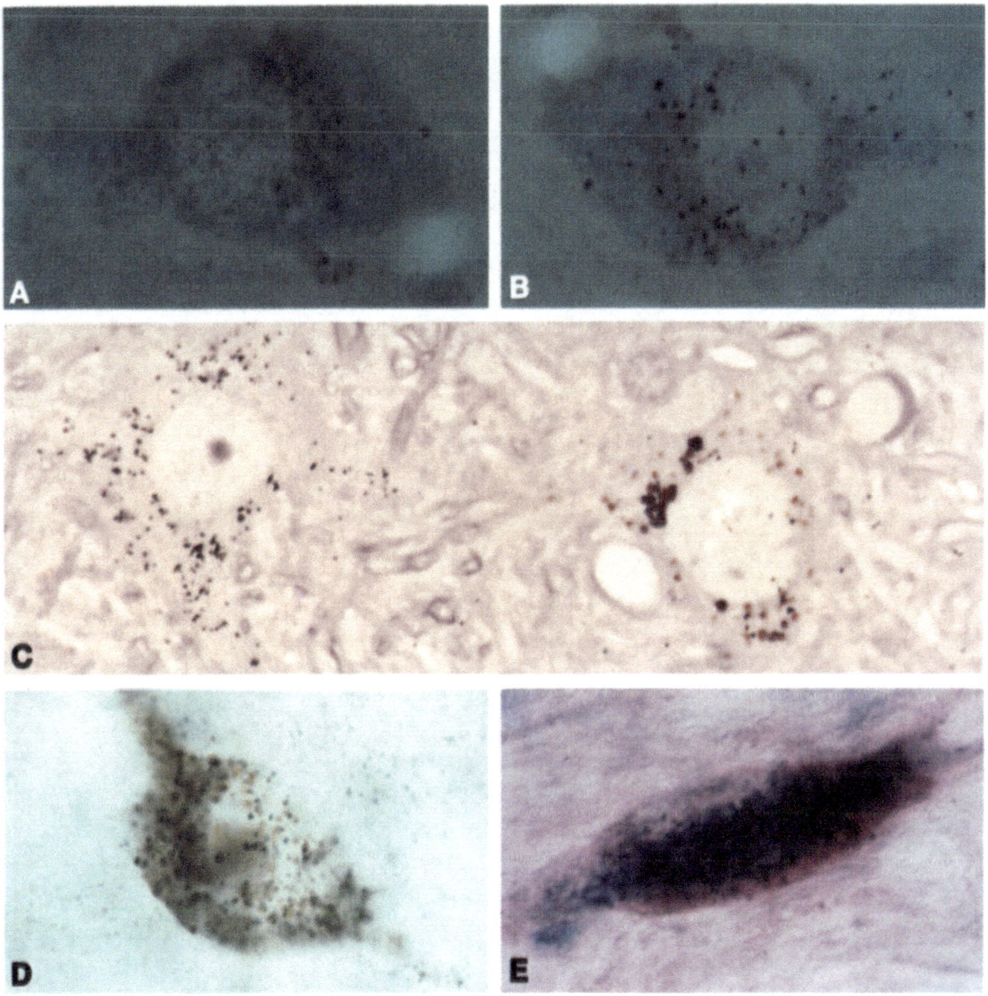

Figure 2

grains indicating the presence of [^{3}H]-BSA are mutually distinguishable at the ultrastructural level. Both labels are concentrated in lysosomelike bodies that appear quite electron-dense when filled with HRP and somewhat more electron-lucent when filled with [^{3}H]-BSA. In addition, the HRP reaction product may be found in what appear to be the cisternae of the smooth endoplasmic reticulum, and isolated silver grains, indicating the presence of [^{3}H]-BSA, are sometimes apparent in the cytoplasm where it is impossible to associate them with any single cytoplasmic organelle.

As in the light microscopic material, the cells appear labeled predominantly with one or the other retrograde marker. For example, Fig. 4A illustrates a portion of cell cytoplasm heavily labeled with [^{3}H]-BSA but with no obvious concentrations of HRP.* Figure 4B, on the other hand, illustrates the cytoplasm of a cell that appears heavily labeled with HRP as evidenced both by the large electron-dense lysosomelike bodies and by the collections of electron-dense material in the cisternae of the smooth endoplasmic reticulum. These cells also have a few silver grains indicating the presence of [^{3}H]-BSA.

b. Problems of Interpretation with the [^{3}H]-Protein Methods. As illustrated by these examples, the combination of [^{3}H]-apo-HRP or [^{3}H]-BSA with enzymatically active HRP has the advantage that the cells of origin of the collaterals can be directly visualized at both the light and electron microscopic levels. However, a potential disadvantage is the fact that the autoradiographic method depends on a concentration of silver grains that is clearly above background levels. Because high background levels are found for a considerable distance surrounding the injection sites, at least in the [^{3}H]-BSA procedures, the usefulness of the method is limited to relatively long axon systems.

Related to the problem of "background" is the phenomenon of the "halo" that seems to surround collections of labeled cells in the tritiated protein procedures (Geisert, 1976; Steward and Scoville, 1976). Unlike the HRP molecule, which is enzymatically inactivated by proteolysis within the cell, the tritium label remains after degradation of the labeled proteins and can be reutilized for protein synthesis. Perhaps such reutilization and subsequent transport of the newly labeled protein into dendrites account for the halo. Indications that dendritic labeling might account for the halo come from observations in plastic-embedded material, where collections of grains can frequently be associated with dendritic profiles (Fig. 2). The labeled amino acid breakdown products could also diffuse out of the cells into neighboring neurons or glia. The possibility of such secondary effects should be kept in mind, particularly if the labeling of a particular neuron is near background level.

As noted above, the label that is detected autoradiographically need not

*It cannot be excluded that the dark lysosomelike bodies contain small quantities of HRP reaction product.

be associated exclusively with the injected protein but may also be associated with amino acids incorporated into other proteins or with fragments of the original polypeptide. The question thus arises, what exactly is being transported? In the preceding discussion, the assumption was that the transported material was the protein subjected to the labeling procedure. However, it is possible that some limited portion of the BSA molecule is in fact the only portion transported. Indeed, gel electrophoretic analysis of our own samples of [^{3}H]-BSA labeled by New England Nuclear indicate that most of the protein is in the form of small protein fragments with molecular weights ranging from 3500 to 25,000,* rather than in the form of the intact BSA molecule. This does not affect the utility of the double retrograde labeling method, but it does indicate the possibility for the development of more sensitive tracers that might be transported in greater quantities than the BSA. There are reports in the literature that BSA is not effectively transported by some neurons, at least not to the degree that these neurons can transport other proteins (Stoeckel *et al.*, 1977). This apparent difference may represent differences in the ability of various neurons to transport certain protein moieties or differences in the ability of given synaptic populations to degrade the injected substances into transportable fragments. These questions will, it is hoped, be answered through future investigations.

While the preceding discussion is related to [^{3}H]-BSA and [^{3}H]-apo-HRP, the utilization of other proteins presumably should offer similar advantages and limitations. In the future, more efficient methods of labeling the proteins may be introduced, and it is possible that these methods may remove the potential confounding problems of the autoradiography.

c. Technical Considerations. All procedures combining HRP histochemistry with autoradiography share certain technical problems. One important problem is the appearance of false negative results. In both of the

* Analysis courtesy of Dr. Lynn Churchill, University of Wisconsin.

Figure 3. (A) Neurons labeled with true blue. (B) Retrobulbar neurons in the rat labeled with granular blue and nuclear yellow following injections of the fluorescent substances into the two olfactory bulbs. The neurons to the left are labeled with granular blue, whereas the neuron to the right is labeled with both granular blue and nuclear yellow. (C) and (D) illustrate dorsal raphe cells by the aid of two different filter systems providing 550-nm and 360-nm excitation wave lengths. The cell to the left in the two photographs is labeled only with propidium iodide, whereas the cell to the right is triple-labeled with propidium iodide, granular blue, and nuclear yellow. Propidium iodide was injected into the ipsilateral septum, granular blue into the ipsilateral mediodorsal thalamus, and nuclear yellow into the ipsilateral olfactory cortex of the rat. (E) Double-labeled cells in the substantia nigra, pars compacta, following injection of primulin in the neostriatum. The section was processed for indirect immunohistochemistry using tyrosine hydroxylase (TH) antiserum as a marker for dopamine neurons. The primulin-containing clear yellow fluorescent granules can be observed in many cells. (F) Same as in E with higher magnification. Yellow primulin granules are present in many of the cells that also contain green fluorescent TH immunoreactivity. [Courtesy of Dr. Kuypers (A), Drs. de Olmos and Heimer (B, C, and D), and Dr. Hökfelt (E and F)]

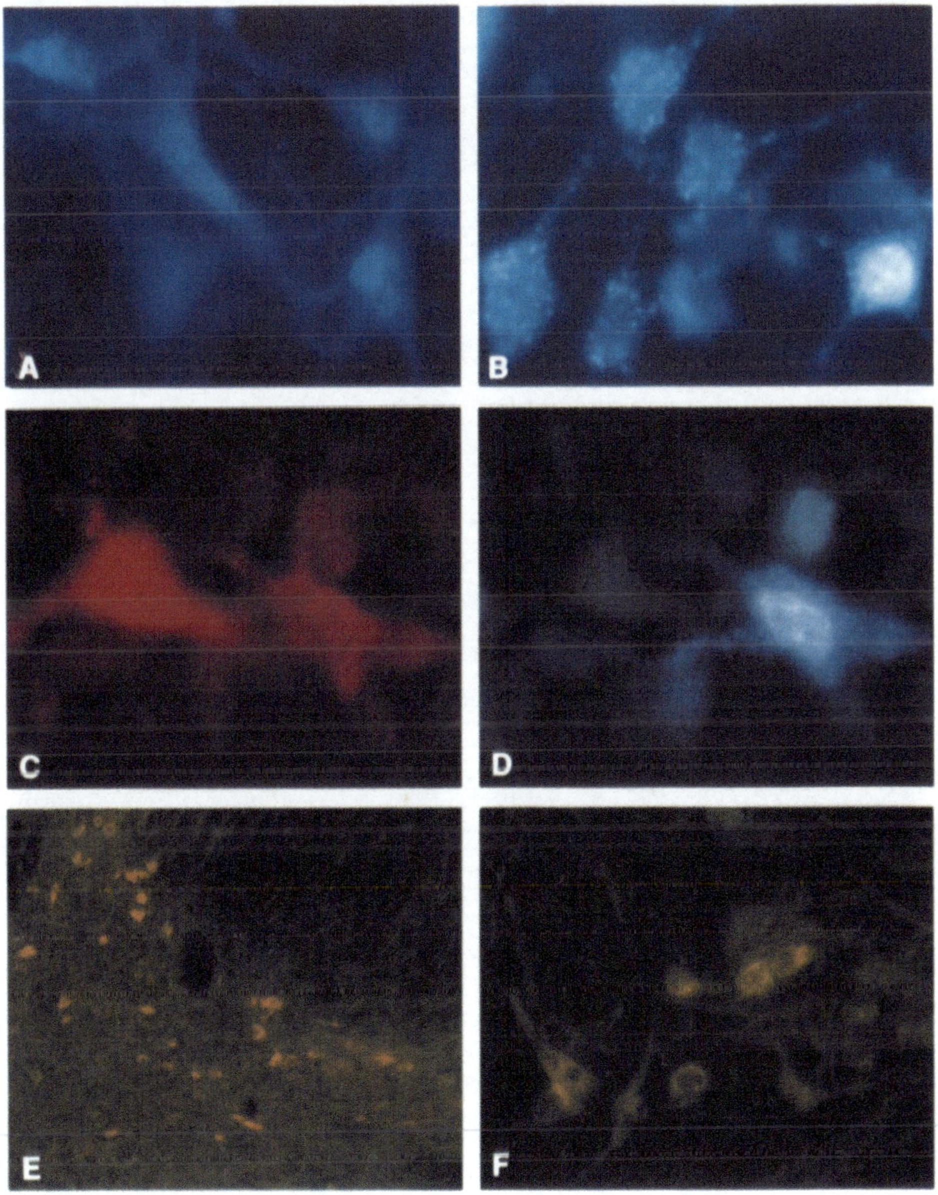

Figure 3

abovementioned methods utilizing labeled proteins, thin sections were used. The reason for this is that the β emissions of tritium have a very limited ability to penetrate the tissue. Tritium sources more than a few micrometers from the section–emulsion interface will therefore not expose the emulsion. This potential risk of false negative results can be largely avoided if appropriate care is taken to insure that some part of the cell in question lies adjacent to the emulsion. In most cases reasonable assurance would be provided even with the relatively thick 15 μm sections obtained from material embedded in paraffin, since any cell with a visible nucleus in the plane of the section has a high probability of being large enough to have some portion of its cytoplasm abut the section–emulsion interface (Rustioni, 1978; Hayes and Rustioni, 1979).

A second difficulty arises with regard to the illustration of the labels at the light microscopic level. The HRP–[^{3}H]-BSA double-labeling method utilizes autoradiography in combination with a DAB histochemical procedure or a procedure utilizing the Hanker–Yates reagent.* One is therefore left with a brown HRP reaction product and black silver grains. In the DAB procedures, the two types of particles can be distinguished in the microscope, but photographic discrimination requires color photography or special illumination techniques (see Appendix, Section VII. A.2). The use of the Hanker–Yates reagent (Hanker *et al.*, 1977), however, offers some advantage in this regard, since the Hanker–Yates product, unlike the DAB reaction product, is not highly reflective in dark-field illumination. Therefore, the combination of the Hanker–Yates reagent with autoradiography facilitates discrimination, since both the silver grains and Hanker–Yates reaction product are visible with bright-field illumination, while only the silver grains are highly reflective with dark-field illumination.

B. Double Labeling with Fluorescent Substances

One disadvantage of the double-labeling technique utilizing HRP and [^{3}H]-proteins is that the methods are rather time-consuming. Recently, however, a simple double-labeling method utilizing fluorescent substances has been described (Van der Kooy *et al.*, 1978). Whereas Evans blue, primulin, and DAPI were initially used (Kuypers *et al.*, 1977, 1979; Bentivoglio *et al.*, 1979a,b, 1980a,b), several other substances have recently been introduced (see Appendix, Section VII.C) (Bentivoglio 1980a,b). Each substance requires a particular wavelength of light for excitation and also emits light at characteristic wavelengths. Thus, by judicious use of excitation and emission filters on a fluorescence microscope, it is possible to single-, double-, and even triple-label cells in a single preparation. Figure 3A illustrates cells la-

*The reason for choosing DAB or the Hanker–Yates reagent as a chromogen is that other chromogens such as *o*-dianisidine, tetramethylbenzidine, and benzidine salts give rise to less stable products that have so far proved difficult to preserve during autoradiographic processing.

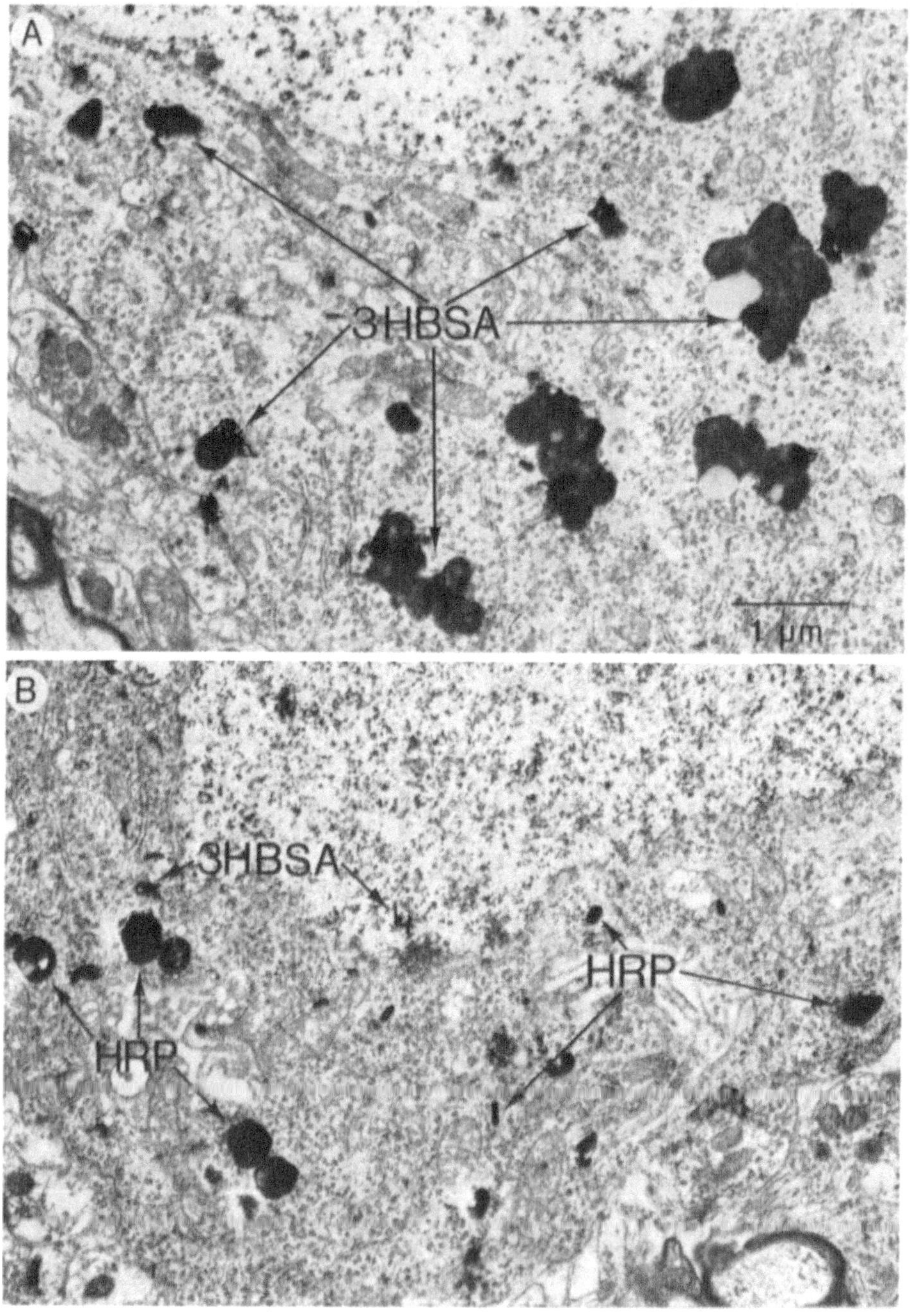

Figure 4. Electron microscopy of cells labeled with HRP and [^{3}H]-BSA. (A) The cytoplasm of a cell labeled with [^{3}H]-BSA exhibits silver grains overlying lysosomelike bodies that are relatively electron lucent. (B) In a double-labeled cell, a few silver grains are also present along with electron-dense accumulations of HRP reaction product in lysosomelike bodies and in what appear to be the cisternae of smooth endoplasmic reticulum. Layer II stellate cells from the entorhinal cortex of a rat with a contralateral entorhinal lesion.

beled with true blue (blue fluorescent cytoplasm and nucleus). Figure 3B shows several cells labeled with granular blue (blue cytoplasm with silvery granules) and one cell double-labeled with granular blue and nuclear yellow (yellow–green nucleus). Figures 3C and D illustrate a triple-labeled cell. The two photographs show the same cells through different excitation filters. The cell to the left in the two photographs is labeled with propidium iodide, whereas the cell to the right is triple-labeled with propidium iodide, granular blue, and nuclear yellow. Interestingly, as was the case with the HRP–[^{3}H]-BSA method, cells that are double- or triple-labeled usually exhibit less intense fluorescence for each of the labels than do cells that are singly labeled.

1. Advantages and Disadvantages of the Fluorescent Probes

The fluorescence methods are simple, fast, and quite sensitive. In addition, the possibility is offered for the development of triple labeling utilizing other fluorescent compounds, and it is also possible to combine the fluorescent retrograde markers with immunohistochemical methods for specific transmitters (see Appendix, Section VII.F). In addition, since it is not necessary to react the tissue in a histochemical medium, the fluorescent probes may be quite useful in combinations with 2-deoxyglucose autoradiography.

There are, however, some disadvantages. Some of the fluorescent probes have a tendency to fade, particularly when illuminated. The fading can be kept to a minimum, however, if the sections are stored in a refrigerator. Furthermore, the methods seem to be relatively incompatible with other standard neurohistological procedures including Nissl staining and silver staining, and it is difficult to envision how they might be adapted for electron microscopic studies unless the fluorescent substance is combined with an electron-dense substance. Finally, in our own work with Evans blue-labeled BSA (Steward and Scoville, 1976), the documentation of the extent of the injection site has proved difficult. This problem, however, is by no means unique to the fluorescent probes.

C. Collateral Transport of HRP

Although the double-labeling methods described above can provide important information regarding collateral projections, they do have their shortcomings. For example, double-labeling studies are unlikely to provide information on the site of origin of axon collaterals, the route of the different collateral branches, or the terminal distribution of each of the collateral systems. Recently developed sensitive HRP methods, however, offer certain advantages in this regard (de Olmos and Heimer, 1977). These methods depend on the fact that neurons apparently can transport HRP bidirectionally under appropriate circumstances (Robertson, 1977). Therefore, injections of HRP into the terminal fields of a given group of cells may result not only in the retrograde transport of the protein back to the cell body but also in anterograde transport into collateral axonal projections of the labeled cell population. In this situation, the objective evidence for the presence of collateral projections is the appearance of the tracer protein in the terminal field of the presumed collateral (Fig. 1). To date, evidence for such collateral transport of marker protein has been obtained in hippocampal pathways and in the mammillary projection systems by de Olmos and Heimer (1977) and in the projections of the trigeminal nucleus to the superior colliculus and inferior olive by Frankfurter *et al.* (1978).

The phenomenon of collateral transport of HRP may reveal several aspects of collateral projections that cannot be defined by the double-labeling methods. The course of the axons as well as the distribution of the collateral terminal field can be revealed. The method may also be capable of revealing the point of collateralization. One potential difficulty relates to the nature of the evidence, however. As noted above, the evidence for collateral projections is the appearance of "terminal" labeling in the projection area of the proposed collaterals. It is essential to exclude the possibility that such terminal labeling results from the direct anterograde transport of HRP from neuronal cell bodies in the area of injection or from axons passing through the injection area. These possibilities, however, can be evaluated by analyzing the anterograde projections of the neurons in the area of the injection through use of silver methods and autoradiography.

III. HRP AND ANTEROGRADE TRACING METHODS

The neuroanatomical study of interneuronal connections has now reached the point where the mere tracing of pathways between cell groups A and B must be considered insufficient. In order to make the tracing of central nervous pathways more meaningful, information is needed not only with regard to the type of synaptic complexes established by a given pathway but also with regard to the nature and specific efferent connectivity of the postsynaptic neuron. In other words, to appreciate the real significance of a specific input to a heterogeneous neuronal pool, the postsynaptic neuron must be classified in terms of its total shape, including not only the cell body and its dendritic arborization but also its axonal projections. Although the combined Golgi–EM technique (Chapter 12) can provide much of this information, the combination of the HRP method with orthograde tracing methods would seem to offer an additional advantage in the sense that a long-axon postsynaptic neuron might be classified in regard to its area of termination.

The HRP method has been successfully combined with both the Fink–Heimer method for silver impregnation of degenerating terminals and the autoradiographic method for the anterograde tracing of neuronal pathways. For example, following dorsal rhizotomy in the cat and HRP injections into the nucleus ventralis posterior lateralis thalami, Blomqvist and Westman

(1975) processed sections of lower brainstem and spinal cord for HRP histochemistry and subsequent Fink–Heimer staining. Argyrophilic debris could be seen in areas of the gracile nucleus which also contained HRP-labeled neurons. Colwell (1975) studied thalamocortical reciprocity by injecting solutions of both HRP and [^{3}H]-leucine into the visual cortex. The anterograde projections from the area of the injection were revealed with the autoradiographic procedure, whereas the cells of origin of the afferent projections to the injected area were revealed by the retrograde labeling of cell bodies with HRP.

The information provided by these combined procedures is limited unless the synaptic relations can be studied at the electron microscopic level. This is a difficult problem, since the afferents often terminate on dendrites that are not easily labeled with the retrograde methods. Until it is possible to retrogradely label the full extent of a neuron at the ultrastructural level, the combination of retrograde and anterograde tracing procedures would seem to offer little more than convenience. The information obtained by the combined approach on the light microscopic level can usually be gained in a sequence of experiments in which each method is used separately.

IV. HRP AND TRANSMITTER-RELATED HISTOCHEMICAL PROCEDURES

Retrograde transport methods using HRP or fluorescent substances (Appendix, Section VII.F) have been successfully combined with various histochemical procedures capable of demonstrating transmitter-related enzymes in single neurons. The motivation for these studies derives from observations that particular brain regions contain neurons that apparently utilize a particular transmitter or at least contain the transmitter-related enzyme. The question then arises whether the cells identified on the basis of their enzyme contents are also the cells that can be retrogradely labeled following injections of HRP or a fluorescent substance into a specific target area.

To date, the HRP method has been combined with monoamine oxidase histochemistry (Satoh *et al.*, 1977), formaldehyde-induced fluorescence techniques (Berger *et al.*, 1978; Blessing *et al.*, 1978; Bjorklund and Skagerberg, 1979), immunofluorescence microscopy for tyrosine hydroxylase (Ljungdahl *et al.*, 1975; and see Section VII.F.), and the histochemical method for acetylcholinesterase (Mesulam, 1976; Hardy *et al.*, 1976). Each of these transmitter-related enzymes serves as a more or less specific marker for neurons that presumably utilize the relevant transmitters.

The major progress to date seems to be in the combination of the acetylcholinesterase (AChE) histochemical methods with HRP histochemistry. Hardy *et al.* (1976) have demonstrated that some of the basal forebrain cells that contain AChE are retrogradely labeled following HRP injections into the mediodorsal thalamic nucleus, and Mesulam (1976) has demonstrated that both AChE-negative as well as AChE-positive neurons of the basal forebrain are labeled following HRP injections into the olfactory bulb and neocortex. Both the AChE histochemical reaction product and the HRP reaction product can be simultaneously visualized in single neurons (see Fig. 2D,E). Detailed protocols for the combined methods for AChE histochemistry are presented in Section VII.D.

At this time, the major weaknesses of the combined procedures for HRP and transmitter-related enzymes are related to the specificity of the histochemical methods for the transmitters in question. Although the accumulation of enzymes in specific cell types may provide a convenient marker for that cell population, it does not necessarily indicate the transmitter utilized by that cell. In this context, it is worth mentioning that degradative enzymes (AChE and MAO) are typically more ubiquitous and less specific for neuronal systems utilizing specific transmitters than are synthetic enzymes (e.g., tyrosine hydroxylase).

There are some technical problems associated with the combination of HRP histochemistry with methods for revealing transmitter-related enzymes. One relates to the difficulties of distinguishing the two reaction products in the tissue, whereas the other concerns the development of procedures that are appropriate for each method with the same piece of tissue. As is the case with the double retrograde labeling studies, the deposition of the reaction product from one histochemical reaction often obscures the reaction product from the other, so that special modifications of one or both methods are necessary. For example, Hardy *et al.* (1976) found that substituting sodium sulfite for sodium sulfide in the AChE histochemical procedure results in an AChE reaction product that appears black instead of brown and, therefore, is more readily distinguishable from the brown HRP reaction product. In the most difficult situations, e.g., those involving fluorescent labeling of tyrosine hydroxylase, it may be necessary to process the tissue for each method sequentially, photographing an identified region at each stage to reveal the neurons labeled with each treatment (see Ljungdahl *et al.*, 1975). Recent technical refinements have permitted the simultaneous visualization of retrograde and transmitter-related fluorescent markers, however (see Section VII.F. and Fig. 3,E,F).

V. HRP AND 2-DEOXYGLUCOSE PROCEDURES

The newest addition to the neuroanatomical armory is the 2-deoxyglucose (2-DG) method for analyzing ongoing metabolic activity in specific brain regions (Chapter 15). To date, the bulk of these studies have involved measurements of uptake of ^{14}C-labeled 2-DG. The resolution provided by ^{14}C autoradiography permits an analysis of patterns of uptake in nuclei or specific laminae but does not permit the analysis of uptake in single neuronal somata. For example, it has been shown in a variety of situations that increasing or decreasing "activity" in some neuronal population by modifying the activity along afferent pathways to the region results in increased uptake of 2-DG and thus in increased concentration of radioactive 2-DG in autoradiographic preparations (Sokoloff, 1977). Further attempts have been made to define the specific neuronal populations involved or the neuronal or glial compartments in which the changes in 2-DG uptake associated with activity are occurring. These efforts have resulted in the development of the [^{3}H]-2-DG method, which permits the resolution of the concentration of 2-DG even at the level of the single cell (Sharp, 1976; Basinger *et al.*, 1979). Given this capability, it is of interest to define 2-DG uptake in neurons that are identified on the basis of their patterns of connectivity.

Our own interest in this problem arose since we were interested in whether cells that give rise to lesion-induced sprouts increase their metabolic activity during the sprouting response as measured by changes in 2-DG uptake. In an attempt to answer this question, HRP was injected into the dentate gyrus to identify entorhinal cells sprouting in response to a contralateral entorhinal lesion (see Steward and Vinsant, 1978a). Just prior to sacrifice, [^{3}H]-2-DG was injected as described in Section VII.E. The animal was perfused with aldehydes, and the tissue was removed and processed for both [^{3}H]-2-DG and HRP as described in the Appendix, Section VII.E. In such material silver grains may be seen overlying the tissue with a distribution comparable to that in tissue processed without perfusion fixation and utiliz-

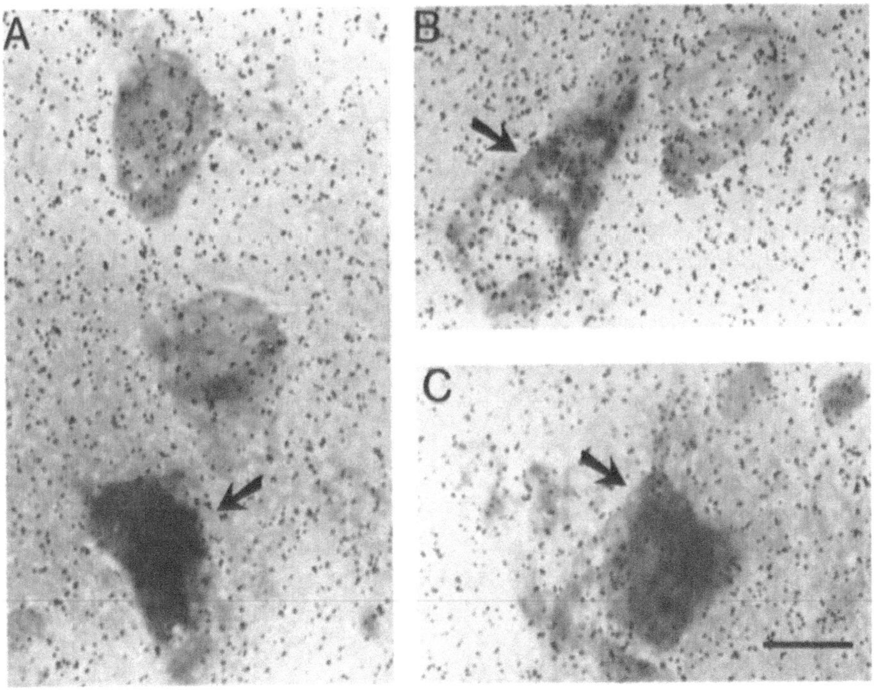

Figure 5. HRP and [^{3}H]-2-deoxyglycose autoradiography. (A), (B), and (C) illustrate 3 cells with HRP reaction product (arrows) in the hilus of the fascia dentata after HRP injections into the contralateral fascia dentata. Cells not labeled with HRP are also present. Silver grains from the [^{3}H]-2-deoxyglucose may be seen to overlie both the cell bodies and the surrounding neuropil.

ing routine procedures for [^{3}H]-2DG autoradiography.* These routine procedures have been quite useful for studies of activity-related changes in the uptake of 2-DG in the brainstem auditory nuclei of the chick (Lippe *et al.*, 1980) and in the study of changes in 2-DG uptake associated with denervation and reinnervation of dentate gyrus (Steward and Smith, 1980). In addition, HRP-labeled cells are quite evident in tissue reacted for HRP histochemistry after autoradiographic development for 2-DG (Fig. 5). While it is too early to determine whether this combined procedure will offer useful information, it is clear that the combination of the two procedures is possible.

VI. PROSPECTS FOR THE FUTURE

The evolution of the neuroanatomical methodology over the past decade has been astounding, and there is no evidence of a slowdown in the near future. One can envision a number of additional applications of the methods described in this chapter. For instance, it should be possible to utilize the autoradiographic retrograde methods on thin plastic sections of Golgi-stained material in which the silver grains from the autoradiographic procedure could be shown, with appropriate illumination techniques, to overlie Golgi-impregnated cells.

The flexibility of the HRP method for various combined procedures has also been greatly enhanced as a result of the possibility for *in situ* perfusion of the histochemical substrate (Hendry, 1978). Until now, it was necessary at some stage to slice the brain into pieces thin enough for histochemical processing for HRP. An *in situ* protocol eliminates this step and makes it possible to treat the brain in any fashion including direct embedding in plastic or paraffin subsequent to the *in situ* reaction. Further advances in the basic HRP methodology have the potential to revolutionize our experimental approaches. If it becomes possible, for example, to obtain "solid HRP labeling" of all the processes of retrogradely labeled cells at the electron microscopic level, the potential will exist for actually defining the synaptic relations between afferents to a region and the cells of origin of long efferent projections (see Chapter 7, Section V.B).

VII. APPENDIX

A. Procedures for Retrograde Double Labeling with HRP and [^{3}H]-BSA

Oswald Steward

For the double injections, HRP is prepared as for routine histochemical procedures. At the moment, no commercial source of [^{3}H]-BSA† of suffi-

*Recent studies indicate that perfusion does not alter tissue concentration or distribution of the 2-DG (Des Rosiers and Descarries, 1978).

† 125-iodine-labeled BSA is available from New England Nuclear, Boston, Mass. We have not tried the substance, but it might be useful for double-labeling procedures.

ciently high specific activity is available, and thus it must be custom-made. Our own material is prepared from Sigma BSA by the tritium gas exposure method (New England Nuclear). The cases illustrated in Figs. 2 and 6 in the present chapter were injected with [^{3}H]-BSA prepared by exposing 100 mg of dry BSA to 6 Ci of tritium gas for 3 weeks. This exposure results in a specific activity of approximately 1.0 mCi/mg. It has been our practice to request that the BSA material be sent frozen in dry form, since this prevents bacterial contamination. The material typically appears as a thin sheet coating the edges of the vial. The material can be dissolved with 0.3 M NaOH in half of the volume eventually desired for the dilutions, and the pH of the solution can be returned to neutral by addition of an equivalent quantity of equimolar HCl. The material thus solubilized remains in solution through several freezings and thawings. Once the [^{3}H]-BSA has been solubilized, it may be injected in whatever volume is desired. We have found adequate transport following injections of 1 μl of the material (with 20 Ci of [^{3}H]-BSA).

The survival times of the animals used in the study of the entorhinal projections were 24 and 48 hr. Both survival times resulted in adequate transport of the [^{3}H]-BSA from the hippocampus to the entorhinal cortex. However, the "halo" of silver grains surrounding cell groups retrogradely labeled with the [^{3}H]-BSA seemed somewhat less following 24 hr than 48 hr.

1. Plastic Sections

Animals are perfused transcardially with a mixture of 2% glutaraldehyde and 2% paraformaldehyde in cacodylate buffer. Following the removal of the brain from the skull and postfixation of the brain in fixative for 1 hr the appropriate part of the brain is dissected by hand utilizing a razor blade. Slabs of approximately 250 μm thickness are rinsed in Tris buffer for approximately 45 min and are then reacted in an HRP histochemical medium with 15 mg of 3,3'-diaminobenzidine and 0.02 ml of 3% H_2O_2 in 30 ml of Tris buffer. Slabs to be processed for electron microscopic and autoradiographic studies are transferred directly from the rinse to a 2% buffered OsO_4 solution for 45 min, dehydrated through a graded ethanol series, and embedded in epon–araldite. The remaining portions of the brain, containing the injection sites, are stored overnight in the cacodylate buffer and cut the following day at 30 μm on a freezing microtome. The frozen sections are stained with the DAB method and mounted on microscope slides for subsequent autoradiographic processing.

Autoradiography. Semithin plastic sections of approximately 1 μm are heat-stretched and air-dried on glass slides cleaned in alcohol. For good results, we have found that it is necessary to remove the plastic from the sections by soaking the slides in a 10% solution of ethoxide in absolute alcohol for 10 min prior to dipping. The deplasticized sections are rinsed in three changes of absolute EtOH and then in flowing water for 20 min. When the slides are dry, they are coated with Kodak NTB-2 emulsion, exposed for 4–12 weeks, and developed in D-19®. Counterstaining can be done with toluidine blue, azure blue, or cresyl violet. The photographs in Fig. 2 are from material stained with cresyl violet.

2. Frozen Sections

In a considerably less time-consuming modification of the double-labeling methods, we inject and perfuse the animals as described above and then section the brains at 5 μm on a cryostat. The cryostat sections are thaw-mounted on microscope slides and reacted in a histochemical medium containing the Hanker–Yates reagent (see Appendix, VII.B). After histochemical reaction, the slides and sections are defatted through an ascending alcohol series, rehydrated to water through a descending alcohol series, and coated with NTB-2 photographic emulsion. From this point, the slides with sections may be treated exactly as any other autoradiographic preparation.

3. Photography for Light Microscopy

Although it is not difficult to distinguish silver grains from HRP reaction product in the microscope, the distinction can be facilitated considerably by viewing the preparations with phase contrast illumination, particularly at higher power. This, however, does not improve the photographic distinction between the two markers on black-and-white film. This distinction can best be obtained utilizing a 100 × Planapochromat® objective with a variable aperture and a dark-field oil-immersion condenser. When viewed in bright-field, both the silver grains and the HRP reaction product appear dark. Similarly, when viewed with dark-field illumination with the numerical aperture set at the highest value on the variable aperture objective, both the silver grains and HRP reaction product again appear dark, but the image is of poor quality and appears as a bright-field image. Closing down the variable aperture results in a transition to a dark-field image, and at the lowest numerical aperture, the silver grains are considerably more reflective than the

---

Figure 6. HRP- and [^{3}H]-BSA-labeled cells in bright-field, dark-field, and transition forms of illumination in a rat with unilateral entorhinal lesion. Four stellate cells in layer II of the intact entorhinal area are illustrated following an HRP injection into the contralateral and a [^{3}H]-BSA injection into the ipsilateral dentate gyrus. Four cells are apparent in the bright-field image in (A). The two on the left contain HRP reaction product, and the two on the right are labeled with silver grains. In the dark-field image in (B), the silver grains are somewhat more reflective than the HRP reaction product, and a few such grains can be seen over the two cells on the left. In dark-field illumination (C) with variable objective aperture, the silver grains reflect light, whereas the HRP reaction product, which is out of the plane of focus, does not. Thus, it can be seen that the cells on the left labeled with HRP are in fact also labeled with [^{3}H]-BSA. The calibration bar indicates 10 μm. [From Steward and Vinsant (1978b).]

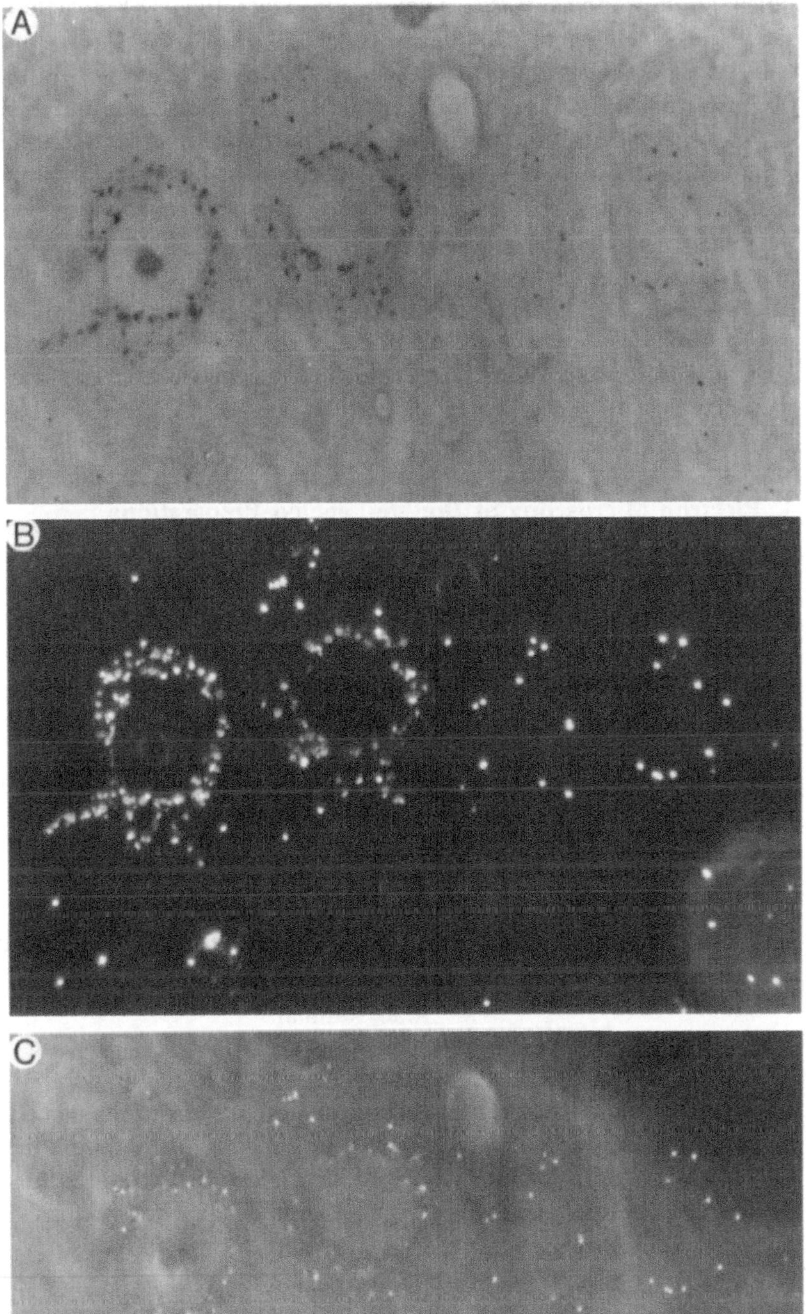
A
B
C

HRP reaction product (Fig. 6). At this point, it is usually possible to distinguish photographically the silver grains from the HRP reaction product, since the silver grains are highly reflective and thus appear quite bright, while the HRP reaction product is less reflective. If it is still difficult to distinguish the two markers photographically, the distinction may be enhanced by opening the variable aperture condenser approximately to the point at which the dark-field image begins to turn to a bright-field image. At this setting, the reflective objects in the plane of focus appear as they would in dark-field, i.e., they reflect light, while objects out of the plane of focus are less reflective. Thus, one may focus on the silver grains which appear as bright spots. The HRP reaction product that is out of the plane of focus appears as dark spots. This method provides a particularly striking documentation of the difference between the silver grains and the HRP reaction product (Fig. 6).

4. Electron Microscopy of Double-Labeled Preparations

From the same blocks that are prepared for light microscopy, thin sections are also taken for electron microscopy after appropriate trimming. Sections (75–80 nm thick) are mounted on 150-mesh copper grids that are then dipped in a freshly made 0.05% solution of sodium lauryl sulfate to enhance the hydrophilia of the grids and sections. The grids are then coated with Ilford L4 emulsion utilizing the loop method. After exposure for 2–6 months, the grids are developed in D-19® or Microdol X®, fixed in Kodak Rapid Fix®, rinsed in running distilled water, and stained with uranyl acetate and lead citrate.

B. Procedures for Retrograde Double Labeling with HRP and [^{3}H]-apo-HRP

Aldo Rustioni*

Tritiated apo-HRP is currently produced by New England Nuclear on an experimental basis and may be ordered directly from the company. Deactivation of the enzyme is accomplished by cleavage of the heme moiety, producing an apoenzyme. Optimal results have been obtained with [^{3}H]-apo-HRP exhibiting 0.44×10^{-4} activity, or 0.43×10^{-3} units of enzyme/mg and having a specific activity near 0.6 mCi/mg. The solution may be stored at 4°C prior to use. After desiccating and redissolving the material to the desired concentration (for example, 20 μCi/μl), the material may be injected into one site, and enzymatically active HRP into another. To date, survival times of 24 and 48 hr have been employed in experiments with rats, cats, and monkeys.

*Departments of Anatomy and Physiology, University of North Carolina, Chapel Hill, North Carolina 27514.

To prepare tissue, animals are perfused with 2.5% glutaraldehyde and 0.5% paraformaldehyde in 0.1 M phosphate buffer at pH 7.2. After perfusion, blocks are taken (2-mm-thick slabs) and processed in a histochemical reaction medium comprised of 50 ml of 0.1 M Tris buffer (pH 7.2) and 75 mg of paraphenylene diamine pyrocatechol (the Hanker–Yates reagent, Hanker *et al.*, 1977) for 2 hr. After the 2-hr pretreatment, the slabs are transferred to a solution of fresh substrate with 0.5 ml of 1% H_2O_2 and reacted for an additional 2 hr. The blocks can then be embedded in paraffin and sections prepared for autoradiography in routine fashion. The autoradiographs may be stained after development in D-19® with a number of counterstains including cresyl violet, thionin, etc. Silver grains and HRP reaction product may be distinguished by depth of focus and color and reflectance when viewed with dark-field illumination.

C. Procedure for Retrograde Double Labeling with Fluorescent Substances

H. G. J. M. Kuypers*

A number of different substances may be used for retrograde tracing. These include:

1. Evans blue (EB), Merck 3169
2. DAPI, 4′-6-diamidino-2-phenylindol · 2HCl, Serva 18860
3. Primuline (Pr), Eastman 1039
4. Bisbenzimide (Bb), Hoechst 33258
5. Nuclear yellow (NY), Hoechst S769121†
6. Propidium iodide (PI), Sigma P-5264
7. True blue (TB), trans-1,2-Bis(5 amido 2 benzofuranyl) ethylene-dihydrochloride‡
8. Granular blue (GB), 2-[4-(4-amidinophenoxy)phenyl]indol-6-carboxamidin-dihydrochloride‡
9. Fast blue (FB), trans-1-(5-amidino-2-benzofuranyl)-2-(6-amidino-2-indolyl) ethylene-dihydrochloride‡

For the study of collateral pathways, two tracers may be chosen so that each fluoresces in a different color at a different excitation wavelength. They can then be identified in one and the same cell by illuminating the tissue with the respective excitation wavelengths. For this purpose, EB and

*Department of Anatomy II, Erasmus Universiteit, Rotterdam, Postbus 1738, Rotterdam, The Netherlands.

†Nuclear Yellow is not commercially available yet but can be obtained for research purposes from Dr. H. Loewe, Hoechst Aktiengesellschaft, Pharma-Synthese, D-6230 Frankfurt-am-Main 80, Postfach 800320, Germany.

‡TB, GB, and FB can be obtained for research purposes from Dr. Otto Dann, Institut für Pharmazie und Lebensmittelchemie der Friedrich-Alexander-Universität, Postfach 2044, D-8520 Erlangen, Germany.

DAPI–Pr (Kuypers *et al.*, 1977, Van der Kooy *et al.*, 1978, Bentivoglio *et al.*, 1979a,b) and PI or EB and Bb (Kuypers *et al.*, 1979) have been used with success in rats and cats. However, it is advantageous to choose two substances that fluoresce at the same excitation wavelength but label in different colors different features of the cell. This eliminates the necessity of continuously having to illuminate the tissue with two excitation wavelengths. The following sets of compounds fulfill this ideal: true blue, fast blue, or granular blue, which at 360 nm excitation wavelength give a blue fluorescence of the cytoplasm (GB with many fluorescent granules), in combination with Bb or NY which, in 1% concentration and after very short survival times (see below), mainly produce a green (Bb) or a golden-yellow (NY) fluorescence of the nucleus at 360 nm excitation wavelength. Double-labeled neurons then display a blue fluorescent cytoplasm with a green or yellow fluorescent nucleus (Fig. 3 B).

The use of the various tracers is basically very simple. They are dissolved or suspended in distilled water (for concentrations, see Table 1) and injected in quantities of 0.1 to 1.0 μl into the brain or spinal cord. The various characteristics of the different retrograde tracers are summarized in the table. In rat, the fluorescence of EB and PI in retrogradely labeled neurons tends to decrease with survival times longer than 4 days. However, in the case of DAPI–Pr, true blue, granular blue, and fast blue, prolonging the survival time results in an increase in the intensity of the fluorescence. After longer survival times, TB and FB diffuse out of the retrogradely labeled neurons as indicated by a halo of blue fluorescent neuropil around the labeled cells including some blue fluorescent glial nuclei. In this regard, special attention should be paid to Bb and NY, since they easily diffuse out of the cells as indicated by fluorescence of many surrounding glial nuclei. This diffusion occurs gradually both *in vivo* during the survival time and, in cat material, also *in vitro* when the sections are kept for a long time in water. After short survival times, Bb and NY only give retrograde labeling of the neuronal nucleus—8 hr for Bb and 16 to 20 hr for NY from rat C2 to cerebral cortex. After slightly longer survival times (Bb, 12 hr and NY, 29 hr), dull fluorescent glial nuclei appear around retrogradely labeled neurons. After still longer survival times (Bb, 24 hr and NY, 44 hr), the neuronal cytoplasm also becomes labeled, and the glial nuclei become as brilliantly fluorescent as the neuronal nuclei, indicating extensive diffusion of the tracers. Therefore, *in vivo* migration of the tracers may be prevented by using short Bb and NY survival times so that the neuronal nuclei become brilliantly fluorescent while surrounding glial nuclei are either not fluorescent or only slightly fluorescent. The *in vitro* migration may be prevented by rapid mounting of the sections after they have been cut (Bentivoglio *et al.* 1980b).

The animals are sacrificed under deep anesthesia by transcardial perfusion with saline followed by 10% formalin. The brain and/or spinal cord are removed and soaked in cacodylate-buffered 10% sucrose (pH 7) until the tissue sinks. Frozen sections 25–35 μm thick are cut, rapidly mounted

Table I. Properties of Fluorescent Compounds Used for Retrograde Tracing Studies

Compound	Cocn.	Fluorescent color seen with Leitz filter–mirror combinations			Cellular fluorescence			Glia around retrogradely labeled neurons	Long-distance transport		Fading	Combined with in double-labeling combinations
		N2	D	A	Cytoplasm	Nucleus	Nucleolus		Rat	Cat		
Evans blue	10%[a]	Red	—	—	Bright	Bright	Bright	—	±	±	—	DAPI–Pr, Bb
DAPI	2.5%	—	Dull green-blue	Blue	Soft + granules	Bright	Bright	+	+	±	—	Pr, EB
Primuline	10%	—	Golden	Golden	Many bright granules	—	—	—	?	?	—	DAPI
Propidium iodide	3%	Orange	—	—	Soft + many bright granules	—	Bright	+	+	+	—	Bb
Bisbenzimide	1–3%	—	Yellow-green	Yellow-green	Soft + yellow-green granules	Soft + bright membrane	Soft + bright ring around nucleolus	+++[c]	+++	+++	—	PI, EB, TB, GB
True blue	5%	—	Dull green	Blue	Bright	Soft	Bright	±	±	?	+	Bb, NY
Granular blue	5%	—	Dull green	Blue	Bright + bright granules	—	—	±	+	+	+	Bb, NY
Nuclear yellow	0.5–1%	—	Yellow	Yellow	—	Yellow	?	+++	+++	+++	—	TB, GB
Fast blue	2%	—	Green-blue	Light blue	+	±	—	—	++	++	±	Bb, NY

[a]Solution in 1% poly-L-ornithine.

from distilled water, and air-dried. Recently, in cats, soaking of the material in 30% sucrose has been replaced by transcardial perfusion with 8% to 10% cacodylate-buffered sucrose, after which the brains are immediately transferred to the microtome. They may be studied under a Leitz Ploemopack® fluorescence microscope equipped with three-filter–mirror systems N2 (excitation filter 2 mm BG36 + KP560 + K530, mirror TK580, Sperrfilter K580 + KP610), D [excitation filter 3 mm BG3 + KP425 (=KP410), mirror TK455, Sperrfilter K469], and A (excitation filter 2×2 mm UG1, mirror TK400, Sperrfilter K430) providing excitation light of 550 nm, 390 nm, and 360 nm wavelengths, respectively. Comparable Zeiss filters are No. 487715EM (Rhodamin) for 550 nm, No. 48770EM (Quinacrine mustard) for 390 nm, and No. 487702EM (U.V.) for 360 nm.

Although the sections can be studied in the microscope without coverslip, the image is improved if the sections are coverslipped using DPX* as mounting medium. If the sections are not coverslipped, emulsion oil is required for medium- and high-power magnification. The oil is placed directly on the air-dried sections. Ultimately the oil seems to make the fluorescence less brilliant, and it is therefore advisable to keep some sections without oil for final photomicrography. Regardless of whether or not coverslips are used, the sections should be studied soon after mounting and drying, since the fluorescence has a tendency to fade with time. However, the fluorescence can be well-preserved for several weeks, especially if the material is stored in a refrigerator.

Some examples of retrogradely labeled cells are illustrated in Fig. 3. Cells labeled with true blue show a deep blue cytoplasm and a bright blue fluorescent nucleolus in a lightly labeled nucleus (Fig. 3A). Granular blue-labeled neurons are characterized by a blue fluorescent cytoplasm with silver to gold fluorescent granules. The nucleus is generally unlabeled or slightly labeled (Fig. 3B). Nuclear yellow, which is transported quite rapidly, primarily labels the nucleus which exhibits a bright yellow color. In the combination of true blue or granular blue and nuclear yellow, the cytoplasm of a double-labeled neuron is blue and the nucleus yellow or greenish-yellow (Figs. 3B and D). In addition, in the case of granular blue, the silver-to-gold fluorescent cytoplasmic granules remain even if some of the blue fluorescent color fades under the microscope. Propidium iodide gives a brilliant orange labeling, especially of the cytoplasm (Fig. 3C).

Recently, it appeared that Bb may also be combined with HRP (Kuypers, unpublished observations). This was concluded from the observation that neurons containing both tracers could be seen in the center–median parafascicular nucleus and substantia nigra after combined injections of Bb and HRP in the caudate nucleus of the rat followed by paraformaldehyde–glutaraldehyde fixation. However, the procedure has the disadvantage that during the soaking of the sections in the HRP incubation medium the Bb or NY may diffuse out of the cell (Bentivoglio *et al.*, 1980b)

*DPX Mountant, Edward Gurr Ltd., London SW 14, England.

and result in false double labeling. This could perhaps be prevented by using the Hardy procedure (see Section VII.D.1) followed by rapid histological processing of the material.

D. Procedures for Simultaneous Demonstration of HRP and AChE

The following procedures allow the simultaneous visualization of the HRP reaction product and the fine granular AChE reaction product in single neurons. In general, injections, survival times, etc., may be optimized for the HRP transport in each chosen system. With the first protocol, the HRP reaction product is brown, and the AChE reaction product black. With the second procedure, the HRP reaction product is blue and the AChE product brown. Suggested protocols for the combined methods are described below.

1. Protocol Published by Hardy *et al.* (1976)

a. Perfusion. Wash briefly with 0.05 M phosphate buffer, pH 7.4. Perfuse with 4% glutaraldehyde in 0.05 M phosphate buffer at the same pH. Both solutions should be ice-cold. The brains may be sectioned 2 hr after perfusion. Sections (50 μm thick) may be collected into ice-cold phosphate buffer saturated with sodium sulfate.

b. Acetylcholinesterase. The following protocol reveals the AChE reaction product.

1. Incubate for 30 min in an incubating solution at 40°C. The incubating solution is made by combining equal volumes of Solutions A and B.

Solution A

0.226 gm Acetylthiocholineiodide per 100 ml of distilled water.

Solution B

Stock glycine (80 mM)	25 ml
Stock $CuSO_4$ (16 mM)	25 ml
0.2 M Acetate buffer (30 ml of acetic acid, 70 ml of Na acetate, pH 5)	50 ml
Phenergan	12.5 mg

2. Wash six times in distilled water (30 sec each).
3. 1.25% sodium sulfite for 1 min.
4. Wash six times in distilled water.
5. 1% silver nitrate for 1 min.
6. Wash twice.
7. 5% sodium thiosulfate for 5 min.
8. Wash six times in distilled water.

c. Horseradish Peroxidase. The following procedure reveals the HRP reaction product.

1. Incubate for 5 min in a solution of 5% sucrose and 0.03% DAB in 0.05 M Tris buffer.
2. Add 5 ml of 0.3% hydrogen peroxide to each 50 ml of the incubating solution. Shake for 5–20 min or until pink.
3. Wash twice in 0.05 M phosphate buffer and once in distilled water.
4. Collect in 10% formalin, mount on microscope slides, and counterstain with cresyl violet.

2. Protocol Published by Mesulam (1976)

a. Perfusion. Wash with saline at 21°C (50 ml for rats, 200 ml for monkeys). Continue with fixative (500 ml for rat, 2000 ml for monkeys) containing 1.25% glutaraldehyde and 1% paraformaldehyde in 0.1 M phosphate buffer at pH 7.4 and a temperature of 21°C. Follow with an equal volume of 4°C buffer with 10% sucrose. The brains may be stored for 24 hr at 4°C and sectioned at 40 μm on a freezing microtome. The sections may be collected in the phosphate buffer.

b. Acetylcholinesterase. After a distilled water wash for 1 min, incubate the sections for 60 min in the AChE incubation solution described above. Prior to development of the AChE histochemical preparation (step 7), the tissue is processed for HRP.

c. Horseradish Peroxidase. The following steps develop the HRP reaction product.

1. Rinse six times in distilled water (30 sec each).
2. Presoak sections for 5 min. Each 100 ml of solution contains 120 mg of benzidine dihydrochloride (BDHC), 100 mg of sodium nitroferricyanide (SNF), 5 ml of 0.2 M acetate buffer, pH 5.0 (19.1 g of sodium acetate and 3.6 ml of glacial acetic acid/liter), 65 ml of distilled water, and 30 ml of absolute ethanol. The solution must be freshly prepared and requires vigorous mixing.
3. Add 5 ml of 0.3% H_2O_2 to presoak solution and incubate for 4–4.5 min.
4. Stop the incubation by transferring the sections into a stabilizer solution (0–4°C). Each 100 ml of stabilizer contains 9 g of sodium nitroferricyanide, 5 ml of 0.2 acetate buffer (pH 5.0, see step 2), 45 ml of distilled water, and 50 ml ethanol. Since the nitroferricyanide is more soluble in water than in ethanol, it may be dissolved in the water prior to adding the ethanol.
5. Rinse six times in distilled water.

6. Develop in a solution made with 100 ml of water and 10 g potassium ferricyanide.
7. Rinse, mount sections, and air-dry. Slides may be dehydrated and the sections covered with Permount.

E. 2-Deoxyglucose Autoradiography and HRP Histochemistry

Oswald Steward

Animals are injected with HRP 1–2 days prior to sacrifice using standard doses and procedures. On the day of sacrifice, the animals receive injections of [^{3}H]-2-deoxyglucose ([^{3}H]-2-DG) obtained from New England Nuclear (specific activity, 40 Ci/mmole). The material is delivered in ethanol which must be evaporated prior to use. This may be accomplished by passing a stream of nitrogen over the liquid. The dried 2-DG may then be dissolved in 0.9% saline to a concentration of 2 mCi/ml.

For the injections, rats are briefly anesthetized with ether, and the 2-DG is injected via the tail vein at a dose of 150–200 μCi/100 g body weight. After 30–45 min survival, the animal is perfused transcardially under deep anesthesia with a glutaraldehyde–paraformaldehyde mixture (1% and 1.25%, respectively). The brain is removed, mounted on a cryostat stage, and frozen by immersing the brain in methylbutane cooled with liquid nitrogen.

The frozen tissue on the stage is placed in the cryostat and allowed to equilibrate to the temperature of the cryostat (−20° to −25°C) for at least 1 hr. Sections (5 μm thick) are cut and mounted on emulsion-coated slides in the dark. This procedure is greatly facilitated by the use of a Thomas sodium vapor safelight. The emulsion-coated slides are prepared by dipping acid-cleaned slides in Kodak NTB-2 emulsion diluted 1:1 with water. The slides should be thoroughly dried and may be stored in a freezer at −30°C in light-tight boxes with some hydrophilic substance.

To affix a section to a slide, the slide is rapidly rotated onto the section which is lying on the cryostat blade. To avoid compressing the sections against the knife blade, two narrow strips of cellophane tape are placed on the knife so that they will be on either side of the sections as they are cut. These strips of tape thus prevent the slides from actually coming into contact with the blade.

Two methods of affixing the sections have been used in our laboratory. In the first, the emulsion-coated slides are stored in the cryostat so that they are at the same temperature as the cut sections; this prevents thawing of the tissue during mounting. The adherence of all portions of the section to the slide is sometimes a problem with this procedure. After mounting, the slide is placed in a light-tight box with Drierite and stored in a freezer at −30°C. In the second method,* which has given more consistent results, the emul-

*Suggested by Dr. Donald Keefer, Department of Anatomy, University of Virginia.

sion-coated slides are allowed to reach room temperature prior to sectioning, and the cut section is thaw-mounted onto the slide. Because the emulsion underlying the tissue has been thoroughly dried, the water in the tissue is drawn into the immediately underlying area of emulsion. The slide is then kept at room temperature for 1 hr, placed in a light-tight box, and stored in a freezer at −30°C. The slides may be stored in the freezer at −30°C for 1–6 weeks before development. The slides are developed as follows:

1. Chloroform for 1 min.
2. 95% alcohol for 1 min.
3. 70% alcohol for 1 min.*
4. Distilled water (three changes, 1 min each).
5. D-19® for 3 min. The D-19® is kept at 17°C and changed after developing six slides.
6. Rinse briefly in cold water.
7. Kodak Rapid-Fix® (without hardener) diluted 1:1 with water for 1–1.5 hr.

After the autoradiographic development and rinsing, the slides may be reacted for HRP. For this purpose the slides are treated as follows:

1. Rinse 0.1 M phosphate buffer, pH 7.3.
2. Using glass slide racks, transfer the slides to the following incubation solution and react at room temperature for 45 min:

Hanker–Yates reagent	375 mg
0.1 M Tris-HC1 buffer, pH 7.6	250 ml
3% H_2O_2	0.8 ml

3. Rinse twice in phosphate buffer.
4. Rinse in distilled H_2O.
5. The tissue may now be stained with cresyl violet, aqueous safranin, etc., dehydrated through alcohols, cleaned in xylene, and covered.

F. A Note on the Combination of Retrograde Fluorescent Tracers with Transmitter Histochemistry

Tomas Hökfelt†

During recent decades, interest has been directed towards identification of neurons on the basis of their transmitter substance. One example of, such an approach is the histochemical identification of monoamines by means of formaldehyde-induced fluorescence (Chapter 13). Another approach relying on the use of immunohistochemistry is discussed in Chapter 14. These his-

*The alcohol solutions have shown significant amounts of radioactivity and should be disposed of accordingly.

†Department of Histology, Karolinska Institutet, 10401 Stockholm, Sweden.

tochemical methods usually are not suitable for tracing pathways, since axons contain low amounts of the transmitter. Therefore, several attempts have been made to combine the HRP tract-tracing method either with immunohistochemistry (Ljungdahl *et al.*, 1975) or with formaldehyde-induced fluorescence techniques (Berger *et al.*, 1978; Blessing *et al.*, 1978). However, as these procedures do present some difficulties, we have recently replaced the HRP with a fluorescent dye as the retrograde tracer (Kuypers *et al.*, 1977). Briefly, our procedure is to inject the dye into the neostriatum. Following a few days' postoperative survival, the animal is perfused with formalin, the brain is removed, and frozen sections are cut on a cryostat. Thereafter, the sections are processed with the aid of the indirect immunofluorescence technique (Chapter 14).

Many of the fluorescent dyes used in retrograde tracing experiments have a tendency to diffuse out of the cell during the immunohistochemical procedure, in which case the retrogradely labeled cells have to be photographed before the immunohistochemistry is done. However, with some of the dyes, leakage from the cells is not a problem; primuline, for example, has been used with good success. This substance produces yellow fluorescent granules that are easily distinguished from the diffusely localized clear green color of fluoresceinisothiocyanate, the compound routinely used for labeling antibodies in immunofluorescence studies. The photographs in Figs. 3E and F show double-labeled cells in the substantia nigra following injection of primuline in the neostriatum. The section was processed for indirect immunohistochemistry using tyrosine hydroxylase (TH) antiserum as a marker for dopamine neurons. The primuline-containing clear yellow fluorescent granules can be observed within numerous zona compacta cells of the substantia nigra that also contain green fluorescent TH-like immunoreactivity. Note that several TH immunoreactive cells lack primuline. Several pathways have now been demonstrated by combining fluorescent dyes with transmitter histochemistry (Hökfelt *et al.*, 1979, 1980a), and the technique of combining immunohistochemistry and retrograde tracing is described in detail elsewhere (Hökfelt *et al.*, 1980b).

Björklund and Skagerberg (1979) have recently combined fluorescent retrograde tracers with the formaldehyde-induced fluorescent technique. Since this technique is based on freeze-drying of tissue, the problem of diffusion is less important. True blue, propidium iodide and Evans blue were found to have the most suitable properties for such combined studies. Finally, Streit and his colleagues (1979) have traced transmitter-specific pathways by using autoradiography to visualize retrogradely transported radioactive transmitters.

In sum, many successful attempts have been carried out to combine retrograde tracer techniques with different types of methods for visualizing transmitters or related substances. Monoamine neurons can be traced by combining the formaldehyde-induced fluorescence technique or immunohistochemistry with antisera either to enzymes involved in the monoamine synthesis or to the transmitter itself (e.g., 5-hydroxytryptamine). The immunohistochemical approach offers the further advantage of tracing other types of neurons such as those containing peptides.

ACKNOWLEDGMENTS. The author extends special thanks to L. Heimer for his helpful criticisms and contributions to this manuscript, and to Drs. T. Hökfelt, H. G. J. M. Kuypers, M.-M. Mesulam, and A. Rustioni for their contribution of material to this chapter. Thanks are also due to E. Geisert for his helpful advice during the early stages of our own work on the development of double retrograde labeling techniques, and to S. A. Scoville, S. L. Vinsant, and L. Smith for their excellent technical assistance. The author is supported by USPHS-NINCDS Grant 5 RO1 NS 12333 and USPHS-NINCDS RCDA Grant 5 KO4 NS 00325.

REFERENCES

Basinger, S. F., Gordon, W. C., and Lam, C. M. K., 1979, Differential labelling of retinal neurons by ^{3}H-2-deoxyglucose, *Nature* **280**:632–684.

Bentivoglio, M., Kuypers, H. G. J. M., Catsman-Berrevoets, C. E., and Dann, O., 1979a, Fluorescent retrograde neuronal labeling in rat by means of substances binding specifically to adenine–thymine rich DNA, *Neurosci. Lett.* **12**:235–240.

Bentivoglio, M., Van der Kooy, D., and Kuypers, H. G. J. M., 1979b, The organization of the efferent projections of the substantia nigra in the rat. A retrograde fluorescent double labeling study, *Brain Res.* **174**:1–17.

Bentivoglio, M., Kuypers, H. G. J. M., Catsman-Berrevoets, C. E., Loewe, H., and Dann, O., 1980a, Two new fluorescent retrograde neuronal tracers which are transported over long distances, *Neurosci. Lett.* **18**(1):25–30.

Bentivoglio, M., Kuypers, H. G. J. M., and Catsman-Berrevoets, C. E., 1980b, Retrograde neuronal labeling by means of bisbenzimide and nuclear yellow (Hoechst S 769121). Measures to prevent diffusion of the tracers out of retrogradely labeled neurons, *Neurosci. Lett.* **18**(1):19–24.

Berger, B., Nguyen-Legros, J., and Thierry, A. M., 1978, Demonstrations of horseradish peroxidase and fluorescent catecholamines in the same neuron, *Neurosci. Lett.* **9**:297–302.

Björklund, B., and Skagerberg, G., 1979, Simultaneous use of retrograde fluorescent tracers and fluorescence histochemistry for convenient and precise mapping of monoaminergic projections and collateral arrangements in the CNS, *J. Neurosci. Methods* **1**:261–277.

Blessing, W. W., Furness, J. B., Costa, M., and Chalmers, J. P., 1978, Localization of catecholamine fluorescence and retrogradely transported horseradish peroxidase within the same nerve cell, *Neurosci. Lett.* **9**:311–315.

Blomqvist, A., and Westman, J., 1975, Combined HRP and Fink–Heimer staining applied to the gracile nucleus in the cat, *Brain Res.* **99**:339–342.

Colwell, S. A., 1975, Thalamocortical–corticothalamic reciprocity: A combined anterograde–retrograde tracer technique, *Brain Res.* **92**:443–449.

de Olmos, J., and Heimer, L., 1977, Mapping of collateral projections with the HRP method, *Neurosci. Lett.* **6**:107–114.

Des Rosiers, M. H., and Descarries, L., 1978, Adaptation of the deoxy-glucose method for use at the cellular level, *Neurosci. Abstr.*, **4**:330 (abstr. 1050).

Fillenz, M., Gagnon, C., Stoeckel, K., and Thoenen, H., 1976, Selective uptake and retrograde axonal transport of dopamine-β-hydroxylase antibodies in peripheral adrenergic neurons, *Brain Res.* **114**:293–303.

Frankfurter, A., Persing, J., and Steward, O., 1978, Collateral afferents to the inferior olive and superior colliculus originating in the spinal trigeminal nucleus, *Neurosci. Abstr.* **4**:551.

Geisert, E. E., Jr., 1976, The use of tritiated horseradish peroxidase for defining neuronal pathways: A new application, *Brain Res.* **117**:130–135.

Hanker, J. S., Yates, P. E., Metz, C. B., and Rustioni, A., 1977, A new specific, sensitive and noncarcinogenic reagent for the demonstration of horseradish peroxidase, *Histochem. J.* **9**:789–792.

Hardy, H., Heimer, L., Switzer, R., and Watkins, D., 1976, Simultaneous demonstration of horseradish peroxidase and acetylcholinesterase, *Neurosci. Lett.* **3**:1–5.

Hayes, N. L., and Rustioni, A., 1979, Dual projections of single neurons are visualized simultaneously: Use of enzymatically inactive (^{3}H)-HRP, *Brain Res.* **165**:321–326.

Hendry, I.A., 1978, Localization of horseradish peroxidase by *in situ* perfusion with substrate mixture, *Brain Res.* **156**:97–101.

Hökfelt, T., Terenius, L., Kuypers, H. G. J. M., and Dann, O., 1979, Evidence for enkephalin immunoreative neurons in the medulla oblongata projecting to the spinal cord., *Neurosci. Lett.* **14**:55–60.

Hökfelt, T., Phillipson, O., and Goldstein, M., 1980a, Evidence for a dopaminergic pathway in the rat descending from the AII cell group to the spinal cord, *Acta Physiol. Scand.* **107**(4):393–395.

Hökfelt, T., Kuypers, H. G. J. M., Bentivoglio, M. Catsman-Berrevoets, C. E., Van der Kooy, D., Dann., O., Philipson, O., Goldstein, M., Steinbusch, H., Verhofstad, A., and Nilsson, G., 1980b, A combined method for transmitter histochemical identification of retrogradely traced neurons. Descending spinal systems characterized by immunohistochemistry using antisera to catecholamine synthesizing enzymes, serotonin and substance P, *Neuroscience,* in press.

Kristensson, K., 1970, Transport of fluorescent protein tracers in peripheral nerves, *Acta Neuropathol. (Berlin)* **16**:293–300.

Kuypers,.H. G. J. M., Catsman-Berrevoets, C. E., and Padt, R. E., 1977, Retrograde axonal transport of fluorescent substances in the rat's forebrain, *Neurosci. Lett.* **6**:127–135.

Kuypers, H. G. J. M., Bentivoglio, M., Van der Kooy, D., and Catsman-Berrevoets, C. E., 1979, Retrograde transport of Bisbenzimide and Propidium iodide through axons to their parent cell bodies, *Neurosci. Lett.* **12**:1–7.

Kuypers, H. G. J. M., Bentivoglio, M., Catsman-Berrevoets, C. E., and Bharos, A. T., 1980, Double retrograde neuronal labeling through divergent axon collaterals, using two fluorescent tracers with the same excitation wavelength which label different features of the cell, *Exp. Brain Res.* **40**:383–392.

Lippe, W. R., Steward, O., and Rubel, E. W., 1980, The effect of unilateral basilar papilla removal upon nuclei laminaris and magroccUularis of the chick examined with [^{3}H]-2-deoxy-D-glucose autoradiography, *Brain Res.* **196**:43–58.

Ljungdahl, Å., Hökfelt, T., Goldstein, M., and Park, D., 1975, Retrograde peroxidase tracing of neurons combined with transmitter histochemistry, *Brain Res.* **84**:313–319.

Mesulam, M.-M., 1976, A horseradish peroxidase method for the identification of the efferents of acetylcholinesterase-containing neurons, *J. Histochem. Cytochem.* **24**:1281–1286.

Robertson, R. T., 1977, Bidirectional movement of horseradish peroxidase and the demonstration of reciprocal thalamocortical connections, *Brain Res.* **129**:538–544.

Rustioni, A., 1978, Double-labeling technique. Neuroanatomical techniques, 8th Annual Meeting of the Society for Neuroscience, 73–80.

Satoh, K., Tohyaina, M., Yamamoto, K., Sakumoto, T., and Shimizu, N., 1977, Noradrenaidine innervation of the spinal cord studied by the horseradish peroxidase method combined with monoamine oxidase staining, *Exp. Brain Res.* **30**;175–186.

Sharp, F. K., 1976, Relative cerebral glucose uptake of neuronal perikarya and neuropil determined with 2-deoxyglucose in resting and swimming rat, *Brain Res.* **110**:127–139.

Sokoloff, L., 1977, Relation between physiological function and energy metabolism in the central nervous system, *J. Neurochem.* **29**:13–26.

Spreafico, R., Hayes, N. L., and Rustioni, A., 1979, Collateral projections from the thalamus to SI and SII of the cat as demonstrated by HRP and ^{3}H-apo-HRP double-labeling technique, *Anat. Rec.* **193**:691.

Steward, O., and Scoville, S. A., 1976, Retrograde labeling of central nervous pathways with tritiated or Evans blue-labeled bovine serum albumin, *Neurosci. Lett.* **3:**191–196.

Steward, O., and Smith, L. K., 1980, Metabolic changes accompanying denervation and reinnervation of the dentate gyrus of the rat measured by [^{3}H]-2-deoxyglucose autoradiography, *Exp. Neurol.* **69:**513–527.

Steward, O., and Vinsant, 1978a, Identification of the cells of origin of a central pathway which sprouts following lesions in mature rats, *Brain Res.* **147:**223–243.

Steward, O., and Vinsant, S. L., 1978b, Collateral projections of cells in the surviving entorhinal area which reinnervate the dentate gyrus of the rat following unilateral entorhinal lesions, *Brain Res.* **149:**216–222.

Steward, O., Cotman, C. W., and Lynch, G. S., 1976, A quantitative autoradiographic and electrophysiological study of the reinnervation of the dentate gyrus by the contralateral entorhinal cortex following ipsilateral entorhinal lesions, *Brain Res.* **115:**181–200.

Steward, O., Scoville, S. A., and Vinsant, S. L., 1977, Analysis of collateral projections with a double retrograde labeling technique, *Neurosci Lett.* **5:**1–5.

Stoeckel, K., Schwab, M., and Thoenen, H., 1977, Role of gangliosides in the uptake and retrograde axonal transport of cholera and tetanus toxin as compared to nerve growth factor and wheat germ agglutinin, *Brain Res.* **132:**273–285.

Streit, P., Knecht, E., and Cuenod, M., 1979, Transmitter-specific retrograde labeling in the striato–nigral and raphe–nigral pathways, *Science* **205:**306–308.

Van der Kooy, D., Kuypers, H. G. J. M., and Catsman-Berrevoets, C. E., 1978, Single mammillary body cells with divergent axon collaterals. Demonstration by a simple, fluorescent retrograde double labeling technique in the rat, *Brain Res.* **158:**189–196.

The Golgi Methods 9

O. EUGENE MILLHOUSE

I. INTRODUCTION

A full account of the history of the Golgi method has not been written. In his *Recollections of My Life,* Ramón y Cajal (1937) tersely summarized the discovery of this method when he wrote: "It was discovered by C. Golgi, the famous histologist of Pavia, through the favor of chance, the muse who inspires great discoveries." A bit more detail is given in a brief passage in Retzius' autobiographical reflections (Ramón-Moliner, 1970). According to Retzius, Golgi was investigating the pial covering of the brain and studied sections of brains that had been hardened in potassium dichromate and subsequently put in a solution of silver nitrate. Golgi's genius consisted in the fact that, at a time when knowledge about neuronal structure was very imperfect, he recognized nerve cells in his preparations.

Following his original publication in 1873, Golgi modified the dichromate fixative by adding osmic acid. Today this technique is referred to as the "rapid Golgi method." Golgi also formulated a way of impregnating nerve cells by fixing tissue in a solution of mercuric salts (1879). This method, as further developed by Cox (1891), is known as the "Golgi–Cox method."

Relying on the German edition of Golgi's works (Teuscher, 1894), Strong (1896) translated Golgi's explanation of his method into English.

O. EUGENE MILLHOUSE • Department of Neurology, College of Medicine, University of Utah, Salt Lake City, Utah 84132.

Apart from its obvious historical significance, this article deserves close reading because in it Golgi carefully considers the problems that still confront anyone who uses his method or one of its variations. A valuable compendium and evaluation of many of the variations of the Golgi method is provided by Ramón-Moliner (1957).

With the successful application of Golgi techniques, and particularly the rapid Golgi method, a very few neurons and glial cells fill with a dense precipitate of silver chromate.* The impregnated cells stand out against a relatively clear background, thus permitting direct visualization of cellular shapes and contours. Herein lies the uniqueness of the method: it demonstrates the shape and surface characteristics of an entire neuron, including its cell body and all of its processes. In short, Golgi methods are especially suited to answer questions about the shape and articulation of neurons, to study the local circuitry of an area, and to explain how afferents ramify and terminate in a region of the central nervous system.

The major drawback of Golgi methods is their unpredictability. In most preparations, only a few neurons in any region of the central nervous system (CNS) are impregnated, and it is not possible to predict which neurons will be filled with the dense precipitate. Furthermore, neurons in one area may be impregnated while those in an adjacent region are not. In view of this, it is necessary to study numerous Golgi preparations of an area to be certain that all or most of the structural patterns have been revealed. The point has been made that the limited number of impregnated cells is not the weakness of this method but is its very strength. If the majority of the elements in a region were engorged with the dense precipitate, it would be impossible to visualize individual cell patterns.

II. THE RAPID GOLGI METHOD

The rapid Golgi method has been used for nearly a century to study the architecture of the CNS. It could be argued that no other single technique has been as useful in understanding the geometry and organization of local neuronal connections. Moreover, as pointed out by Scheibel and Scheibel (1970, 1978), this potent technique can be most effective in interpreting data gathered by a variety of other techniques, including intracellular recording and electron microscopy.

Briefly, the method (Fig. 1) consists of fixing small pieces of CNS in a solution of potassium dichromate and osmic acid. After a period of time, the tissue is transferred to a weak solution of silver nitrate. Finally, the pieces

*Although the "Golgi-precipitate" in the rapid Golgi method has long been regarded to be silver chromate, this has only recently been demonstrated to be the case by Fregerslev *et al.* (1971) and Chan-Palay (1973) who examined the precipitate with X-ray diffraction techniques. It is also clear that the precipitate fills the interior of the cells with the exception of the nucleus and is not a dense coat applied to the external cellular surface.

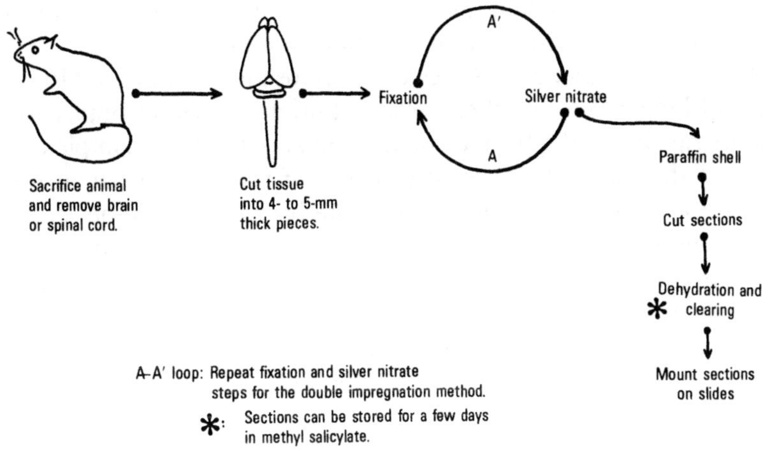

Figure 1. Outline of the proceaure for the rapid Golgi method.

are cut into sections that are dehydrated and put on glass slides. Each of these steps will be considered in detail.

The technique described in this chapter was published originally by Ramón y Cajal and DeCastro (1933) and is essentially the one used by Scheibel and Scheibel (1955) who applied it in their classical study of the inferior olivary nucleus.

A. Preparatory Steps

1. Selection of Tissue

It has been assumed that the brains and spinal cords of only young animals are suitable for the rapid Golgi method. This assumption, which is not entirely warranted, rests on three impressions. First, the impregnation of cells in mature brains does not appear to be as complete as that in immature brains. In particular, axons and axon collaterals are not seen as clearly or as fully as they are in young material. Second, fewer cells seem to be impregnated in older brains than in young brains. Finally, the background that occurs in all sections is darker in mature than in immature brains, and it obscures some of the fine detail.

One possible reason that so few axons are impregnated in the adult brain is that the myelin sheaths retard or prevent the Golgi reaction. Indeed, areas of neuropil in mature brains that are not crowded with myelinated axons—e.g., the cerebral and cerebellar cortices—can be exquisitely impregnated; yet, it is difficult to achieve impregnations of similar quality in the heavily myelinated reticular formation.

It is important to consider that if only immature brains (e.g., 1- to 2-

week-old mouse, rat, kitten) are used, significant structural features of a region may be overlooked because they have not developed.

As a general rule, 3- to 4-week-old brains are suitable for visualizing the courses of long axons and their collateral branches. Older brains are suitable for studying the configurations of the somata–dendritic complex. To begin a study of the salient features of a region of the CNS, it has proved useful to use 3- to 6-week-old rodent brains. The background in the sections is not dense enough to obscure the impregnated cells, and the majority of the neuronal and glial population has probably developed sufficiently so that none of the structural patterns is likely to be ignored. Also, the brains are small enough to permit sagittal and horizontal sections to be made through the entire brain; thus, rostrocaudally directed axons can be followed for considerable distances in single sections.

2. Dissection of Tissue

Young rats and mice can be decapitated, and sharp, pointed scissors used to reflect the skin and remove the calvarium. In the case of adult animals, rongeurs or wire cutters will be needed to open the skull. Once the skull cap has been removed, insert curved forceps between the brain and skull; use the forceps to gently pry (or scoop) the brain out of the cranial cavity. When the brain is lifted from the base of the skull, cut the optic and other cranial nerves with curved, sharp-pointed scissors. Place the brain on either a piece of glass or beeswax and, with a single-edged razor blade, cut it into 4- to 5-mm-thick pieces that are immediately immersed in the fixing solution (see Section II.B.1). Sacrifice one animal at a time to insure that as little time as possible intervenes between death and the beginning of fixation. With practice, this entire procedure will take no more than 2–3 min.

The fresh brain is typically cut in either the transverse, the horizontal, or the sagittal plane. An attempt should be made to cut the brain to take advantage of the course of the axons or dendrites that are to be studied. For example, if an axonal group courses between the dorsal and ventral surfaces at an oblique angle, its full course will not be appreciated in true transverse or sagittal sections. However, if the brain is cut at about the same oblique angle that the axons travel, their course can be traced more completely in a single section.

B. Fixation

1. Fixative Solution

This solution is made by dissolving 8 g potassium dichromate in 300 ml distilled water. Use a wide-mouth jar. The dichromate will dissolve quickly with a little agitation. One gram of osmium tetroxide (which is packaged in individual glass vials) is then added to the solution. *Osmium is toxic, and its fumes are hazardous and can damage the cornea. Skin contact with this chemical must be avoided. All handling of osmium should be done in a hood.* The glass vial that contains the 1 g osmium should be broken *inside* the jar that is used for making the fixative. Because osmium dissolves slowly, the fixative should be made the day before it is to be used, or the solution can be placed on a laboratory shaker for a few hours until the osmium crystals have dissolved.

The fixative can be kept for several days at room temperature or for several weeks if stored in a refrigerator. Although it was once thought that the solution had to be kept in a dark bottle, this is not necessary.

When the fixative is to be used, it should be at room temperature. Pour fixative into small, wide-mouth bottles that hold 100–125 ml; they need to be filled with only 90–100 ml of fixative. As a convenient rule of thumb, six pieces of adult mouse brain, 4–5 mm thick, can be fixed in one small bottle. Once the tissue is in fixative, the solution should be kept at a relatively constant temperature of 21–24°C.

This same fixing solution can be used to obtain Golgi impregnations in brains that have been perfused with 1% glutaraldehyde and 1% formaldehyde in 0.12 M phosphate buffer (see Appendix, Section VII.C).

2. Fixation Time

There is no fixed answer to the question of how long the tissue should remain in fixative. At least three factors have to be considered: the age of the animal, the area of CNS to be studied, and the neuropil component to be emphasized (e.g., axonal patterns *vs.* dendritic trees).

a. The Age of the Animal. Older brains are usually left in fixative for 2–8 days. Tissue of animals up to 3 weeks old is generally fixed 10–12 days. The longer the tissue is left in the fixative, the more brittle it will get.

b. The Area of CNS to be Studied. The fixing times indicated above are suitable for impregnating the cerebral cortex and hypothalamic neuropil, but the cerebellum probably will not be well-impregnated. Upon undertaking a study of the particular region of the CNS, the best way to proceed is initially to immerse several pieces of brain in fixative and to remove them one piece at a time at 24-hr intervals. Put each piece of tissue in silver nitrate (see Section II.C) for 18–20 hr. By comparing such a series of Golgi impregnations, it will be possible to arrive at the overall optimal fixation time.

c. The Neuropil Component to be Emphasized. The longer the material is kept in fixative, the "harder" it gets. "Hard" blocks produce impregnations of axons, axon collaterals, and terminals, but fewer examples of somata and dendrites. The latter usually occur in greater abundance in tissue that is fixed for shorter time periods. For example, exquisite impregnations of only a few axons and their collaterals can be obtained in brains fixed for several weeks. In these preparations there probably will not be any example of cell bodies and dendrites, and the background in the section will be lighter. Unfortunately, it is difficult to process the very hard blocks because they tend to break.

C. Silver Impregnation

The silver solution is made by dissolving 6 g silver nitrate in 800 ml distilled water. The silver nitrate crystals, as well as the solution, have to be handled carefully or skin and clothes will be covered with black spots by contact with the silver nitrate.

Small jars similar to the ones used in fixation can be used for this step. Use the same tissue-to-solution ratio as in the fixing step. Although it has been recommended that the tissue be blotted before being put in the silver solution, this can cause mechanical disruption of the tissue. Instead, carefully lower the tissue into the silver solution. This immediately produces a cloudy precipitate red–brown in color. Gently swirl the tissue around in the cloudy solution and then pour off the silver nitrate. Add fresh solution and keep the tissue in it for 18–20 hr at room temperature. The longer the blocks are kept in the silver solution, the more fragile they become. Longer impregnation time also produces a darker background and an increased amount of extraneous precipitate in the sections.

As a matter of course, the silver nitrate solution and the tissue once it is in it are stored in the dark. It is not altogether clear that darkness is essential for the impregnation reaction to proceed, but the question has not been critically considered.

D. Sectioning the Tissue

The tissue is cut into sections that are 100–200 μm thick. Thick sections are needed in order to follow the processes of an impregnated neuron without interruption. However, the thicker the sections are, the less light will be transmitted. Consequently, thick sections are darker, and there is less contrast between the impregnated cells and the background. On the other hand, the thinner the sections are, the more difficult it will be to visualize the geometry of a neuron and to appreciate its structural relationships unless the cell is reconstructed from several serial sections. Yet, if the primary purpose is to photograph segments of a dendrite for the purpose of studying spine detail, thinner sections (50–90 μm) are required.

1. The "Paraffin Shell" Technique

A variety of methods are available for sectioning the brain tissue; however, the "paraffin shell" technique is the quickest, easiest, and most inexpensive method (Figs. 2–7). In this technique, the brain is supported by a shell of paraffin while it is being sectioned.*

a. Equipment. Take a 2- to 3-cm cube of solid paraffin and melt one of its surfaces with a hot spatula. Attach it to a wooden chuck (or fiber block or cork block) that fits in the specimen holder of a sliding microtome. Several of these paraffin blocks should be made in advance and kept on hand for use.

The sections are cut using an "a-profile" knife (Fig. 8). On crosssection, most microtome knives have a triangular shape; in the case of an "a-profile" knife, one of the sides of the triangle is slightly concave. The knife is put into the microtome so that the concave side faces up. The concavity helps to retain both the cutting solution and the cut section on the knife.

To make the paraffin shell, have available a pair of stainless steel blunt forceps (about 12 mm long), a dissecting needle, a stainless steel teaspoon, and a small Stender dish (frequently listed as "low Stender dish" in supply catalogues) filled with warm paraffin. The paraffin should be the low-melting-point variety so that it will not harden quickly when taken out of the paraffin oven. The paraffin is kept hot by heating the teaspoon in a flame and then putting it in the paraffin dish.

b. The Shell. Pour the tissue and silver solution into a petri dish. Carefully pick up the tissue, one piece at a time, with the forceps and put it on a paper towel. Capillary action will draw off the excess silver solution. Put the tissue in the warm paraffin. Bubbles will form around the tissue, but they will dissolve. When moving the tissue into and out of the paraffin, the tips of the forceps should be warm or the paraffin will harden around them, and they will be difficult to manipulate. Several pieces of tissue can be placed in the paraffin dish and left there while making the individual shells.

Heat the forceps' tips in a flame and use them to make a well in one of the paraffin blocks. The well serves as a base for the tissue; therefore, it should be shaped so that one of the surfaces of the tissue rests securely in it. While the paraffin at the base of the well is still melted, orient the tissue in it. Once the tissue is in place, spoon warm paraffin over it. Let it cool slightly, then add more warm paraffin. In this way, the shell is built up around the brain block. With a little practice, it will be possible to determine how thin the shell can be. The shell should not be so thick that it entirely obliterates the contours of the block. If air bubbles occur in the shell, break them with a heated tip of a dissecting needle. When the paraffin shell has

*An alternative method for dehydrating and embedding the tissue blocks in low viscosity nitro cellulose is described in Section VII.B.

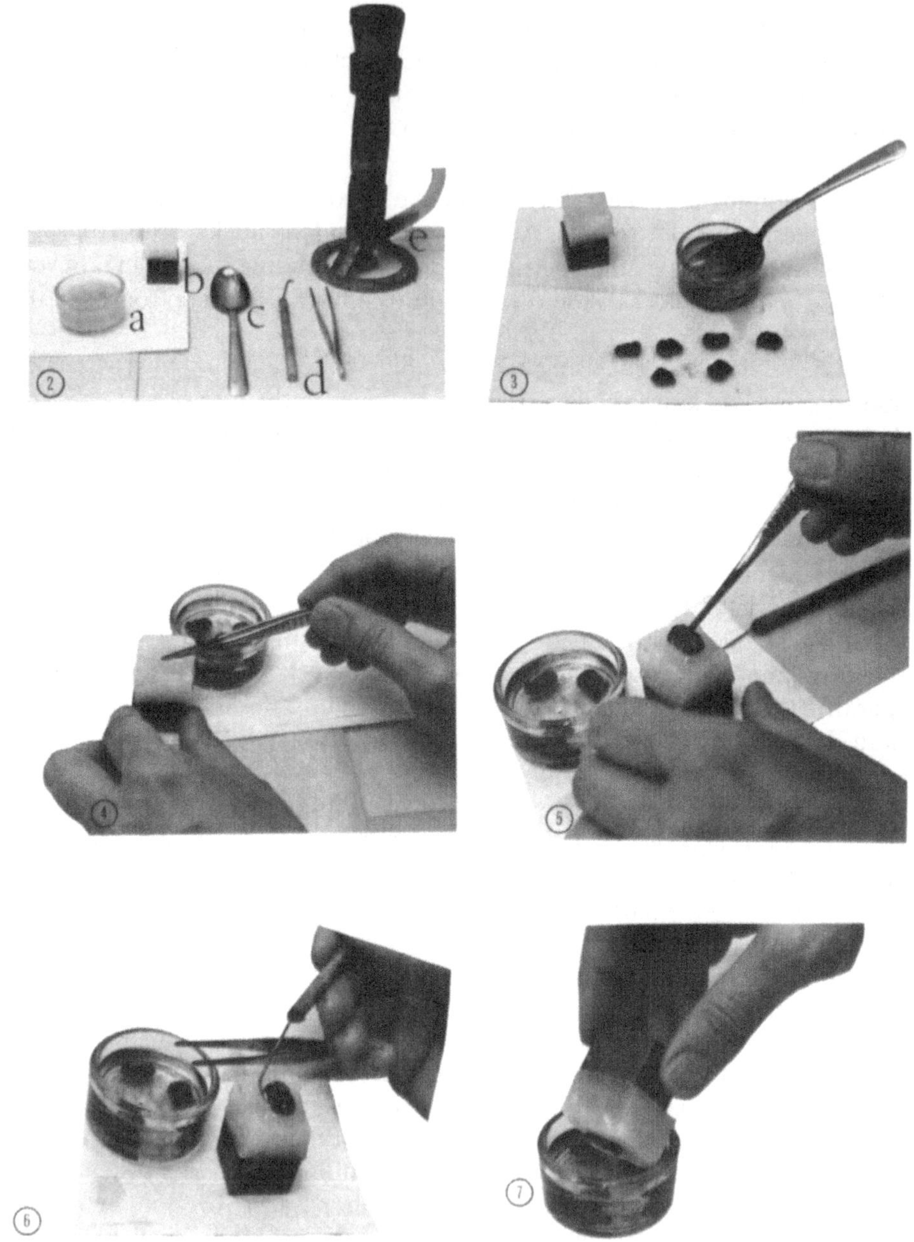

Figures 2–7. Steps in making a paraffin shell.

Figure 2. Equipment needed for making a paraffin shell includes (a) Stender dish filled with warm paraffin, (b) paraffin block attached to a wooden chuck, (c) spoon, (d) dissecting needle whose tip is bent or curved, forceps, and (e) bunsen burner. The spoon is heated in the Bunsen burner flame and put in the Stender dish to keep the paraffin melted.

Figure 3. Placing the tissue in paraffin. Several pieces of tissue can be kept in the paraffin dish while paraffin shells are being made. Before putting the tissue into the paraffin, it should be put on a paper towel so that excess fluid can be absorbed.

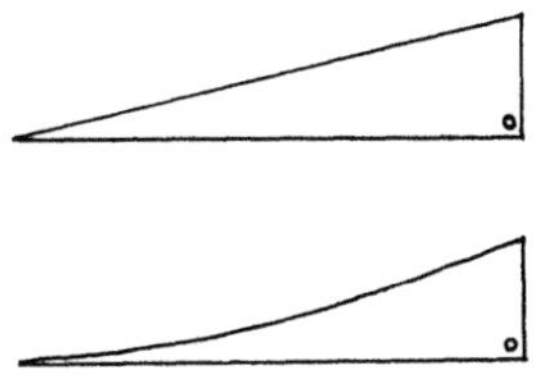

Figure 8. Cross-sections of a typical (top) and an "a-profile" (bottom) microtome knife.

cooled (within 10 min), the block can be sectioned. Several blocks can be prepared at the same time. It is not necessary to prepare the shell immediately before cutting the tissue; however, once prepared, the blocks should be cut within a couple of hours or the tissue will get dry and crumble when it is sectioned.

c. Sectioning. The block is fastened securely in the specimen holder of a sliding microtome, and the knife is aligned parallel to the long axis of the tissue (Fig. 9). The cutting angle of the knife, as indicated on the scale of the microtome knife holder, should read "zero." Keep the knife and the tissue block wet with absolute alcohol. The part of the knife edge that is not used to cut the tissue poses a hazard. Pieces of cork or rubber tubing slit along its length can be used to cover the exposed cutting edge of the knife. This will reduce the chance of cutting the hand or fingers when sectioning the tissue.

As the knife is drawn through the tissue, the section will slide onto it. The section can easily be removed with a flat brush (12–15 mm for rodent brains) that is soaked in absolute alcohol. The same brush is used to put alcohol on the knife. Beginning at the top of the knife, i.e., the "knife back," gently stroke the blade to bring the brush down over the section. Do not press very hard. The even, smooth stroke of the brush will carry the section off the knife onto the brush (Fig. 10).

If the sections curl or break during sectioning, the rim of paraffin around the tissue is too thick. Very carefully trim away a bit of paraffin from around the entire block with a single-edge razor blade. However, enough of the shell must remain to keep the tissue firmly in place. Extreme caution has to be taken while trimming the block because the fingers will be working very close to the edge of the knife.

Figure 4. The tips of the forceps are warmed in the flame and then used to make a "well" or depression in the paraffin block that will hold the tissue.

Figure 5. The tissue is oriented in the well so that it is snug and lies flat.

Figure 6. The heated tip of the dissecting needle melting the edge of paraffin around the tissue. Make sure that there are no air bubbles between the tissue and paraffin. If they are present, break them with the heated needle.

Figure 7. Spoon warm paraffin over the tissue to make a shell, allow it to cool, and spoon more paraffin over it. Then dip the block into the paraffin dish. Do this three or four times over a 5-min period.

(9)

(10)

Figure 9. A 180-μm-thick section cut with an "a-profile" knife. If the paraffin shell is not too thick, the section will remain intact while being cut; otherwise, it will break into several pieces. Also, if the rim of paraffin is very thick, the section will curl and, consequently, be hard to mount on a slide.

Figure 10. A section is removed from the knife with a brush that is wet with 100% ethanol. As the brush moves down over the section, the section will adhere to it.

Figure 11. Sections can be stored and processed in Stender dishes that have slits drilled along their sides and bottoms. The slits allow fluids to enter and leave the dish without causing excessive agitation of the fragile sections.

Figure 12. Stender dishes are placed in a pan that can be conveniently filled and drained with dehydrating and clearing agents. By keeping the dishes in order, the sections can be kept in serial sequence.

E. Dehydration and Clearing

As each section is cut, it is put into absolute ethyl alcohol. Dip the brush that holds the section into the alcohol bath; the section will float away from it. Stender dishes that have slits drilled in the rim along their bottom are arranged in a pan that contains the alcohol (Fig. 11).* The sections are put in the Stender dishes in serial order (Fig. 12). To change solutions, drain the pan, keep the dishes in order, and pour on the next solution. Do not pour the solution directly into the Stender dishes but into the pan, thus filling the dishes from the bottom. This will minimize the risk of damaging the fragile sections while changing the solutions.

The sections are kept in three changes of absolute ethyl alcohol for 10–15 min each, then bathed in three changes of methyl salicylate (oil of wintergreen) for 15 min each. The first oil bath is a 1:1 mixture of absolute alcohol and oil. The remaining two oil baths are full-strength methyl salicylate. The sections can be left in an oil bath, preferably the last one, for a few hours provided the pan is covered so that the oil does not evaporate. It is important that the sections be completely dehydrated†; otherwise, excessive background precipitate develops, and the preparations will be useless in a few weeks' time.

*This method of processing individual Golgi sections is adapted from a method devised (but not published) by Dr. A. Globus.

†The last change of alcohol should be fresh. The first change can be used as cutting alcohol for the next day; thus alcohol II will be used as alcohol I, and III will become II. The oil solutions can be used every day for at least a month provided that they are filtered after each use.

The sections are cleared in either xylene or toluene. Because the sections get brittle if they are left very long in either xylene or toluene, put only the sections of a couple of blocks in xylene at one time and immediately begin to mount them on slides.

F. Mounting

The sections are removed (one at a time) from the xylene solution by being brushed onto a slightly curved spatula from which they are then brushed onto a dry microscope slide. With gentle action of the brush, the sections can be oriented as desired. Any type of coloring brush is adequate. It is not necessary to pretreat the microscope slides with albumin; they can be used directly from the box.

Once the slide is filled with sections, blot the sections with several (2–3) layers of paper towel that have been trimmed to fit the slide. Lay the paper on top of the sections and rub a thumb along the length of the slide. Lift the paper away, and the sections will adhere to the slide unless the rubbing action has been too swift. Immediately cover the blotted sections with Permount or a similar type of synthetic resin. There usually is no need to dilute the resin if it is used directly from its original container. Occasionally the sections will drift on the slide after the resin is applied. If this happens, wet the tip of the brush with xylene and carefully press the sections in place. Use just enough resin to cover the sections. As the resin dries, some of the sections or small parts of them may be left exposed, in which case one can add a little more resin to these areas. An excessive amount of resin will interfere with the use of the high-power microscope objectives.

The resin will dry rather evenly in a matter of 4–5 days. If ripples develop on the surface, brush them with xylene until the surface is again smooth. Over a period of months, the resin may become opaque. The opacity can be dissolved by immersing the slide in xylene until the resin is clear. If the sections come off the slide during the process, remount them and add fresh resin.

Golgi sections prepared according to this method can be used for some time. Although some of the sections may gradually fade, others remain good for a number of years. If the sections are coverslipped, a precipitate may form and the background become more dense. Consequently, the slides are not routinely coverslipped unless they are to be studied with an oil-immersion objective. In this case, soften the resin (if it has dried) with xylene and add the coverslip.

G. A Note on Perfusion Fixation

Immersion fixation has been emphasized in this chapter because a large number of brains can be fixed quickly and conveniently with no special equipment needed. The alternative method is to perfuse the CNS with rapid Golgi fixative. However, the amount of time required to do a single perfusion limits the number of preparations that can be made. Also, perfusions are not always successful because the fixative may fail to reach some areas of the CNS. The practical details regarding the perfusion technique and the solutions for rapid Golgi perfusion are described in the Appendix (Section VII.A).

III. ANALYSIS OF THE DATA

A critical interpretation of rapid Golgi material requires more than merely looking at a few sections in a microscope and recording the shapes of cells. It is essential to examine numerous sections that have been cut in several planes and to study sections that have been prepared by other methods, such as the Nissl and myelin stains, before any careful interpretation can be made. The following discussion of the identification and location of neurons and their processes will emphasize these two points.

A. Cell Location

A prepared mind is a prerequisite for interpreting rapid Golgi sections. Nissl- and myelin-stained sections that have been cut in the transverse, sagittal, and horizontal planes should be studied intensively before a Golgi study of a particular region is undertaken. The reason for this is obvious. The nuclei and tracts of the CNS, which are clearly labeled in an atlas, are easily identified in Nissl- and myelin-stained sections but will not be as apparent in Golgi sections in which only a few cells are impregnated. Therefore, the landmarks available for identifying a nucleus and its boundaries in Golgi sections will be few.

The following questions might serve as a guide while studying Nissl- and myelin-stained sections. Can two or more neuronal populations be distinguished on the basis of their cytological features and soma size? If so, how are the cells distributed in the nucleus? What proportion of the neuronal population does each type represent? Is there a remarkable or constant pattern of blood vessels that can serve as landmarks? What major fiber groups occur in the area? The answers to these and similar questions will be of immense help both in studying the Golgi sections and in organizing the data collected from the rapid Golgi material.

Golgi–Cox preparations are also invaluable for the purpose of orientation. Golgi–Cox impregnations do not usually furnish the exquisite dendritic and axonal detail that are characteristic of the rapid Golgi technique, but they are adequate for indicating the neuronal patterns that occur in an area. Furthermore, Golgi–Cox sections can easily be counterstained with thionine or cresyl violet so that the nuclear groups and cytoarchitectonic features can be recognized. Recently, Geisert and Updyke (1977) have developed a method for counterstaining rapid Golgi sections (See Appendix, Section VII.D).

If a nucleus is identified in a Golgi section, and impregnated neurons seem to be present in it, it is important to determine whether the cells or other impregnated elements, e.g., afferents, are actually within the nucleus or are superimposed on it. A nucleus (or fiber tract) may be grazed in a thick (150–200 μm) section, and instead of looking at its contents, one may be peering into it through an overlying part of CNS. The study of serial sections and of sections cut in different planes can circumvent this possible source of error.

B. Cell Processes

It is not ordinarily difficult to distinguish between dendrites and axons (Figs. 13–17). The dendrites, of which there are usually several per neuron,

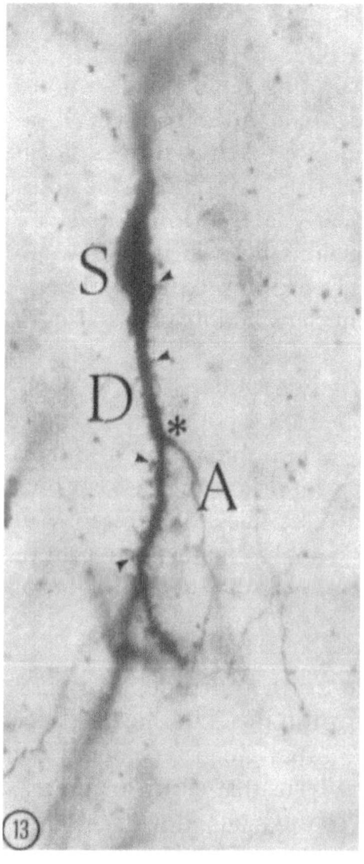

Figure 13. The three regions of a neuron as seen in a rapid Golgi preparation. S, soma; D, dendrite; A, axon. The soma and dendrites may have small projections (arrowheads) that are called spines. Compare the diameter of the dendrites and axon. At the point where the axon originates from a dendrite or soma, there is usually a small protuberance or elevation of the cell's surface (*). Three other axons can be seen in the field, but their origins and terminations cannot be specified. The small black dots that are visible throughout the picture are ubiquitous in Golgi sections; some of them may be extraneous precipitate or cross-sections of axons and dendrites (750 ×, reproduced at 65%).

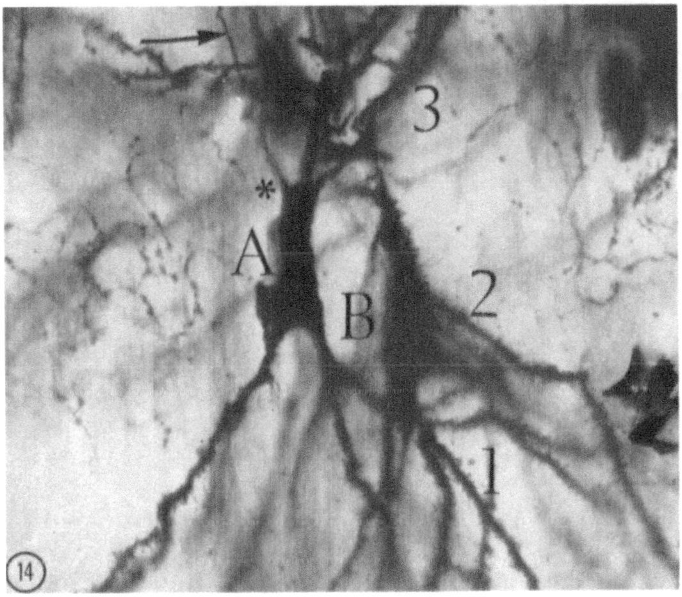

Figure 14. Two pyramidal neurons, A and B, in the pyriform cortex. The axon of neuron A originates (*) from the apical part of the soma. The axon appears out of focus at the arrow. Although dendrites 1, 2, and 3 seem to belong to cell B, 1 is really a process of A, and 2 and 3 come from other neurons. Unless the section is studied by continuously focusing up and down with the microscope, the dendritic pattern of cell B could be misrepresented. For this reason, it is imperative that sections that are to be photographed must first carefully be studied with the microscope; otherwise, the images may be misinterpreted (750×, reproduced at 65%).

taper perceptibly as they extend away from the cell body. This feature, in addition to their thickness and great number, makes them easily recognizable. Axons, of which there are only one per neuron, have to be sought. In comparison with dendrites, axons have a smaller diameter and a more even caliber along their length. Furthermore, the axonal surface is more regular in spite of varicosities and short side branches and collaterals.

The origin of an axon from either the cell body or proximal portion of a dendrite is marked by a conical or differently shaped protuberance (Figs. 13 and 14). If an axon is thought to originate from a specific neuron, but this specialization is not evident, there may be a case of mistaken identity. It is possible that an axon close to a neuron has been cut so that it appears to arise from a cell body that it only passes by or crosses over. This is the problem of superimposition, a source of interpretative error that should always be considered in studying Golgi preparations.

Dendrites and axons, including collaterals and terminals, commonly appear in Golgi sections as discrete and separate entities (Fig. 16). The pre- and postsynaptic configurations are visualized as though they are isolated from one another. The challenge to the Golgi anatomist is to understand how the two structural patterns articulate. Each Golgi section usually sup-

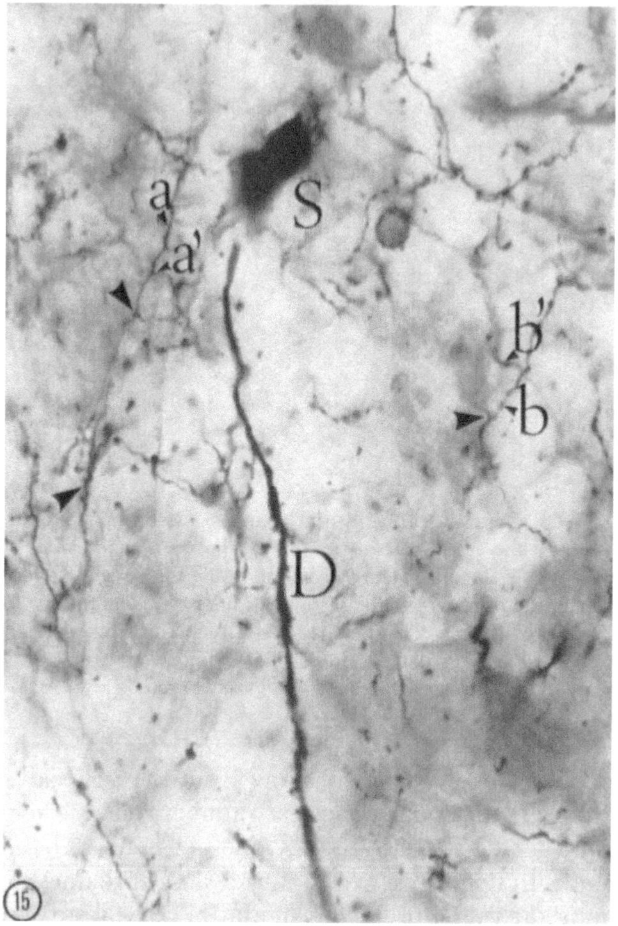

Figure 15. The short axon segments a′ and b′ appear to be collaterals of a and b; but, if the section is carefully examined by focusing the microscope up and down, it will be evident that a′ and b′ merely cross a and b and are entirely separate axon systems. The origin of a collateral is usually marked by a swelling or slight increase in the diameter of the axon (arrowhead). The dendrite, D, is close to, but clearly detached from, the neuronal soma, S. Proximity is not sufficient evidence that these two are parts of the same cell. In this case, it was concluded after examining adjacent sections that the dendrite belongs to a neuron in the preceding section, and, furthermore, some of the dendrites of S are in the following section. Notice that there is not a uniform distribution of spines along the length of D; its proximal portion is devoid of these processes (750 ×, reproduced at 65%).

plies a clue that has to be confirmed in other sections that will, it is hoped, provide additional data. It is only through the close scrutiny of numerous preparations that the pattern of organization and the structural relationships between the pre- and postsynaptic elements begin to be apparent.

Axons that curl around or cross over cell bodies and dendrites do not necessarily establish synaptic contact with these structures. The superimposition of neuronal processes in a thick Golgi section can be mistaken for a functional contact. However, if axons consistently demonstrate structural specializations such as swellings, small protrusions, or mossylike endings at the point where they contact a postsynaptic surface, it is likely that they represent synaptic regions. The parallel fibers in the cerebellar cortex, for instance, have small nodes that have been demonstrated to be synaptic terminals. This was established by examining Golgi preparations in the electron microscope. However, it is not reasonable to assume that all beads or swellings that occur along axons are axon terminals. In this context, it is well to remember that final proof of synaptic contact can only be obtained by the aid of electron microscopic technique.

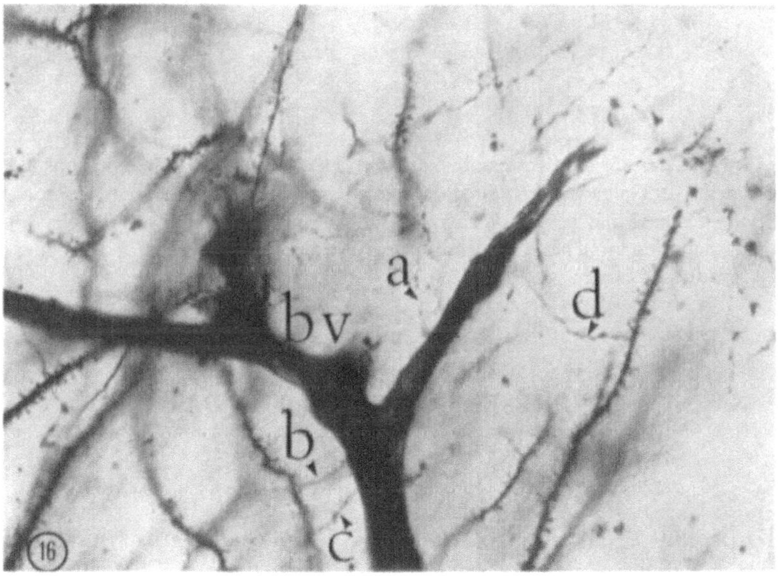

Figure 16. In addition to neurons and glia, blood vessels (bv) are often impregnated in Golgi sections. They pose a handicap because they obscure so much detail. If an axon under study disappears behind a thick vessel, it is nearly impossible to follow it further because it could be confused with other axons. For example, is axon a continuous with b, c, or d? Axons seldom assume a straight course. Also, fine capillaries can be mistaken for thick dendrites. Some regions of the CNS, including the hypothalamic paraventricular nucleus, have unique blood vessel patterns that can assist in identifying a nucleus or area, especially when one is studying sections cut in different or oblique planes (750×, reproduced at 65%).

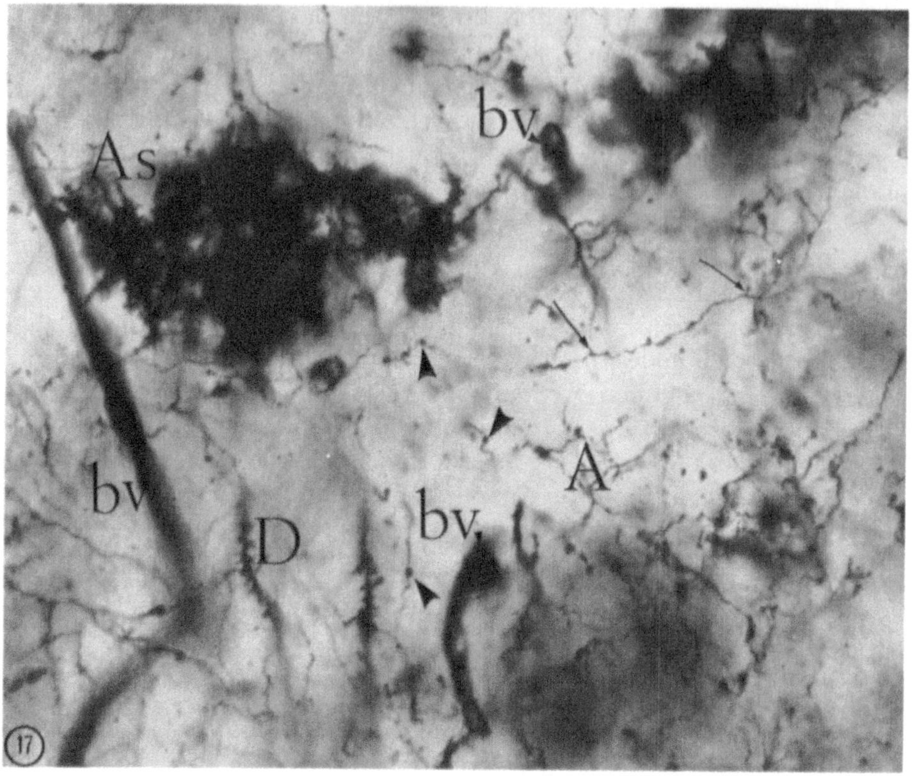

Figure 17. Examples of blood vessels (bv), astrocytes (As), dendrites (D), and axons (A) as they typically appear in Golgi sections. Small swellings (arrowheads) often occur along axons. Although some of these may be axon terminals, this can only be verified with the electron microscope. The small side branches or short arms (arrows) are generally regarded as being boutons (750 ×, reproduced at 65%).

IV. PRESENTATION OF THE DATA

A. Golgi Drawings

To present their data, the majority of Golgi anatomists have relied on pen-and-ink drawings made with a camera lucida, i.e., a drawing tube that is attached to the microscope. The process is laborious and tedious, and it may initially be difficult to adapt to the camera lucida. However, after some practice with adjusting the light, learning how to follow a pencil point around a neuron, and coming to trust the fingers and hand to move independently of direct visual guidance, the camera lucida will almost become an extension of one's self. The original drawings can be made with a hard lead pencil on a sheet (or several attached sheets) of bond paper. The pencil should have a sharp point, but the point and the line it makes have to be seen in the camera lucida while tracing the cells. Certain mechanical pencils have such fine points and make so fine a line that their image is not projected by the camera lucida. This makes it difficult to visualize how the tracing is progressing and whether or not it faithfully reflects the impregnated cells. The illustrations intended for publication can be made by laying a sheet of good quality tracing paper, e.g., linen tracing cloth, over the original drawing and tracing it in ink.

To allow the drawing to continue without the paper being stopped by the side of the microscope, suspend the microscope above the table top. Dr. J. B. Angevine has devised a simple way of accomplishing this (Fig. 18). A board about 1 m long and 14–15 cm wide is firmly attached at right angles to the wall next to the back of the drawing table. The microscope sits on the board, which is 0.5 cm above the table surface; thus, the drawing paper is free to slide under the microscope.

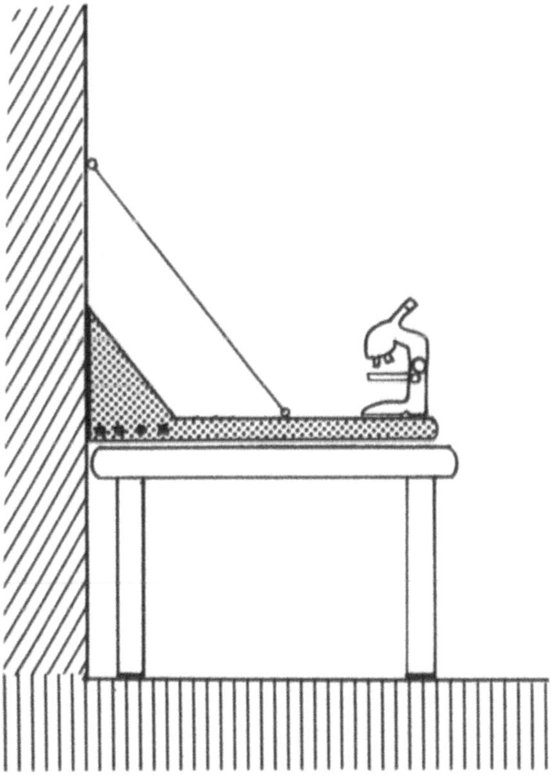

Figure 18. Diagram of the apparatus used to suspend a microscope above a table top.

B. Photography

If it requires considerable time and patience to do a Golgi drawing, why not photograph the section or field of interest and forego all of the effort? It is difficult to photograph thick Golgi sections. First, there are multiple focal planes in a single section. Second, the neurons and glia are not arranged so that they occur entirely within a single plane: they are three-dimensional cells that have extensive processes. Thus, to photograph only a single neuron requires making several photographic exposures at different focal levels. After the film is developed and printed, a number of photographic prints have to be pasted together to reconstruct the cell. In addition, published photographs are not easy for a reader to interpret because blood vessels, background precipitate, and pieces of cells obscure the essential features of the picture. This sort of extraneous material should be deleted from a drawing. Ramón y Cajal (1960) summarized this when he wrote, "The best histological photographs offer only a pale reflection of the preparations. Save in the case of sections that present only a single focal plane . . . they have value only for those who have made them and are understood only by those who have studied the original preparations." He argued, further, that the "scientific artist" should omit useless details: "And we must declare that if our drawings at times omit details, in compensation, all details that they show have been copied with the utmost fidelity. . . . We prefer to sin by omission rather than by excess. . . ."

Photography, however, is important for illustrating specific points such as the shapes and varieties of dendritic spines and axon terminals. Lower-magnification photographs can also be effective in conveying an impression of how the dendrites are arrayed in a particular region of the CNS.

V. VARIATIONS OF THE GOLGI METHOD

There are a number of variations of the Golgi method, and once the fundamental steps described in this chapter have been learned, the modifications can easily be adopted.

A. Double and Triple Impregnations

Although cell bodies and dendrites appear to be more abundant in single- than in double- or triple-impregnated sections, a relatively greater number of axons and their collaterals are visible following double and triple impregnations. Therefore, repeated impregnations have been used to bring out or emphasize axonal details, particularly in the brain stem (Ramón y Cajal, 1960; Valverde, 1965).

Double or triple impregnations are achieved by repeating the fixation and silver impregnation steps a second or third time before the blocks are sectioned (Fig. 1). However, the blocks become increasingly brittle in the fixative and silver nitrate solutions, and they can be difficult to section. The sections often crack or break into several pieces. They may also be quite dark and hard to study, especially if subjected to a triple impregnation.

There are no well-defined rules about how long the blocks should be left in the fixing and silver solutions a second or third time. On a first attempt, leave the tissue in the initial fixing solution for 1 week and in silver nitrate 48 hr. Return the tissue to fresh fixative for an additional week and keep it in the second silver bath for 24 hr. For a triple impregnation, repeat the above steps one more time.

Rapid Golgi blocks that become hard and brittle because of, for example, repeated fixation and silvering steps and are difficult to section with the paraffin shell technique can be dehydrated and embedded in low-viscosity nitrocellulose. Other celloidin-embedding media such as Parlodion® are not entirely effective because they do not completely infiltrate the osmium-treated Golgi blocks and they are not dense enough to support the Golgi blocks during the sectioning process. A method for embedding rapid Golgi blocks in nitrocellulose is described in the Appendix, Section VII.B. This method has been used for several years by Dr. Morest and has been published by Palay and Chan-Palay (1974).

B. Golgi–Kopsch Method

There are a dozen recommended ways for using the Golgi method on formalin-fixed tissue (for a nearly complete list, see Lee, 1950). The methods vary only in the relative percentage and strength of potassium dichromate used and in the length of time the tissue is kept in the fixative. These differences aside, the method consists of putting formalin-fixed material into a potassium dichromate solution without the addition of any osmic acid and then into silver nitrate. The name most often associated with this modification is the Golgi–Kopsch method. Kopsch (1896) originally suggested fixing fresh tissue in one part formalin to four parts of 3.5% potassium dichromate for 24 hr and continuing fixation for 2–6 days in only 3.5% potassium dichromate.

One advantage of the Kopsch variation is that formalin-fixed material obtained from other experiments or processed according to other histological procedures can be used. Second, because osmic acid is not required, this method is considerably less expensive than the rapid Golgi technique. Also the background is much clearer than it is in osmium-treated tissues. Finally, the method works equally well for very young or very old brains.

There are limitations to the Golgi–Kopsch method. Fewer cells are impregnated than in rapid Golgi sections. The capillary bed is usually more completely impregnated in these than in rapid Golgi preparations. Finally, as Braitenberg *et al.* (1967) point out, "The constancy of the results, in our experience, is . . . slightly inferior to that of the rapid Golgi method."

Two ways of proceeding with this variation of the Golgi method are described in the Appendix, Section VII.E.

C. Golgi–Cox Method

A well-known and frequently used modification of the Golgi technique is the Golgi–Cox method. Whereas a larger percentage of neurons is impregnated in Golgi–Cox than in rapid Golgi preparations, the Golgi–Cox impregnations are not as complete. The fine dendritic branches, spines, axon collaterals, and terminals are not as fully revealed in the Golgi–Cox material. The utility and popularity of the Golgi–Cox method stem from the fact that it provides a panorama of the neuronal field; the topographic details are filled in with observations made on rapid Golgi material. Furthermore, because the background is very clear, it is easier to photograph Golgi–Cox sections.

The Golgi–Cox method and suggestions for processing the material are described in the Appendix, Section VII.F.

D. Golgi Method for Embryonic Tissue

Golgi methods that employ osmic acid in the fixative are not entirely satisfactory for impregnating the CNS of embryos, particularly of mammals (Stensaas, 1967). To study developing neurons, and especially axonal patterns, Stensaas (1967) recommended a modified Golgi fixative that was described by Klatzo (1952) and was, in turn, derived from a recipe originally published by Del Rio-Hortega.

The details concerning this method are given in the Appendix, Section VII.G.

VI. SUMMARY OF ADVANTAGES AND LIMITATIONS

A. Advantages

1. The rapid Golgi method (or one of its several modifications) permits the direct visualization of neuronal and glial cells, including their numerous processes. Therefore, the shapes of the cells and their surface characteristics can be studied and drawn or photographed. This means that Golgi methods are invaluable in studying the somatic and dendritic architecture of the CNS.
2. The Golgi methods can demonstrate the origins and courses of axons and their collaterals. In addition, these methods illustrate how axons terminate. Most tract-tracing methods indicate the relative density

and distribution of axon terminals in a particular area but do not usually reveal the pattern of termination of individual axons.

3. It follows from (1) and (2) that Golgi methods are particularly important in studying local circuit patterns in the CNS.

B. Limitations

1. Because it cannot be predicted which cells (or how many) will be impregnated, a large number of Golgi preparations have to be made and studied in order to understand the structural features of an area of CNS. Making the preparations is both time-consuming and expensive.
2. Not all of the axons of the impregnated neurons can be followed to their sites of termination. Nor can all of the axons that are impregnated in a preparation be followed directly to their origin. Therefore, a considerable amount of specific detail remains locked up, as it were, in Golgi sections because of the inability to identify precisely the origin and termination of most of the impregnated axons. That is to say, Golgi methods are particularly advantageous for tracing the short axonal projections but not usually effective for following the long projections of a specific neuron. Although long axon paths, such as the fornix, medial lemniscus, and corticospinal tracts are revealed, their individual axons can very rarely be traced directly to the cell of origin.
3. Two separate and unrelated cells or processes of cells may be superimposed on one another in the thick Golgi sections and may mistakenly be taken to be structurally related.

VII. APPENDIX

A. Recipe for Perfusion Technique

A method for perfusing the brain with rapid Golgi fixative has been published by Morest and Morest (1966), Morest (1968), and Palay and Chan-Palay (1974).

The animal is anesthetized and artificially respirated. By means of transcardial perfusion, the vascular system is flushed out with a 0.1 M phosphate buffer wash that contains 0.5% polyvinylpyrrolidone (PVP-40). The latter is employed to prevent excessive swelling of the brain. The washout solution is made in two steps as follows. First dissolve 2.65 g of sodium phosphate (monobasic) and 14.00 g of potassium phosphate (dibasic) in enough distilled water to make a total volume of 1000 ml. Then dissolve 1 g PVP-40 in 200 ml of the first solution. Filter using Whatman paper #3.

Immediately following the washout, perfuse the rapid Golgi fixative. In this method, the fixative is made by dissolving 1 g PVP-40 in 150 ml 4% potassium dichromate. The dichromate solution should have a pH of 6.8 (the pH can be adjusted with 40% sodium hydroxide). The PVP-40 should be completely dissolved prior to adding 50 ml 1% osmium tetroxide. Filter the fixative using Whatman paper #3.

Once the brain has been perfused, remove it from the skull and slice it into 4 to 5-mm-thick pieces. These pieces are immersed in rapid Golgi fixative and processed according to the rapid Golgi technique. The blocks can be sectioned either by the "paraffin shell technique" or by embedding them in low-viscosity nitrocellulose.

B. Embedding Rapid Golgi Blocks in Nitrocellulose

The following procedure is based on that used by Dr. Morest and published by Palay and Chan-Palay (1974).

1. Dehydration of the Tissue

Transfer the tissue from the silver solution into a silver–alcohol solution that is made by dissolving 5 g silver nitrate in 50 ml distilled H_2O. Use a 1000 ml graduated cylinder (preferably one that is fitted with a glass stopper). Add distilled water to make 450 ml. Next add 500 ml 95% ethyl alcohol. Keep the tissue in this solution for about 1 hr; then proceed with the dehydration according to the following schedule:

70% Ethyl alcohol	2 hr
90% Ethyl alcohol	1 hr
Absolute alcohol	two changes, 1 hr each

Before infiltrating with nitrocellulose, expose the tissue to ether by immersion in an alcohol–ether solution that is made by adding one part absolute alcohol to two parts diethyl ether. Initially keep the tissue in this solution for 3–4 hr, then put it in a fresh change of solution and keep it there overnight.

2. Infiltration

Infiltrate the tissue with nitrocellulose by placing it in 5% and 10% nitrocellulose for 24 hr each and in 20% and 30% nitrocellulose for at least 48 hr each. The wide-mouth jars that contain the tissue and nitrocellulose solution have to be made air-tight so that the solvent for the nitrocellulose does not evaporate.

The nitrocellulose infiltration and embedding media are made by dissolving various amounts of nitrocellulose in alcohol–ether. *Dry nitrocellulose (gun cotton) is explosive and must be handled with extreme caution. When weighing it out and putting it in alcohol–ether, work in a static-free area. Electic sparks and flames must be avoided. Work in a chemical hood.*

To make 5% nitrocellulose, add 254 ml absolute ethyl alcohol to 254 ml diethyl ether. To this, add 40 g nitrocellulose, not all at once but in small amounts. After adding a small amount of nitrocellulose, stir it in the alcohol–ether with a glass rod in order to thoroughly wet it. Once all 40 g have been added, pour in an additional 254 ml diethyl ether. Use large, wide-mouth jars that can be tightly sealed because it is important that the alcohol–ether not evaporate. Adhesive or bandage tape can be used to seal the jars.

The same general procedure is followed in making 10%, 20%, and 30% nitrocellulose.

For 10% nitrocellulose, combine 240 ml absolute alcohol with 240 ml diethyl ether and gradually add 80 g nitrocellulose. When this is dissolved, add an additional 240 ml diethyl ether.

For 20% nitrocellulose, the alcohol and ether volumes are 214 ml, and 160 g nitrocellulose is used. For 30% nitrocellulose, the fluid volumes used are each 186 ml, and 240 g of nitrocellulose is dissolved. Allow 2 weeks or more for the nitrocellulose, especially the 30% solution, to dissolve completely. If a laboratory shaker is available keep the jars on it; otherwise, lay the jars on their side and rotate them each day.

3. Embedding the Blocks

Fold thick white paper or index cards to make a paper boat. The tissue should sit in the boat without touching its sides. Pour enough 30% nitrocellulose into the boat to form a thick layer. Orient the tissue on the layer of nitrocellulose and then fill the boat with 30% nitrocellulose. There should be enough nitrocellulose around the tissue so that, as the block hardens, none of the tissue will be exposed. The tissue should have a thick rim (4–5 mm) of nitrocellulose all around it.

Put the boats in a closed container, such as a desiccator, that contains chloroform. The boats should not sit directly in the chloroform but rest on a shelf or platform that allows the chloroform vapors access to the boats. As the boats sit for several hours in the vapors, the nitrocellulose will become hard. Once it is hard, peel away the paper and put the blocks in a covered dish or jar of 70% alcohol where they can be stored (in a refrigerator) for several days.

As the alcohol–ether evaporates and the nitrocellulose hardens, bubbles develop. They ordinarily do not pose any problem; however, it is frequently recommended that before the boats are exposed to chloroform vapors they sit in a covered dish, either at room temperature or in a refrigerator, for about an hour. This will allow the gas bubbles to escape from the block before being trapped in the hardening nitrocellulose.

4. Mounting the Blocks on Chucks

Trim the firm nitrocellulose block to fit a microtome chuck. Wet one face of the chuck and the base of the nitrocellulose block in alcohol–ether (a 1 : 1 mixture) for 2–3 min. Cover the wet chuck face with fresh 30% nitrocellulose and press it against the base of the block. Let this ensemble stand for 1–2 min and then drip 30% nitrocellulose around the chuck–block junction. Put the whole assembly in chloroform vapors until the nitrocellulose is hard and store it in 70% alcohol until the block is sectioned.

5. Sectioning and Processing

An ordinary microtome knife will be sufficient for sectioning the block, but if an "a-profile" knife is available, it is more convenient because it helps to retain the cutting solution and the section on the knife. Be careful in bringing the knife to the nitrocellulose block because if the knife hits the block too fast, the block may come off the chuck. Also make certain that the chuck is fastened securely in the microtome holder.

Keep the knife and block wet with 75% alcohol. The sections, which are cut 100–150 μm thick, can be removed from the knife with a brush. The sections are put in 75% alcohol in serial order. There are several ways of keeping the sections in serial order. One is to put each section on lens paper that can be numbered with an indelible pencil. Another is to have a series of small dishes that are kept in order.

Dehydrate the sections in three changes each of 85% and 95% ethyl alcohol for 15 min each change. The final dehydration occurs in butyl alcohol, three changes of 15 min each. It is imperative that *butyl alcohol* be used; otherwise the nitrocellulose becomes soft and dissolves, causing the sections to stick together. Clear the sections in cedarwood oil. The sections will float on the cedarwood oil when they first contact it. After the sections sink to the bottom, they are briefly rinsed in two changes of xylene and mounted on slides. The sections may be stored for several days in cedarwood oil but will become brittle and wrinkle and thus be difficult to mount if allowed to sit in xylene for more than a few minutes. The slides are coverslipped.

C. Rapid Golgi Method for Use on Aldehyde-Fixed Material

The Golgi method can be applied to tissue that has been perfused with an aldehyde solution for an electron microscopic study. The advantage of this is obvious.

Perfuse the CNS with 1% gluteraldehyde and 1% formaldehyde in 0.12 M phosphate buffer (Wuerker and Palay, 1969; Palay and Chan-Palay, 1974). Remove the brain; cut it into 4 to 5-mm-thick pieces and put them in rapid Golgi fixative. Process the tissue as described for the rapid Golgi method. Because the aldehyde-treated tissue is hard and brittle, the blocks should be embedded in low-viscosity nitrocellulose.

The question again arises of how long the tissue should remain in fixative and silver nitrate. On the whole, perfused tissue does not have to remain in fixative as long as does fresh tissue. Palay and Chan-Palay (1974) suggest that the quality of the impregnation can be judged by removing a block of tissue from the silver nitrate solution and, with a single-edge razor blade, cutting a thin slice from the surface of the block. Float the slice in a drop of 70% alcohol on a slide, coverslip the slide, and assess the quality of the impregnation. Although this is a rational way to proceed, it should be pointed out that the Palays examined cerebellar tissue. Any block of cerebellar cortex, for example, consists of the same few cell types that are present throughout, and it is likely that superficial sections are typical of the impregnations that occur deeper in the block. Yet if other, less homogeneous areas of the CNS are studied, this may not be true. The examination of more than a thousand blocks of rat and mouse diencephalon has demonstrated that the quality of impregnation near the surface of the block does not always accurately reflect the state of the impregnation towards the center of the block. For example, the first two or three sections may be disappointing, even blank, but the next ones, proceeding in serial order, may display magnificently impregnated neurons.

D. Stabilizing and Counterstaining Rapid Golgi Sections

Geisert and Updyke (1977) have recommended a procedure for stabilizing the Golgi precipitate in order for the sections to be counterstained. The tissue is embedded in celloidin or low-viscosity nitrocellulose and cut into 100-μm-thick sections. The impregnation is stabilized by converting the silver chromate precipitate to silver bromide and then reducing the silver. The reduction of the silver is thought to stabilize the impregnation, thus preventing it from fading. This, in turn, allows the Golgi sections to be counterstained with aqueous cresyl violet without diminishing the quality of the impregnation.

These authors point out that there is an increase in the opacity of the counterstained sections, and that it is advisable to study and draw or photograph the sections prior to counterstaining them.

The following steps result in stabilizing sections.

1. Collect the 100-μm-thick sections in 70% alcohol.
2. Rinse the sections in two changes of distilled water.
3. Put the sections in 0.4 M sodium bromide and gently agitate them for about 5 min; the sections should be uniformly yellow.
4. Rinse in two changes of distilled water for 5 min each.
5. Reduce the sections in D-19® developer for 5 min; the sections will become gray.

6. Rinse briefly in distilled water.
7. Treat the sections for 5 min with 0.5 M sodium thiosulfate; the sections will become less opaque.
8. Rinse twice with distilled water for 5 min each rinse.

The sodium bromide and sodium thiosulfate solutions need to be renewed frequently.

All of the above steps are performed at room temperature in a lighted room.

At this point, the stabilizing sections can be counterstained with aqueous cresyl violet according to any of the methods that are commonly used for celloidin sections.

E. Variations of the Golgi–Kopsch Method

1. The Method of Braitenberg *et al.*

Braitenberg *et al.* (1967) obtained delicate impregnations in the brains of a variety of species, including insects, fish, amphibians, and mammals, that had been stored in formalin for more than a year. The tissue is placed in the following solution for 8–10 days, and the temperature is maintained at 27°C:

3% Potassium dichromate	100 ml
10% Formalin	5 ml
Sucrose	12.5 g

The tissue is next put into 0.75% silver nitrate for at least two days and then rinsed in distilled water and sectioned. If the tissue blocks are soft, embed them in celloidin.

According to Braitenberg *et al.* (1967), the longer the tissue is fixed in formalin, the better the results will be. In their experience, the quality of the formalin is not critical, and it made no difference whether or not the formalin was buffered. They think that the significance of the sucrose is that it acts as a reducing agent.

2. The Method of Fox *et al.*

Fox *et al.* (1951) fixed brains in formalin and postfixed them in zinc chromate. They emphasized that the tissue must be stored in formalin for at least 2–3 months (and 2–3 years is much better) before the chromating step is carried out. The authors point out that their method is not effective for studying embryonic or newborn brains.

The fixed brain (or spinal cord) is cut into 2- to 5-mm-thick slices and immersed for 2 days in the following chromating solution:

Zinc chromate	6 g
Formic acid	4 ml
Distilled water	98 ml

Make the solution by adding the zinc chromate to the formic acid and 48 ml of the water. Stir until the zinc chromate has dissolved and then add the remaining water.

The tissue is subsequently put in a large volume of 0.75% silver nitrate solution after the first day. Fox *et al.* (1951) recommend using a fine thread to suspend the tissue in the silver nitrate, but if the bottles are placed on a laboratory shaker or rotated several times or have their bottoms lined with glass beads, it will not be necessary to use thread. The authors also suggest that the excess silver chromate crystals that coat the surface of the tissue be brushed away.

The tissue can be sectioned either by the "paraffin shell" technique or by embedment in celloidin.

3. The Method of Colonnier

Colonnier (1964) perfused the CNS with buffered gluteraldehyde, sectioned the brain, and put the sections in a gluteraldehyde–potassium dichromate solution. This solution is made by adding 100 ml of 25% gluteraldehyde to 400 ml of 2.5% potassium dichromate. The tissue is left in this fixative for 5–6 days at room temperature and then in 0.75% silver nitrate for 5–6 days. These times cannot be considered appropriate for impregnating all areas of the CNS and will have to be adjusted (empirically) for the area that is being studied. Double and triple impregnations may also prove useful.

F. The Golgi–Cox Procedure

There are several Golgi–Cox recipes (Ramón-Moliner, 1970). The following procedure, which was published by Ha (1966), is simple and produces relatively good impregnations in the brain and spinal cord. Although Ha originally recommended that the brain be perfused with buffered formalin and then stored in formalin for several weeks, this is not necessary.

1. Animals

The brains of young mammals, not older than 3 months, are usually used in this method. However the method has been used successfully with avian, reptilian, and amphibian brains.

2. Tissue

The brain or cord is removed as rapidly as possible from the animal, cut into 4- to 5-mm-thick blocks, and immersed in the Golgi–Cox fixative.

An alternative method involves perfusing the brain or cord with buffered formalin and, after keeping the tissue in formalin for about 2 months, putting it in Golgi–Cox fixative.

3. Fixing Solutions

Three separate solutions are made and then mixed together in a particular sequence immediately before being used.

Solution A is mercuric chloride, 10 g, dissolved in 450 ml of distilled water.

Solution B is potassium dichromate, 10 g, dissolved in 450 ml of distilled water.

Solution C is potassium chromate, 3 g, dissolved in 100 ml of distilled water.

Solutions A and B can be made and stored for an indefinite time, but solution C must be made just before it is used. The fixative is made by adding 450 ml solution A to 450 ml solution B. Stir the mixture and, while continuing to stir, slowly add 100 ml solution C. It is critical that C be added to A and B slowly and with constant stirring, or a precipitate will form.

The tissue is kept in the fixative in the dark, at room temperature, for 2 months. Glass wool or beads should line the bottom of the fixing jars to allow all surfaces of the tissue to be exposed to the fixative.

4. Dehydration

Dehydrate the tissue in 50% and 70% ethyl alcohol for 15 min each and in 95% and 100% alcohol for 30 min each. Then put the tissue in a solution of one part absolute alcohol to one part diethyl ether for 40–60 min. Do not prolong the time the tissue is in the alcohol–ether solution or the tissue will become very brittle.

5. Infiltration, Embedding, and Cutting

The tissue is next infiltrated with celloidin, e.g., Parloidin®, by putting it in 2% and 4% celloidin for 24 hr each and in 6% and 12% for 48 hr each. Celloidin is usually purchased as dry chips that have to be dissolved in an alcohol–ether solution. For example, to make 6% celloidin, dissolve 6 g celloidin chips in a solution of 50 ml absolute alcohol and 50 ml diethyl ether. This has to be done in a glass container that can be tightly sealed to prevent evaporation of the alcohol–ether solution. It will take several days for the celloidin to dissolve; the process can be speeded up if the bottles are put on a laboratory shaker.

Cover the bottom of a paper boat with fresh 12% celloidin, orient a piece of tissue on it, and cover with fresh 12% celloidin. Harden the blocks in chloroform vapors, store them in 70% alcohol, and then attach them to microtome chucks (see Section VII.B.4).

6. Processing the Sections

Cut the sections 80–150 μm thick and collect them in serial order in 70% alcohol. No special-profile knife is needed to cut celloidin sections; a standard microtome knife is sufficient. Wash the sections briefly in distilled water and "develop" them in 5% ammonium hydroxide made in 95% ethyl alcohol. Keep the sections in the ammonium hydroxide until they turn dark, about 3–5 min. If the sections are left in this solution too long, extraneous precipitate may occur in the sections and decrease their usefulness. Put the sections in butyl alcohol for 5 min to dehydrate them and then clear them in cedarwood oil. They are left in the oil until they sink to the bottom of the dish. Next, the sections are mounted on glass slides and covered with a coverslip.

7. Counterstaining

An advantage of the Golgi–Cox method is that the sections can be conveniently counterstained to show the cell bodies or nuclei of unimpregnated neurons and glia cells. Depending on the recipe of the Golgi–Cox method used, counterstaining may cause a reduction in the impregnation. Initially counterstain, at most, only every other section and compare the impregnations with those of the adjacent, nonstained sections.

The sections are counterstained after they are developed in ammonium hydroxide. Following the development, wash the sections for only a short time in distilled water before putting them in the staining solution.

The recommended counterstain is 0.25% aqueous cresyl fast violet. Its pH must be 3.7. Before using it, add 5 drops of 10% acetic acid to the stain, filter the stain, and put it in a dish on a 45°C hotplate. Allow the stain to reach 45°C and then, keeping the staining dish on the hotplate, stain the sections (only a few at a time). The sections should be stained within 30 sec. If within this time the sections become too dark, i.e., gray and white matter cannot be distinguished, decrease the pH of the staining solution.

The sections are differentiated in 95% ethyl alcohol. Keep the sections in this alcohol until the gray and white matter can easily be distinguished and no more stain comes out of the section. Dehydrate the sections in butyl alcohol, clear them in cedarwood oil, and mount and coverslip them.

G. Golgi Method for Embryonic Tissue

Stensaas (1967) used the following procedure to impregnate developing neurons in the rabbit cerebral cortex. He used two different fixing solutions to emphasize either axonal patterns or soma–dendritic arrays. The first fixing solution mentioned below, which was developed by Klatzo (1952), is very strongly recommended by Scheibel and Scheibel (1978) to study glial cells in young adult brains.

1. Fixation

The fixative Stensaas (1967) considered especially effective in studying developing axonal systems consists of the following:

Potassium dichromate	5 g
Distilled water	99 ml
Chloral hydrate	3 g

The potassium dichromate should be dissolved before adding the chloral hydrate. The solution must be made just before being used.

Fix the tissue for 24–48 hr. The solution should be changed after 24 hr.

Stensaas (1967) developed a slightly different solution to study developing somata and dendrites. This solution must be made immediately before it is to be used. It consists of the following:

Potassium dichromate	5 g
Chloral hydrate	5 g
Formaldehyde (used from the bottle and not diluted)	5 ml
25% Glutaraldehyde	5 ml
Distilled water	100 ml

Stensaas originally added 5 drops dimethyl sulfoxide, but he currently thinks that it has no utility in fixation.

The tissue is kept in this solution for 24–48 hr. Change the solution after 24 hr.

Embryonic tissue is fragile; therefore, it should be put into fiber bags ("tea bags") that can be suspended in the bottles containing the fixative. The bags are convenient for transferring the tissue from solution to solution.

2. Silver Impregnation, Dehydration, and Clearing

After the tissue is fixed in either of the above solutions, briefly rinse it in 1.5% silver nitrate and then suspend it in fresh 1.5% silver nitrate for 18–24 hr. Rinse the tissue in tap water to remove the excess silver nitrate and dehydrate it in alcohol by transferring the fiber bags to 40%, 75%, 95%, and 100% (2 changes) ethyl alcohol for 15 min each. Clear the tissue in xylene for 10–15 min.

3. Paraffin Infiltration and Cutting

Transfer the tissue from xylene to a paraffin mixture and keep it in the paraffin for 2–3 hr in a 60°C oven. The paraffin mixture is made by mixing one part rubber cement to two parts melted paraffin (use the low-melting-point variety). Keep this mixture uncovered overnight in a 60°C oven. The solvent that is contained in the rubber cement will evaporate, and the paraffin mixture will become sticky. The tissue is infiltrated with this sticky mixture.

After 2–3 hr, remove the tissue from the paraffin bath and let it cool to room temperature. Remove it from the fiber bags. Mount the tissue on a wooden or cork chuck whose face has been covered with several layers of paraffin. Use hot forceps (as in making a paraffin shell) to make a small well to hold the tissue block. Pour a small amount of warm paraffin (not the sticky mixture) over the block and let it cool.

Because the tissue is not myelinated and has not been exposed to osmium, it will be soft and pliable. Therefore, it will be hard to cut sections unless pressure is kept on the block. To do this, firmly put the left thumb on the tissue block and *very* carefully draw the knife through the block, being careful to section only the tissue. Put the sections on glass slides that have been coated with a thick layer of an albumin–glycerine mixture. Make this mixture by mixing 3 g albumin powder with 1 ml glycerine. Add 0.5 ml distilled water. The mixture will become cloudy with the addition of the water. Stir thoroughly and add a few thymol crystals to prevent decay. The mixture is spread on the slides with a small piece of sponge or powder puff.

When the slides are filled with sections, put them in a low petri dish containing xylene. This will remove the paraffin. Take the slides out of the xylene and, without allowing them to dry, cover the sections with a synthetic resin. The sections will fade after 6–8 months. However, once the tissue is embedded in paraffin and attached to a chuck, the blocks can be kept for an indefinite period of time and be used to cut additional sections.

ACKNOWLEDGMENTS. The author wishes to thank Dr. Diane Smith for advice concerning the Golgi-Cox method and Dr. L. J. Stensaas for his help regarding the processing of embryonic tissue.

REFERENCES

Braitenberg, V., Guglielmotti, V., and Sada, E., 1967, Correlations of crystal growth with the staining of axons by the Golgi procedure, *Stain Technol.* **42:**277–283.

Chan-Palay, V., 1973, A brief note on the chemical nature of the precipitate within nerve fibers after the rapid Golgi reaction: Selected area diffraction in high voltage electron microscopy, *Z. Anat. Entwickl.-Gesch.* **139:**115–117.

Colonnier, M., 1964, The tangential organization of the visual cortex, *J. Anat.* **98:**327–344.

Cox, W., 1891, Impregnation des centralen Nervensystems mit Quecksilbersalzen, *Arch. Mikr. Anat.* **37:**16–21.

Fox, C. A., Ubeda-Purkiss, M., Ihrig, H. K., and Biagioli, D., 1951, Zinc chromate modification of the Golgi technic, *Stain. Technol.* **26:**109–114.

Fregerslev, S., Blackstad, T. W., Fredens, K., and Holm, M. J., 1971, Golgi potassium-dichromate silver nitrate impregnation. Nature of the precipitate studied by X-ray power diffraction methods, *Histochemie* **25:**63–71.

Geisert, E. E., Jr., and Updyke, B., 1977, Chemical stabilization of Golgi silver chromate impregnations, *Stain Technol.* **52:**137–246.

Golgi, C., 1873, Sulla struttura della sostanza grigia dell cervello, *Gass. Med. Ital. Lombarda* **33:**244–246.

Golgi, C., 1879, Di una nuova reasione apparentemente nera dell cellule nervose cerebrali ottenuta col bichloruro di mercurio, *Arch. Sci. Med.* **3:**1–7.

Ha, H., 1966, A modified Golgi–Cox method with counterstain for the study of synapses, *Anat. Rec.* **155:**59–64.

Klatzo, I., 1952, A study of glia by the Golgi method, *Lab. Invest.* **1:**345–350.

Kopsch, F., 1896, Erfahrungen über die Verwendung des Formaldehyds bei der Chromsilberimprägnation, *Anat. Anz.* **11:**727–729.

Lee, A. B., 1950, *The Microtomist's Vade-Mecum,* 11th ed. (J. B. Gatenby and H. W. Beams, eds.), Blakiston, Philadelphia.

Morest, D. K., 1968, The collateral system of the medial nucleus of the trapezoid body of the cat, its neuronal architecture and relation to the olivo–cochlear bundle, *Brain Res.* **9:**288–311.

Morest, D. K., and Morest, R. R., 1966, Perfusion-fixation of the brain with chrome–osmium solutions for the rapid Golgi method, *Am. J. Anat.* **118:**811–832.

Palay, S. L., and Chan-Palay, V., 1974, *Cerebellar Cortex: Cytology and Organization,* Springer-Verlag, New York.

Ramón y Cajal, S., 1937, *Recollections of My Life* (E. H. Craigie, trans.), MIT Press, Cambridge.

Ramón y Cajal, S., 1960, *Studies on Vertebrate Neurogenesis* (L. Guth, trans.), Charles C Thomas, Springfield, Illinois.

Ramón y Cajal, S., and DeCastro, F., 1933, *Elementos de Tecnica Micrografica del Sistema Nervioso,* Tipografia Artistica, Madrid.

Ramón-Moliner, E., 1957, A chlorate–formaldehyde modification of the Golgi method, *Stain Technol.* **32:**105–116.

Ramón-Moliner, E., 1970, The Golgi–Cox technique, in: *Contemporary Research Methods in Neuroanatomy* (W. J. H. Nauta and S. O. E. Ebbesson, eds.), pp. 32–55, Springer-Verlag, New York.

Scheibel, M. E., and Scheibel, A. B., 1955, The inferior olive. A Golgi study, *J. Comp. Neurol.* **102:**77–131.

Scheibel, M. E., and Scheibel, A. B., 1970, The rapid Golgi method. Indian summer or renaissance? in: *Contemporary Research Methods in Neuroanatomy* (W. J. H. Nauta and S. O. E. Ebbesson, eds.), pp. 1–11, Springer-Verlag, New York.

Scheibel, M. E., and Scheibel, A. B., 1978, The methods of Golgi, in: *Neuroanatomical Research Techniques* (R. T. Robertson, ed.), pp. 89–114, Academic Press, New York.

Stensaas, L. J., 1967, The development of hippocampal and dorsolateral pallial regions of the cerebral hemisphere in fetal rabbits. I. Fifteen millimeter stage, spongioblast morphology, *J. Comp. Neurol.* **129:**59–70.

Strong, O. S., 1896, Review of the Golgi method, *J. Comp. Neurol.* **6:**101–127.

Teuscher, R., 1894, *Untersuchungen uber den Feineren Bau des Centralen und Peripherischen Nervensystems,* G. Fischer, Jena.

Valverde, F., 1965, *Studies on the Piriform Lobe,* Harvard University Press, Cambridge.

Wuerker, R. B., and Palay, S. L., 1969, Neurofilaments and microtubules in anterior horn cells of the rat, *Tissue Cell* **1:**387–402.

Electron Microscopy 10

Preparation of Neural Tissues for Electron Microscopy

VICTOR L. FRIEDRICH, Jr., and
ENRICO MUGNAINI

I. INTRODUCTION

The electron microscope is essential in experimental neuroanatomy for the detailed study of synaptic organization. Its great value is based on the fact that it allows the recognition of synaptic junctions and often the identification of the neurons participating in the synapses. However, the identification of different types of synaptic junctions and the classification of neurons usually depends on the regular occurrence of subtle differences in the shapes and internal structures of processes from different types of neurons. These differences are observed consistently only in specimens that are well-preserved; poor preservation obscures cytological detail and renders results less reliable and interpretation more difficult. Thus, good preservation of tissue structure is a necessity for electron microscopic neuroanatomy.

This chapter deals primarily with details of fixation and embedding;

VICTOR L. FRIEDRICH, Jr., and ENRICO MUGNAINI • Laboratory of Neuromorphology, Department of Biobehavioral Sciences, University of Connecticut, Storrs, Connecticut 06268.

many other aspects, such as thin sectioning, are not covered here but are treated extensively in more general reviews such as the series edited by Hayat (1970–1978) and by Glauert (particularly Agar *et al.*, 1974; Glauert, 1975; Reid, 1975).

II. BASIC PROCEDURES FOR FIXATION AND EMBEDDING

The brain has a high rate of metabolism and reacts quickly to anoxia. In addition, fresh brain tissue is soft and highly susceptible to mechanical damage. Satisfactory fixation of the central nervous system thus requires rapid fixation *in situ.* Areas at the surfaces of the ventricles may best be fixed by intraventricular perfusion of fixative, and at some pial surfaces by surgical exposure and local application of the fixative. Perfusion of fixative through the vascular system, however, is the method of choice for most areas of the brain and spinal cord (Palay *et al.*, 1962). After the initial fixation, the nervous tissue is dissected out, fixed further, and embedded in a material suitable for the cutting of ultrathin sections. Although the general procedures for the preparation of nervous tissue for electron microscopy are similar in most laboratories, there are many variations in the different steps. The following descriptions of methods for fixation and embedding are representatives of common practice, and they have been used with success for various species such as frog, turtle, chicken, mouse, rat, cat, and monkey. The method of fixation is based on vascular perfusion with buffered aldehydes followed by postfixation of dissected slices or blocks of tissue with buffered OsO_4. It can be used for all areas of the forebrain, the brainstem, and the cervical spinal cord.

A. Anesthesia

Different anesthetics are appropriate for different species. For turtles, we use 0.5–1.0 ml/kg of 6.5% sodium pentobarbital intraperitoneally. For birds, we inject 0.3 ml/kg of 6.5% sodium pentobarbital slowly into the subular vein, or we use methoxyflurane or halothane vapors.* For mice and rats, we use 350 mg/kg of chloral hydrate intraperitoneally (3.5 or 35% aqueous solutions, the latter concentration for terminal surgery only). Larger mammals like cats and monkeys can be anesthetized with intramuscular injections of 10 mg/kg of Ketaject® (ketamine hydrochloride) followed after 10–15 min by 0.3 ml/kg of 6.5% sodium pentobarbital intraperitoneally. The Ketaject® sedates the animal in a few minutes; it also reduces the dose of barbiturate necessary for deep anesthesia, eliminates the excitation phase preceding sedation, and minimizes vascular reaction to the bar-

*These inhalation anesthetics should be stored in the refrigerator, and the bottle should be used only when it is recently opened; toxicity that develops with storage causes mitochondrial condensation.

biturate. Phencyclidine ("angel dust") is a good anesthetic for primates and produces moderate vasodilation. For newborn mammals, methoxyflurane or halothane vapors are effective. Some combinations of anesthetic, dose, and species (for instance, low doses of sodium pentobarbital in cats) result in cerebral vasoconstriction and yield erratic and inferior results, as judged by reduced flow of the fixative, poor clearing of blood from the brain and meninges, and narrowing of blood vessels.

The depth of anesthesia is an important variable. Animals that are not deeply anesthetized often perfuse poorly, whereas too much anesthetic, enough to cause gasping, also leads to erratic results. In assessing the degree of anesthesia, one cannot simply measure the dose administered because of individual differences in sensitivity and rate of absorption. Rather, one must attend carefully to physiological clues such as the diameter of the pupils, the rate and depth of respiration, and the activity of reflexes such as closure of the eyelid after a light touch to the cornea and withdrawal of the foot after it is pinched.

B. Surgical Procedure

The anesthetized animal is placed on its back on a perfusion table (Fig. 1) with the four extremities tied down. The skin on the abdomen and thorax is cleared of hair or feathers and incised along the midline with scissors and scalpel from the hypogastric region to the neck. Along the extent of this incision, the skin is separated by blunt dissection from the underlying musculature and pulled laterally. The abdominal wall is then opened longitudinally along the midline. Two additional cuts are made from the midline laterally along the caudal border of the rib cage, to expose the diaphragm. These cuts sever the superior epigastric arteries; the arteries may be clamped before making the mediolateral cuts to prevent bleeding.

The diaphragm is then cut away from the ribs on both sides, and the thoracic wall is cut at the osteocartilagenous border on both sides, beginning at the diaphragm and ending near the first rib. These cuts define a central slab of thoracic wall, containing the sternum, that is free caudally and fixed rostrally. The caudal end (the xiphoid process of the sternum) is lifted ventrally and rostrally, exposing the heart and lungs. A clamp is placed across the slab at its rostral end to close the internal thoracic (mammary) arteries; the part caudal to the clamp is then cut away.

Ribs are tough, and it is difficult to cut them quickly; we use large paper cutting shears for chickens and cats, surgical scissors for rats and mice. One should avoid nicking the lungs while cutting the diaphragm and ribs.

Following the exposure of the thoracic activity, the pericardium is lifted with fine, toothed forceps and opened with sharp scissors. After the prediaphragmatic descending aorta has been located, the right atrium is cut, and the descending aorta clamped with a hemostat to save fixative. The tip of the left ventricle is then opened with scissors to admit the cannula of the

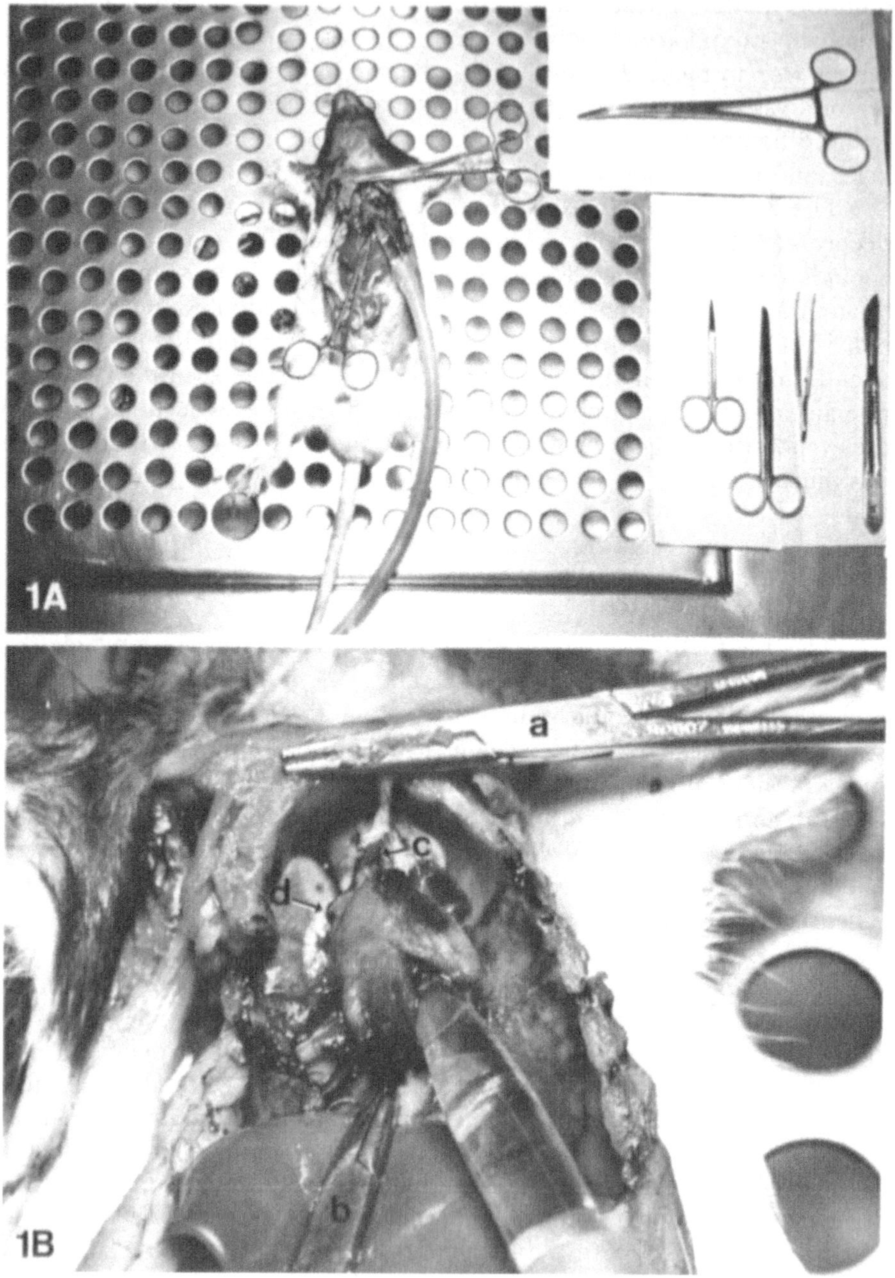

Figure 1. Rat after perfusion. (A) The surgery is done in a well-lighted area that is cleared except for the few surgical instruments that are necessary. Here, the work surface covers a sink which receives the outflow of fixative. A fan draws air from above through the perforated work surface and exhausts it from the building, thus limiting exposure to strong fixative fumes. (B) Close view of the thorax. The upper clamp (a) compresses the internal mammary arteries, and the lower clamp (b), which is angled, the descending aorta. The tip of the tapered glass cannula

perfusion apparatus (see Fig. 2). Since air bubbles tend to collect at the tip of the cannula, it must be inspected immediately before insertion. If bubbles are present, they should be expelled with a brief burst of fluid before the cannula is passed through the left ventricle and its tip inserted into the aorta. Perfusion with vascular rinsing solution (Section VII.A) is begun as the cannula is inserted. Unless it impedes insertion, flow should actually be started immediately *before* the cannula is inserted to further guard against bubbles. This is followed after 30 to 60 sec by the perfusion fixative (Section VII.C). The cannula may then be secured in place with a small clamp, a thread, or a string around the heart or around the aorta. This securing can be omitted if the cannula is well placed and of the right size and shape, in which case the fixative itself constricts the heart and aorta around the cannula, and minimal effort is required to maintain it in position.

Blood and perfusates exit via the cut in the right atrium. Initial flow should be about 0.05 to 0.5 ml/min per g body weight, at pressures from 40 to 90 mm Hg (55–120 cm H_2O), but flow usually slows with time.

In mammals, the interval from first disruption of the diaphragm to onset of fixation should not be more than 60 to 90 sec; consequently, the procedures described above must be performed in rapid succession. During the surgery, one must avoid pulling the heart out of its natural position, since this may reduce perfusion of ascending arterial branches; moreover, excessive manipulation induces vasoconstriction. The cannula and its associated tubing must be arranged so that they do not strain the vasculature when the cannula has been secured and the experimenter releases it. This is critical and often difficult to do with small animals. We prefer to perfuse for at least 1 hr, but this is impractical for animals larger than a rat. The perfusion pressure can be reduced after 15 min to insure that the fixative is not exhausted prematurely. Reasonable volumes of fixative are from two times body weight in cat and monkey to 20 times body weight in the mouse.

Animals are left undisturbed for several hours after the perfusion. The atlantooccipital membrane is then exposed by removing skin and muscle from the dorsal surface of the neck, and a scalpel is inserted to sever the spinal cord. This procedure prevents the strain on the brainstem and spinal cord that would accompany inadvertent movement of the neck. Following this, the head is cleaned of skin and muscle, and the mandible removed. The head is then separated from the rest of the body, immersed in the perfusion fixative, and left in the refrigerator overnight. The trunk is saved if the spinal cord is the focus of investigation. An alternative procedure,

---

is inserted through the left ventricle and into the aorta (c). The right atrium (d) has been incised to allow outflow of fixative. A channel has been cut in the rib cage on the animal's right to drain fixative away from the heart; this makes it easier to see the aorta and the tip of the cannula. The placement of the tip of the cannula is important, and its position should be monitored with an operating microscope in small animals. It is best if the tip is just within the aorta and large enough to fit snugly; if the cannula is pressed too far into the aorta, perfusion may be impaired.

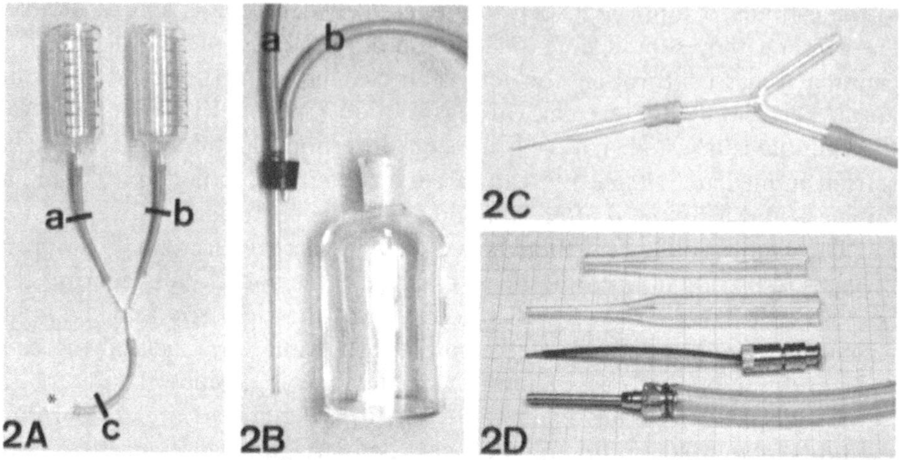

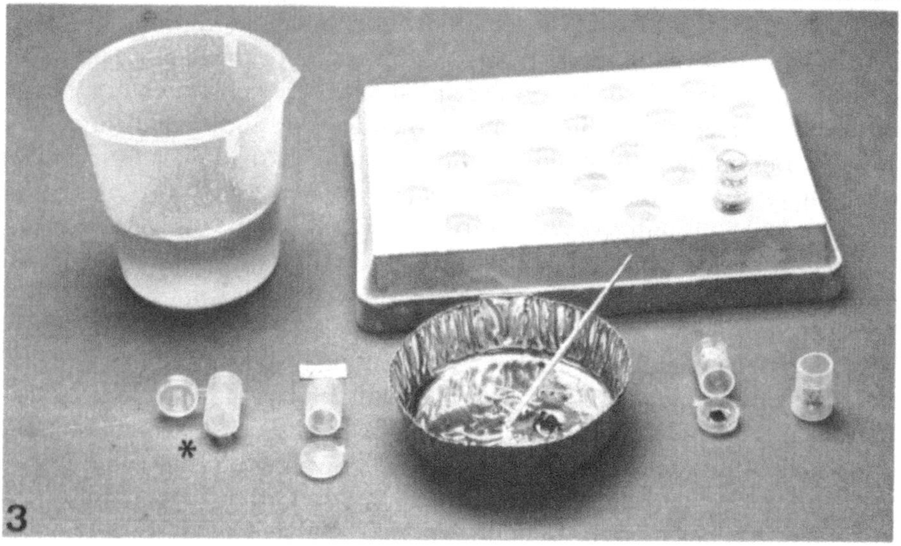

Figure 2. Devices for perfusion fixation. (A) Gravity perfusion apparatus. The two flasks allow delivery of two fluids, such as vascular rinse and fixative. The graduated bottles are hung from pegboard to allow adjustment of their height; alternatively, they can be suspended with twine and a pulley. The height of the bottles determines the perfusion pressure (1 cm $H_2O = 0.74$ mm Hg). Clamps at a, b, and c control the flow; the clamp at c is released at the beginning of the perfusion to initiate flow, and a single clamp is moved quickly from a to b to change from one perfusate to the other. The tubing, which can be gum rubber or clear plastic, has been shortened for the purpose of photography; a and b should be near the Y-tube and c near the cannula (*). The tubing should be free of bubbles after it is filled. Bubbles can be dislodged and driven to one end of the tubing or the other by tapping or pinching the tubing while holding it vertical. The small volumes of vascular rinse appropriate for very small animals, such as mice, can be siphoned directly into the cannula after the latter and the tubing have been filled with fixative. (B) Compressed air perfusion device. After the bottle is filled, it is closed with the rubber stopper, which is secured in place with adhesive tape. Compressed air (0–3 lb/in.^{2}; 1 lb/in.^{2} = 51.7 mm Hg = 70.3 cm H_2O) is delivered through line b to the short glass tubing in the stopper, and fixative is expressed through line a from the long glass tubing. The

attractive for small animals such as mice and rats, is to put the whole body in the refrigerator overnight and perform the cleaning and separation of the head the next day.

C. Dissection and Postfixation

To maximize hardening of the tissue by the fixative, dissection and postfixation are performed the day after the perfusion. It is advisable, however, not to wait longer than 1 day to initiate postfixation in order to prevent the formation of whorls of membrane and other artifacts.

Dissection of the cranium and spinal column must be done with care to avoid accidental compression or stretching of the neural tissue. Dissection of small animals is best done under an operation microscope with a working distance long enough to allow free manipulation of dissecting instruments. Safety glasses should be worn, particularly when one is dissecting large animals, to prevent damage to eyes by flying bone chips.

The cranium is usually best entered from the ventral side. The skull should already be stripped of skin and muscle, the mandible removed, and the head separated from the body. Starting at the foramen magnum and progressing forward to beyond the optic chiasm, the bones comprising the

---

air pressure must be controllable and steady, as from a small compressor pump with a bleeder valve (for example, #54906-001, VWR Scientific), and should be monitored with a mercury manometer. (C) A simple device for artificial respiration. The gas mixture, 95% O_2–5% CO_2, is delivered to the tubing on the right; the cannula, on the left, is inserted into the trachea as described in the text. When the opening of the Y-tube is obstructed, gas is forced into the lungs; when it is open, gas can escape from the lungs. (D) Cannulas for the aortas of animals of different sizes. The upper two cannulas were drawn from glass tubing; the cannula at the bottom was made from a hypodermic needle and is shown with tubing attached. The other cannula, which is used for mice, consists of a length of thin polyethylene tubing with a short piece of blunted hypodermic needle at one end and a fitting to mate a hypodermic syringe at the other. The cylinder of a syringe is then fitted to the relatively large tubing of the perfusion apparatus. The cannulas are photographed on graph paper with the small squares measuring 1 mm. For best results, the cannula should be as large as can pass into the aorta, and the bore of each cannula should be as large as possible.

Figure 3. Embedding tissue slices. Specimens are often prepared as thin slices and must be embedded for sectioning parallel to the large faces of the slices. This can be accomplished using plastic embedding capsules manufactured by BEEM (Better Equipment for Electron Microscopy, Inc., Bronx, N.Y.) and available from many suppliers. The capsule (*) is modified by cutting off the conical end with a razor blade and separating the flat cap from the cylinder. Tissue slices infiltrated with epoxy monomer are centered in the caps, and paper labels are placed in the empty cylinders. Sometimes it matters which face of the slice is down; if necessary, the slices can be placed in a weighing dish and examined under a dissecting microscope before being transferred to the embedding capsules. After the tissue is in place in the caps, the cylinders are carefully pressed into the caps so as not to compress the tissue. The capsules are then placed in special holders (upper right) or fastened to index cards with double-sided adhesive tape. The cylinders are filled with epoxy monomer by using a disposable syringe or by pouring directly from the disposable beaker in which the monomer is prepared (upper left).

base of the cranium are broken away. The channel so formed is widened laterally and dorsally until the brain is freed from the skull. One must avoid pulling the cranial nerves and the dura. The dura presents a special problem as it is fused in places with the periosteum of the cranial vault, and one may distort the brain severely when breaking and removing the cranial bones. This problem is minimized if the dura is removed as it is exposed; its attachments to the arachnoid must be cut without pulling on either membrane.

The spinal cord is exposed by removing the dorsal musculature, successively breaking each vertebral arch in several places, and removing the pieces. Again, the nerve roots and the dura must not be pulled. It is extremely important that the tissue not be compressed or stretched during the dissection.

The exposed parts of the brain and spinal cord must be kept thoroughly wet with fixative or buffer wash (Section VII.E) throughout the dissection.

A well-fixed brain and spinal cord should appear relatively hard, free of blood, and pale yellow to pale orange in color. If the tissue is soft and white or pink, and the meningeal vessels are darkened by blood, it has not been perfused and should be discarded. Some blood almost always remains in the venous sinuses, which should, therefore, be ignored in this evaluation. The color of a well-fixed brain depends on the composition of the primary fixative; with the recipe given here, it should be pale yellow.

After the brain is dissected out, it is marked on one side with a shallow incision, and the region of interest is divided into serial slices with a razor blade. Each slice, 1–1.5 mm thick, is collected in a separate vial. If slices are too thin, the cells are often swollen and extracted, whereas slices too thick often suffer from uneven postfixation. The slices may also be trimmed in the other dimensions to reduce the amount of osmium required for postfixation; however, they should always be shaped to preserve necessary information about orientation. The slices are immersed in washing solution (Section VII.E) for 10–30 min and postfixed with osmium tetroxide (Section VII.H) for 2–4 hr. The washing and osmication may be performed at room temperature or on ice. Chilling seems to reduce extraction of tissue components and to stabilize the osmium. Osmium tetroxide reacts with aldehydes still in the tissue. If that happens, the solution turns red or brown and should be changed to insure complete osmication.

Fiber tracts and nuclei become markedly visible upon osmication, and drawings or photographs of the cut surfaces of each slice can therefore be made after osmication. The slices are then trimmed to their final size, and the modifications are recorded on the drawings or photographs. The final slices should be small enough that they can be sectioned "semithin" across the entire face. This usually means 3–4 mm on each side, unless extra-wide knives are available. Highly accurate characterization of the areas finally selected for thin sectioning is usually possible if this procedure is followed.

Mechanical distortion of the tissue is, in our opinion, the source of most tissue artifacts, and even if the tissue has been thoroughly reacted with aldehydes, it is still unstable and responds to mechanical stress. Dissection and slicing, therefore, are among the most critical steps in the fixation procedure. A well-known artifact that is common when tissue is not adequately fixed before it is dissected is the dark neuron (Cammermeyer, 1962; Mugnaini, 1965). Other changes that can occur from distortion, even of well-fixed tissues, include swelling of cells and cell processes and splitting of myelin lamellae. Swelling of dendrites gives tissue a spongy appearance that is easily seen in semithin sections.

D. Dehydration and Embedding

The trimmed slices are dehydrated in graded ethanols according to the following sequence: 50%, 65%, 80%, 5 min each; 95%, 10 min; 100%, 15 min, twice. They are then transferred to propylene oxide (15 min, twice, in a fume hood), and finally to a mixture of equal parts of propylene oxide and an epoxide monomer mixture (Section VII.L). Absolute alcohol and propylene oxide are hygroscopic, and the vials must therefore be capped during those steps. The tissue is left in the last mixture overnight in tightly capped vials. All steps are performed with continuous mechanical agitation at room temperature. The diluted alcohols, however, may be chilled.

The following day, the slices are transferred to fresh monomer mixture, left for 4–8 hr on a rotator at room temperature, and then embedded (Fig. 3). The blocks are polymerized at 60°C for 2 days.

III. VARIATIONS

A protocol insuring good results for all areas of the central nervous system in all animals does not exist. The difficulty in establishing such a standard procedure results from the lability of the vascular system and its variability within the nervous system as well as among species. The most critical parameters of an intravascular perfusion are the pressure and flow of fixative in the capillaries of the area of interest; unfortunately, these parameters cannot be measured directly during a perfusion.

A considerable drop in pressure normally occurs in the circulation between the aorta and the capillaries. The drop reflects the resistance to flow in the arteries and arterioles that feed the capillaries. The flow resistance in these vessels changes markedly in response to the surgery related to the perfusion, and even if it is possible to determine the pressure delivered to the aorta during each perfusion, it is impossible to control or even determine the pressure delivered to the capillaries during the perfusion. In addition, only part of the fixative reaches the brain, while the rest is delivered to other parts of the body. The flow resistance of the vessels supplying the central nervous system varies independently of that of other parts of the circulation, and the division of the fixative between the nervous system and the rest of the body is thus both variable and unknown. Thus, even if two perfusions are characterized by similar aortic pressure and the same rate and total volume of flow, they may result in fixations of radically and unpredictably different quality.

The greatest sources of variability in fixation quality are the manipulations that lead to the perfusion and the dissection. Good results are likely when the surgery is performed with minimal bleeding and trauma, when the interval between interruption of respiration and beginning of fixation is brief, and when dissection does not traumatize the tissue. Complete failure of perfusion may occur when awkward surgery or slow cannulation induce total vascular collapse or when bubbles of air are introduced from the cannula into the aorta. Erratic and poor preservation of well-perfused tissue results from accidental mechanical straining of tissue during dissection and later processing.

A. Artificial Respiration

The use of artificial respiration with a mixture of 95% oxygen and 5% carbon dioxide (Palay *et al.*, 1962) can contribute substantially to a reliable and reproducible perfusion of mammals, in part because it allows more time for surgery, which can therefore be performed with great care, and in part because of the vasodilating action of the carbon dioxide. However, artificial respiration is not necessary, and, as it complicates the procedure and requires additional personnel, we have not included it in our basic fixation protocol. Nevertheless, artificial respiration is necessary in certain instances to insure good results. This is the case when the animal's ribs are too large to be severed quickly or when the aorta must be inspected before the size of the cannula can be determined. Artificial respiration is also recommended whenever very expensive or rare mammals are used. Artificial respiration is not necessary for frogs, since they respire in part through the skin, or for turtles, which are highly resistant to anoxia. Nor is it required for birds, which have a peculiar respiratory system.

Artificial respiration requires special equipment, and the surgical procedure must be modified. The head is fixed to the table with a string or a rubber band across the upper jaw. The initial midline incision of the skin is carried nearly to the jaw, and the skin is then separated by blunt dissection from the underlying structures and pulled laterally. The layers of connective tissue and muscle overlying the trachea are each split down the midline by blunt dissection using two pairs of toothed forceps to pull the two halves of each layer apart. The exposed trachea is incised between two rings of cartilage with a sharp scalpel, and a pair of sharp scissors is then used to cut through the cartilage in order to make a window large enough to admit the cannula. The cannula, with as large a bore as possible, is inserted and held by an assistant seated at the head of the animal. The assistant must watch the cannula carefully during the artificial respiration to prevent it from being pulled out of the trachea or pushed in too far.

The operation must be done with great care to prevent damage of the thyroid arteries, which lie immediately adjacent to the trachea. Mechanical irritation of the trachea and bronchi, furthermore, causes an accumulation of fluid in the lungs that may significantly reduce the effectiveness of respiration. This problem is minimized if the trachea is not pulled or twisted during its exposure and if sharp instruments are used to open it.

The artificial respiration, which is controlled by the assistant, must start immediately after the cannula has been inserted into the trachea. With the apparatus illustrated in Fig. 2, respiration is effected by alternately covering and uncovering the open end of the Y-tube at regular intervals. Gas is forced into the lungs under pressure when the open end of the Y-tube is covered and escapes from the lungs when the opening is uncovered. Alternatively, a small animal respirator may be used.

The depth of respiration is gauged by cues that vary during subsequent surgery. At first, movements of the abdominal wall are observed, then those of the diaphragm and adjacent viscera, and finally of the lungs themselves. Respiration should be rapid and deep enough to suppress reflex contractions of the diaphragm and the intercostal muscles. Darkening of the liver indicates inadequate oxygenation.

After initiation of artificial respiration, the abdomen and thorax are dissected as already described. Respiration should be continued until the heart is incised.

B. Ventilation with O_2–CO_2

An alternative to artificial respiration is to let the animal breath the mixture of 95% oxygen and 5% carbon dioxide. This is accomplished simply by sliding around its head a beaker into which tubing from the gas cylinder has been inserted and through which the mixture is flowing. The beaker is left in place through the surgery and removed after the thorax is opened or when cannulation is completed.

We strongly recommend this procedure for mammals, as it seems to improve the general quality of the perfusion and reduces variability without complicating the procedure. The improvement may be related to the vasodilation caused by the carbon dioxide, to the more efficient respiration, or both.

C. Pressure of Perfusion

The optimum delivery pressure of the perfusion apparatus should, as a rule, be similar to the physiological pressure in the aorta. However, it is also dependent on the size and shape of the cannula used, on constrictions in the tubing, and also, to a certain extent, on the region of interest. If the pressure is too low, the flow will be slow, and the tissue inadequately fixed. If it is too high, tissue disruptions may occur. Most areas of the central nervous system contain capillaries without fenestration, and these areas will tolerate a high perfusion pressure that causes the liver and intestines to swell grossly. Certain areas, however, do have fenestrated capillaries and must be perfused under lower pressure to reduce bulk filtration of fixative into the intercellular space.

We usually perfuse at pressures from 70 to 90 mm Hg (95 to 120 cm water) for birds and mammals.

D. Temperature of Perfusates

Although satisfactory results can be obtained with perfusates at room temperature or chilled in ice, we recommend that they be adjusted to body temperature, which is about 37–41°C for most mammals and birds. In our experience, isothermal perfusates yield more reproducible results and require somewhat less pressure than those that are colder than body temperature.

The temperature of the fluid delivered during the first few minutes is critical in establishing good perfusion, and the tubing with the fluid, as well as the large reservoirs, should be warmed. For mice, only the filled tubing need be warmed, since flow is comparatively slow. After 5–10 min of perfusion, the perfusate and the animal may be chilled with ice to reduce possible extraction of tissue components.

E. Vascular Rinsing Solution

Any balanced salt solution that is isosmotic to the animal's blood and at neutral pH can be used for the washing of the vascular tree. Calcium salts should be omitted if the perfusion fixative contains phosphate to prevent precipitation of calcium phosphate in the tubing of the perfusion apparatus. Solutions with carbon dioxide–bicarbonate buffer (Section VII.A) have an advantage because carbon dioxide is a cerebral vasodilator, and its inclusion in the vascular rinse may act to increase the flow of fixative to the brain.

It is essential to flush the blood from the vascular system before introducing a primary fixative that contains a potent coagulant such as osmium tetroxide. Glutaraldehyde and formaldehyde, however, act more slowly on the blood, and it is often possible to omit the rinse when they are used for primary fixation. We routinely use a vascular rinse buffered with bicarbonate, as it seems to contribute to reproducibility.

F. Composition of the Primary Fixative

Although osmium tetroxide was originally the primary fixative of choice, it has largely been replaced by formaldehyde and glutaraldehyde, which penetrate tissues more quickly than does osmium. The use of aldehydes also results in better fixation and preservation of microtubules. Formaldehyde and glutaraldehyde are commonly used in combination (Karnovsky, 1965); formaldehyde insures rapid penetration, and glutaraldehyde improves preservation. The proportion of the two aldehydes can vary within a wide range; for example, 4% formaldehyde (Pease, 1962); 4% formaldehyde and 0.5% glutaraldehyde (Vaughn and Peters, 1967); 1% formaldehyde and 1% glutaraldehyde (Wuerker and Palay, 1969); or 2.5% glutaraldehyde (Larramendi *et al.*, 1967). High concentrations of aldehydes are useful for fixation of white matter but may cause shrinkage of neuropil. The degree of overall shrinkage can be estimated during dissection by noting the amount of space between the brain and the cranial wall.

The exact proportion and concentrations of the two aldehydes are usually not the most critical factors for the quality of the fixation, provided that the combined concentrations are in the reasonable range of about 2–4%. The beginner is well advised to choose one formula and stick to it. Comparing results from different aldehyde mixtures is possible only when the quality of perfusions and dissections is consistently good, and this usually comes only after considerable experience.

Formaldehyde and glutaraldehyde cross cell membranes rapidly, and they do not exert strong osmotic effects on cells provided the total concentration of the aldehydes is less than 3–5%. The osmotic strength of a fixative containing dilute aldehydes is thus, in effect, nearly independent of the aldehyde concentration (the particle strengths of the aldehydes themselves) and depends only on the concentrations of other, osmotically active particles such as buffer and sucrose. For good results, the tonicity of the fixative, excluding the aldehydes, should approximate that of the animal's plasma or interstitial fluid. This value is about 320 mOsm for mammals and birds, but it varies by an order of magnitude among other vertebrates and the invertebrates. The nonreactive components of the fixative are called collectively the "vehicle" of the fixative (Maunsbach, 1966).

Areas with fenestrated capillaries, such as the area postrema, are not fixed well by the fixative mixture given in the Appendix (Section VII.C). Better fixation for such areas is obtained with a perfusion fixative whose total osmolarity, including the aldehydes, is near the tonicity of the blood, such as 2% glutaraldehyde in 0.06 M phosphate buffer, or by addition of colloidal agents such as dextran or polyvinylpyrrolidone to a fixative like that in the Appendix (Maunsbach, 1966).

A different aldehyde fixative is obtained by mixing glutaraldehyde and hydrogen peroxide (Peracchia and Mittler, 1972), a procedure that yields improved preservation of membranes. This fixative can be perfused, although oxygen bubbles may collect if the red cells are not thoroughly washed from the vasculature before introduction of the fixative (Williams and Jew, 1975). Formaldehyde cannot be included in this fixative.

G. Double Perfusion

Many workers have obtained good results from preparations perfused with a combination of two aldehyde fixatives, such as the one given in the

Appendix (Section VII.C), followed by a fixative containing aldehyde concentrations of 4–5%. As with so many procedural variants, it is difficult to know when double perfusion is appropriate.

We often use a second fixative that is isotonic to the first fixative and contains glutaraldehyde but not formaldehyde (Section VII.D). Our impression is that this leads to fewer membrane whorls, better retention of cytoplasmic ground substance, and more uniform distribution of cytoplasmic organelles; in addition, the brains also seem harder and more resistant to the stresses of dissection. While this procedure may make good material a little better, it will not make poor material good. For this reason and for simplicity, it is not included in the basic procedure of the previous section.

H. Buffer Wash and Postfixative

The buffer wash, which may be omitted for small blocks or very thin slices, removes free aldehydes that would otherwise react with the osmium in the postfixative. The type of buffer and the osmotic strengths of the buffer wash and postfixative markedly influence the shapes of synaptic vesicles (Valdiva, 1971). With the recipes given in the Appendix, mossy, climbing, and parallel fibers in the cerebellar cortex contain almost exclusively round synaptic vesicles, while Purkinje, stellate, and Golgi cell axons have many flattened vesicles. Optimal differentiation of the synaptic vesicles in another area might require a buffer wash and postfixative of a different tonicity. In our experience, the buffer wash and postfixative should be hypertonic to the vehicle of the perfusion fixative and isotonic to the total particle strength of the fixative, including the aldehydes. When the buffer wash and postfixative are isotonic only to the vehicle of the perfusion fixative, the tissue obtained appears swollen and extracted.

I. Stabilization with Uranyl Acetate

In some cases it is desirable to treat the tissue with uranyl acetate after osmication but before dehydration. This procedure yields enhanced membrane contrast and also stabilizes some membrane structures, including gap junctions (Brightman and Reese, 1969) and myelin (Revel and Hamilton, 1969). However, it also extracts glycogen (Vye, 1970) and cytoplasmic ground substance and should, therefore, be used with caution.

The phosphate of the postfixative must be washed out of the tissue before it is exposed to the uranyl acetate; otherwise, uranyl phosphate precipitates. If the washing is incomplete, a crust of white precipitate forms around each piece of tissue during its exposure to uranyl. In less severe cases, the crust may not be grossly evident but will be noticed in semithin and thin sections. Sometimes dense, rhomboidal crystals appear within inadequately washed tissue, especially in the astrocytes. After exposure to uranyl acetate, the tissue pieces must be washed again to remove free uranyl, which will otherwise inhibit polymerization of the epoxy embedding medium. Uranyl acetate is photoreactive and must, therefore, be protected from light. It should also be kept at or below room temperature, both during preparation and use, as the uranyl is unstable when warmed.

A typical schedule following postfixation in osmium tetroxide (Section VII.H) is: acetate buffer wash (Section VII.J), 10 min, five times; uranyl acetate treatment (Section VII.K), 1–2 hr; acetate buffer wash, 10 min, three times; dehydration and embedding as already described. These steps are all carried out with constant mechanical agitation, and the bottles should be covered or wrapped with foil to exclude light when they contain the uranyl. The procedure can be performed at room temperature but is better done on ice to reduce extraction. Tissue can be left overnight in the acetate buffer wash if desired.

J. Phosphate Precipitate

If phosphate ions are present during osmication, a fine granular precipitate sometimes forms on membranes and on the walls of blood vessels (Gil and Weibel, 1968). The precipitate, which is electron-dense, has about the same size as those that sometimes contaminate lead-staining solutions. The phosphate precipitate is easily diagnosed, as it is visible in unstained sections. It can be prevented by washing the tissue slices for 30 min in 0.8% NaCl with 8% glucose between osmication and dehydration. Such washing is not necessary if the slices are treated with uranyl acetate before dehydration.

K. A Procedure for Myelin

The animal is perfused with the vascular rinse (Section VII.A) and the first perfusion fixative (Section VII.C) for 10–15 min, followed by second perfusion fixative (Section VII.D) for 15 to 45 min. The initial perfusate is warmed to body temperature, but the second fixative is chilled with ice. Tissues are dissected out about 2 hr after the perfusion and are postfixed with osmium tetroxide and uranyl acetate as already described (Sections III.H,I). The animal is kept cold before and during the dissection, and all solutions in the postfixation and dehydration are chilled in ice. The tissue is brought to room temperature in 100% ethanol. Osmium ferrocyanide (Section VII.I) may be substituted for osmium tetroxide in the postfixation; in this case, slices should be thin (0.2–0.5 mm), as the osmium ferrocyanide penetrates poorly, and the uranyl block treatment can be omitted.

IV. EVALUATION OF RESULTS WITH THE LIGHT MICROSCOPE

The results of perfusion fixations are always somewhat variable, and it is difficult to obtain consistently good material for the electron microscopic analysis before one has learned to reject the bad and the mediocre material following light microscopic inspection. Therefore, tissue processed for electron microscopy should always be sectioned "semithin" at 1–2 μm, stained (Section VII.M), and examined in the light microscope for quality of fixation before time is invested in thin sectioning. Semithin sectioning can be done with great rapidity, and it is therefore advisable to section all or many of the blocks available from a perfusion, check them in the light microscope, and select the best-fixed ones for thin sectioning. All semithin sections should be saved. They can be stored without coverslips, or with coverslips mounted with immersion oil or epoxy-embedding medium. Permount may cause fading and should therefore not be used.

Fixation is never uniform within a tissue slice. Zones of poor fixation always extend 25 to 50 μm or more into the block from every cut face. Areas adjacent to natural surfaces formed by the pia or the ventricular system also often show artifacts. These effects, which are probably caused by mechanical or osmotic damage during dissection and subsequent manipulations, manifest themselves most consistently as diffuse swelling. For best results, it is essential that all sections be taken at least 50 μm from surfaces cut before embedding. When tissue appears poorly fixed, we cut deeper.

Some fixation artifacts such as separation of tissue constituents by artifactual spaces (Fig. 4) and dark cells (Fig. 5), are immediately evident in the light microscope. Other artifacts are less conspicuous, at least in the semithin sections, and it requires some experience to detect them. The perikarya of neurons should be uniformly filled with organelles. Pale spots within or halos around cell bodies indicate focal displacement or swelling of organelles or gross shrinkage of the perikaryon. The nuclear chromatin, which is usually finely dispersed and featureless in well-preserved tissue, has a tendency to form clumps in poorly fixed material. Fixation of neuropil is judged chiefly by the texture and intensity of staining. Poor fixation generally results in the swelling of some or all cell processes, causing reduced staining. If different cell types are differentially affected by the fixation, the tissue often appears spongy (Fig. 6).

Dendrites, which are especially sensitive indicators of fixation quality, may either swell or shrink as a result of poor fixation. Although uniform swelling or shrinkage may occur, both fixation artifacts are usually present and give the dendrites a beaded appearance. This artifact is well known to practitioners of the Golgi method, and it can usually be detected in the semi-thin sections.

In some areas, for example, the cerebellar cortex, the large dendrites have about the same density as the surrounding neuropil; the small dendrites are not apparent, although they are revealed by their elongated mitochondria. Pallor or darkness of the larger dendrites suggests swelling or shrinkage. Dendrites in other areas, including the cerebral cortex, are often pale. However, they may look different, and the significance of any particular appearance must be established by coordinated light and electron microscopy.

The preservation of myelin sheaths presents a special problem. The

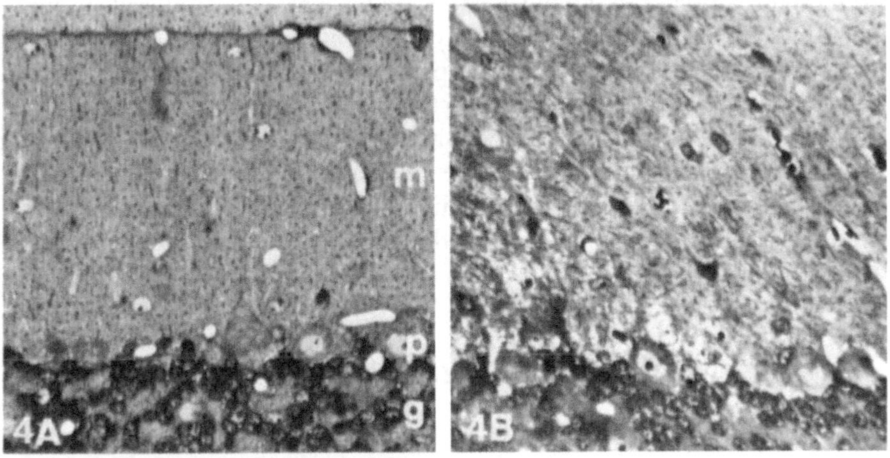

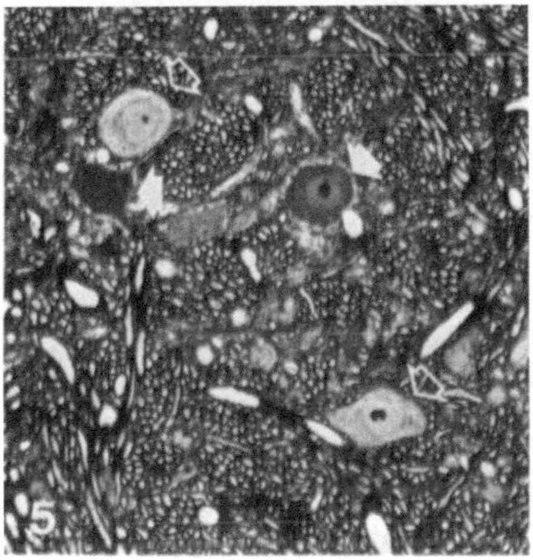

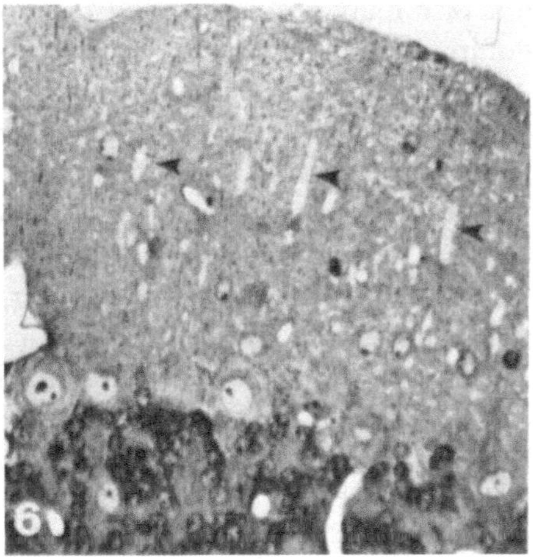

Figure 4. Mouse cerebellar cortex, 1.5 μm sections. (A) In well-fixed specimens, the neuropil in the molecular (m), Purkinje cell (p), and granular (g) layers appears more or less uniformly dense. The small dark dots and rods are mitochondria. The blood vessels are empty and appear white in the picture. (B) Perfusion was poor in this specimen, as indicated by the presence of red blood cells within the vessels. Apparently empty spaces are prominent next to the Purkinje cells and also occur elsewhere; they represent the severely swollen perikarya and processes of astrocytes. Such swelling is common when the fixative does not reach the tissue quickly.

Figure 5. Section of mouse cerebellar nucleus, 1.5 μm thick. The nuclei and perikarya of neurons in the cerebellar nuclei usually appear pale or moderately stained (open arrows); dark-staining neurons (filled arrows) result from trauma during the dissection. Dark neurons appeared in several of our specimens from normal mice, even though the tissue was well-perfused and not dissected until the day following the perfusion. We eliminated the dark neurons by changing the order of the final slicing during the dissection and the manner in which we held the tissue during the slicing.

Figure 6. Section of mouse cerebellar cortex, 1.5 μm thick. Several large, pale processes are visible in the molecular layer (arrows); these are the swollen dendrites of Purkinje cells. The perikarya of the Purkinje cells, however, are not conspicuously swollen. The molecular layer in the upper part of the picture has a spongy appearance because of the presence of many small, swollen profiles (compare with Fig. 4A).

membrane lamellae often separate from one another, even in otherwise well-preserved material. On the other hand, we have also encountered specimens whose myelin sheaths were reasonably well preserved, but in which the neuronal elements were poorly fixed. Apparently the myelin sheaths differ from other tissue components in their reaction to fixation, and they may require special conditions for optimal preservation. Improvements in the technology of fixation may eventually eliminate this disparity, but for the present it seems sensible to discount the condition of myelin sheaths in evaluating fixed material unless the myelin sheath itself is the main target of the investigation.

V. CUTTING AND STAINING OF ULTRATHIN SECTIONS

After having selected blocks of well-fixed tissue, one should trim them to a size manageable for thin sectioning, usually 1 mm^2 or less. Exact positioning and orientation of the trimmed block face in relation to structures in the tissue is usually essential. If one has carefully recorded the process of dissecting and slicing tissue before embedding, it remains only to control and document the further trimming.

One begins with the semithin section of the full block face, which is drawn or photographed. The desired position of the reduced face for thin sectioning is marked on the drawing or photograph, which is then used as a guide during the trimming itself.

The block is viewed under a dissecting microscope. Such features as blood vessels, fiber tracts, and neuropil are usually revealed by slight changes in the elevation of the cut surface or by differences in the texture of the surface. If such landmarks cannot be identified, the block should be replaced in the microtome and sectioned semithin with a knife already overdulled by use. This roughens the surface and enhances the visibility of landmarks. The landmarks are then used to find exactly the area desired for thin sectioning, the sides of the desired reduced face are marked on the block face with a pencil or razor blade, and the trimming is performed.

Trimming can be done freehand with a razor blade or with glass knives on a microtome. It may also be accomplished with the LKB pyramotome, an excellent but expensive device. An especially useful technique is to cut away with a glass knife about 25 to 100 μm of tissue and plastic around the area for thin sectioning. This creates a shallow "mesa" for thin sectioning; after sections have been collected and photographed, one may collect another semithin section of the full piece of tissue, destroying the mesa in the process, and trim down to another area from the same block. This process may be repeated many times and allows sampling over a wide area of a tissue slice but with small block faces. A full discussion of the trimming of blocks and of ultrathin sectioning is presented by Reid (1975).

A semithin section is collected from the reduced face immediately after trimming, and the exact position of the pyramid is determined by comparing this section with the previously cut full-face section. The position is recorded on a drawing or photograph. Other semithin sections should be collected immediately before and after thin sectioning; these are used as aids for orientation during electron microscopy as described in Chapter 11. The semithin sections should be saved and carefully labeled so that their sequence can be kept.

Obstruction of thin sections by grid bars is minimized or eliminated by using large mesh, slot, or single-hold grids. These must be covered by a supporting film before the sections are picked up (see, e.g., Pease, 1964). Preparing good films is very tricky, yet critical; it is best that a person who does it successfully routinely prepare films for everyone in a laboratory.

Recipes and a protocol for staining thin sections are given in Section VII.N.

VI. SYNTHESIS

The interactions between fixative and tissue are complex and poorly understood, and much basic information is lacking; protocols for preparing tissues for electron microscopy are derived empirically with no detailed theoretical basis. Inevitably, every microscopist tries to improve his protocol. In comparing the procedures used in different laboratories, one finds bewildering variation in the composition of fixatives and postfixatives, in the schedules for postfixation and embedding, in the dehydrating agents and embedding resins, and in other features. The procedure and formulas given here are representative of the common practice and have produced good results; but we do not suggest that they are better than others for any particular application. The common procedural variations often have little effect on the quality of the results; but to be sure of the best, one must sooner or later try them. However, we strongly urge that this step be taken later rather than sooner. Inadequate skill at perfusing and dissecting causes large variability in results, and valid comparisons among fixatives of different composition are difficult if not impossible in the face of this underlying variability. Furthermore, improvements in results may wrongly be ascribed to some minor variation in procedure, when the real cause is an improvement in the experimenter's skill.

Mastery of the techniques of perfusion and dissection develops with practice. It is best, at the very beginning, to perfuse a number of animals with a simple protocol with no vascular rinse and one perfusion fixative at room temperature. Discard the animals until the operation is performed smoothly and quickly, with each step in the right order and a minimal interval of anoxia. The cranium of each perfused animal should be opened; the brain should appear cleared of blood and, after 10 min of perfusion with fixative, should be noticeably firmer than fresh brain tissue. When the surgery runs smoothly, a specimen should be dissected, and a number of pieces should be postfixed, embedded, and sectioned for light microscopy. The quality of fixation should be evaluated with the light microscope by the criteria given above, and the best-fixed piece should be examined in the electron microscope. The results can be compared with published electron micrographs, for example in Peters *et al.* (1976).

Perfusions under the same conditions should routinely give about the same results. If they do not, one should include a vascular rinse in the procedure, adjust the rinse and fixative to body temperature, and, for mammals, add ventilation with O_2–CO_2 or artificial respiration. These elaborations should only be included when the surgery itself has become routine. The most likely causes of poor reproducibility are poor health of the animals, variations in the tempo and precision of the surgery, and inadvertent stressing of tissue during dissection. The last two problems are corrected by practice and intense, careful attention to detail.

When a reasonable degree of reproducibility has been achieved, variations in procedure should be tried and evaluated. There are many potential variables, but the following seem to us most important: procedure for dissection, delivery pressure of the fixative during the perfusion, composition of the fixative and the use of a second perfusion fixative, and duration of fixation.

VII. APPENDIX

A. Vascular Rinse

1. Composition for Mammals, Birds, and Reptiles

NaCl	8.5 g/1000 ml
KCl	0.25 g/1000 ml
$NaHCO_3$	0.20 g/1000 ml

2. Composition for Amphibians

NaCl	6.5 g/1000 ml
KCl	0.14 g/1000 ml
$NaHCO_3$	0.20 g/1000 ml

3. Preparation

The ingredients are dissolved in water and the solution made to volume. The pH should then be about 7.7. The pH is adjusted to 6.8–7.0, 10 min or less before the perfusion by bubbling carbon dioxide or a mixture of carbon dioxide and oxygen through the solution. The carbon dioxide passes out of the solution with time, and the pH is not stable.

B. 0.4 M Phosphate Buffer Stock

1. Composition per Liter

$NaH_2PO_4 \cdot H_2O$	15.9 g
Na_2HPO_4	40.5 g

2. Preparation

The ingredients are dissolved in water, and the solution brought to volume. pH is 7.2.

C. First Perfusion Fixative

1. Composition per Liter

Formaldehyde (paraformaldehyde, depolymerized)	10 g
Glutaraldehyde (25% stock solution)	40 ml
Phosphate buffer (Section VII.B)	300 ml

2. Preparation

1. Paraformaldehyde (trioxymethylene) is an insoluble polymer of formaldehyde. It is depolymerized, yielding formaldehyde, as follows. For *W* grams paraformaldehyde, heat about 10*W* ml water to steaming (75°C) in a covered beaker; remove from heat and add the paraformldehyde (it remains in suspension). Add 1–2 drops 5 N NaOH per 100 ml of suspension with stirring. The solution clears within 1 min (slight cloudiness persists). The hot solution must be handled in a fume hood and cooled rapidly after the depolymerization; the cooling can be done by proceeding immediately to the following step.
2. Mix formaldehyde, buffer stock, and water to about 75% of final volume in a graduated cylinder of volumetric flash; add the glutaraldehyde. Add water to about 95% of final volume.
3. The pH should be 7.2 to 7.4; if not, adjust with NaOH or HCl. Add water to make final volume.

4. Filter the fixative with a fine paper (e.g., Schleicher and Schuell No. 576); this can be done rapidly with suction and a Büchner funnel. If the fixative is noticeably cloudy before filtration or filtration is very slow, the fixative should be filtered first with a coarse paper and then with a fine one.

3. Remarks

Formaldehyde is prepared fresh from paraformaldehyde because commercial formaldehyde contains methanol. Glutaraldehyde should be of high purity. Glutaraldehyde can be purified by filtration with activated charcoal (Fahimi and Drochmans, 1965; Anderson, 1967); the pure solution is colorless.

Fixative should be prepared the day it is used or the day before; for overnight storage, it should be refrigerated.

If $CaCl_2$ is desired, it should be added as a dilute (0.05%) solution in Step 2. Calcium precipitates with phosphate.

Caution: Formaldehyde and glutaraldehyde are volatile, toxic, and allergenic. Make the fixative in a fume hood and wear gloves. Be especially careful of the concentrated glutaraldehyde and formaldehyde solutions. The fixative should be made and used only in areas with brisk exhaust ventilation; this includes the perfusion and dissection procedures. Clean up spills and surfaces contaminated with fixative thoroughly and immediately. Carelessness in these precautions endangers the health of others.

D. Second Perfusion Fixative

1. Composition per Liter

Glutaraldehyde (25% stock solution)	172 ml
Phosphate buffer (Section VII.B)	300 ml

2. Preparation

Mix glutaraldehyde and buffer stock and water to 95% of final volume; check and adjust pH; make to final volume; filter. Refrigerate for overnight storage.

Caution: As for first perfusion fixative, make and use the fixative only with adequate ventilation, and clean contaminated surfaces thoroughly.

E. Phosphate Buffer Wash

1. Composition per 100 ml

Glucose	8.0 g
Phosphate buffer (Section VII.B)	30 ml

2. Preparation

Mix ingredients and make to volume. Final pH is 7.2. The solution keeps for weeks in the refrigerator.

F. Double-Strength Osmication Buffer

1. Composition per 100 ml

Glucose	14 g
Phosphate buffer (Section VII.B)	60 ml

2. Preparation

Mix ingredients and make to volume. The solution keeps for weeks in the refrigerator.

G. 4% Osmium Tetroxide Stock Solution

The solution is prepared by dissolving 1 g osmium tetroxide in 25 ml of water. Prepare the solution in a glass-stoppered, wide-mouth glass bottle. The bottle and stopper are washed with detergent or cleaned in acid and thoroughly rinsed with running tap water followed by distilled water. If a sonicator is available, the bottle should be filled to the brim with distilled water, the stopper emplaced, the two sonicated for a few minutes, and the water discarded. These measures help to prevent spoilage of the osmium by contamination in the bottle. The vessel used for measuring the distilled water should be treated in the same way. Osmium tetroxide is solid at room temperature and is supplied in sealed glass ampules. To prepare the solution, ampules are cleaned thoroughly, broken, and dropped into the appropriate amount of distilled water in the clean bottle. To assure complete dissolution and uniform dispersal of the osmium, it may be necessary to fragment the body of the ampules as well as to break their necks. The bottle is capped and sonicated or left standing until the osmium is dissolved.

The stock solution, which should be protected from light, can be stored for weeks at room temperature or in a refrigerator if the container is clean. The top of the bottle should be sealed by wrapping with plastic film, and the bottle should be kept within a larger sealed container; even so, fumes will escape and blacken nearby objects.

Caution: Osmium tetroxide is volatile, highly reactive, and especially dangerous to the eyes and respiratory system. The stock solution and fixative should be prepared and used only in a fume hood, and one should wear gloves. Bottles and vials containing this reagent should be sealed well and wiped on the outside if they are removed from the hood. When poured down a sink, solutions containing osmium tetroxide should be followed by abundant running water; the sink itself should be within a fume hood.

H. Osmium Tetroxide Postfixative

Combine 5 ml each of 4% osmium tetroxide solution (Section VII.G) and double-strength osmication buffer (Section VII.F). Prepare this solution immediately before use and protect from light.

Caution: Make and use the postfixative in a fume hood as described in Section VII.G.

I. Osmium Ferrocyanide Postfixative

1. Composition per 10 ml

OsO_4 (Section VII.G)	2.5 ml
H_2O	2.5 ml
$K_4Fe(CN)_6 \cdot 3H_2O$	150 mg
Double-strength osmication buffer (Section VII.F)	5 ml

2. Preparation

Mix osmium tetroxide stock solution, water, and ferrocyanide; add double-strength buffer. Protect the solution from light.

Caution: Make and use the solution in a fume hood as described in Section VII.G.

J. Acetate Buffer Wash

1. Composition per 100 ml

Glucose	8 g
Sodium acetate (0.1 M)	70 ml
Acetic acid (0.1 M)	30 ml

Mix ingredients; pH is 5.0.

K. Uranyl Acetate Block Treatment

1. Composition per 100 ml

Glucose	7.5 g
Uranyl acetate	2 g
Sodium acetate (0.1 M)	100 ml

2. Preparation

Protect the solution from light while mixing and using. pH is 5.0–5.2; adjust with NaOH or HNO_3 if necessary.

L. Epoxide Embedding Mixture

1. Composition

LX-112	55g
DDSA	15 g
NMA	30 g
DMP-30	2 ml

2. Preparation

A disposable 250-ml beaker is placed on a balance, and the first four reagents added. The DMP-30 is added in a fume hood with a syringe; disposable plastic syringes are satisfactory and can be discarded without cleaning. The necks of the bottles should be wiped before the bottles are recapped to prevent the caps from sticking later; use separate paper towels for each bottle.

The ingredients are mixed by stirring with wooden sticks. Stirring should be continued for at least twice the time necessary for the mixture to appear uniform (about 10–15 min). Incomplete mixing is a principal source of embedding failure.

The mixture should be prepared immediately before making the mixture with propylene oxide and again when the tissue is first immersed in the undiluted monomer, because its viscosity increases progressively. Low viscosity is thought to minimize damage to specimens during infiltration.

Blocks are hardened at 60°C for 2–3 days.

3. Comments

The suggested embedding mixture has given us good results when sectioned with glass and diamond knives. Its hardness can be varied by changing the relative amounts of NMA and DDSA, keeping their total weight constant. Many other embedding mixtures are in use (e.g., Glauert, 1975). This mixture is from Mascarro *et al.* (1979).

4. Reagents

DDSA is dodecenylsuccinic anhydride, NMA is nadic methyl anhydride, and DMP-30 is tri-(dimethylaminomethyl)-phenol. LX-112 is made by Ladd Research Industries, Inc. (P.O. Box 901, Burlington, Vermont 05402); the other reagents are sold by a number of dealers in electron microscopy supplies.

The embedding reagents deteriorate on exposure to the moisture in air, yielding blocks that are too soft. Bottles in use should be kept at room temperature. If they are stored cold, they should be warmed to room temperature before opening to prevent condensation of water. Bottles in current use should be discarded in favor of fresh stock if they yield soft blocks. The DMP-30 is a frequent offender; it should be stored in lots of 10 to 20 ml, and a new lot opened every month or two in dry weather and much more often in humid weather.

Caution: Some embedding reagents give off toxic fumes, and others cause contact dermatitis; they all should be handled in a fume hood if possible and always with gloves. Surplus resin should be hardened before discarding. Contaminated gloves and paper towels should be deposited in plastic bags, and the bags closed before discarding. Contaminated surfaces can be cleaned with 95% ethanol.

M. Mounting and Staining Semithin Sections

1. Preparation

Dissolve toluidine blue O (0.1%) and sodium borate (1%) in water. The stain should be filtered after it is made. For optimal result it should also be filtered again immediately before use, but this second filtration can usually be omitted for routine work. Penetration of the stain into sections 2 μm or thicker is incomplete but can be increased by adding dimethylsulfoxide to the mixture.

2. Procedure

Glass slides are cleaned by dipping them in 95% ethanol and wiping dry immediately before use. Sections 0.5 μm or thicker are cut, picked up dry with forceps, and dropped onto small pools of distilled water on the slides. The slides are then placed on a hot plate at 60–80°C and allowed to dry. The heat causes the sections to spread on the water, so they dry flat. However, bubbles will form and be trapped between the sections and the slide if the temperature is too great.

The dried slides are cooled to room temperature, the sections covered with stain from a dropper or syringe, and the slides replaced on the hot plate. They are removed when the stain is steaming, the stain is rinsed from the slide with a stream of distilled water, and the slides are dried on the hot plate. Sections sometimes detach from the slides during staining and washing; in this case, the drying time should be increased. However, drying at high temperature or for too long a time reduces staining.

The sections can be examined immediately in the microscope; a coverslip of thickness correct for the microscope's lenses should be dropped over the sections to reduce spherical aberration.

Permanent mounting of coverslips can be achieved by putting a drop of epoxy embedding mixture (Section VII.L) over the sections, adding the coverslip, and leaving the slide overnight at 60°C. This procedure removes nonspecific stain from the section and conceals surface roughness. If too much clearing occurs, slides should be kept at 60°C for shorter periods and then left at lower temperature for the epoxy to harden. The epoxy embedding mixture is preferable to other mounting media, some of which cause fading with time. The excess mixture from embedding tissue can be saved in small vials in a freezer, warmed as needed, and transferred with toothpicks to the slides. Only the amount needed for spreading under the coverslip should be applied.

N. Staining Thin Sections

1. Solutions

Uranyl Acetate

Uranyl acetate	0.4 g
50% Ethanol	10 ml

Lead Citrate

Lead citrate	10 mg
NaOH (10 N, carbonate-free)	0.1 ml
H_2O	9.9 ml

2. Preparation

Put the uranyl acetate and 50% ethanol in a centrifuge tube wrapped with foil and shake it for 5 to 10 min; all but a trace of uranyl should be dissolved. Put the lead citrate and water in a centrifuge tube, add the sodium hydroxide, cap, and shake until the lead citrate is dissolved (about a minute). The strains are centrifuged at about 2000 rpm (about 400 *g*) for 15 min immediately before use and are kept wrapped in foil or otherwise protected from light.

Uranyl acetate decomposes at 37°C, and one must take care not to warm the solution with one's hand while shaking it. The 50% ethanol should be at room temperature when the uranyl acetate is added and must therefore be made in advance or with cold alcohol or water, since alcohol and water are heated by mixing.

The natural mixture of uranyl isotopes is radioactive, and the stain should be handled in accordance with radiation safety regulations unless special nonradioactive uranyl acetate is used.

The stains can be kept for a few weeks if refrigerated and protected from light.

3. Procedure

Grids are floated or immersed (edge first) in drops of stain on clean, disposable surfaces such as Parafilm®: uranyl acetate, 5 to 10 min followed by lead citrate, 2 to 5 min. The drops are covered with inverted petri dishes, and the uranyl stain with an opaque cover as well. The grids should be rinsed and dried after each of the two stains. Rinsing is done by immersing the grids in a succession of several beakers of distilled water or in a gentle stream of water directed down the tines of the forceps. The fluid between the tines of the forceps should be removed with wedges of filter paper at least once during each rinsing and again immediately before each drying. Grids are dried by drawing the water off them with filter paper. Excessive rinsing may remove stain.

The lead stain draws carbon dioxide from the air and forms lead carbonate precipitate, which contaminates grids. To prevent this, the tube should be kept capped except when withdrawing stain (never return either stain to its tube), staining should be done in still air, the drops of lead stain should be covered with a petri dish as soon as they are deposited, and one should avoid breathing on the stain. The presence of a globular precipitate, 100 nm or more in diameter, on sections indicates that the stain in the tube is bad or that the drop absorbed carbon dioxide. A smaller precipitate of diameter 10 nm or less and distributed on membranes and other dense structures indicates overstaining with lead; the staining time or concentration of lead citrate should be reduced. If the small precipitate persists, it may be from the phosphate buffer, as discussed earlier.

Angular crystals come from the uranyl stain. These sometimes occur within tissue treated with uranyl during postfixation; in this case, the crystals are present in unstained sections. If angular crystals are present only on stained grids, then try more thorough rinsing; if that fails, make a new batch of stain.

Except for the specific patterns just described, contamination of sections usually comes from sources other than the stains, most commonly the water on which the sections are floated during thin sectioning. Troughs, even of glass knives, should be rinsed before use with a stream of distilled water directed to prevent its running over the cutting edge. Forceps and other instruments for manipulating sections and grids should be cleaned in acetone or ethanol before use and stored on a clean paper towel. Grease from fingers is transmitted to the surface of the water if one accidentally touches the water or if one touches parts of instruments which later contact the water. Another source of contamination is the razor blade used for trimming blocks and, especially, pyramids; razor blades are often supplied coated with oil and should be cleaned in acetone before use. Otherwise, the oil is transferred from razor blades to the specimen and thence to the water and the sections floating on it.

The surface of the water should be cleaned often during sectioning, and especially whenever a film appears or fragments of sections are present; draw a piece of filter paper or a hardened tissue across it. This soaks up the water and contamination at the surface. In addition, it is often advisable to change the water in the trough during sectioning. This can be done without removing the knife from the microtome or changing its position by using two syringes. First, move section pieces away from the cutting edge with a hair and clean the surface of the water with paper. Then, remove the old water with a syringe used only for removing water, and refill the trough with another syringe used only for adding clean water.

Much attention and care should be devoted to keeping the water in the trough clean; sections taken from dirty water are inevitably contaminated.

ACKNOWLEDGMENTS. Supported by USPHS Grant NS 09904. The authors wish to thank Ms. Mary Wright Goss and Ms. Anne-Lise Dahl for their skillful assistance.

REFERENCES

Agar, A. W., Alderson, R. H., and Chescoe, D., 1974, *Principles and Practice of Election Microscope Operation,* North-Holland, Amsterdam.

Anderson, P. J., 1967, Purification and quantitation of glutaraldehyde and its effect on several enzyme activities in skeletal muscle, *J. Histochem. Cytochem.* **15:**652–661.

Brightman, M. W., and Reese, T. S., 1969, Junctions between intimately apposed cell membranes in the vertebrate brain, *J. Cell Biol.* **40:**648–677.

Cammermeyer, J., 1962, An evaluation of the significance of the "dark neuron," *Ergeb. Anat. Entwickl.-Gesch.* **36:**1–61.

Fahimi, H. D., and Drochmans, P., 1965, Essais de standardisation de la fixation au glutaraldehyde. I. Purification et determination de la concentration du glutaraldehyde, *J. Microscopie* **3:**725–736.

Gil, J., and Weibel, E. R., 1968, The role of buffers in lung fixation with glutaraldehyde and osmium tetroxide, *J. Ultrastruct. Res.* **25:**331–348.

Glauert, A. M., 1975, *Fixation, Dehydration and Embedding of Biological Specimens,* North-Holland, Amsterdam.

Hayat, M. A., 1970–1978, *Principles and Techniques of Electron Microscopy, 8 Vols.,* Van Nostrand-Reinhold, New York.

Karnovsky, M. J., 1965, A formaldehyde–glutaraldehyde fixative of high osmolality for use in electron microscopy, *J. Cell Biol.* **27:**137A.

Larramendi, L. M. H., Fickenscher, L., and Lemkey-Johnston, N., 1967, Synaptic vesicles of inhibitory and excitatory terminals in the cerebellum, *Science* **156:**967–969.

Mascarro, J. A., Knox, C. A., and Underwood, E., 1979, Morphological evaluation of tissues embedded in LX-112 (an Epon replacement) and in LX-112/Araldite combination, *37th Annual Proceedings of the Electron Microscopy Society of America* (G. W. Bailey, ed.), p. 346.

Maunsbach, A. B., 1966, The influence of different fixatives and fixation methods on the ultrastructure of rat kidney proximal tubule cells. II. Effects of varying osmolality, ionic strength, buffer system and fixative concentration of glutaraldehyde solutions, *J. Ultrastruct. Res.* **15:**283–309.

Mugnaini, E., 1965, "Dark cells" in electron micrographs from the central nervous system of vertebrates, *J. Ultrastr. Res.* **12:**235–236.

Palay, S. L., McGee-Russell, S. M., Gordon, S., and Grillo, M. A., 1962, Fixation of neural tissues for electron microscopy by perfusion with solutions of osmium tetroxide, *J. Cell Biol.* **12:**385–410.

Pease, D. C., 1962, Buffered formaldehyde as a killing agent and primary fixative for electron microscopy, *Anat. Rec.* **142:**342.

Pease, D. C., 1964, *Histological Techniques for Electron Microscopy, Second Edition,* Academic Press, New York.

Peracchia, C., and Mittler, B. S., 1972. Fixation by means of glutaraldehyde–hydrogen peroxide reaction products, *J. Cell Biol.* **53:**234–238.

Peters, A., Palay, S. L., and Webster, H. DeF., 1976, *The Fine Structure of the Nervous System: The Neurons and Supporting Cells,* W. B. Saunders, Philadelphia.

Reid, N., 1975, *Ultramicrotomy,* North-Holland, Amsterdam.

Revel, J.-P., and Hamilton, D. W., 1969, The double nature of the intermediate dense line in peripheral nerve myelin, *Anat. Rec.* **163:**7–16.

Valdivia, O., 1971, Methods of fixation and the morphology of synaptic vesicles, *J. Comp. Neurol.* **143:**257–273.

Vaughn, J. E., and Peters, A., 1967, Electron microscopy of the early postnatal development of fibrous astrocytes, *Am. J. Anat.* **121:**131–152.

Vye, M. V., 1970, The morphological alteration of particulate glycogen by *en bloc* staining with uranyl acetate, *J. Ultrastruct. Res.* **33:**291.

Williams, T. H., and Jew, J. Y., 1975, An improved method for perfusion fixation of neural tissues for electron microscopy, *Tissue Cell* **7:**407–418.

Electron Microscopy 11

Identification and Study of Normal and Degenerating Neural Elements by Electron Microscopy

ENRICO MUGNAINI and VICTOR L. FRIEDRICH, Jr.

I. INTRODUCTION

The electron microsope is now an established tool in experimental neuroanatomy, and many laboratories have long used it routinely to complement studies on connectivity based on light microscopic degeneration methods. The results of these studies have contributed significantly not only to the definition of cytological changes accompanying anterograde degeneration in general but also to the knowledge of the synaptic organization of specific systems.

It seems appropriate to point out that the use of electron microscopy in experimental neuroanatomy is not restricted to the study of neuronal degeneration provoked by experimental lesions. Valuable information relating to

ENRICO MUGNAINI and VICTOR L. FRIEDRICH, Jr. • Laboratory of Neuromorphology, Department of Biobehavioral Sciences, The University of Connecticut, Storrs, Connecticut 06268.

neuronal connectivity can, for instance, be obtained if electron microscopy is used to localize radioisotopes in autoradiographic studies or to identify neuronal profiles and cellular organelles filled with horseradish peroxidase in studies where this enzyme is used for tracing pathways. More recently, electron microscopy has also been used for identifying neurons impregnated by the Golgi method or "stained" by intracellular injected tracers or by immunocytochemical methods. Since the application of electron microscopy to autoradiography, Golgi impregnation, and immunocytochemistry will be dealt with in Chapters 5, 9, and 14, the present chapter will be concerned primarily with the identification of degenerating nervous tissue elements. We will not deal specifically with degeneration of neural cells after administration of neurotoxins; however, insofar as neurons degenerating for whatever reason often become electron-dense, the present review also applies to that experimental neuropharmacological approach.

To appreciate the ultrastructural characteristics of degenerating nerve cells, one must first have a good understanding of the criteria used in the identification of normal nervous tissue elements. Many excellent descriptions of the ultrastructural appearance of normal nervous tissue are available, and we would urge the novice to consult a recent volume on the subject (Peters *et al.*, 1976).

II. BRIDGING THE GAP BETWEEN LIGHT AND ELECTRON MICROSCOPY

After tracing a fiber projection to its termination by means of a light microscopic method, the next logical step is to confirm the termination electron microscopically and to describe the nature of the synaptic contacts in detail. Before this can be done properly, the target area should be examined in a variety of other light microscopic preparations to gain familiarity with its topography and composition. For instance, Nissl-stained material provides information regarding cell size and shape, nucleus-to-cytoplasm ratio, amount of Nissl substance, number of nucleoli, and presence of any nuclear indentations. Myelin-stained preparations offer a good picture of the distribution of myelinated fibers in the region. Golgi-impregnated material offers an unexcelled view of whole neurons and their dendritic pattern.

The task of analyzing experimentally altered neuronal profiles with the electron microscope is severely restricted by sampling problems unless the target area is reasonably large or the experimental lesion has produced widespread and abundant degeneration in the target tissue. Unless the target area for analysis is carefully selected, one may select blocks for electron microscopy only to find that the ultrathin sections contain either scanty degeneration or none at all.

In the past, ultrastructural studies of silver-impregnated frozen sections have been performed to demonstrate the location of silver precipitates previously identified by light microscopy. Usually sections stained by one of the silver methods are postfixed with osmium and embedded in resin. Unfortunately, the presence of nonspecific silver deposits throughout the tissue and unsatisfactory preservation of the ultrastructure have limited the value of this approach merely to the identification of those cytological elements that are argyrophilic. Synaptic detail is often difficult to assess in these preparations.

Considerable improvement in bridging the gap between light and electron microscopy was achieved with a modified Fink–Heimer method in which semithin resin-embedded sections are silver impregnated directly (Heimer, 1969). The advantages of the latter technique have remained largely unexploited. However, since the search for degenerating elements has to be done on the blocks of tissue embedded for electron microscopy, this method does not solve the sampling problem completely, and another procedure was therefore introduced by Brand and Mugnaini (1975). Following an appropriate survival time determined by previous light microscopic analysis, the experimental animal is perfused according to the procedure recommended in Chapter 10. After 12–24 hours, the brain is dissected out and sliced on a Vibratome (Oxford Instruments, Inc., San Mateo, California). With this instrument, thin slices of tissue are sectioned in a bath of cold buffer (with or without formaldehyde), thus eliminating freezing artifacts. The sections are cut by a vibrating razor blade that is slowly advanced through the slab of tissue with minimal mechanical damage. The sections are then carried through the various steps of one of the Fink–Heimer procedures (see Chapter 3) and examined wet, one at a time, in the light microscope. Selected areas are dissected out or punched away, osmicated, and embedded for ultrathin sectioning (see Chapter 10) and meticulous prepara-brain tissue are scanned, and appropriate areas, no matter how small, can be selected for electron microscopy. If one perfers not to have silver precipitates in the ultrathin section, alternate sections can be used for light and electron microscopy. The method can also be used for other procedures, such as autoradiography.

As a matter of routine, appropriate care in collecting the tissue to be embedded for ultrathin sectioning (see Chapter 9) and meticulous preparation and study of toluidine-blue-stained semithin sections prior to thin sectioning are sufficient to provide an adequate transition between light and electron microscopic analysis. A general rule, to which one should strictly adhere at the onset of a study, is to make a semithin section of the whole tissue block and then trim a smaller area using the mesa technique (Heimer, 1970). A second semithin section is then cut, stained, and photographed or drawn with attention to landmarks such as fiber bundles, capillaries, or cell groups. Ribbons of ultrathin sections are collected on serially numbered grids. Since mesh-type grids may obscure some of the landmarks, it is often advisable to use single-hole grids covered by a formvar film and reinforced with carbon. The whole section is then easily surveyed in the electron microscope, and the position of individual electron micrographs can then be noted directly on the drawing or on the light photomicrograph of the semithin section. Since as many as 10–15 thin sections occupy only 1 μm of the tissue, any semithin section preceding or following the series of thin sections will contain the same cell bodies, large cell processes, and blood vessels that are visible in the electron microscope. If possible, the block should be left mounted in cutting position until the ultrathin sections have been analyzed in the electron microscope. If serial electron micrographs are deemed necessary, one can thus proceed with the thin sectioning without having to realign the block.

III. PRACTICAL GUIDELINES FOR ELECTRON MICROSCOPY

The importance of photography in electron microscopy cannot be overemphasized. Since thin sections are progressively degraded by the electron beam, micrographs provide the only lasting record of electron microscopic experiments. Furthermore, the micrographs reveal details not easily visible on the relatively dim viewing screen of the microscope itself.

Thus, while it is certainly not practical to photograph everything you see in the electron microscope, a comprehensive record is essential for evaluating your material. Only photographs allow unhurried study. The beginning electron microscopist is best counseled to err in the direction of taking too many rather than too few photomicrographs. Large montages of electron micrographs (4000–10,000 ×) are often indispensable, especially in experimental studies. Magnifications in the lower range are appropriate for studying neuronal cell bodies and processes, whereas those in the higher range (10,000–20,000 ×) are necessary for defining structural details of synaptology.

It is important to remember that the larger the section is exposed to the electron beam, the less suitable it will be for photography. The destructive effect of the electron beam increases proportionately with increasing beam intensity and magnification. Scanning should therefore be done at the lowest intensity and magnification practical, reserving high powers for detailed study of the finest ultrastructural features. A good rule of thumb is to exhaust the resolving power of low and intermediate magnification before embarking on high-power electron microscopy. An anticontamination device that cools the specimen area with liquid nitrogen should always be used to reduce sublimation of the specimen.

A. Selecting Sections

Select the best section for photography by reviewing all the stained grids. Photographs are expensive and should be taken only of sections that are free or nearly free of such artifacts as holes, dirt, scratches, and chatter. Although thicker sections (70–140 μm gold interference color) are useful for low magnification survey pictures (1000–8000 ×), photography at higher magnification (10,000–25,000 ×) requires progressively thinner sections. Try to choose an intermediate section thickness that will serve your requirements for both low- and high-power photographs.

In examining each grid, begin by finding the number and position of the sections on it. With some electron microscopes, this can be done at a very low power (about 100 ×). Each section is then checked for holes and dirt at about 1000–5000 × at low beam intensity. If a very low magnification is not available, examine the grids by eye by holding them with forceps, section side down, under a strong light. When the grid is held at the correct angle, the sections reflect the light, and their position and color become visible.

B. Topography

It is essential to perform electron microscopic studies with certain knowledge of how the highly magnified image relates to the gross topography visible at a light microscopic level. You must know the exact location of a target area within the section and its relationship with such features as cortical laminae, major fiber bundles, or oriented processes. To establish orientation and define the areas of interest, a low-power photograph of the section (ca. 100 ×) can be compared with notes taken during slicing and embedding of the tissue and with a semithin section of the block face taken just before or after thin sectioning. Ideally, this photomicrograph can be taken on the electron microscope itself (Fig. 1). The location of fields photographed can be marked directly on the print. If possible, the low-power micrograph should be of the same section used for further work, for knife marks, stain contamination, and grid bars provide exceptionally useful, if unwelcome, landmarks. It may be necessary to print the negative of the lower-power micrograph upside down to give the print the same orientation as the electron microscope's image at some magnifications.

Some microscopes do not operate at very low power (e.g., 100 ×). On those that do, contrast may be inadequate. Contrast is maximized by using the lowest accelerating voltage available (for the survey photograph only) and by inserting, when possible, a contrast aperture. We use 600-μm single-hole grids in our Zeiss EM10 and Hitachi HU-12 microscopes. If a suitable picture cannot be obtained, then a light micrograph or drawing of a nearby semithin section of the block face should be used instead. Useful initial landmarks are cell bodies, large processes, blood vessels, and fiber tracts; the position of grid bars and others features may be filled in during low-power scanning.

It is important to remember that several days may intervene between photography and the inspection of the finished prints, and many details of the grid and photographs will undoubtedly be forgotten in the meantime. Therefore, a scrupulous record of the individual micrographs must be kept, including the numbering of the mechanical stage of the electron micrograph

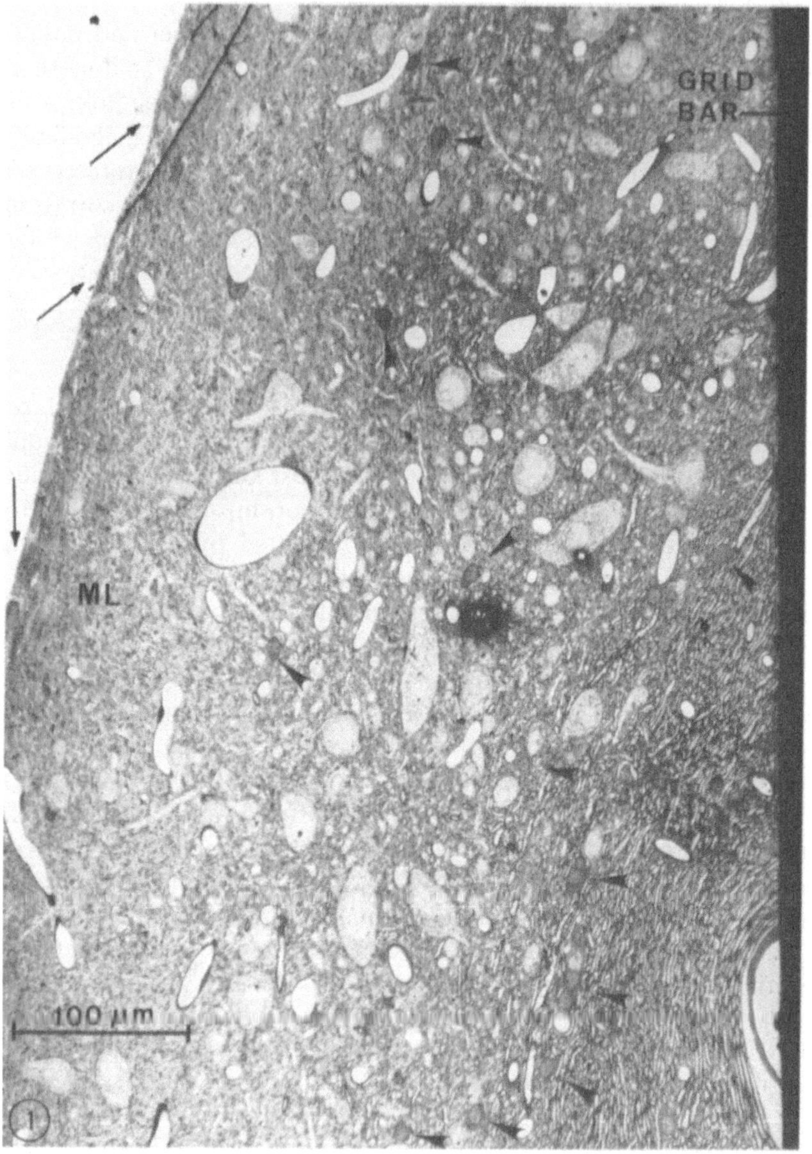

Figure 1. Electron micrograph of the rat dorsal cochlear nucleus at 100× primary magnification. The cilia of ependymal cells (arrows) are barely visible. In the molecular layer (ML) one can recognize the cell bodies of stellate neurons and the large dendrites of fusiform cells. Other large neurons in deeper layers are obvious. Oligodendrocytes (arrowheads) are evident throughout, as are blood vessels, capillaries, and myelinated axons.

(see, e.g., Alksne *et al.*, 1966) in relation to the 100× electron micrograph or a light photomicrograph of the accompanying semithin section.

C. Scanning at Low Magnification (1000–4000×)

1. Recognize the plane of section and the orientation of the section on the grid.
2. Verify the landmarks noted on the semithin section. In laminated structures, recognize the successive layers.
3. Try to distinguish electron-dense glia as a preliminary stage in the classification of glial and neuronal cell bodies (Figs. 2–4).
4. Identify bundles of myelinated fibers and blood vessels. These structures will continue to serve as landmarks in facilitating orientation at higher-magnification scanning, although the image will rotate substantially as magnification is increased.
5. Note gross features of nuclear morphology, cytoplasmic density, and nucleus-to-cytoplasm ratio. These latter features may serve as a clue to distinguish the cell bodies of intrinsic neurons from those of principal neurons. Unlike principal neurons, which project their axons out of the region of study, the efferent processes of intrinsic neurons are not long-traveling, and their somata may possess features, such as lobulated nuclei and scant Nissl substance, that facilitate their identification at low-power electron microscopic scanning.

D. Scanning at Intermediate Magnification (5000–12,000×)

1. Continue the classification process of neuronal and glial elements, concentrating on such details as cytoplasmic character and intracellular organelles (e.g., amount and type of endoplasmic reticulum, ribosomes, mitochondria, Golgi apparatus).
2. Recognize synaptic arrangements (i.e., glomeruli, en passant and terminal boutons, symmetric or asymmetric synaptic junctions, round or polymorph synaptic vesicles).
3. Identify axonal and dendritic processes and define the postsynaptic membrane, when possible, as somatic, dendritic, or axonal.

E. Scanning at High Magnification (15,000–25,000×)

Use high magnification only for details of conspicuous cytological interest. Primary magnifications higher than 20,000× are seldom necessary, and printing the negatives to a final magnification of 100,000× is sufficient even for quantitative analysis of synaptic vesicle size and shape. Scanning at 25,000–30,000× is necessary only to identify gap junctions.

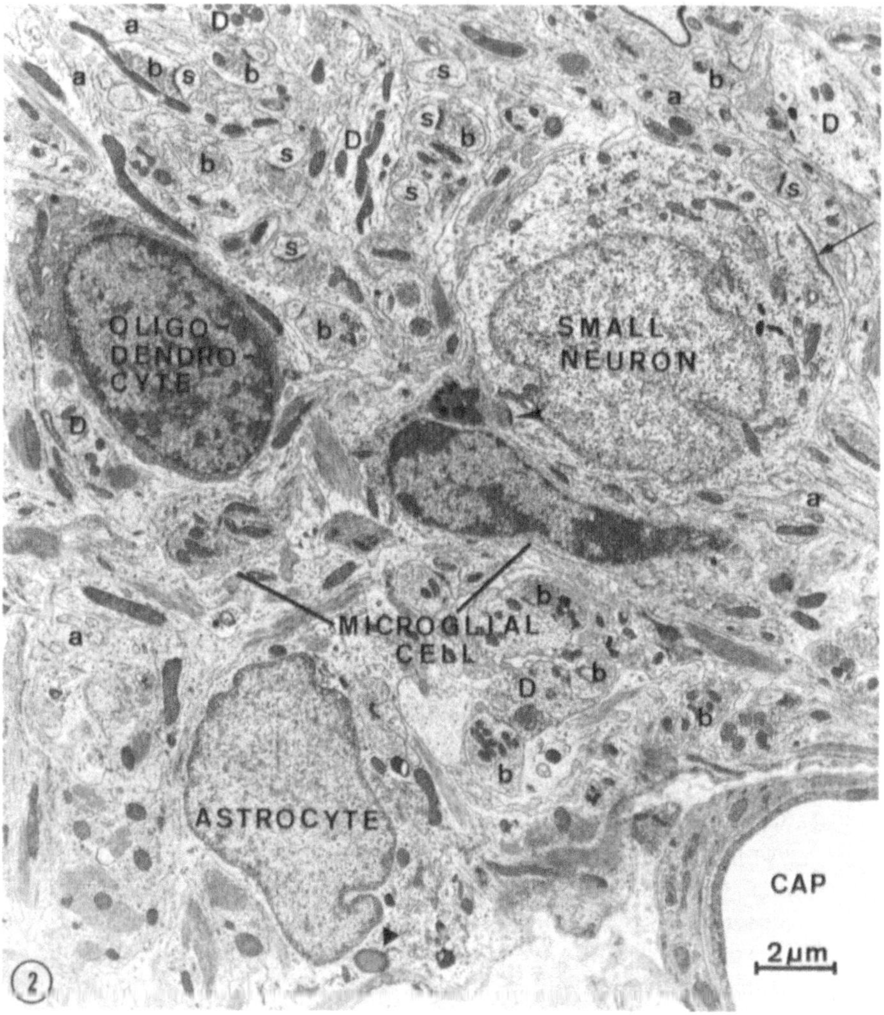

Figure 2. Low-power electron micrograph illustrating the three types of glial cells and a small neuron in the cat dorsal cochlear nucleus. The astrocytic nucleus has evenly dispersed chromatin with few and small peripheral clumps. The oligodendrocytic and microglial nuclei are more electron-dense, and the chromatin clumps are larger, especially in the microglial cell. The nuclear envelope of the oligodendrocyte is slightly dilated. The small neuron has a lighter nuclear content, forms one axosomatic synapse (arrowhead) and displays a hypolemmal cistern (arrow). The triangle points to a mitochondrion with matrix filaments in the astrocyte. Note also the bundles of intermediate filaments (unlabeled) in the astrocytic cell body and processes. Some of the dendrites are labelled D; boutons, b; unmyelinated axons, a; and dendritic spines, S.

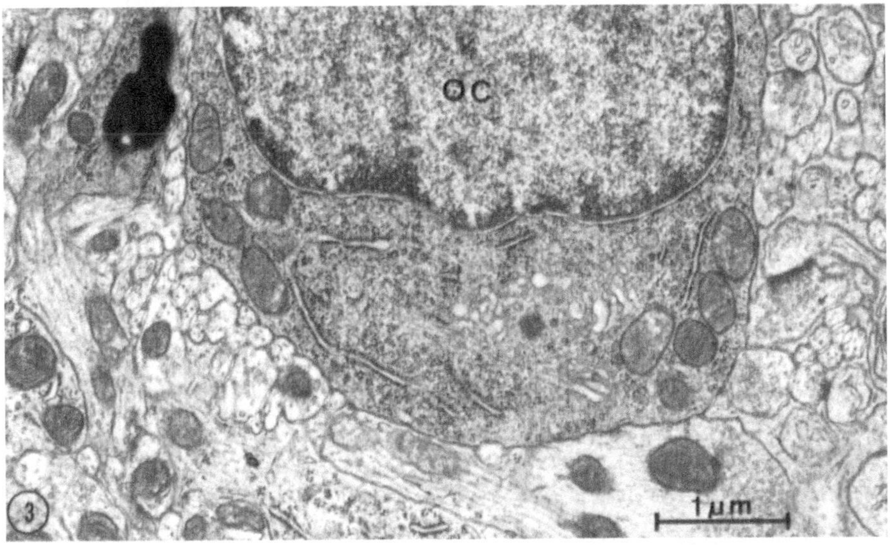

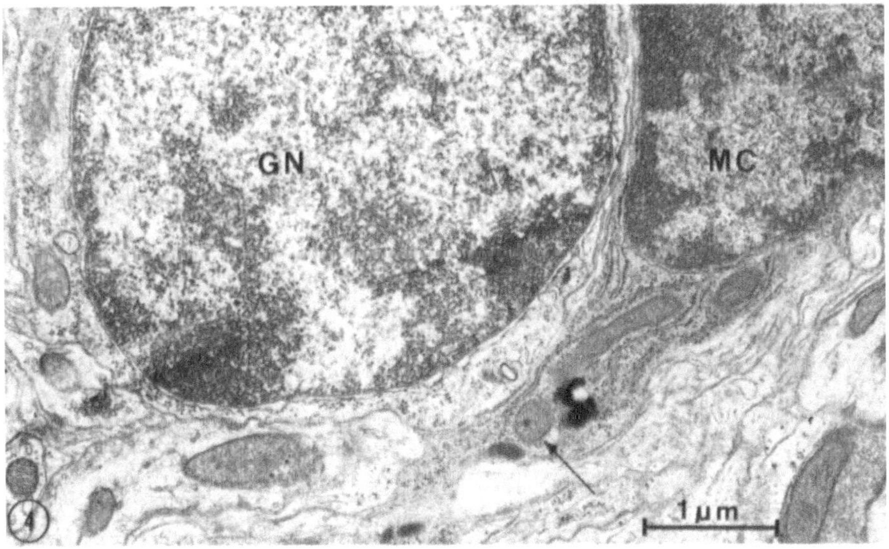

Figures 3 and 4. Medium-power electron micrographs illustrating portions of an oligodendrocyte (OC), a microglial cell (MC), and a granule cell neuron (GN). The endoplasmic reticulum and nuclear envelope of the oligodendrocyte are slightly dilated and have a transparent lumen, whereas those of the microglial cell are thinner and have a flocculent content. The microneuron has large clumps of chromatin, but the cytoplasm is relatively electron-lucent. The arrow points to a mitochondrion with a matrix dense granule in the microglial cell.

Table I. Summary of Glial Cell Features[a]

Cell	Cytoplasm	Nucleus	Processes	Mitochondria	Granular endoplasmic reticulum
Astrocyte Protoplasmic[b]	Relatively clear	Various shapes, often oval; chromatin evenly dispersed, very few peripheral clumps; medium-size to large nucleolus	Numerous, with thin side arms, pericapillary endfeet	Numerous, pleiomorphic, sometimes with matrix filaments and angular cristae	Present in cell body; some also in processes and endfeet
Fibrous[b]	Relatively clear	Various shapes, often oval; chromatin evenly dispersed, very few peripheral clumps; medium-size to large nucleolus	Less numerous and smoother Pericapillary endfeet	Less numerous, sometimes with matrix filaments and angular cristae	Present in cell body and in some processes and endfeet
Oligodendrocyte	Usually dense	Round or oval; chromatin in clumps; swollen nuclear envelope; small nucleolus	Few to numerous, often thin	Few and small	Abundant in cell body but not in distal processes; short cisterns with wide transparent lumen
Microglial cell	Dense or moderately dense	Oval or indented; chromatin in large clumps; small to large nucleolus	Few, both thin and large, with thin side arms	Few and small, with matrix and dense granules	Few cisterns with electron-dense content; some of these are long with narrow lumen

[a] Modified from Mugnaini and Walberg (1964–). Golgi apparatus, multivesicular bodies, and centrioles are not included in the table since they are present in all types of cell and present no useful differential features.

Intermediate filaments	Microtubules	Heterogeneous dense bodies	Glycogenlike granular material	Cilium
Scanty	Present but scanty	Present (some probably specific)	Abundant, especially in processes and endfeet	Present
Numerous, especially in processes and around the nucleus	Present but scanty	Present (some probably specific)	Present, especially in some endfeet	Present
Usually not present	Numerous both in cell bodies and processes	Present (some specific)	Not observed	Not observed
Usually not present	Present but scanty	Present, often numerous, some containing lipid droplets	Not observed	Not observed

[b] Protoplasmofibrous astrocytes have properties intermediate between those of the protoplasmic and the fibrous astrocytes.

IV. IDENTIFICATION OF NEURAL ELEMENTS

A neuronal cell body is easily recognized by its large nucleus with little heterochromatin and a large nucleolus and the presence of an extensive Golgi apparatus and of Nissl bodies. Most neurons have at least a few synapses on their perikaryal surface. When the synapses are absent and the neuronal nuclear and cytoplasmic features are not pronounced, as in the case of microneurons, the presence of hypolemmal cisterns (see Peters *et al.*, 1976), one of the special forms of endoplasmic reticulum in neurons, may serve as a reliable criterion to distinguish neurons from glia. The main features of the various kinds of glial cells are summarized in Table I.

The most common glial cell bodies in the gray matter are astrocytic.

The most common cell processes in the neuropil are axons, dendrites, and astrocytic processes. Although, at times, even the most experienced electron microscopist may occasionally have difficulty distinguishing between an axon and a dendrite, it is often possible, even for the beginning student, to identify the two types of processes by applying the following general criteria.

A. Axons

1. With few exceptions, myelinated processes are axons.
2. Small-diameter (i.e., 0.2–1.0 μm) profiles tightly packed in bundles usually are axons.
3. Profiles with synaptic vesicles are generally preterminal axons or boutons (see, however, Section IV.C).
4. Small axons usually contain some neurofilaments.

B. Dendrites

1. Dendrites usually do not occur in bundles except in synaptic glomeruli, they are often larger than 0.5 μm in diameter, contain polyribosomal arrays, and are postsynaptic to axonal boutons.
2. Small dendrites usually contain one or more mitochondria and a few microtubules. Neurofilaments are rarely observed.
3. Spines or other kinds of evaginations, most of which bear at least one synapse, are often seen arising from dendritic trunks.

The recognition of spines sometimes presents difficulties if no association with a parent dendrite is evident. The most valuable criteria for the identification of an isolated dendritic evagination are the absence of microtubules and the presence of a spine apparatus (cerebral cortex, corpus striatum) or some agranular reticulum and floccular cytoplasm (see Peters *et al.*, 1976).

C. Axoniform Dendrites

Local-circuit neurons may have axoniform dendrites that subserve both presynaptic and postsynaptic functions. These have been observed in various brain regions, including the diencephalon, the retina, the olfactory bulb, and the cortex. Principal neurons, with the exception of the mitral cells of the olfactory bulb, lack this feature. When presynaptic dendritic profiles have scattered polysomal arrays in the vicinity of synaptic areas, their dendritic nature may be established with reasonable confidence, but when the presynaptic site is located on a dendritic appendage, the nature of the profile can be assessed unequivocably only with the use of serial sections and the Golgi–EM procedure (Chapter 12).

D. Synthesis

A crucial step in the electron microscopic analysis is the production of good electron micrographs. Especially at the onset of a new investigation, thin sections should be collected on single-hole grids, and large montages of electron micrographs should be made from areas of the thin section that can easily be recognized on a preceding or successive semithin section. One should quickly learn to establish landmarks in the thin section before taking photographs. Each micrograph should include the record of the object's location in the thin section. One should first establish orientation with respect to gross features visible in the light microscope, then expose the cell bodies, then the cell processes and the synaptic details.

Micrographs at a primary magnification of 1000–4000× are best suited for cell bodies, at 8000–12,000× for cell processes, and at 15,000–20,000× for synaptic junctions and synaptic vesicles. The careful study of a textbook on ultrastructure of the nervous tissue is an essential first step toward learning one's way into the neuropil. The distinction between axons and dendrites is not always easy. Although short series of sections may facilitate identifying small cell processes and cell junctions, long series of sections or Golgi–EM (Chapter 12) may be required to identify axoniform dendrites with certainty.

V. ULTRASTRUCTURE OF DEGENERATING NERVE FIBERS

It is difficult to overestimate the value of the electron microscope as a tool in the tracing of interneuronal connections. Light microscopic studies can, at best, outline the area of termination of a fiber system. It remains for the electron microscopist to identify the synaptic endstructure of a specific pathway and clarify its relation with other neurons. It is important in this context to reiterate that as soon as the objective of an electron microscopic study goes beyond the mere confirmation of the existence of degenerating terminals, it becomes essential to gain a good understanding of the normal ultrastructure in a given area. For example, if one wants to know whether a certain kind of bouton present in normal tissue belongs to the experimentally damaged fiber system, the experimental study must be preceded by a thorough qualitative and quantitative analysis of the synaptology in normal material.

A few general points are worth remembering before one embarks on an extensive analysis of the pattern of degeneration.

1. Subtle degeneration changes can be properly evaluated only in first-rate material.
2. The existence of scanty degeneration is of little experimental value unless it is given proper statistical evaluation. Degeneration may occur as a consequence of pathological alterations unrelated to the experiment. Various forms of dystrophic axons and boutons have been described in several publications (see especially Sotelo and Palay, 1971).
3. Profiles with high electron density can be portions of artifactually generated "dark neurons" (Chapter 4) or may be electron-dense types of glial cells, either oligodendrocytes or microglia.
4. The degeneration induced by the experiment may not only proceed in anterograde or retrograde directions but may also be transneuronal (Cowan, 1970; Jean-Baptist and Morest, 1975). As a rule, retrograde and transneuronal degeneration have a longer time course than anterograde degeneration, although cases in which the transneuronal degeneration occurs together with the anterograde degeneration may well exist (Heimer and Kalil, 1978).
5. The time course of degeneration depends not only on the species but also on the fiber system being studied. Furthermore, the distance between the site of the lesion and the target area is of great importance in estimating the optimal survival time for terminal degeneration. Fiber diameter may also influence the choice of survival time (Chapter 4).
6. It is especially important to analyze carefully early stages of degeneration, especially because the internal structure of the degenerating elements is often quickly deranged, making their classification difficult. Although similar structures within a given system have a similar rate of degeneration, it is nevertheless true that profiles at different stages of degeneration will be found even in the same electron micrograph.
7. If one wants to determine the percentage of the total synapses in an area belonging to the fiber system under study, it is obviously important to make sure that the lesion is total or near total. If, on the other hand, the focus of interest is differential distribution of boutons on the receptive surfaces (i.e., dendritic shafts, spines, soma, etc.), a partial lesion may still be useful.

8. If the operation does not completely interrupt an afferent pathway, degeneration of neuronal cell bodies and passing fibers may continue in the area of the scar and around it. Thus, dystrophic changes and early stages of degeneration in the terminal area should be expected even several days after the experimental lesion. Furthermore, in cold-blooded animals, the degeneration processes may be very slow (Kruger and Maxwell, 1969).

A. Patterns of Degeneration

Although a great deal of attention has been paid to the description of various degeneration patterns at the EM level, much of the apparent variation may be attributed to temporal factors. Nonetheless, it would be an oversimplification to assert that there exists only one main pattern of degeneration whose early stages may vary among fiber systems and species.

We will discuss five degenerative patterns that have been reviewed in previous publications (Guillery, 1970; Raisman and Mathews, 1972; Westrum, 1973). These five kinds of degeneration are electron-dense, filamentous, flocculent, watery, and pinocytotic.

1. Electron-Dense Degeneration

Electron-dense degeneration is the most common type of Wallerian axonal degeneration and may also occur retrogradely (Grant, 1970). This type of degeneration is characterized foremost by an increased electron opacity of the axoplasm (Fig. 5). However, since electron opacity may vary depending on fixation and handling of the specimens (Chapter 10), other changes in cell organelles that usually accompany, or sometimes even precede, the increase in electron density of the axoplasm should be identified if possible.

Early changes are seen especially in mitochondria, axoplasmic reticulum, and synaptic vesicles. The mitochondria lose their common rodlet shape and become globoid and then crenated. The mitochondrial cristae undergo a rearrangement and often become stacked. Occasionally, a needle-shaped precipitate (hydroxyapatite crystals) and osmiophilic bodies are found in the mitochondrial matrix (Mugnaini *et al.*, 1967). The axoplasmic reticulum tends to swell, and the synaptic vesicles become larger (Akert *et al.*, 1971; Kawana *et al.*, 1971), possibly as consequences of an imbalance of water and ion exchanges in the severed axon. A flocculent appearance of the axoplasm may precede the conspicuous darkening observed at later stages. At the time when the electron density of the axoplasm has become conspicuous, the boutons usually appear collapsed and may present a stellate configuration with tiny projections "caught" either among surrounding normal boutons and dendrites, or between astrocytic and microglial cell processes. It appears probable, judging from these features, that water moves

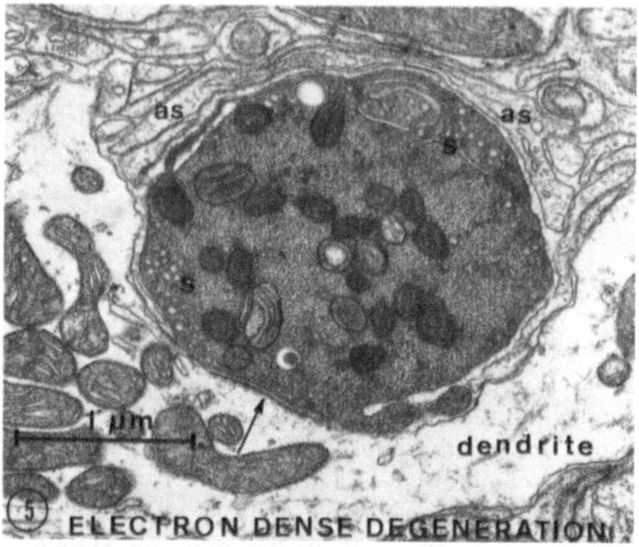

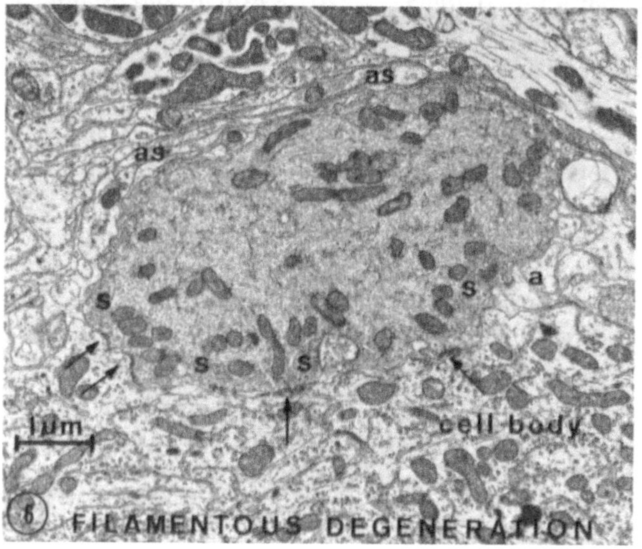

Figure 5. High-power electron micrograph illustrating an axon terminal in the cat cochlear nucleus undergoing electron-dense degeneration. Arrows indicate synaptic junctions; s, synaptic vesicles or their remnants; as, astrocyte processes; mg, microglial cell processes.

Figure 6. Axon terminal of the cat cochlear nucleus undergoing filamentous degeneration. Abbreviatins as in Fig. 4.

from the darkened, degenerating elements into the extracellular space. Thus, the increase in electron density might be related not only to chemical changes within the axoplasm but also to a relative decrease of water in the degenerating element.

Glial changes are usually evident where dark degeneration takes place.

Within a short time, sometimes only 12–24 hr following the operation, astrocytic processes and microglial cells invade the area, presumably under the stimulus of some chemotactic agents leaking from the degenerating axons (Mugnaini *et al.*, 1967).

During this process, the degenerating fibers lose their continuity and form chains of fragments. It is unknown whether the fragmentation is the result of axonal degeneration or whether it is partly induced by the glial cells.

The darkened synaptic boutons remain attached to the postsynaptic elements during the early stages of degeneration. At this point it is often still possible to distinguish between symmetric and asymmetric synapses and puncta adherentia, although the synaptic vesicles may be removed from the presynaptic densities. Short series of thin sections (10–20) are usually helpful in this classification by furnishing electron micrographs of the same junction in successive planes.

The relationships between pre- and postsynaptic elements may become altered in different ways during the last stages of the degeneration process (e.g., Wisniewski *et al.*, 1972; Gentschev and Sotelo, 1973). The degenerating bouton may detach from the postsynaptic surface and be engulfed by the phagocyte, and the vacated postsynaptic half of the synaptic junction may be covered by glial cell processes. The postsynaptic membrane may gain contact with a newly sprouted axon terminal, or it may be internalized and digested away by the postsynaptic neuron itself.

In certain cases a fragment of the presynaptic membrane has been reported to remain attached to the postsynaptic membrane for several days, while the rest of the degenerating bouton is phagocytized. The opposite phenomenon may also take place; that is, the postsynaptic membrane, or even part of the postsynaptic neuron, such as a spine, may be phagocytized together with the degenerating bouton.

Although the progressive steps of degeneration and phagocytosis have been well described, little is known about what triggers the glial cell mobilization or what determines which pre- and postsynaptic elements become detached and what will be the fate of the postsynaptic membrane. Nor do we have a clear understanding of the respective roles played by astrocytes and microglial cells. In some cases, oligodendrocytes have also been reported to participate in the removal of degenerating boutons and axons (Spencer and Thomas, 1974).

An instructive example of the reaction of glial cells to anterograde degeneration is found in the cerebellar cortex. The electron-dense, degenerating rosettes of severed mossy fibers lose their contact with surrounding granule cell dendrites while they are ensheathed by both astrocytic and microglial cell processes. Fragments of mossy terminals completely engulfed by a single astrocytic process are more rarely encountered than those surrounded by a microglial element. The fragments apparently engulfed by single astrocytic processes usually have identifiable internal components, whereas most of the fragments taken up by microglial cells contain severely altered organelles. Moreover, while fragments in the microglial cytoplasm are usually rounded, degenerating fragments contacted by astrocytic processes often have a stellate configuration with tiny arms invaginating the astrocytic processes. These arms are most frequently seen in cross section and appear as small, electron-dense bodies surrounded by a double membrane.

These observations support the contention that the two chief kinds of glia—astrocytes and microglia—enter into separable phases of phagocytosis and may be mobilized by different aspects of the degeneration process. Astrocytes, which are present in greater number than microglia, normally surround the glomerulus, thus placing them in a more strategic position than the microglia to initiate the phagocytic response. Astrocytic processes seem very active, both in the disorganization of the glomerular array and in the engulfment of the degenerating rosettes, but the trigger that activates this process is unclear and is probably distinct from the one that mobilizes the microglial elements later. Microglial cells are typically seen in mitosis 1–4 days after the lesion, and they give the appearance of being most active in the digestion of the degenerating fragments. Microglial cell division occurs in response to retrograde as well as anterograde changes and accompanies loss of axosomatic and axodendritic contacts by axotomized neurons (Blinzinger and Kreutzberg, 1968; Torvik and Skjørten, 1971).

2. Filamentous Degeneration

The most conspicuous feature of the filamentous degeneration is the increase in the number of neurofilaments (Fig. 6). This kind of degeneration, however, is also characterized by other features that might reflect less striking but more relevant changes of axonal functions than the accumulation of filaments. The most notable of these changes are a sharp decrease in the number of synaptic vesicles and hypertrophy of the axoplasmic reticulum. Less reliable signs of degeneration may include swelling of the fibers and the boutons and changes in the shape, size, and distribution of the mitochondria. Accumulation of glycogen particles has occasionally been noted (Szentagothai *et al.*, 1966). Boutons undergoing filamentous degeneration are usually not phagocytized, and, therefore, remained attached to the postsynaptic structure.

The mechanism responsible for the accumulation of neurofilaments is unknown. The neurofilaments may polymerize from soluble precursors or accumulate either through continuing proximodistal transport or slowing down of the retrograde transport. If the filaments are transported to the terminals, their number should decrease in more proximal portions of the severed axon.

There is substantial evidence that filamentous degeneration is but an early stage of electron-dense degeneration peculiar to certain fiber systems and to certain species. For instance, severed retinogeniculate fibers and Purkinje axons in the rat undergo electron-dense degeneration, whereas in the cat and the monkey they go through a stage of filamentous hypertrophy, after which they become progressively more electron-dense and are eventually removed by astrocytes and microglial cells (Mugnaini and Walberg, 1967; Wong-Riley, 1971).

A peculiar situation has been described in the medial trapezoid nucleus of the cat after cochlear ablations (Jean-Baptist and Morest, 1975). The calycine fibers undergo filamentous degeneration in their calyciform ending but electron-dense degeneration in the small appendages of the calyx. Moreover, the calyciform ending may be phagocytized while undergoing filamentous degeneration, or its axolemma may disappear, leaving behind an extracellular mass of neurofilaments that progressively disappears.

3. Flocculent Degeneration

This type of degeneration (Fig. 7) has recently been shown by Kane (1974) to occur extensively in the cat cochlear nuclei after destruction of the membranous cochlea and the spiral ganglion cells. In addition, it has been observed in some of the boutons of afferents undergoing the electron-dense type of degeneration. It is characterized by disintegration of neurofilaments, swelling, packing, and depletion of the synaptic vesicles, mitochondrial degeneration, and a flocculent appearance of the axoplasm. The flocculent degeneration seems to be part of a more diversified degeneration process (Westrum, 1973). At the same time that many of the boutons are characterized by a flocculent degeneration, others show electron-dense degeneration, while still others may display the so-called watery type of degeneration (see Section V.A.4). Whereas the flocculent terminals are usually removed by phagocytes, there seem to be a few instances in which the flocculent content is dissolved, leaving a watery ghost. It is also possible that in some of the boutons the flocculent reaction, like the filamentous degeneration, may represent an early stage of electron-dense degeneration. The existence of a large number of neurofilaments in normal axon terminals may be involved in the formation of the flocculent precipitate during the degeneration process.

4. Watery Degeneration

Watery degeneration is characterized by clumping and depletion of synaptic vesicles and mitochondrial degeneration similar to that observed in terminals undergoing electron-dense degeneration. The axoplasm, however, becomes electron-transparent instead of electron-dense (Fig. 8). Watery terminals are engulfed by glial elements, indicating that this degeneration process activates phagocytes. These features have been observed in the cochlea (Smith and Rasmussen, 1965) and cochlear nuclei (Kane, 1974; Gentschev and Sotelo, 1973), in the cerebellum (Fox *et al.*, 1967), in preganglionic (Raisman and Matthews, 1972) and postganglionic (Thomas and King, 1974)

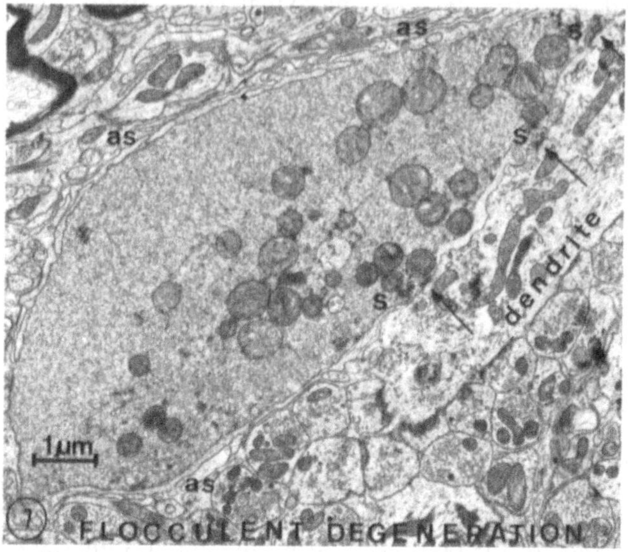

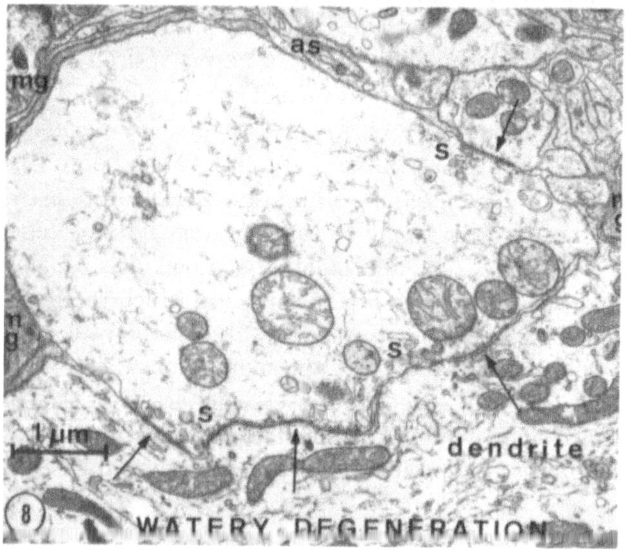

Figure 7. Axon terminal of the cat cochlear nucleus undergoing flocculent degeneration. Abbreviations as in Fig. 4.

Figure 8. Axon terminal of the cat cochlear nucleus undergoing watery degeneration. Abbreviations as in Fig. 4.

autonomic fibers, in the olfactory system (Westrum, 1969; Berger, 1971), and in a few other locations (Westrum, 1973; Korthals *et al.*, 1974). The recent studies of Arnesen *et al.* (1978) suggest that in the anterior cochlear nucleus the large majority of the acoustic nerve terminals undergo the watery type of degeneration. Swelling seems to be the most conspicuous first sign. Many of the calyciform endings become so ballooned that they can easily be seen in the semithin sections. They then shrink and can be studied only with the electron microscope.

The degeneration process is not simultaneous in the whole terminal area of the VIII nerve but follows a cochleotopic sequence. Swelling starts 48 hr after the lesion in the fibers from the basal coil and after 72 hr in those from the apical coil. An earlier-than-usual breakdown of the ion pumps may be responsible for the watery appearance of the degenerating terminals. Other factors, such as the proximity of the terminal area to the lesion site, may also be involved.

5. Pinocytotic Degeneration

A peculiar postoperative change has recently been seen in a minority of degenerating axon terminals (Gentschev and Sotelo, 1973; Jones and Rockel, 1973). This is referred to as "pinocytotic" degeneration. The pinocytotic terminals appear moderately swollen and are depleted of their synaptic vesicles. The axoplasm is electron-transparent, and the boutons are filled with an unusually large number of coated vesicles and "empty baskets." Based on the effect of stimulation on the muscle end plates Heuser and Reese (1973) and Jones and Rockel (1973) have suggested that these boutons are the terminals of fibers that have been stimulated above normal range during the lesioning. Whether the boutons underoing "pinocytotic degeneration" in the early stages become electron-dense later is not known.

6. Degenerative Changes in Nerve Fibers

Usually, one would expect to find in the parent fiber essentially the same type of degeneration observed in the terminal aborization (see Morest and Jean-Baptist, 1975). This seems true in most but not all instances. For example, the retinal fibers in cat and monkey show filamentous degeneration in their terminal aborization, but the initial part of the severed axon shows loss of neurofilaments and microtubules and undergoes an electron-dense degeneration without going through a filamentous stage. "Pinocytotic" degeneration of nerve fibers has not been described.

In general, degenerating fibers show changes not only in the electron density of their axoplasm but also in their axonal organelles. Microtubules decrease in number, ultimately to disappear entirely. Mitochondria alter their shape and crystal pattern. The number of neurofilaments declines, culminating in their dissolution. Finally, there are changes in the number and shape of the tubules of axoplasmic reticulum. These changes have been observed in practically all degenerating fiber systems.

Wallerian degeneration of myelinated fibers involves changes of both axonal and glial sheaths. Whereas in mammals the axonal changes precede or at least become evident earlier than those in the surrounding myelin, in anamniotes the opposite seems to be true (Turner and Singer, 1975; Turner and Glaze, 1977). Early degenerative changes in the myelin sheaths are more difficult to study in birds and mammals than in fishes, amphibians, and reptiles, since fixation of myelin is often unsatisfactory in the former classes. There seems to be agreement, however, that in both groups the outer loops of the myelin sheath remain continuous with the compact myelin for several days after the operation, indicating that continuity of the sheath with the parent glial cell is maintained long after the onset of axonal degeneration. According to Cook and Wisniewski (1973), the inner loop increases in size during degeneration and may show vacuolization. Disruption of the compact myelin is initiated from the inner lamellae, and the outer loop disappears only when alteration of myelin lamellae is well advanced. Although splitting of the myelin lamellae is often seen in mammals at early degenerative stages, one should be aware that it may be an artifact of fixation.

Fragmentation of anterogradely degenerating axons has often been described in experimental material and is accompanied by fragmentation of the myelin sheath. Phagocytized fragments of axons may still have a myelin sheath, although it may be altered. How fragmentation of the myelin tube takes place along the internodes is unclear, but it has been suggested that Schmidt–Lanterman incisures may be involved in the fragmentation process. On occasion, cell bodies may be seen in the process of phagocytosing the degenerating axon inside the myelin tube itself, an occurrence well known to neuropathologists (Cook, 1974). Empty myelin tubes are observed frequently, suggesting that intratubal phagocytosis is not a rare phenomenon. The intratubal phagocytes are of the microglial type. Astrocytes have never been reported inside the myelin sheath, although their processes may penetrate the myelin tube for short distances. The phagocytes enter the tube from Ranvier nodes or, presumably, from the ends of the fragments.

In general, Wallerian degeneration of myelinated fibers proceeds at a faster rate in peripheral nerves than in the CNS. There is evidence that in certain pathways coarse myelinated fibers degenerate at a different rate than the thinner ones, although within the target areas, some fine fibers degenerate earlier than the coarser ones. It has been suggested that the early degenerating fibers within the target area are the preterminal branches of thicker axons. In general, distal portions of the axon degenerate before the proximal portion. There are reports indicating that in tracts consisting of both unmyelinated and myelinated axons, the unmyelinated ones are removed first.

Whereas Schwann cells are evidently involved in phagocytosis of degenerating myelin and axons in peripheral nerves, participation of oligodendrocytes in the removal of the CNS myelin fragments has not been established. Most authors have been unable to detect unequivocal changes in the oligodendrocytes, whereas astrocytes and, especially, cells considered by most investigators to be the equivalent of microglia at the light microscopic level are involved. Of the mitotic cells in the fields of degeneration, a few have been identified as astrocytes, and most as microglial cells. The cells considered to be mitotic astrocytes have been identified on the basis of the presence of bundles of cytoplasmic filaments, a criterion that may not be fully acceptable since reactive cells are concerned (see, e.g., Bignami and Ralston, 1960; Mathews and Kruger, 1973a,b).

In considering the different reaction of Schwann cells and oligodendrocytes to degeneration of their related axons, one should keep in mind that whereas one Schwann cell is involved in only one segment of the myelin sheath, each oligodendrocyte may provide a sheath to a number of different axons. Thus, one could expect different reactions in individual oligodendrocytes when only some or all of their target axons undergo degeneration.

B. Synthesis

Despite the variety of changes observed in different severed fiber systems in several animal species, the predominant end product of Wallerian degeneration is the conspicous collapse and fragmentation of the axonal element which stands out among the normal components of the tissue because of its deranged internal structure. Depending on the fiber system and the animal species, anterograde degeneration may take several forms that the beginning student should try to identify and distinguish from naturally occurring dystrophic changes. Since time may also change the appearance of degeneration in a given system, several survival times should be examined. The most common pattern of degeneration is the electron-dense type. Electron density alone, however, is a misleading criterion, since it may be the result of fixation artifact. Thus, changes in axonal organelles should receive special attention. When virtually all the fibers of a homogeneous pathway—that is, a pathway where all axons are derived from the same class of neuron—show other types of degenerative changes, then experimental parameters such as topographic factors and the distance of the degeneration field from the lesion site should be investigated with special care. On the other hand, when aberrant patterns of degeneration are observed only in a small number of the degenerating terminals, as in the case of the pinocytotic degeneration, then variations in the lesion method may be tried.

The possibility of the early occurrence of retrograde and transneuronal changes should always be kept in mind.

With respect to the process of Wallerian degeneration in general, the exact sequence of events is still unclear, especially in the early stages. It may be of some contingent value to subdivide the continuum of the process of degeneration arbitrarily into a probable series of morphologically identifiable steps or stages, as proposed below.

1. Stage 1: initial changes primarily involve internal constituents of the axon and the ionic pumps. Impulse conduction and synaptic transmission may persist for several hours after severance of the stem fi-

ber (Pecci-Saavedra *et al.*, 1970). Swelling of the axon and of the synaptic vesicles and the endoplasmic reticulum and disruption of microtubules may be the first morphological alterations. Depletion and clumping of the synaptic vesicles may follow. Secondary changes may be dissolution of neurofilaments and mitochondrial degeneration.

2. Stage 2: permeability of the axolemma becomes severely altered, and chemotactically active substances leak from the axolemma. Water is lost to the extracellular spaces. The fibers collapse inside their sheaths, and the terminals shrink, assuming an irregular configuration. These changes begin to affect the surrounding tissue. Synaptic contacts are still maintained, especially if the terminals are involved in glomerular arrays, where cell junctions are numerous and direct apposition to glial elements is limited.
3. Stage 3: degenerating boutons show increased contact with glial elements and loss of synaptic connections. The parent fibers become fragmented. Myelin sheaths begin to show reactive changes and alterations.
4. Stage 4: boutons are engulfed by glial elements and phagocytosed. Axonal organelles are still recognizable.
5. Stage 5: the lysosomes of phagocytes fuse with the phagosomes. Digestion proceeds and culminates in the formation of residual bodies. Removal of the degenerating myelin sheath continues for longer periods.

VI. MORPHOMETRY

As detailed in the preceding sections, degenerating synaptic terminals can be identified by the occurrence of specific changes within them and in their surroundings. In addition to the presence of degenerating synapses, it is usually possible to discover something about the pattern of their distribution, an important feature in unraveling local circuitry. Such information is most useful when it is quantitative.

The general problem of characterizing three-dimensional structures from their appearance in two-dimensional sections is treated in a substantial literature, and we urge interested readers to consult systematic reviews (e.g., DeHoff and Rhines, 1968; Elias *et al.*, 1971; Underwood, 1970; Weibel, 1969).

A. Measures

A quantitative description of degeneration may address any of three broad goals. First, one may simply measure the total amount of degeneration. This might be used to compare different regions, such as areas of cerebral cortex, or parts within a region, such as cortical layers. Second, one may define the relative amounts of degeneration associated with different postsynaptic sites at a given region. Third, one may focus on a particular postsynaptic site and ask what fraction of the total input to that site is comprised by the degenerating elements.

The most obvious measure of the total amount of degeneration in a region is the number of degenerating profiles observed in the thin sections, usually expressed as the number per unit area (N_A). Intuitively, the more degenerating profiles visible in thin sections, the more degenerating boutons are present in the tissue. However, the size and shape of the boutons also influence the number of their profiles in thin sections. Thus, differences between specimens in N_A for degenerating profiles may be caused by differences in the size or shape of the boutons rather than in their actual number.

For instance, a large bouton clearly has a higher probability of appearing in any random section than does a small bouton; because of its convoluted surface, a single mossy fiber synaptic swelling may appear as several isolated profiles rather than only one.

Two other measures are also useful. They are the volume density and the surface density of the boutons. The volume density (V_V) is the total volume of degenerating boutons in a region divided by the volume of the region. It is exactly equal to the area fraction of degenerating profiles as determined by adding the areas of degenerating profiles in micrographs and dividing by the area of the micrographs. The areas can be determined planimetrically, or the very rapid method of point hit counting may be used (see Weibel, 1969). The surface density (S_V) is the integrated surface areas of degenerating terminals in a region divided by the volume of the region. It is determined by measuring the total circumference of the plasmalemmas of degenerating profiles in micrographs, dividing by the area of the micrographs, and multiplying by $4/\pi$ (path lengths and areas should be corrected for magnification). This measure is particularly meaningful when one measures only plasmalemma apposed to other neurons or that in synaptic adhesions. The method described here for determining surface density assumes random orientation of the objects measured and should not be used when the objects are highly oriented (for example, cerebellar parallel fibers).

For assessing the distribution of degenerating elements one may, by counting, find the proportion of all degenerating profiles associated with each type of postsynaptic site. Alternatively or in addition, one might find the volume density or surface density separately for profiles associated with each type of site.

A useful measure for comparing the amount of input from degenerating terminals with that from nondegenerating elements is the "synaptic covering," the fraction of surface of postsynaptic neurons in direct apposition to synaptic boutons (Conradi, 1969); in this case, one compares the synaptic covering by degenerating boutons with that by nondegenerating boutons. Synaptic covering is found by measuring the length of postsynaptic surface apposed to boutons and dividing by the total length of postsynaptic surface, as seen in thin sections. Somas may be treated separately from dendrites. It is essential that the denominator include even those profiles of the postsynaptic neuron (or part, as soma) that are not apposed to any synaptic boutons.

The measures N_A, V_V, and S_V evaluate different properties of the specimen and may vary independently. Furthermore, they are likely to be maximally sensitive to different aspects of the degenerative process. For example, if the reaction includes loss of water with darkening and crenation of boutons, then the surface density of the degenerating boutons might change little during this phase, while the volume density would change substantially. On the other hand, the separation of terminals from the postsynaptic surface by glial cells would have maximal effect on the observed surface density of terminal membrane apposed to postsynaptic sites or the surface density of synaptic adhesions. Thus, since terminals may be in different stages or modes of degeneration, it is well to use more than one measure in comparing regions or distribution on postsynaptic sites and to make use of qualitative observations in interpreting the quantitative ones.

One effect of the degeneration of afferents may be a reduction in the volume of a region, particularly if transneuronal atrophy occurs. Such changes are not likely to be substantial in short-survival experiments but might well be in long-survival ones, such as those for analyzing plasticity (synaptic rearrangement). Shrinkage could cause an apparent increase in any of the measures described here without actual growth in the tissue.

B. Sampling

Perhaps the most critical step in performing a quantitative study of any kind is the collection of the micrographs. Given human fallibility, it is not good practice for an investigator to examine a specimen in the microscope and then photograph fields he feels are representative. Instead, fields must be selected according to a precise and explicit protocol that eliminates investigator bias. This step is central; to be meaningful, quantitative data must refer to a well-defined region, and the micrographs used for measurement must be representative of that region. Achieving these conditions is a particular challenge in the nervous system, which is dramatically nonhomogeneous and anisotropic, and in which substantial changes in local structure may occur over a few tens of micrometers or less.

The obvious method for sampling is to construct montages of micrographs of contiguous fields. This approach is particularly attractive for laminated structures where narrow rectangular montages are oriented to run across the layers. The structure can either be studied as a whole, or divided by layers. Unfortunately, such montages often require a large number of pictures (for example, at 5000 × magnification, it would take more than 50 negatives 4 inches long to span the 1 mm thickness of mouse neocortex). Furthermore, significant variation in structure may occur in directions parallel to laminae, such as orientation and ocular dominance columns in the visual cortex. In these cases, a narrow montage could be unrepresentative of the structure as a whole, whereas a sufficiently wide montage might be truly prohibitive (3000 negatives 3.25 by 4 inches to cover 1 mm^2 at 5000 × magnification).

Because of the limited area that can practically be included in them, it is generally not desirable to construct montages for quantitation but rather to scatter single micrographs over as wide an area as available.

A simple scheme uses mesh grids. Sections are mounted on the grids, and micrographs are taken in fixed positions (for example, one or more corners) in grid squares distributed across the specimen. The exact position of each field can be determined later using a 100× picture of the same section. This scheme samples from large areas of section and produces a manageable number of micrographs. It has the drawback that parts of the specimen are obscured by grid bars and are unavailable for sampling.

An alternative scheme, more elaborate but offering great flexibility, uses microscope stage counters to specify position of micrographs. Sections are mounted on very-wide-mesh, single-hole, or slot grids, and a whole section is photographed at 100× in the electron microscope. Regions to be studied are marked on the print, and the stage coordinates of their boundaries are determined by moving the specimen in the microscope to center the boundaries on the screen, using the print of the low-power micrograph as a guide. Each region to be studied is then represented on graph paper by plotting the stage coordinates of its boundaries, and the stage coordinates of fields to be photographed are determined by choosing points within the region. A reasonable plan is to distribute fields regularly across the area to be studied. A better one is to combine regular spacing with some degree of randomness to prevent interaction between regular spacing of the micrographs and possible repeating structure in the specimen. This can be done by dividing the region to be studied into rectangles or squares and placing one field at a random position within each, using a table (e.g., Snedecor and Cochran, 1967; Rand, 1955) or some other source of random numbers.

This scheme may be executed with little fuss. For example, in a study of cerebellar cortex, we omit the low-power picture because we are able to orient without it. We determine the coordinates of three strategic points in a section and enter them into a calculator that is programmed to reconstruct the boundaries of the region we want and to generate random points within a series of rectangles covering that region. With little trouble we obtain a sample of micrographs that combines randomness and regularity and from which we are certain our own bias has been excluded. The key is the use of the microscope stage counters to specify position.

The sampling schemes discussed so far are useful only with a sufficient density of degenerating elements. When degeneration is sparse, it may be necessary to hunt for it in the microscope over units of known area and clear boundaries, such as grid squares. This approach should be avoided when possible, because results may be influenced by the operator and cannot easily be checked. On the other hand, it is certainly acceptable to search systematically for cell somas or other structures that are prominent but may be sparse.

Because of the need for precise localization, careful and well-documented dissection, slicing, embedding, and trimming are important. Particular attention should be paid the orientation of the blocks, for neural tissues are usually highly anisotropic, and the values obtained for N_A, S_V, and synaptic covering may vary with the orientation of the sections. Obviously, this possibility can be checked by sectioning at different orientations, but the effect cannot be controlled unless the orientation of the blocks is known.

C. Synthesis

In planning a morphometric study of any kind, one must establish clear goals and choose appropriate measures and a suitable sampling plan to meet them. In making these choices, one should obtain and use information on the amount of time different measuring and sampling schemes entail. Thus, as a practical matter, planning should include trial runs of all steps, with a careful accounting of the time invested. This information can be used to optimize design (Shay, 1975) and to choose parameters such as magnification and number of pictures.

Number of profiles per unit area of section, volume density, and surface density can readily be determined from electron micrographs and are useful in characterizing degeneration quantitatively. Valid results require good fixation and sectioning, precise localization of block faces, adequate sampling, and correct identification of degenerating profiles.

Acknowledgments. Supported by USPHS Grant NS 09904. The authors wish to thank Ms. Annie-Lise Dahl and Ms. Mary Wright Goss for their skillful assistance.

REFERENCES

Akert, K., Cuénod, M., and Moor, H., 1971, Further observations on the enlargement of synaptic vesicles in degenerating optic nerve terminals of the avian tectum, *Brain Res.* **25**:225–263.

Alksne, J. F., Blackstad, T. W., Walberg, F., and White, L. E., Jr., 1966, Electron microscopy of axon degeneration: A valuable tool in experimental neuroanatomy, *Ergeb. Anat. Entwickl.-Gesch.* **39**(1):1–31.

Arnesen, A. R., Osen, K. K., and Mugnaini, E., 1978, Temporal and spatial sequence of anterograde degeneration in the cochlear nerve fibers of the cat. A light microscopic study, *J. Comp. Neurol.* **178**:679–696.

Berger, B., 1971, Formes divèrses de degenerescence des boutons synaptiques dans le glomeru le olfactif de lapin après lesion du nerf olfactif, *Brain Res.* **33**:218–222.

Bignami, A., and Ralston, J. H., 1960, The cellular reaction to Wallerian degeneration in the central nervous system of the cat, *Brain Res.* **13**:444–461.

Blinzinger, K., and Kreutzberg, G., 1968, Displacement of synaptic terminals from regenerating motor neurons by microglial cells, *Z. Zellforsch. Mikr. Anat.* **85**:145–157.

Brand, S., and Mugnaini, E., 1975, Electron microscopy of light microscopically identified degenerating axons and dendrites selectively impregnated with a Fink–Heimer method, *Anat. Rec.* **181**:318.

Conradi, S., 1969, Ultrastructure and distribution of neuronal and glial elements on the motoneuron surface in the lumbosacral spinal cord of the adult cat, *Acta Physiol. Scand.* [*Suppl.*] **332**:5–48.

Cook, R. D., 1974, Observations on glial cells within myelin sheaths in degenerating optic nerves, *J. Neurocytol.* **3**:737–751.

Cook, R. D., and Wisniewski, H. M., 1973, The role of oligodendroglia and astroglia in Wallerian degeneration of the optic nerve, **61**:191–206.

Cowan, W. M., 1970, Anterograde and retrograde transneuronal degeneration in the central and peripheral nervous system, in: *Contemporary Research Methods in Neuroanatomy* (W. J. H. Nauta and S. O. E. Ebbeson, eds.), pp. 217–251, Springer-Verlag, Berlin.

DeHoff, R. T., and Rhines, F. N., 1968, *Quantitative Microscopy,* McGraw-Hill, New York.

Elias, H. A., Hennig, A., and Schwartz, D. E., 1971, Stereology: Applications to biomedical research, *Physiol. Rev.* **51**:158–200.

Fox, C. A., Hollman, D. E., Siegesmund, K. A., and Dutta, C. R., 1967, The primate cerebellar cortex: A Golgi and electron microscopic study, Prog. Brain Res. **25**:174–225.

Gentschev, T., and Sotelo, C., 1973, Degenerative patterns in the ventral cochlear nucleus of the rat after primary deafferentation. An ultrastructural study, *Brain Res.* **62**:37–60.

Grant, G., 1970, Neuronal changes central to the site of axon transection. A method for the identification of retrograde changes in perikarya, dendrites and axons by silver impregnation, in: *Contemporary Research Methods in Neuroanatomy* (W. J. H. Nauta and S. O. E. Ebbeson, eds.), pp. 173–186, Springer-Verlag, Berlin.

Guillery, R. W., 1970, Light and electron microscopical studies of normal and degenerating axons, in: *Contemporary Research Methods in Neuroanatomy* (W. J. H. Nauta and S. O. E. Ebbesson, eds.), pp. 77–104, Springer-Verlag, Berlin.

Heimer, L., 1969, Silver impregnation of degenerating axons and their terminals on Epon–Araldite sections, *Brain Res.* **12**:246–249.

Heimer, L., 1970, Bridging the gap between light and electron microscopy in the experimental tracing of fiber connections, in: *Contemporary Research Methods in Neuroanatomy* (W. J. H. Nauta and S. O. E. Ebbesson, eds.), pp. 162–172, Springer-Verlag, Berlin.

Heimer, L., and Kalil, R., 1978, Transneuronal degeneration and death of cortical neurons following removal of the olfactory bulb in adult rats, *J. Comp. Neurol.* **178**:559–609.

Heuser, J. E., and Reese, T. S., 1973, Evidence for recycling of synaptic vesicle membrane during transmitter release at the frog neuromuscular junction, *J. Cell Biol.* **57**:315–344.

Jean-Baptist, M., and Morest, D. K., 1975, Transneuronal changes of synaptic ending and nuclear chromatin in the trapezoid body following cochlear ablations in cats, *J. Comp. Neurol.* **162**:111–134.

Jones, E. G., and Rockel, A. J., 1973, Observations on complex vesicles, neurofilamentous hyperplasia and increased electron density during terminal degeneration in the inferior colliculus, *J. Comp. Neurol.* **147**:93–118.

Kane, E. S., 1974, Patterns of degeneration in the caudal cochlear nucleus of the cat, *Anat. Rec.* **179**:67–92.

Kawana, E., Akert, K., and Bruppacher, H., 1971, Enlargement of synaptic vesicles as an early sign of terminal degeneration in the rat caudate nucleus, *J. Comp. Neurol.* **142**:297–308.

Korthals, J. K., Wisniewski, H. M., Ghetti, B., and Cook, R. D., 1974, The fate of the axon and its terminal in the Pacinian corpuscle following sciatic nerve section, *J. Neurocytol.* **3**:385–403.

Kruger, L., and Maxwell, D. S., 1969, Wallerian degeneration in the optic nerve of a reptile. An electron microscopic study, *Am. J. Anat.* **125**:247–270.

Mathews, M. A., and Kruger, L., 1973a, Electron microscopy of non-neuronal cellular changes accompanying neuronal degeneration in thalamic nuclei of the rabbit. I. Reactive hematogenous and perivascular elements within the basal lamina, *J. Comp. Neurol.* **148**:255–312.

Mathews, M. A., and Kruger, L., 1973b, Electron microscopy of non-neuronal cellular changes accompanying neuronal degeneration in thalamic nuclei of the rabbit. II. Reactive elements within the neuropil, *J. Comp. Neurol.* **148**:313–346.

Morest, D. K., and Jean-Baptist, M., 1975, Degeneration and phagocytosis of synaptic endings and the axon in the medial trapezoid nucleus of the cat, *J. Comp. Neurol.* **162:**135–156.

Mugnaini, E. and Walberg, F., 1964. Ultrastructure of neuroglia, *Ergebn. Anat. Entwicklungsgesch.* **37:**193–236.

Mugnaini, E., and Walberg, F., 1967, An experimental electron microscopical study on the mode of termination of cerebellar corticovestibular fibres in the cat lateral vestibular nucleus (Deiter's Nucleus), *Exp. Brain Res.* **4:**212–236.

Mugnaini, E., Walberg, F., and Brodal, A., 1967, Mode of termination of primary vestibular fibres in the lateral vestibular nucleus, an experimental electron microscopical study in the cat, *Exp. Brain Res.* **4:**187–211.

Pecci-Saavedra, J., Vaccarezza, O. L., Reader, T. A., and Pasqualini, E., 1970, Synaptic transmission in the degenerating lateral geniculate nucleus, an ultrastructural and electrophysiological study, *Exp. Neurol.* **26:**607–620.

Peters, A., Palay, S. L., and Webster, H. deF., 1976, *The Fine Structure of the Nervous System,* Saunders, Philadelphia.

Price, D. L., and Porter, K. R., 1972, The response of ventral horn neurons to axonal transection, *J. Cell Biol.* **53:**24–37.

Raisman, G., and Mathews, M. R., 1972, Degeneration and regeneration of synapses, in: *The Structure and Function of Nervous Tissue, Vol. 5* (G. H. Bourne, ed.), pp. 61–104, Academic Press, New York.

Rand Corporation, 1955, *A Million Random Digits With 100,000 Normal Deviates,* Free Press, Glencoe, Illinois.

Shay, J., 1975, Economy of effort in electron microscope morphometry, *Am. J. Pathol.* **81:**503–511.

Smith, C. A., and Rasmussen, G. L., 1965, Degeneration in the efferent nerve endings in the cochlea after axonal section, *J. Cell Biol.* **26:**63–77.

Snedecor, G. W., and Cochran, W. G., 1967, *Statistical Methods, Sixth Edition,* Iowa State University Press, Ames, Iowa.

Sotelo, C., and Palay, S. L., 1971, Altered axons and axon terminals in the lateral vestibular nucleus of the rat, possible example of axonal remodelling, *Lab. Invest.* **25:**653–671.

Spencer, P. T., and Thomas, P. K., 1974, Ultrastructure of the dying-back process. II. The sequestration and removal by Schwann cells and oligodendrocytes of organelles from normal and diseased axons, *J. Neurocytol.* **3:**763–783.

Szentagothai, J., Hamori, J., and Tömböl, T., 1966, Degeneration and electron microscopic analysis of the synaptic glomeruli in the lateral geniculate body, *Exp. Brain Res.* **2:**283–301.

Thomas, P. K., and King, H. M., 1974, The degeneration of unmyelinated axons following nerve section: An ultrastructural study, *J. Neurocytol.* **3:**497–512.

Torvik, A., and Skjørten, F., 1971, Electron microscopic observations on nerve cell regeneration and degeneration after axon lesions. II. Changes in the glial cells, *Acta Neuropathol. (Berl.)* **17:**265–282.

Turner, J. E., and Glaze, K. A., 1977, The early stages of Wallerian degeneration in the severed optic nerve of the newt, *Anat. Rec.* **187:**291–310.

Turner, J. E., and Singer, M., 1975, The ultrastructure of Wallerian degeneration in the severed optic nerve of the newt, *Anat. Rec.* **181:**267–286..

Underwood, E. E., 1970, *Quantitative Stereology,* Addison-Wesley, Reading, Massachusetts.

Weibel, E. R., 1969, Stereological principles for morphometry in electron microscopic cytology, *Int. Rev. Cytol.* **26:**235–302.

Westrum, L. E., 1969, Electron microscopy of degeneration in the lateral olfactory tract and plexiform layer of the prepyriform cortex of the rat, *Zellforsch. Mikr. Anat.* **98:**157–187.

Westrum, L. E., 1973, Early forms of terminal degeneration in the spinal trigeminal nucleus following rhizotomy, *J. Neurocytol.* **2:**217–235.

Wisniewski, H. M., Ghetti, B., and Horonpian, D. S., 1972, The fate of synaptic membranes of degenerating optic nerve terminals, and their role in the mechanism of trans-synaptic changes, *J. Neurocytol.* **1:**297–310.

Wong-Riley, M. T. T., 1971, Terminal degeneration and glial reactions in the lateral geniculate nucleus of the squirrel monkey after eye removal, *J. Comp. Neurol.* **144:**61–92.

Tract Tracing by Electron Microscopy of Golgi Preparations 12

THEODOR W. BLACKSTAD

I. INTRODUCTION

A. Purpose and Potentialities

The unique contribution of the Golgi methods is their well-known ability to demonstrate soma, axons, and dendrites together and thereby to permit classification of neurons at the light microscopic level. Chapter 9 deals with various aspects of this subject. With the aid of electron microscopy (EM), on the other hand, details of synaptology and other ultrastructural features can be demonstrated. A combined Golgi–EM strategy makes possible ultrastructural analysis of synaptic relations where one of the components of a synapse is classified by means of Golgi impregnation. The impregnated component may be either post- or presynaptic. A few such situations suitable for Golgi–EM analysis will be mentioned (see Fig. 1).

1. When the Postsynaptic Neuron is Impregnated

If the unimpregnated presynaptic terminal has a special appearance (Fig. 1A, a) that permits its identification and classification without a special

THEODOR W. BLACKSTAD • Institute of Anatomy, University of Oslo, Oslo 1, Norway.

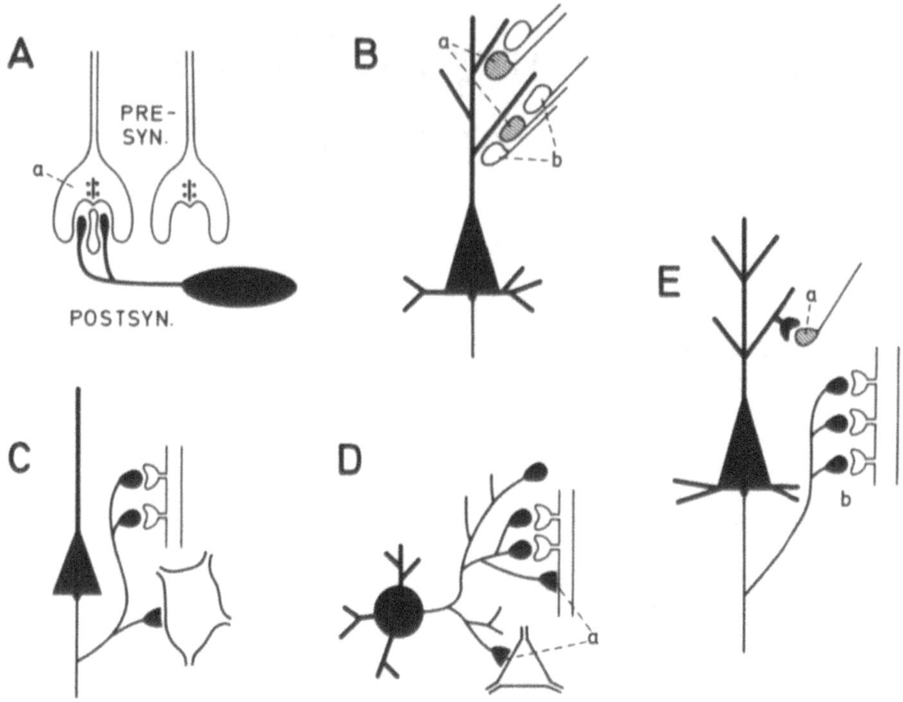

Figure 1. Diagrammatic representation of some situations where the Golgi–EM strategy can give valuable information. Impregnated structures (can be deimpregnated), solid black; degenerated axon terminals, hatched; unimpregnated structures, open. (A) The case when the normal presynaptic component (a) is readily recognizable because of its special configuration; (B) when some of the afferent terminals (a) in contact with an identified, Golgi-impregnated neuron are distinguishable from other terminals (b) because they show degeneration provoked by interruption of their parent axons; (C) when both a parent cell and collaterals of its axon are Golgi-impregnated; the synaptic contacts formed by the axon collaterals can then be analyzed; (D) similarly, the mode of contacts made by the axon ramifications of neurons with very short axon (a) can be studied when cell and axon ramifications are impregnated; (E) combining severance of afferents (a) to an impregnated cell with a study of the relationships of its axon collaterals (at b) allows description of the synapses connecting the neurons in a three-neuron chain.

"tagging" procedure, the Golgi–EM strategy can reveal specific synaptic relations in normal material. This is exemplified by studies on the retina, exploring the synaptic relations between the short photoreceptor axon, with its peculiar spherule bouton, and impregnated horizontal cells (Stell, 1965, 1967; Stell *et al.*, 1975).

In the majority of situations, presynaptic components of interest must be tagged (Fig. 1B, a) by some experimental procedure in order to be discriminated from other afferent systems (Fig. 1B, b). Up to the present time, only transection of afferent pathways has been used for this purpose (e.g., Blackstad, 1965, 1975b; Somogyi, 1978; White, 1978). As described in Chapters 4 and 11, this provokes axonal degeneration demonstrable with the electron microscope. With patience, one may succeed in obtaining ultrathin sections of degenerating axonal terminals in contact with impregnated nerve cells.*

2. When the Presynaptic Neuron is Impregnated

The synaptic relations of short axons and collaterals are difficult to study with degeneration methods. Light and electron microscopy of Golgi-impregnated structures of this kind in normal animals can provide information on their relations in the tissue (Figs. 1C,D). Examples include studies on cells in the visual cortex by LeVay (1973), Somogyi (1978), Peters and Fairén (1978), and Fairén and Valverde (1980).

3. When Both Dendrites and Axons are Impregnated

In this case pre- and postsynaptic relations of the same cell may be elucidated (Fig. 1E, a,b; Somogyi, 1978).

A further important refinement is Golgi impregnation of HRP material as done by Somogyi and Smith (1979); HRP is injected in an area of termination of axons far away from the parent cells. The region containing these cells is first incubated to reveal HRP distribution and is then Golgi-impregnated. The parent cells may demonstrate retrogradely transported HRP even when Golgi-impregnated, particularly if the Golgi impregnation is reduced by deimpregnation. If degeneration of afferents to these cells has first been provoked (Somogyi *et al.*, 1979, 1981), further detailed information on connectivity is obtained.

B. Technical Approach

Prior to a more detailed discussion of technical aspects, the general steps involved in a Golgi–EM study will be briefly previewed. The basic requirement, obviously, is that the tissue be subjected to Golgi impregnation. On the whole, this is done as for light microscopy (Chapter 9) by treating the tissue with dichromate followed by silver nitrate.† However, to secure good

*Presynaptic elements can also be labeled by injecting tritiated amino acids into the area of origin of a pathway (Chapter 5). However, no Golgi–EM study of autoradiographically labeled material has yet been published. This combination would be useful in order to show the exact origin of the boutons ending on the impregnated neurons. This may have remained unsettled after anterograde degeneration studies (see Chapter 4) because these may fail to differentiate between passing fibers and true cell bodies of origin. Moreover, it would demonstrate the true appearance and contact mode of the terminals, whereas anterograde degeneration might lead to serious distortion of these features. After the recent demonstration (Somogyi and Smith, 1979) that tissue incubated to show retrogradely transported horseradish peroxidase (HRP) may successfully be subjected to Golgi impregnation, it seems certain that even a combined anterograde HRP transport–Golgi–EM strategy must be possible.

†So far only silver chromate impregnation has been successfully combined with electron microscopy, and the present chapter will therefore deal with this method only and not with the mercury method.

preservation of ultrastructural details, the treatment must be preceded by or combined with fixation for electron microscopy. It is usually an advantage to perfuse the animal with aldehyde and subsequently transfer the tissue to dichromate. In order to stabilize the tissue structure, and also, in part, to favorably influence the impregnation process, a treatment with osmium tetroxide must be added. The osmium treatment can precede the dichromate, or the two agents may be used together. Then the tissue is impregnated with silver nitrate.

The impregnated tissue may now be embedded in plastic, and ultrathin sections can be cut. That is, it is feasible to go directly from the silver nitrate, through an ascending series of dehydrating baths, to the resin which is then polymerized (see Chapter 10). As a rule, however, some additional treatments are necessary. Thus, further division into smaller parts and sectioning of the impregnated tissue are required before embedding. These are necessary both in order to identify interesting impregnated structures in the light microscope and to obtain samples that are sufficiently small for ultramicrotomy and electron microscopy. This topic is dealt with in Section II.D. Another treatment that is highly recommended for most purposes is a partial "deimpregnation" of impregnated structures, that is, a removal of silver chromate in order to prevent the latter from obscuring ultrastructural details. Furthermore, following deimpregnation, the impregnated structures become easier to section with the ultramicrotome, and they do not rupture in the electron beam.

When the different technical processes listed above have been performed, the investigator may see, in the electron microscope, pictures like those shown in Figs. 2–12.

II. GENERAL DESCRIPTION OF TECHNIQUES

A. Fixation

No specific recipe need be given here, since any good aldehyde fixation for electron microscopy can be used prior to Golgi impregnation. Neither

Figure 2. Normal bouton from the molecular layer of the rat fascia dentata (ib) deimpregnated with the UV photochemical method. d, unimpregnated dendrite; g, glycogen in astroglia; s, spine (45,000 ×).

Figure 3. Impregnated dendrite (id) with spines from rat hippocampal cortex. Asymmetric synaptic contact between an impregnated spine (is) and a bouton is seen (similar contacts with unimpregnated spines (s) are visible). Deimpregnation by means of the photochemical method with gold. b, unimpregnated axon terminals (boutons); s, unimpregnated spines (45,000 ×).

Figure 4. Preterminal and terminal portion (ib) of an impregnated axon forming a synapse with the cell body (C) of a stellate cell in layer III of the visual cortex. The symmetric type of synapse can be recognized. Deimpregnation by gold toning. (Courtesy of A. Peters and C. C. Proskauer). b, unimpregnated axon terminal (bouton); d, unimpregnated dendrite; s, spine (27,800 ×).

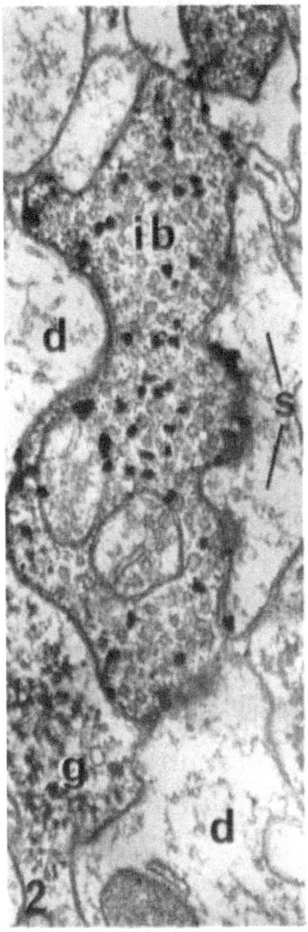
ib
d
s
g
d
2

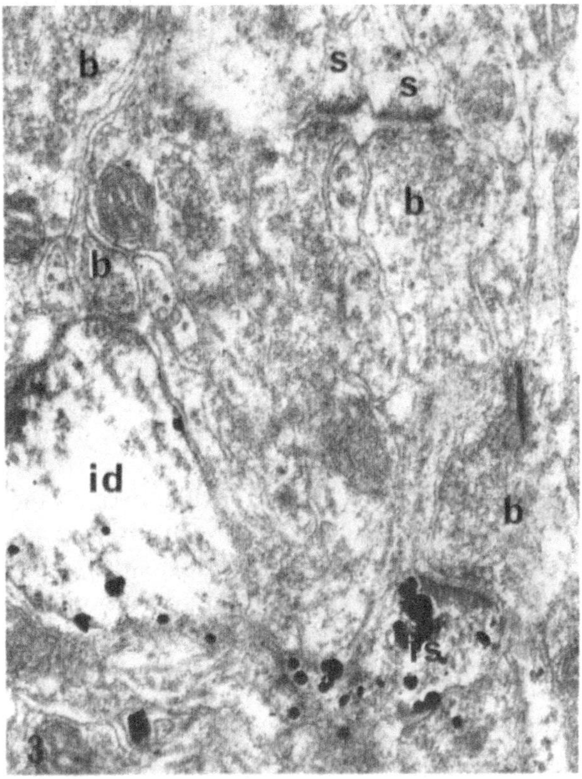
b
s
s
b
b
id
b
is
3

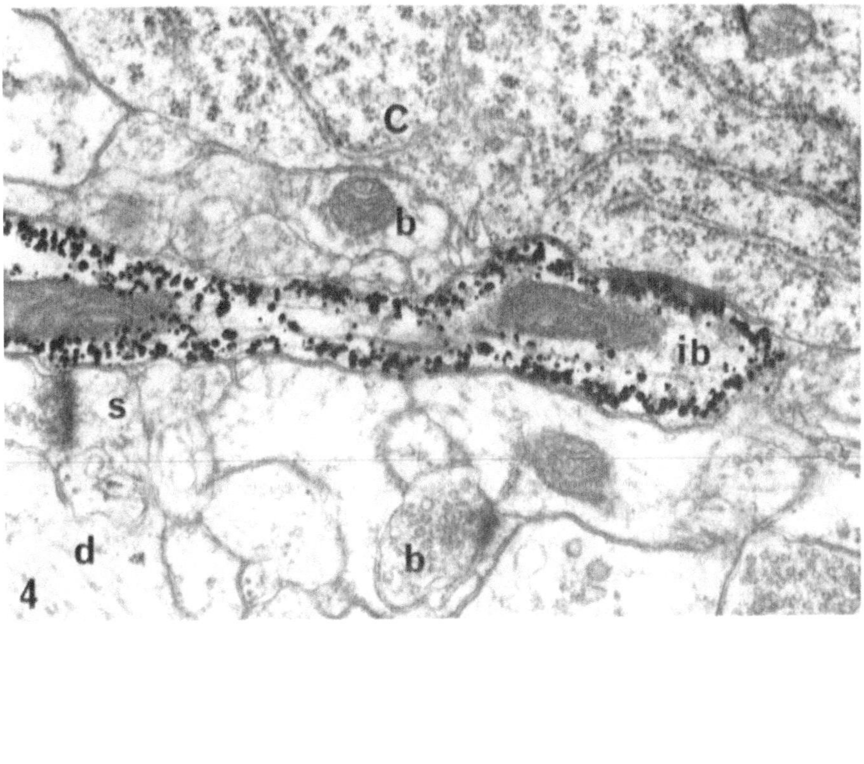
c
b
ib
s
d
b
4

the aldehyde composition nor the choice of buffer (e.g., phosphate, cacodylate, etc.) seems to be critical for Golgi impregnation. The use of a buffer wash after fixation is optional. However, in an attempt to favor impregnation of specific structures in a particular species, one may experiment with different fixatives.

Perfusion and subsequent quality control of the tissue must follow the same strict rules as for electron microscopy in general (Chapter 10). At all stages, from fixation to embedding, the tissue should be manipulated minimally and treated gently; osmotic shocks and violent diffusion currents must be avoided, and after impregnation as much of the processing as possible should be carried out in ice baths.

B. Osmification and Chromation

Good results can be obtained by treating the aldehyde-fixed tissue with osmium tetroxide, OsO_4, before proceeding to the potassium dichromate (Section V.A.1). But most investigators have preferred to include OsO_4 in the dichromate solution constituting the initial stage of impregnation, either in full strength (1–2%; see Section V.A.2) or in a lower concentration (0.2%), necessitating further osmification after completed silver impregnation (Section V.A.3).

These three alternatives seem to be fairly equivalent as regards tissue preservation. Examples of particularly good ultrastructure after the second alternative are given by LeVay (1973). The third alternative will cause less

Figure 5. Hippocampal cortex: impregnated bouton (ib). Photochemical method with gold. d, unimpregnated dendrite; s, spine (45,000 ×).

Figure 6. Hippocampal cortex: impregnated dendritic shaft (id) and spine (is). Boutons in contact with the impregnated spine and with unimpregnated spines nearby are seen. Photochemical method with gold. b, unimpregnated axon terminal (bouton); s, unimpregated spine (29,300 ×).

Figure 7. Impregnated axon terminals (ib) in contact of asymmetric type with dendritic spine within layer IVcβ of rhesus monkey visual cortex. A dense plaque is discernible within the synaptic cleft. For comparison see contact between unimpregnated bouton and spine nearby. The impregnated terminals belong to one of the local axon collaterals of a spiny stellate cell seen to receive monosynaptic input from the lateral geniculate nucleus (cf. Figs. 10A,B). Impregnation according to Somogyi (1978). (Courtesy of P. Somogyi and A. Cowey). b, unimpregnated axon terminal (bouton); s, spines (57,800 ×).

Figure 8. Impregnated spine (is) in contact with degenerating bouton (db). Rat. Deimpregnation by means of the UV photochemical method. The spine belongs to a fascia dentata granule cell, and the bouton to the perforant path arising in the entorhinal area. The postsynaptic specialization is discernible in the deimpregnated spine. b, unimpregnated axon terminal (bouton). (33,800 ×).

Figure 9. Same material and methods as in Fig. 8 (both are from the material studied by Blackstad, 1975b). Postsynaptic density in impregnated and unimpregnated spine can be compared. db, degenerating bouton; is, impregnated spine; s, spine. (45,000 ×).

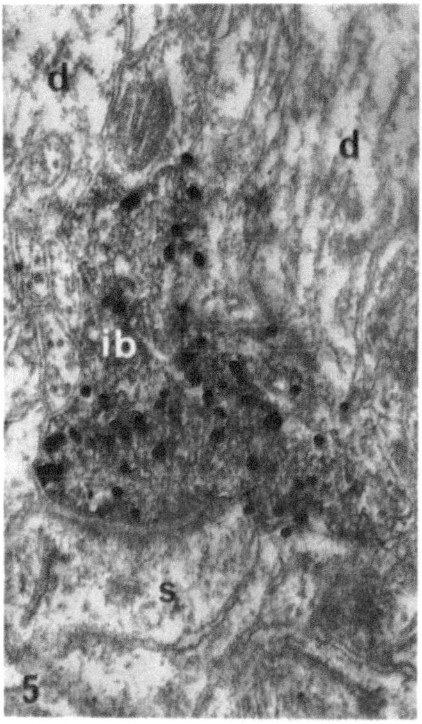
d
d
ib
s
5

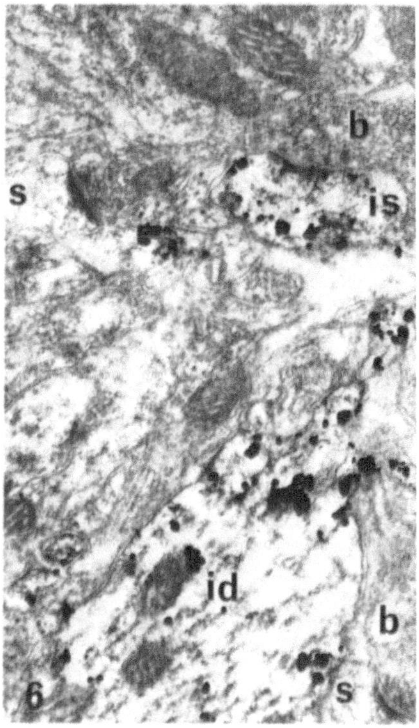
b
s
is
id
b
s
6

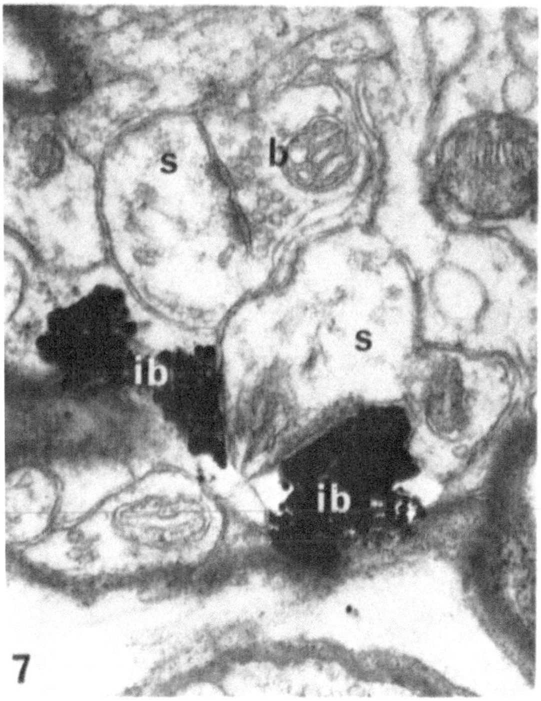
s
b
s
ib
ib
7

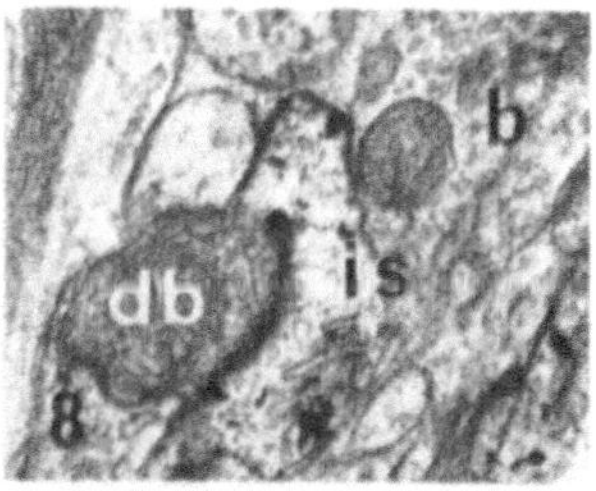
b
db
is
8

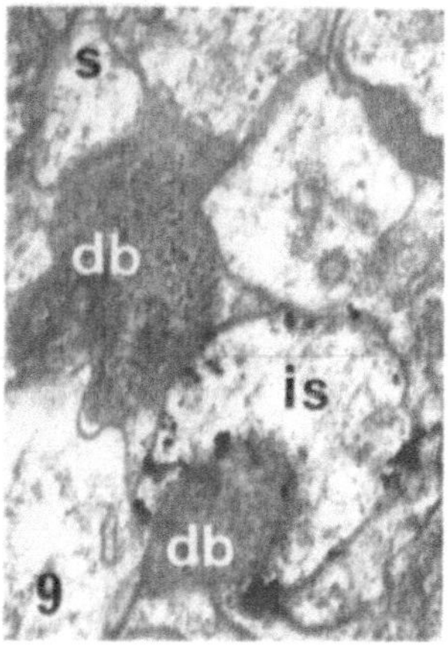
s
db
is
db
9

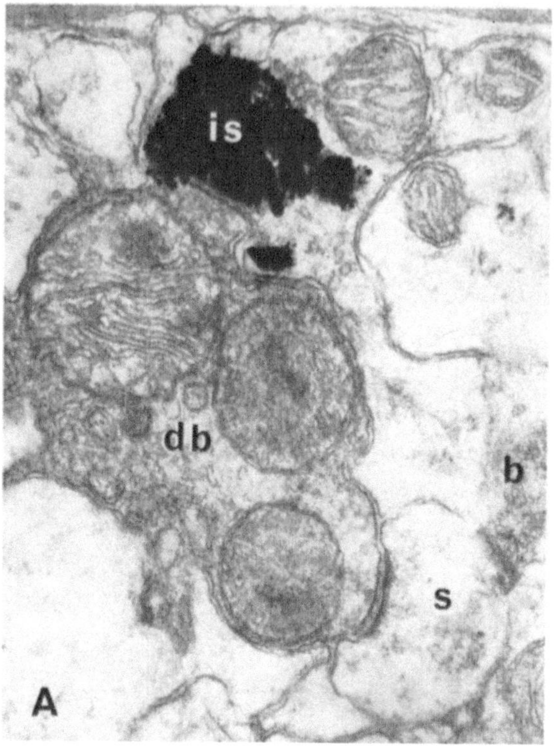
is
db
b
s
A

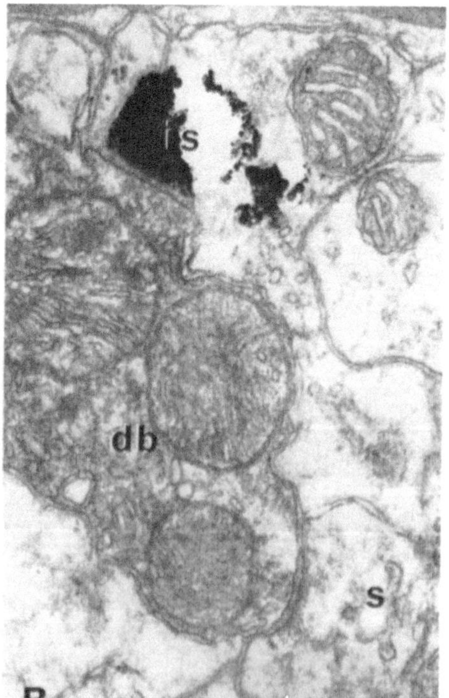
is
db
s
B

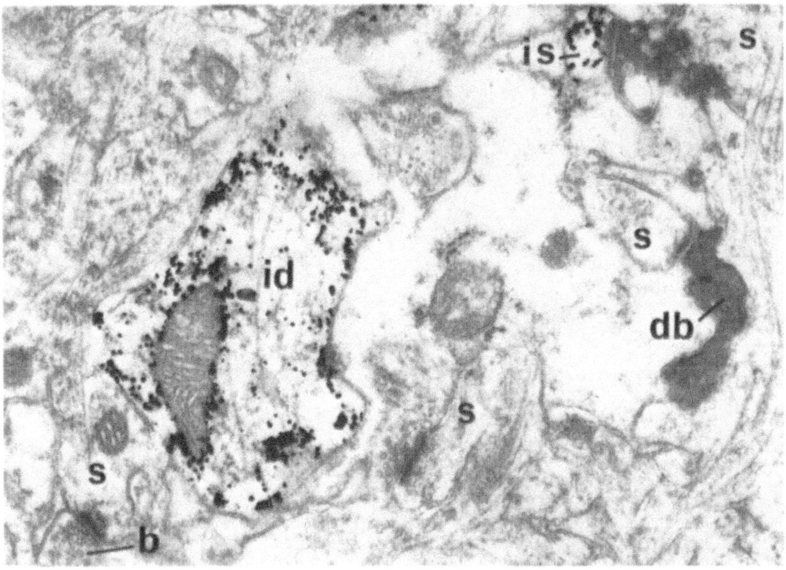

Figure 11. Impregnated apical dendrite (id) of a layer III pyramidal neuron in the rat visual cortex with spine (is) synapsing with a degenerating axon terminal from the lateral geniculate nucleus. In the vicinity, a similar contact with an unimpregnated spine. In both places the postsynaptic specialization of the spine is well seen. Gold toning deimpregnation. [From Peters *et al.* (1979). Courtesy of the authors and the *Journal of Neurocytology* (1980). b, unimpregnated axon terminal (bouton); db, degenerating bouton; s, unimpregnated spine (19,500 ×).

darkening of the tissue because of the lower OsO_4 concentration and may be preferred if thick sections are to be used and these prove to be too dark with other alternatives.

Before the brain or spinal cord is taken to the osmificating or chromating solution, it is sliced into 2- to 4-mm-thick slabs. It is essential in later processing to be aware of the fact that the osmificated zone often extends no more than a few hundred micrometers below the surface; deeper parts of the tissue may contain nicely impregnated structures but will have less-well-preserved ultrastructure. Although it is practical to process osmificated and unosmificated tissue areas together perhaps as far as through the embedding in resin, final selection of tissue for electron microscopy must be done from the osmificated, darker zone.

←

Figure 10. Serial sections from the primary visual cortex of the rhesus monkey. The impregnated spine (is) belongs to a small spiny stellate local circuit interneuron in layer IVcβ, the cell of which impregnated axon collateral terminals are shown in Fig. 7. The degenerating bouton (db) belongs to afferents from the lateral geniculate nucleus where a lesion was placed in the parvocellular layers 4 days before sacrifice. The degenerating bouton also contacts an unimpregnated spine (s), and at both synapses a widened synaptic cleft, electron-dense cleft material, and a dense plaque present within the cleft are seen. In the unimpregnated spine, the postsynaptic density is visible in addition. (Courtesy of P. Somogyi and A. Cowey.) b, unimpregnated axon terminal (bouton) (57,800 ×).

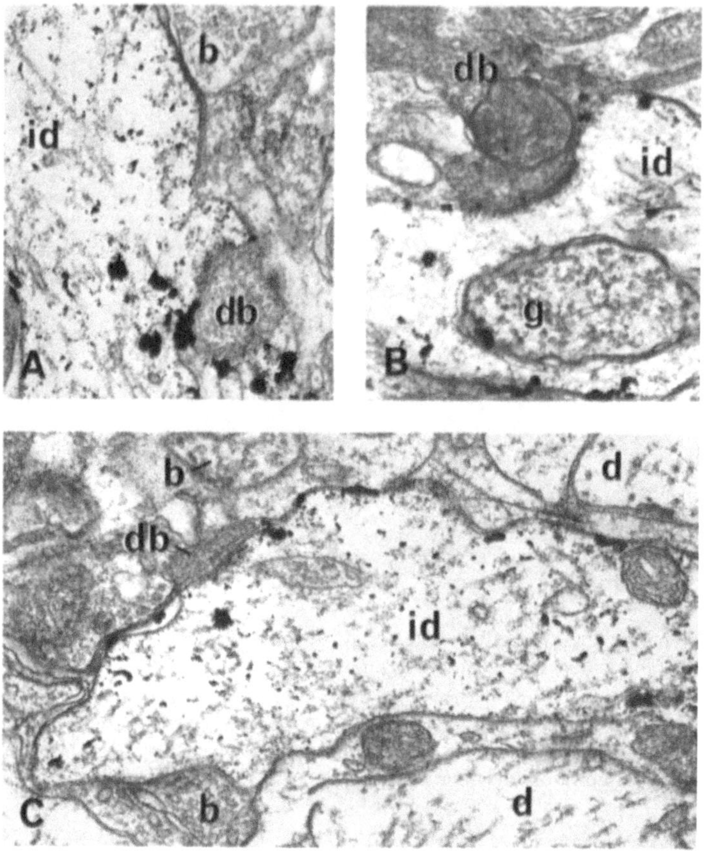

Figure 12. Synaptic contacts between degenerating boutons (db) of rat perforant path and impregnated dendritic stems (id) of fascia dentata granule cells. The postsynaptic density of these asymmetric contacts is well seen. Photochemical UV method without gold. [A and B from Blackstad (1975b).] b, unimpregnated axon terminal (bouton); d, unimpregnated dendrite; g, glycogen in astroglia (45,000 ×).

The duration of chromation can be varied considerably. On the short side, usually 2 or more days are required to obtain satisfactory impregnation. However, even after as long as 4–9 days of chromation, good to excellent ultrastructure will still be found.

C. Silver Impregnation

After an optional brief rinse in distilled water following the dichromate or dichromate–osmium treatment, the tissue is transferred to a solution of 0.75% silver nitrate. Rinsing in water is thought to counteract the formation of a dense crust of silver chromate at the surface of the tissue. A period of 1 to 3 days in silver nitrate is recommended. In order to obtain a more complete impregnation of specific structures, double or triple impregnation may be resorted to and can successfully be combined with electron microscopy (Stell, 1965, 1967). However, it is not recommended as a routine procedure, since protracted impregnation for many weeks may jeopardize the preservation of ultrastructural detail.

After completed silver impregnation, it is useful in many cases to infiltrate the blocks with glycerol, as described in Section V.B. This is a way to make storage of impregnated tissue possible when, for practical reasons, it cannot be processed immediately after impregnation. Extended periods of storage are not recommended, since the fine tissue structure may be endangered. Glycerol-infiltrated tissue is also a suitable basis for various further processing as, for example, shown in Section V.E,F.

D. Primary Thick Sectioning

By merely inspecting impregnated slabs it is not possible to see, to any significant degree, whether adequate impregnation of target areas has been achieved. Therefore, sections must be made of the slabs and these must be checked with the light microscope (Fig. 13A,B). (For the sake of clarity, these

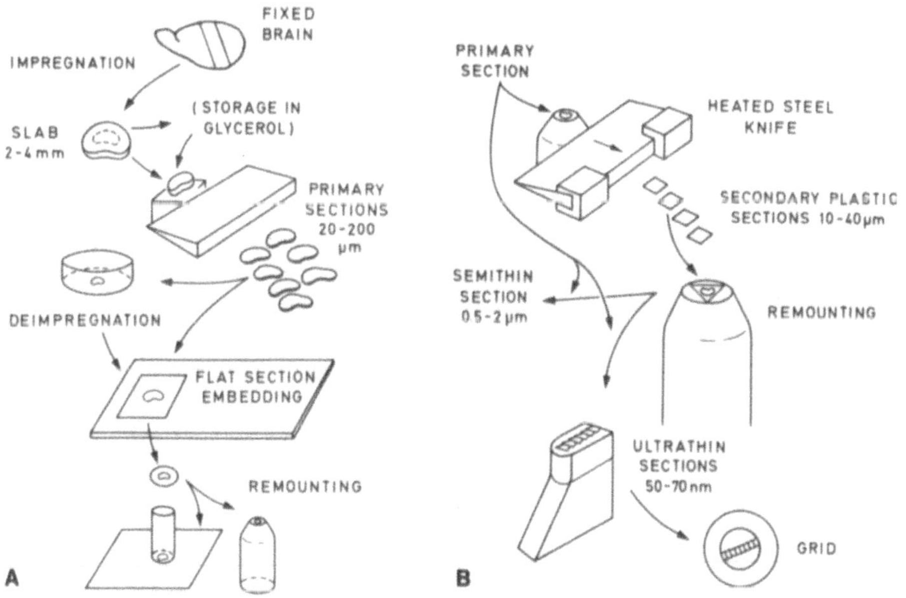

Figure 13. Diagrams showing the main steps in the preparation of material for Golgi–EM. (A) From gross specimen to remounted primary sections; (B) from primary sections to ultrathin sections. Some alternative routes are indicated, e.g., in A that storage in glycerol or deimpregnation may be included or omitted, and in B that primary sections need not always be further split into thinner secondary sections. For the sake of clarity, trimming, staining, and other obligatory or optional steps mentioned in the text have been omitted.

will be called "primary" sections.) First trim away larger irrelevant areas, e.g., cerebral cortex if thalamus is to be studied. Then cut sections from 20 μm to 200 μm thick or more, depending on the purpose of the investigation and the further treatment of the sections. Sectioning can be done with a sliding microtome, a Vibratome, or a tissue chopper, or by hand with a razor. On the whole, the method used is unimportant for the final result. However, to preserve the ultrastructure of the tissue, full embedding (e.g., in celloidin or paraffin) must be dispensed with; moreover, during the sectioning it is of paramount importance that the cut surface be kept thoroughly wet. Some guidelines for "primary sectioning" are given in Section V.C.

E. Deimpregnation

Ultrathin sectioning and EM of tissue impregnated with silver chromate are plagued with some disadvantages as already mentioned; these have been discussed in detail by Blackstad (1965, 1970) and Somogyi (1978). Although valuable information may be obtained through careful study of such material (see, e.g., LeVay, 1973; Somogyi, 1978; Somogyi and Cowey, 1981), it is recommended that some deimpregnation treatment be applied routinely. Deimpregnation is done after the primary sectioning of the impregnated tissue and before the embedding in plastic (Fig. 13A). See Fig. 14 for simplified flowcharts showing the sequence of the main steps in the deimpregnation procedures discussed below.

1. Gold Toning

The most common method of deimpregnation is that of Fairén *et al.* (1977). The sections, which may be up to 200 μm thick, are placed in a weak solution of gold chloride; this leads to the appearance of submicroscopic particles of metallic gold in the impregnated structures. The excess silver chromate is dissolved with thiosulfate, after which the sections are dehydrated and embedded flat. A detailed recipe for gold toning is given in Section V.D. Examples of the results of this method are shown in Figs. 4 and 11. At present, this method must be considered the most generally useful way of deimpregnation.

2. Photochemical Method without Gold

With this method (Blackstad, 1975a,b), the sections are irradiated with ultraviolet light (UV) which produces submicroscopic particles of silver in the impregnated structures (a "latent image" in photographical terminology). Again, the remaining silver chromate is dissolved, and the tissue is

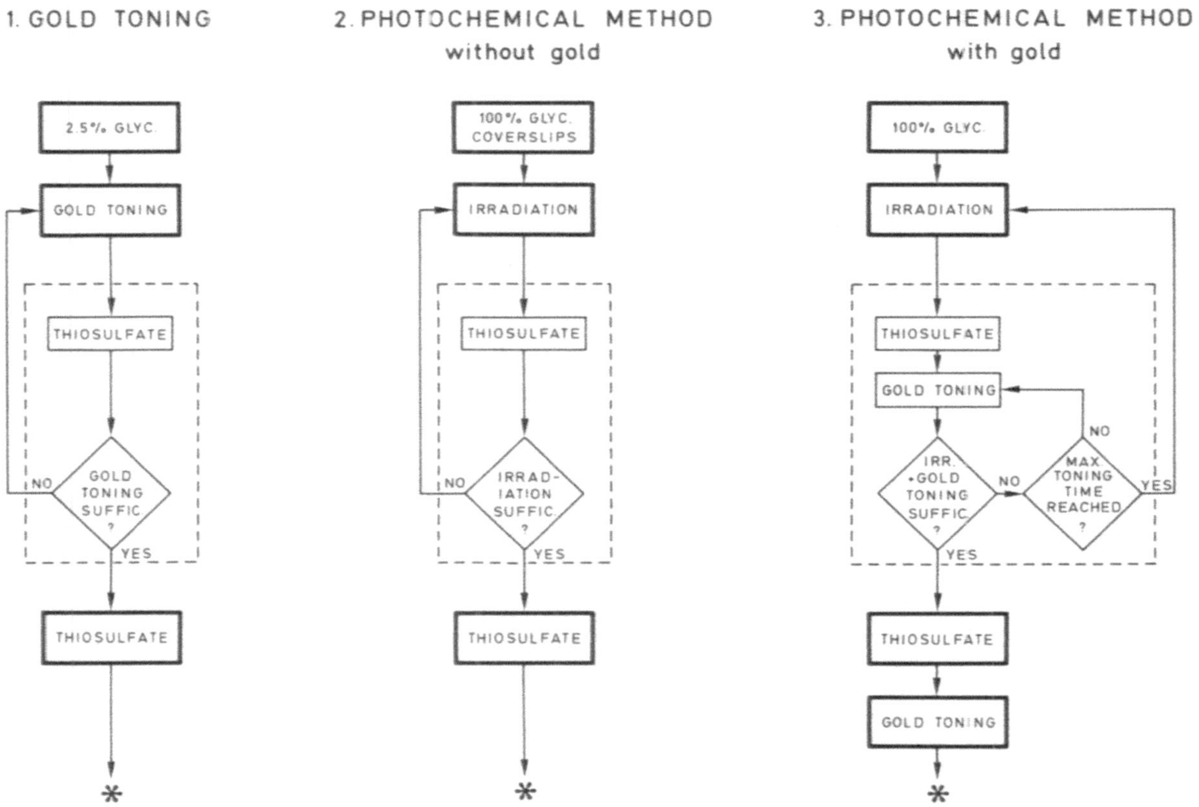

Figure 14. Flowchart diagrams of the essential steps in three procedures (Sections V.D–F) for deimpregnation and partial substitution of the silver chromate of impregnated structures. Heavily outlined rectangles describe the treatment of all main sections. Weakly outlined rectangles and diamond-shaped areas (collectively circumscribed by broken lines) describe treatment of test sections only. Asterisk indicates subsequent steps: Postosmification if needed, dehydration, and embedding.

embedded for EM (Section V.E). Examples of the results of this method are given in Figs. 2, 8, 9, and 12. A drawback of the UV method is that the shallow depth of the photochemical effect limits the useful section thickness to about 25–30 μm. Pilot studies by the author suggest, however, that the UV rays can be replaced by strong cold light which will penetrate to a greater depth of tissue.

3. Photochemical Method with Gold

In this method (Section V.F), the sections are first irradiated with intensive white light which produces a latent image; then the excess silver chromate is dissolved, and, finally, the image is intensified to the desired extent with gold chloride. Examples of the results of this method are shown in Figs. 3, 5, and 6. This method may be tried if difficulties are experienced in obtaining the correct degree of gold toning in the method described in Section V.D.

With all three methods, test sections should be processed. Here, a warning seems justified. It is likely that some capriciousness experienced with deimpregnation methods has resulted from the quite intensive illumination of test sections during microscopic examination. The microscope condenser focuses enough light on a section to produce a latent image (silver particles) in structures containing silver chromate. It can also accelerate growth of gold particles in a section containing gold chloride. The heat of the focused light may enhance these undesirable effects and by itself endanger the fine tissue structure. Accordingly, test sections ought to be examined briefly (for seconds rather than minutes) if they are to be representative of the unilluminated sections selected for EM. In the recipes (Section V.D–F), it is recommended that test sections exposed to light be compared with test sections protected from light. Normal room lighting does not affect the result.

4. Alternative Principles of Deimpregnation

Methods based on the partial substitution of silver chromate for another visible substance, lead chromate, were proposed by Ramón-Moliner and Ferrari (1972, 1976) to whom the reader is referred for details. An interesting feature of this approach is the good visibility of the lead chromate in light microscopy with incident light.

Another principle, here called Interrupted Golgi Impregnation (Scott and Guillery, 1974; W. A. Ribi, personal communication), does not involve deimpregnation in a strict sense but is based on impregnation for such a short time that only scattered speaks of silver chromate are deposited in the impregnated structures. See Section V.G for further information on this approach. Somogyi and Cowey (1981) dissolved some of the silver chromate with 5% ethanol before proceeding to dehydration and embedding for EM.

F. Embedding and Remounting of Primary Sections

After deimpregnation, the "primary" sections are dehydrated and embedded in resin by conventional procedures (Chapter 10). Flatmount embedding is indispensable because it allows examination under the light microscope for drawing and photography of interesting features. Also, after flat-mount embedding, tissue around and close to the interesting structure can readily be trimmed away; unless sufficient trimming is carried out, the ultrathin sections will become to large, serial sectioning become impossible, and searching in the electron microscope become too time-consuming.

Flat embedding can be realized in many ways; two are mentioned here. The primary sections are either placed on microscope slides directly under a coverslip (Section V.H.1) or between acetate foils from which the resin containing the sections can be more easily stripped after polymerization (Section V.H.2). In either case, the investigator will now have a slice or small disk of resin containing the section. After the section has been studied with the light microscope, it must be remounted on a blank plastic base for further sectioning (Fig. 13A). A simple method of doing this is to place the embedded section (the disk) on an acetate foil in a drop of resin, lower a blank cylindrical base down onto it,* and then polymerize the resin at temperatures between 45 and 100°C. It is also possible to fasten the primary section to the base with fast-setting epoxy resin, nail varnish (LeVay, 1973), or cyanoacrylic glue. Another method of remounting, particularly suited for thinner sections, is mentioned in Section V.J.

G. Cutting Thick Sections of Tissue Embedded in Resin

Very often the purpose of a Golgi–EM study will be to analyze individual, specific structures such as single dendrites. In such cases it may prove impractical to proceed directly to ultramicrotomy of thick (e.g., 50–200 μm) primary sections, since this may entail searching for the desired structures through hundreds of ultrathin sections in the electron microscope.† It will normally be more rational to split the embedded primary section into thinner "secondary" sections, perhaps no more than 10–20 μm thick (Fig. 13B). These are then inspected in the light microscope, and the secondary section containing the desired structure is selected for further processing (remounting, ultramicrotomy, and EM). Making sections of plastic for this

*One may follow a procedure described by Holländer (1970) when EM study of a specific minor part of a larger section is needed. It insures that the part of interest comes to be positioned centrally on the plastic cylinder.

†This statement requires qualifications. As discussed in Section V.K, there are methods for monitoring sectioning so closely that it may be feasible—even in thick plastic slices—to locate fine structures within the slice during ultramicrotomy. But such monitoring will also be easier if the slice is not too thick.

and other purposes requires special techniques, some of which are treated in Section V.I.

Remounting plastic sections of this reduced thickness for ultramicrotomy and electron microscopy can be done in the manner already described for resin-embedded primary sections (Section II.F) or in other ways; the method preferred by the author is described in Section V.J.

H. Ultramicrotomy

In some cases trimming and ultramicrotomy of the final, remounted section (block*) for Golgi-EM may be carried out in conventional ways in order to obtain ultrathin sections of structures of interest. The use of semithin sections may be advantageous under certain circumstances (Section V.K.1). Only the investigator will be able to judge when purely conventional methods will suffice.

If a remounted tissue block is very thin (e.g., 5–10 μm), it may be reasonable to cut the entire block on the ultramicrotome, collecting all the ultrathin sections and afterwards locating, with the aid of the electron microscope, the grids and sections that contain the structures of interest. If features of interest, e.g., abundant, uniform axonal or dendritic ramifications, are present throughout the block, the work can further be simplified merely by collecting ultrathin sections from a few levels.

In general, however, Golgi–EM work requires some method of monitoring the ultrathin sectioning, that is, observing in the light microscope how the interesting structures are approached and finally reached by the microtome knife. Several methods have been devised for this; they can be used separately or in combination.

The tissue in semithin sections is lost to electron microscopy, and such checking can, accordingly, not be used in those parts of the block where it is crucial to include virtually every ultrathin section. Therefore, it is likely that the reader will sooner or later find some of the methods mentioned in this section indispensible; they are based on light microscopic inspection of the block itself.

The least sophisticated method, requiring no extra equipment, is to repeatedly detach the block from the microtome holder for microscopic examination (since conventional holders do not permit transillumination). A better solution is to fasten the plastic cylinder with the block rigidly in a metal ring or hollow sleeve that makes an exact fit with a modified microtome chuck. The block plus sleeve are removed for transillumination and examination in the light microscope and can afterwards be reinserted to the same position in the chuck. Descriptions of such systems have been given by

*Although the term block is not very descriptive for thin remounted sections, it is used for the sake of clarity.

LeVay (1973) and by Fairén *et al.* (1977). If equipment for incident illumination of the block is available, the block may be checked in the microscope while remaining in the unmodified holder (see Ramón-Moliner and Ferrari, 1972, 1976, for details). Transillumination of the block without a need to detach it from the holder is made possible if a mirror is placed within the resin, under the tissue (Section V.K.2).

Despite the usefulness of the above methods, the detaching and refastening of the block with its holder nevertheless tend to cause some displacement of the block in relation to the knife. Some readjustment is therefore necessary before ultrathin sectioning can be resumed, entailing a risk of losing some vital material. This is completely avoided if a further mirror is placed in front of the block (Section V.K.3); this external mirror enables the block to be watched with the light microscope during sectioning.

The choice of method for monitoring ultrathin sectioning should depend on the specific needs and available facilities in each case. The least complicated yet efficient method should be selected. Faintly visible, thin structures necessitate more critical methods than more distinct structures.

1. Protection of Silver Chromate

If nondeimpregnated tissue is sectioned on the ultramicrotome, the silver chromate generally disappears from the ultrathin sections and leaves holes with torn edges in the tissue. This can be prevented by picking up the sections so rapidly that the silver chromate is not dissolved (LeVay, 1973) or by filling the trough with water saturated to 80% with silver chromate. In the latter case, rinse the grids with the sections in 90% ethanol (Somogyi, 1978).

Advice on the use of silver chromate for protection of impregnation in whole blocks or thick sections is given in Section V.L.

2. Staining for Electron Microscopy

Some electron optical contrast exists in Golgi material as a result of osmification and impregnation, but additional staining is necessary. Silver or gold particles in deimpregnated tissue will not be attacked by the usual stains for electron microscopy. The silver chromate of tissue that has not been deimpregnated will also usually withstand such treatment. Therefore, specific staining recipes for Golgi–EM need not be given here. Block staining with uranyl acetate or PTA plus uranyl acetate, followed by lead citrate on the grids, has proved adequate. Alternatively, uranyl staining may also be applied to the grids. Staining with lead should always be added. See Chapter 10 as well as standard textbooks for details on staining procedures.

3. Serial Sectioning

The student of Golgi–EM preparations should remember the extreme usefulness of serial sections, not only in visualization of three-dimensional relations but also for the purpose of identifying profiles in electron micrographs. It is a poor strategy to invest time and effort in obtaining electron micrographs of selected impregnated structures if the nature of the adjacent structures is not fully understood. In single sections of neuropil, many profiles cannot be unequivocally classified because they do not display enough ultrastructural criteria (glycogen, synaptic vesicles, tubules, etc.). If a profile is traced through 5–10 serial sections, criteria facilitating identification nearly always appear.

III. SUMMARY OF ADVANTAGES AND LIMITATIONS

Tract tracing by means of electron microscopy of Golgi preparations is a composite procedure that has the advantages and disadvantages of each of such component procedures as Golgi impregnation, electron microscopy, and HRP histochemistry.

A. Advantages

1. Light microscopic examination of a Golgi specimen prior to electron microscopy permits the identification and classification of a studied cell, or portion of a cell, seen to be part of a synaptic complex.
2. The initial stages of the Golgi—EM procedure represent a simple task in the sense that little special equipment is needed. The work may be inserted between other daily occupations, and the mere investment of time and patience will not fail to yield sufficient material for EM processing. Similarly, the following steps (including sectioning, gold toning, and EM) mainly require time, accuracy, and patience for success; no equipment and expertise other than what is nowadays normally available in a department of anatomy or biology are needed.
3. No conflict exists between the pretreatments needed for Golgi impregnation and those for electron microscopy, since the preparatory steps needed to obtain adequate ultrastructure for EM analysis create a good basis for Golgi impregnation.
4. *Specific tract-tracing applications*:
   a. Golgi–EM is particularly useful for studying the terminations of afferent pathways on an impregnated postsynaptic neuron (soma, dendrites, or, more rarely, axonal components). So far, the afferent systems in combined Golgi–EM studies have been marked

primarily with anterograde degeneration. Potentially, anterograde labeling of axon and terminal with a tracer, e.g., HRP, may be combined with Golgi–EM in the same way.

b. Golgi visualization of presynaptic structures (axons, terminals) may compete with other methods for tract tracing in the case of ultrashort axons or local collaterals, where surgical methods or injection of a tracer is less applicable.
c. A procedure of great potential for the study of the cells of origin of a long-traveling pathway is the combination of conventional demonstration of the cells by retrograde filling with tracer (HRP) with further classification of the same cells by means of added Golgi impregnation.
d. The demonstration that it is possible to Golgi impregnate a slice previously subjected to histochemical incubation [cf. (c)] suggests the possibility of combining histochemistry, including immunohistochemistry, with Golgi and Golgi–EM.

B. Limitations

1. The capriciousness of the Golgi procedures also restricts the applicability of Golgi–EM.
2. In order to follow long pathways, Golgi–EM alone is not usable and cannot compete with studies based on HRP transport, which can demonstrate both the structure of terminal ramifications (like Golgi specimens) and the origin of pathways (unlike Golgi specimens).
3. The identification of the postsynaptic structure by means of Golgi impregnation, unlike microelectrode stain injection methods, does not have the potential of revealing anything about the electrophysiological characteristics of the cell.
4. The series of steps, from surgery to collection of successful electron micrographs, is long, and even if valuable information can be expected in the end, many difficulties are likely to arise in the course of the procedure.
5. Although the combination of Golgi impregnation and histochemical treatment is possible, the thinness of the blocks required by the histochemical procedure imposes a considerable limitation on the practical usefulness of such a combination.

IV. CONCLUDING COMMENTS AND TROUBLESHOOTING

Golgi–EM methods have been developed during the last few years, and it should be realized that other varieties than those described so far may well prove better for specific purposes. It is wise, therefore, to be aware of the potential usefulness of developing and testing new modifications. The present chapter should, indeed, be conceived of as an introduction and guide to such work rather than a fixed set of rules to follow rigidly.

A few hints on possible sources of disappointment and technical failure may be useful.

1. *Low quality of the ultrastructure.* If the ultrastructure is found to be unsatisfactory, the reason may be that ultrathin sections have been made of poorly perfused material, nonosmificated parts of the tissue, or mechanically damaged superficial tissue.
2. *Broken sections.* The initial (primary) sections of impregnated tissue slabs may tend to break when being cut. This may be because of the tendency of impregnation to make tissue brittle. If possible, reduce the time in dichromate and silver nitrate and section as small blocks as possible. In addition, better sections may be obtained after adjustments of knife-wetting agent, knife angle or speed, vibration amplitude, or other variables of the cutting instrument.
3. *Problematic transfer of sections.* Sections may rupture if lifted with inappropriate needles or brushes instead of with some type of spatula, net, or other nondamaging device.
4. *Scarcity of structures.* It is normal for impregnated profiles to constitute a very small proportion of the neuropil in the ultrathin sections. If this becomes too pronounced, the reason probably is that the block has not been trimmed sufficiently close to the impregnated structures prior to ultramicrotomy. The time spent in the electron microscope room will similarly be prolonged if too many ultrathin sections without impregnated structures have been collected. Then more critical monitoring of the progress of sectioning is needed.
5. *Lost orientation.* Orientation may be lost at the macroscopic level when the brain slabs are impregnated and turn dark. This source of failure is eliminated by taking notes and making sketches of the pieces of tissue through all of the successive trimming stages.
6. *Failure in general.* Continued direct contact with every aspect of the technical process is needed to maintain skill and discover sources of failure when they occur.

V. APPENDIX

In the following pages, a more detailed description is given of some technical procedures mentioned briefly in the main text. Many of them are modified versions of methods used by the quoted authors. Alternative methods given are largely equivalent; their relative usefulness may vary with local laboratory conditions, the nature of the project, and other factors. Figures 13 and 14 are simplified diagrams dealing with certain aspects of making Golgi–EM preparations.

A. Osmification and Chromation

Three alternatives are given (1–3). In all cases, the pieces or slabs into which the brain is divided after perfusion–fixation are placed on a layer of glass wool or glass beads in the bottom of the jar to facilitate access of the solution to all sides of the pieces. Regarding duration of chromation and silver impregnation, see Sections II.B–C.

1. Osmium Tetroxide before Chromate

1. Osmium tetroxide (OsO_4), 1–2% for 1 to several hours (Somogyi, 1978).
2. Potassium dichromate solution without OsO_4 (see Section V.A.2–3).
3. Silver nitrate, 0.75%.

2. Dichromate with 1% Osmium Tetroxide

1. Dichromate–osmium mixture containing 1% OsO_4, 2.4% potassium dichromate, and 2% sucrose. Adjusting the pH to about 7.0 with KOH has been accompanied by good results (LeVay, 1973; Blackstad, 1975b) but may be unimportant (Fairén *et al.*, 1977). The solution may be used ice-cold during the first 24 hr.
2. Optional: replace this mixture with a similar solution without OsO_4 after 1–3 days and continue the chromation.
3. Silver nitrate solution, 0.75%.

3. Dichromate with 0.2% Osmium Tetroxide

1. Dichromate–osmium mixture containing 0.2% OsO_4 and 2.4% potassium dichromate (Fairén *et al.*, 1977) (the pH will be about 4.6).
2. Silver nitrate, 0.75%.
3. After deimpregnation, osmium tetroxide, 2%, for 30 min.

B. Infiltration of Blocks with Glycerol

The following schedule works well with blocks a few millimeters thick, i.e., the impregnation of even the finest structures (thin axons, spine necks) remains intact after days to weeks of storage. Use ice-cold glycerol solutions made up with distilled water (Fairén *et al.*, 1977) or with distilled water saturated with silver chromate (Section V.L). Take the tissue blocks directly from the 0.75% silver nitrate solution to the first glycerol solution. Keep the blocks in glycerol solutions as follows: 20%, 20 min; 20%, 40 min; 40%, 60%, 80%, and 100% (pure glycerol), 60 min each. Then transfer to fresh

100% glycerol and store in refrigerator. It is likely that traces of silver nitrate remain, but these cause no trouble in later processing.

C. Thick Sectioning of Impregnated Tissue

It is not practical to section whole blocks while they are still in the silver nitrate solution. The blocks are therefore transferred to another medium containing ethanol or glycerol, depending on previous treatment and the sectioning method to be used. The choice of sectioning method may be determined by the available facilities. Another factor is the size of the block; the largest sections can be cut with a sliding microtome.

1. Sectioning with a Vibratome

Fill the trough with ice-cold 50% ethanol saturated with silver chromate (Section V.L). Plane one side of the block and glue it to a clean, flat base, preferably of metal. Use a cyanoacrylate glue, such as Loctite IS 496®, that will set in few seconds. A stable junction results if a very thin layer of glue is applied. Keep the tissue moistened with the trough liquid until it can be fully immersed in the liquid (after about 30 sec). Cut with a thick, single-edged razor blade. Empirically find the best angle of clearance, speed, and amplitude setting for obtaining flat, unbroken sections. The trough liquid will wash out the silver nitrate from the sections. Transfer the sections to vials with ice-cold trough liquid. Test sections may be examined in this liquid under a coverslip. These sections should normally not be used for EM because of possible effects of heat and strong light. To obtain better transparency of the test sections, they may be taken to glycerol or even via 95–100% ethanol to a conventional histological clearing agent.

Instead of ethanol, glycerol may be used in the Vibratome trough, particularly with blocks that have been stored in glycerol (Section V.B).

2. Sectioning with a Tissue Chopper

Chopping can be done without an encasing medium (Somogyi, 1978). Follow the instructions for the particular tissue chopper available. The block can be glued to the base with cyanoacrylate glue as explained in Section V.C.1. Encasing in agar (Section V.C.41) is recommended by Fairén *et al.* (1977). During chopping, keep the block wet with the liquid from which it was taken, e.g., glycerol, and immediately transfer the sections to vials with the same medium.

3. Sectioning with a Sliding Microtome

This is the classical way of making sections of Golgi material (see Chapter 9). The sectioning is done largely as for light microscopy. However, the methods of encasing for light microscopy cannot be transferred directly to Golgi–EM since this might be harsh to the ultrastructure. This is likely to be the case with the common encasing in celloidin. Again, remember the essential precaution of keeping the tissue and the knife edge wet with the infiltrating liquid during the sectioning. Glycerol may be too viscous if relatively thin sections (10–50 μm) are to be cut, in which case an ethanol–glycerol mixture is preferable (e.g., glycerol, 1 part; 75% ethanol saturated with silver chromate, 3 parts). For blocks taken directly to sectioning from silver nitrate, 50% ethanol saturated with silver chromate is a suitable medium. Cut with the knife in an oblique orientation.

4. Supporting the Blocks

Blocks from which primary sections are to be made may be fastened to the supporting base without or with encasing ("superficial embedding") being done.

a. No Encasing. The block may be glued to a base as described for Vibratome sectioning. Holländer (1970) described the use of gum Arabic. The base immediately under the block should be of cork, agar, or some other material that will not damage the knife edge if cut into by accident. Clearly, some tissue must be sacrificed where the block has to be planed. Sectioning without encasing is a quick method but can only be used with tissue blocks of sufficient firmness.

b. Encasing in Agar. Because a melted solution of agar sets quickly at a relatively low temperature (about 42°C), agar encasing will not harm the ultrastructure of well-fixed tissue. To obtain a suitable cutting consistency, dissolve 7% of agar in water (no more than 7.5% can be dissolved). This should be done slowly, over several hours, in a boiling water bath or in an oven at 90–95°C. Avoid evaporation of the water. Too high a temperature or heating for too long will spoil the gelating properties of the agar. The clear, syrupy solution is finally poured into boxes fashioned from aluminum foil and allowed to set. The agar is cut into pieces and stored in closed jars in the refrigerator or freezer.

When the agar is needed for encasing, melt a piece in a flask in a boiling water bath or at 90–100°C in the oven (in a moist chamber). Let the agar cool until it nearly sets before immersing the tissue. After the agar has solidified, shape a block as desired. Alternatively, glue a planed piece of tissue to a base (wood, cork, or metal) and pour a layer of melted agar over the tissue. After the agar has barely set (keep it for a few seconds in the refrigerator), pour on a second layer. Trim away excess agar as needed before sectioning.

After encasing, proceed to sectioning without delay to avoid transfer of moisture from the agar gel to the block, causing the firm union between agar and tissue to loosen.

c. Encasing in Paraffin. Submerge the tissue into a well of melted paraffin wax and harden the paraffin in the refrigerator (A. Fairén, personal communication). Solder the trimmed paraffin block to the base (wood or cork) by swiftly pulling the heated blade of a knife between the block and the base. It does only seem safe to use a low-melting-point paraffin for encasing for EM. Sectioning can be done using glycerol as the wetting medium.

D. Gold Toning

The following recipe is based on Fairén *et al.* (1977), Fairén (personal communication, 1979), and unpublished observations by the present author, particularly on the effects of light.

The times needed in the various solutions assume a section thickness of up to 200 μm. Use all solutions ice-cold.

1. Take the sections to 2.5% aqueous glycerol through a graded series of solutions made up with the medium used during the cutting. For instance, 80%, 60%, 40%, and 20% aqueous glycerol, 3–5 min each, if glycerol was used. Avoid vigorous agitation. Renew the 2.5% glycerol once.
2. Gold toning is done in a 0.04% or 0.05% solution of yellow gold (III) chloride ($HAuCl_4 \cdot 4H_20$ or $AuCl_3 \cdot NaCl \cdot 2H_20$ which are equivalent for this purpose) in 2.5% glycerol. Pour the 2.5% glycerol out of the vials and add the gold chloride solution. Agitate the vials gently, either manually or by means of a laboratory shaker. At this stage and later, protect sections containing silver chromate or gold chloride from strong light.
3. After 10 min, replace the gold chloride with 2.5% glycerol.
4. Take a test section to 1% sodium thiosulfate and keep it there until all silver chromate has disappeared. The time needed can be estimated by light microscopic examination of a fragment cut off the test section (or by examining an extra test section). Examine the fragment under a coverslip in thiosulfate solution. Do not mistake traces of silver chromate for gold. Allow some extra time for complete removal of silver chromate from the main test section; then rinse it in water and examine it in the light microscope. The purpose of these precautions is to prevent traces of silver chromate or gold chloride in a test section from being acted on by the strong light of the microscope illumination, which could result in erroneous inferences about the degree of toning of the sections reserved for EM.
5. Usually sufficient gold toning, as shown by a suitable visibility of the impregnated structures after removal of silver chromate, will not have been obtained after 10 min. In that case, pour the 2.5% glycerol off the sections and gently pour on fresh gold chloride solu-

tion. Repeat the sequence of toning, interruption, and checking described above (Steps 2–4), using 5 min periods of gold toning.

6. When toning is considered adequate, rinse the sections in distilled water, three changes of 5 min each. Oxalic acid treatment previously recommended is superfluous.
7. Dissolve any remaining silver chromate with several changes of freshly made, ice-cold 1% sodium thiosulfate, used for 30–90 min depending on section thickness.
8. Rinse in two changes of distilled water, a few min in each.
9. Postosmificate sections of material previously treated with only 0.2% OsO_4 for 30 min at room temperature in 2% OsO_4 in buffer, pH about 7. Alternatively, use 1% OsO_4 for 60–90 min.
10. Dehydrate and embed in resin. Do not use test sections for EM.

E. Photochemical (UV) Method without Gold

This method requires a source of ultraviolet radiation (UV), the intensity of which, however, is not critical, since the normal time needed for irradiation can be determined empirically. Because the useful photochemical effect of UV on Golgi material only reaches about 10–15 μm under the surface, both sides of a 30 μm section must be irradiated.

The following procedure is based on the author's 1975b version and subsequent unpublished observations.

1. If the sections are not already in glycerol but in another medium used during the cutting, take them to 100% glycerol through a graded series of glycerol solutions made up with that medium. For instance, 20%, 40%, and 60% glycerol in 50% ethanol saturated with silver chromate, and 80% glycerol in 50% ethanol. Renew the final 100% glycerol.
2. Place the sections in drops of glycerol on a large coverslip that is held in a frame of cardboard and cover them with small coverslips.
3. Irradiate with UV (for equipment, see Section V.E.1). Turn the frame with the sections upside down once or twice to obtain an equal amount of irradiation on both sides. This is obviously obligatory when one source of UV is used but is also recommended when two sources of UV are present in order to obtain more uniform irradiation.
4. Interrupt illumination after a period to be determined through experience (e.g., 1 hr) in order to check a test section.
5. Take the test section through 80% and 40% aqueous glycerol to water (a few minutes in each) and then to freshly made 0.2% aqueous sodium thiosulfate. Follow, in the light microscope, the gradual disappearance of silver chromate in a fragment cut from the test section and mounted in thiosulfate under a coverslip. Wait until deimpregnation of cells and processes in the test fragment is com-

plete. Do not wait until larger masses of precipitate in impregnated capillaries have disappeared. Wash the main test section in several changes of distilled water, take it through 50% and 95% ethanol to a clearing agent, and examine it in the microscope. If impregnated cells are discernible as desired (pale, yet distinct; spines and fine axons visible), then proceed to Step 6, otherwise repeat Steps 3 to 5.

6. After completed UV irradiation, transfer the sections through 80%, 60%, 40%, and 20% aqueous glycerol to 0.2% aqueous thiosulfate. Deimpregnate for the length of time indicated to be sufficient by the test section (Step 5). Usually this will be about 30 min. Caution: silver particles are less resistant than gold particles to thiosulfate, and too vigorous thiosulfate treatment may, under certain circumstances, completely remove the silver particles produced by irradiation.
7. Rinse in distilled water, e.g., three changes of 5 min each. Postosmificate if needed (Section V.D, Step 9).
8. Dehydrate, blockstain if desired, and embed in resin.

1. Equipment for UV Irradiation

No specific equipment is obligatory, but three general requirements ought to be satisfied: that the intensity be sufficient to allow reasonably short times of irradiation (1–4 hr); that the heat generated by the source be dissipated to prevent overheating of the specimen; and that the source be shielded to protect the eyes. Three arrangements for irradiation with UV are suggested here, the first two of which have been tested by the author.

In the first arrangement, two 125-W UV bulbs (e.g., Philips HPW 125, with ballast BH 125 L12; radiation mostly at 366 nm) are installed in a box (cabinet) 20–30 cm from each other. The specimens are placed on a thin glass plate (such as a window pane of the smallest thickness available) mounted midway between the bulbs. The box has a fan moving air through both halves of the chamber. When this apparatus is used in a cold room (0–4°C), the specimen will stay at room temperature or below. A door or lid provides access to the interior of the box.

In a second arrangement, fluorescent UV tubes may be similarly mounted in a ventilated box or cabinet, 8–10 cm over and under the central glass plate separating the two half-chambers. Because higher watt values merely mean a corresponding lengthening of such tubes, higher radiation intensity can only be obtained by using more tubes. Two tubes over and two under the specimens are suitable. Appropriate types of tubes are Philips TLA 20W/05 and Osram L20/73 with intensity maxima at 370 nm (340–400 nm) and 350 nm (330–370 nm), respectively. The lower current consumption of these tubes (20 W $\times$ 4 = 80 W) permits use of this equipment outside a cold room; the specimen will still stay approximately at room temperature.

Finally, a single UV bulb may be placed over the specimen in a cold room, refrigerator, or freezer. Find out how close the bulb may be without raising the temperature of the specimen above room temperature. If the specimen is placed over a reflecting surface (aluminum foil), a greater photochemical effect is obtained.

F. Photochemical Method with Gold

1. Bring the sections to 100% glycerol as described in Section V.E, Step 1 and place them between coverslips (Section V.E, Step 2).
2. Illuminate the sections with very strong white light, preferably in a cold room. A suitable intensity is obtained if the outlet of a fiber optic cable is placed 10 mm above the sections, and the light generator is turned to its maximum value. Place the sections over a reflecting aluminum foil. After about half the required illumination time, turn the set of coverslips and sections upside down. Try 5 or 10 min from each side the first time and then examine a test section.
3. Take a test section to aqueous 0.5% sodium thiosulfate through aqueous glycerol solutions (e.g., 80% and 40%). Examine a fragment of the test section in the light microscope under coverslip. Deimpregnate until all silver chromate has disappeared from the impregnated nervous structures in the fragment. The main part of the test section should not be exposed to the microscope illumination during the deimpregnation period (Section V.D, Step 4). The impregnated structures will be invisible or very faintly visible at this stage. Then rinse the test section well in distilled water, five changes, 3 min in each.
4. Subject the test section to gold toning, using 0.04% or 0.05% gold chloride (formula in Section V.D, Step 2).
5. Interrupt the toning after 5 min, rinse in water, check a fragment of the test section, again without subjecting the main test section to the microscope illumination. During the gold toning, the deimpregnated structures will reappear. If the illumination time has been sufficient to allow a suitable revisualization of the deimpregnated structures (this must be finally verified in the main test section), then proceed to the next step; otherwise repeat the gold toning, adding further toning of the remaining test section. If sufficient toning can not be obtained after 20 min ("max toning time reached" in Fig. 14), then continue illumination of all sections and repeat the subsequent test steps. *Warning*: do not compensate for overly short illumination times with very prolonged gold toning. This may result in gold particles excessive in size but few in number.
6. Take all remaining sections through 80%, 60%, 40%, and 20% aqueous glycerol to 0.5% sodium thiosulfate. Treat the sections with the thiosulfate for 10–20 min longer than indicated by the test sections to be required for full deimpregnation. Rinse well in distilled water (five changes, 3–5 min in each).
7. Tone in 0.04% or 0.05% gold chloride for the time shown to be optimal for the test sections; then rinse in distilled water, three changes of 5 min each.

8. Postosmificate sections previously treated with only 0.2% osmium tetroxide (as described in Section V.D, Step 9).
9. Dehydrate and embed in resin. Do not use test sections for EM.*

G. "Interrupted Golgi Impregnation"

Whereas as complete an impregnation as possible is usually desired, one may impregnate for a few hours only and obtain an incomplete type of impregnation in which the silver chromate occurs as scattered specks within the neuron (Scott and Guillery, 1974; W. A. Ribi, personal communication). The picture bears a certain similarity to that characterized by the scattered particles of metal seen after deimpregnation of fully impregnated neurons. The impregnated structures display well-preserved submicroscopic details. Therefore, "interrupted Golgi impregnation" may well be the method of choice for specific purposes. However, one can certainly not expect the completeness of impregnation that generally characterizes the slower methods (cf., Valverde, 1976).

In the interrupted Golgi procedures described so far, fixation is performed with formaldehyde–glutaraldehyde mixtures followed by osmification. Scott and Guillery (1974) then used 3.5% potassium dichromate with 0.2% OsO_4 for 1–2 hr followed by 0.75% silver nitrate for 1–2 hr. In contrast, Ribi used weaker solutions: buffered 1% potassium dichromate overnight at 4°C and 0.5% silver nitrate for 0.5–3 hr, also at 4°C. In both cases the tissue was then embedded in resin. Ribi has obtained impregnation of greater lengths of neuronal processes than Scott and Guillery, and some modification of his procedure seems the more promising.

H. Flat Embedding of Primary Sections

1. Direct Embedding under Coverslip

After infiltration with resin, the section is placed on a slide in a drop of resin and coverslipped. The section is examined with the light microscope before polymerization. At this stage, the coverslip may be removed temporarily in order to cut away uninteresting areas. After polymerization followed by further microscopic study, photography, etc., the coverslip is lifted from the section which is next separated from the slide. This is best done under a dissecting microscope with a razor blade and requires care to avoid rupture of the section. The method works both with epoxy and polyester resins.

When the sections are small and few in number, the definite difficulties

*A simplified method could be used when only a few structures are wanted. Bring to glycerol, place between coverslips, expose to light, and treat with thiosulfate. Then carry out the toning with gold under microscopic control until the desired structure has been revisualized to a suitable degree (P. Hurlen, personal communication). Cold light must be used for the microscope illumination to prevent heat damage to the tissue.

associated with the isolation of the sections from the glass surface can be tolerated, and this direct method is acceptable. In other cases, the method described below is preferable.

2. Embedding between Acetate Foils

Some variety of Holländer's (1970) "Section Embedding Method" may be used. The main features of his method are as follows. Three layers of acetate foil are stacked between glass plates. The middle layer, which roughly has the same thickness as the section, e.g., 50 μm, is provided with a window in which the section and unpolymerized resin are accommodated. After curing, the sheets of foil are stripped from the flat piece of resin with the tissue section, which can now be placed on a slide, coverslipped in unpolymerized resin if needed, and studied in the light microscope.

The present author finds it convenient to embed small sections in a stack of foils placed on an ordinary 1 × 3-in. microscope slide as follows. (The letters *a*–*h* refer to the labeling in Fig. 15.)

Attach a piece of acetate foil *c* onto a slide *a* with double-stick cellophane tape *b*.

Stick a second piece (with a stamped-out window, *d*) to the bottom piece with embedding resin and place the section *e* into the window in a little resin.

Place a third piece of acetate foil *f* on top of the second foil, again applying resin to keep the pieces together.

Fix the stacked pieces to the slide, either by adding some resin and a coverslip *g*, which may be weighted, or by means of cellophane tape *h*, in which there is a window to permit microscopy. Or use both coverslip and tape. Very thin acetate foil must be used for the third layer *f* if high magnification microscopy is needed, since thicker foil causes optical distortions. After polymerization, the section in its layer of resin, forming a small disk, is readily separated from the foils by bending the latter away from the disk.

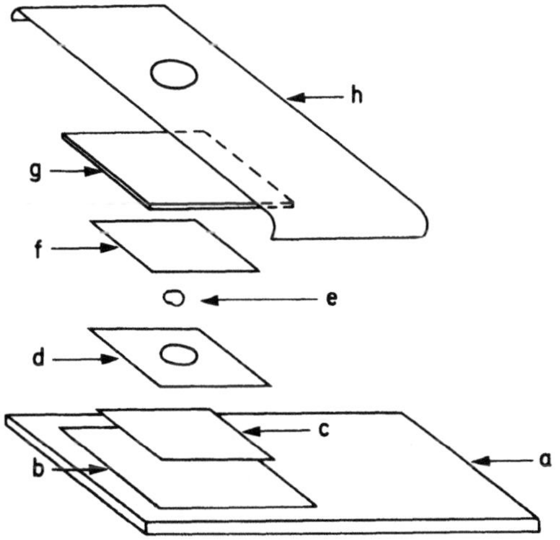

Figure 15. Diagram showing a version of Holländer's Section Embedding Method described in detail in Section V.H.2. Briefly, the section (e) is included in a stack of acetate foils placed on a microscope slide (a).

The embedded section can then be subjected to light microscopy, trimming, and remounting for EM.

I. Cutting Thick Sections of Plastic

Whereas cutting 1- to 4-μm sections with a glass knife on the ultramicrotome offers little difficulty, sectioning in the 5–50 μm range requires other procedures. Four methods that have been used successfully are listed below. Although any good type of microtome for general histology may be used, a sliding microtome is probably preferable. Cutting has been done successfully both with a heavy steel knife and with solid razor blades. See the authors referred to below regarding liquids suitable for moistening the knife during sectioning.

Thick sections may be cut before polymerization is fully completed (Stell, 1965, 1967; Pinching and Brooke, 1973).

Softer resin than normal may be used, such as the mixture suggested by Stell and Witkowsky (1973) or one of those used by Heimer (1970) for silver impregnation of plastic sections.

Normal resin for ultramicrotomy is used, but, following Ribi (1976), it is softened with heat by placing an infrared lamp above the block during sectioning. Ribi used a Philips JR 230- to 250-V, 250-W lamp placed 20–30 cm from the knife while sectioning Araldite blocks.

Heating elements can be mounted on the knife itself which then softens the plastic during the cutting (E. Mugnaini, personal communication). Two soldering iron elements supplied by an autotransformer are used by the writer. The knife is heated until a finger cannot be held on it. The correct degree of softening with different section thicknesses is further regulated by the knife speed and is learned by experience. Sections as thin as 15–10 μm can easily be cut from normal, hard plastic.

J. A Method for Remounting Sections

This method (Fig. 16) is convenient when the specimen is a small section that needs to be studied with the light microscope after remounting. It can be used both with previously unembedded sections and with sections already embedded in plastic (Section V.H,I). The section is immersed in liquid resin before remounting. The top of a blank plastic cylinder (which may contain an internal mirror, cf. Section V.L.2) is ground or filed to form a platform wider than the section. A tiny drop of resin is placed on the platform, and the section is transferred to it with a fine hair of glass or similar tool. So little resin must be used that the section will lie flat in it. The section is covered with a small piece of thin acetate foil held with a fine watchmaker's forceps, and curing is carried out. The purpose of the foil is twofold: to secure a smooth surface appropriate for microscopy, and to keep the section flattened, thus facilitating sampling of tissue during ultramicrotomy. After polymerization, the piece of foil can be stripped off under the dissecting mi-

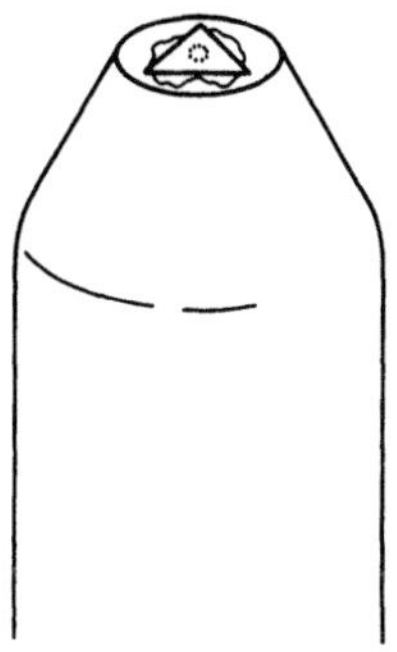

Figure 16. Sketch of a blank plastic cylinder with a small remounted specimen immersed in resin and covered by a piece of acetate foil (here triangular). See the text (Section V.J) for further details.

croscope, but light microscopy of the section is possible both before and after this has been done. The optical quality allows dendritic spines, axons, etc., to be discerned. Removal of the foil may be postponed until it can be done with the glass knife of the ultramicrotome.

K. Monitoring the Ultrathin Sectioning

1. Semithin Sections

The conventional method of cutting and examining "semithin," 0.5- to 2-μm sections (Chapter 10) can be used with Golgi–EM material as long as the sections pass through a part of the block that can be sacrificed (cf. Peters and Fairén, 1978). Deimpregnated structures containing silver or gold particles will not be attacked by normal stains for semithin sections (toluidine blue, paraphenylendiamine). Structures containing unmodified silver chromate may fade. They can, however, be stabilized by reduction with a photographic developer. Another approach is to collect two semithin sections, only one of which is stained. The latter will then show the tissue in general, the other the impregnated structures. Finally, gold toning of Golgi sections to permit counterstaining (Obregia, 1890) is possible.

2. Transillumination with a Mirror in the Block

Plain plastic cylinders for remounting are supplied with an internal slanting mirror, i.e., a spade-shaped strip of aluminum foil embedded in the resin (Fig. 17B). When tissue is later remounted on such a base (Figs. 17B,C), it can easily be transilluminated for light microscopy even if it remains in the ultramicrotome specimen holder. Illumination by means of a fiber optic light guide is ideal.

3. Inspection through an External Mirror

With this method the specimen holder with the block is left untouched in the ultramicrotome during microscopic inspection. A prerequisite for suc-

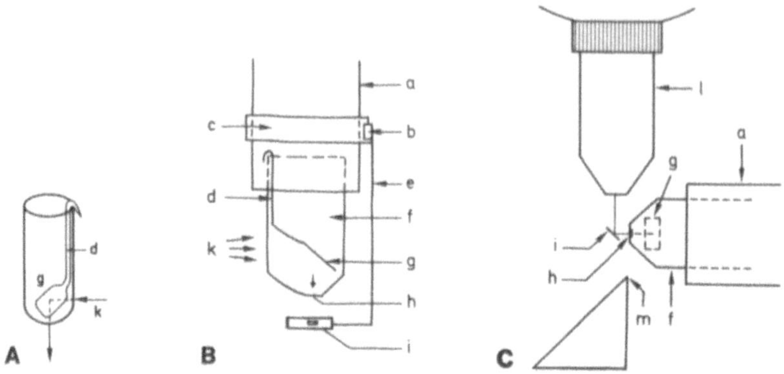

Figure 17. Schematic drawings related to various aspects of embedding, remounting, and ultramicrotomy. (A) Sketch indicating how a spade- or spoonlike piece of aluminum foil (g,d) can be embedded in resin in a gelatin capsule and be used as an "internal mirror." Later, light can be directed through the polymerized block (k), and a specimen may be remounted on it. (B) Drawing of an arrangement for transillumination and observation of a specimen (h) during ultramicrotomy. Seen from above. Light (from k) is reflected by the internal mirror g. The transilluminated specimen is observed in a microscope (not shown) through the external mirror i (the small rectangle in it symbolizing the mirror reflection of the cutting face of the specimen). Here, the mirror is attached to the specimen holder a by a bracket at c and by the arm e which is hinged at b. (C) similar arrangement imagined seen from the side, l being the objective lens of a vertically oriented light microscope placed above the external mirror i and the knife edge m. The meaning of the other symbols is as in (B). Here, the external mirror is mounted on a separate holder (not shown) and remains at the level where it is shown, at the upper limit of the specimen movement, during the actual sectioning. When needed, this microscope and the mirror can be moved to the side to permit inspection of the ultrathin sections in the trough (to the left of m) with the regular microtome microscope.

cessful recognition of details is that the specimen be transilluminated from within, e.g., as described in Section V.K.2.* A small mirror (Figs. 17B,C) is mounted in front of the block. It can be glued to an arm attached to the specimen holder and hinged so that the mirror can be raised and prevented from colliding with the knife during the actual sectioning movements (Fig. 17B) (Blackstad, 1969, 1970). A better solution (P. Hurlen, personal communication) is to mount an arm holding the mirror permanently in front of the highest point (above the knife) to which the block rises during each cutting cycle. With both arrangements it is possible to inspect, through a microscope with normal vertical orientation, the interior of the block through the vertical cut surface. If the mirror is mounted on the specimen holder, the sectioning must be interrupted during microscopy; if the mirror is mounted independently of the moving parts, this is not necessary.

Through the microscope, one can see how far sectioning has proceeded in relation to the structures of interest. The sectioning may be interrupted, and ultrathin sections containing the interesting structure picked up. Afterwards, cutting is resumed without any loss of material. The microscope may

*It is possible that incident illumination through the external mirror can also be used.

be either the regular dissecting microscope of the ultramicrotome or a usual light microscope additionally fitted to the ultramicrotome (Fig. 17C) and permitting inspection at higher magnifications, 50–150 × being suitable.

L. Protection of Impregnation with Silver Chromate

The silver chromate in Golgi-impregnated structures is gradually dissolved if exposed to solutions containing water. One remedy against this is quick passage through such solutions, e.g., during dehydration. Another is to saturate the aqueous solutions with silver chromate. Then, even the finest structures, such as spine necks and thin unmyelinated axons, will remain visible for days.

A suspension of silver chromate can be made by mixing solutions of silver nitrate and potassium dichromate, decanting the supernatant, and then washing the precipitated silver chromate a few times by adding and decanting distilled water. A few drops of the suspension is added to the aqueous solutions of glycerol, ethanol, acetone, etc., needed in the various procedures. (Alternatively, the silver chromate is collected on a filter, dried, and used as powder.) Although this precaution is not always mentioned in the description of procedures in this chapter, it should be adopted whenever necessary.

ACKNOWLEDGMENTS. I am much indebted to Drs. A Fairén, A. Peters, W. A. Ribi, and P. Somogyi for placing unpublished information and pictures at my disposal. For the way their information has been incorporated into the recipes, I am alone responsible. Particular thanks for technical assistance are owed to J. Line Vaaland, K. Ruud Øztürk, and E. Risnes. USPHS Grant NS 07998 was received during earlier stages of this work.

REFERENCES

Blackstad, T. W., 1965, Mapping of experimental axon degeneration by electron microscopy of Golgi preparations, *Z. Zellforsch.* **67:**819–834.

Blackstad, T. W., 1969, Studies on the hippocampus: Methods of analysis, in: *The Interneuron* (M. Brazier, ed.), pp. 391–414, UCLA Forum in Medical Sciences, Los Angeles.

Blackstad, T. W., 1970, Electron microscopy of Golgi preparations for the study of neuronal relations, in: *Contemporary Research Methods in Neuroanatomy* (W. J. H. Nauta and S. O. E. Ebbeson, eds.), pp. 186–216, Springer-Verlag, Berlin.

Blackstad, T. W., 1975a, Golgi preparations for electron microscopy: Controlled reduction of the silver chromate by ultraviolet illumination, in: *Golgi Centennial Symposium, Perspectives in Neurobiology* (M. Santini, ed.), pp. 123–132, Raven Press, New York.

Blackstad, T. W., 1975b, Electron microscopy of experimental axonal degeneration in photochemically modified Golgi preparations: A procedure for precise mapping of nervous connections, *Brain Res.* **95:**191–210.

Fairén, A., and Valverde, F., 1980, A specialized type of neuron in the visual Cortex of cat: a Golgi and electron miscroscope study of chandelier cells, *J. Comp. Neurol.* **194:**761–779.

Fairén, A., Peters, A., and Saldanha, J., 1977, A new procedure for examining Golgi impregnated neurons by light and electron microscopy, *J. Neurocytol.* **6:**311–337.

Heimer, L., 1970, Bridging the gap between light and electron microscopy in the experimental tracing of fiber connections, in: *Contemporary Research Methods in Neuroanatomy* (W. H. J. Nauta and S. O. E. Ebbeson, eds.), pp. 162–172, Springer-Verlag, Berlin.

Holländer, H., 1970, The section embedding (SE) technique. A new method for the combined light microscopic and electron microscopic examination of central nervous tissue, *Brain Res.* **20:**39–47.

LeVay, S., 1973, Synaptic patterns in the visual cortex of the cat and monkey. Electron microscopy of Golgi preparations, *J. Comp. Neurol.* **150:**53–86.

Obregia, A., 1890, Fixirungsmethode der Golgi'schen Präparate des centralen Nervensystems, *Arch. Pathol. Anat.* **122:**387–388.

Peters, A., and Fairén, A., 1978, Smooth and sparsely-spined stellate cells in the visual cortex of the rat: A study using a combined Golgi–electron microscope technique, *J. Comp. Neurol.* **181:**129–172.

Peters, A., Proskauer, C. C., Feldman, M. L., and Kimerer, L., 1979, The projection of the lateral geniculate nucleus to area 17 of the rat cerebral cortex. V. Degenerating axon terminals synapsing with Golgi impregnated neurons, *J. Neurocytol.* **8:**331–357.

Pinching, A. J., and Brooke, R. N. L., 1973, Electron microscopy of single cells in the olfactory bulb using Golgi impregnation, *J. Neurocytol.* **2:**157–170.

Ramón-Moliner, E., and Ferrari, J., 1972, Electron microscopy of previously identified cells and processes within the central nervous system, *J. Neurocytol.* **1:**:85–100.

Ramón-Moliner, E., and Ferrari, J., 1976, Electron microscopy of Golgi-stained material following lead chromate substitution, *Brain Res.* **103:**339–344.

Ribi, W. A., 1976, A Golgi–electron microscope method for insect nervous tissue, *Stain Technol.* **51:**13–16.

Scott, G. L., and Guillery, R. W., 1974, Studies with the high voltage electron microscope of normal, degenerating, and Golgi impregnated neuronal processes, *J. Neurocytol.* **3:**567–590.

Somogyi, P., 1978, The study of Golgi stained cells and of experimental degeneration under the electron microscope: A direct method for the identification in the visual cortex of three successive links in a neuron chain, *Neuroscience* **3:**167–180.

Somogyi, P., and Cowey, A., 1981, Combined Golgi and electron microscopic study on the synapses formed by double bouquet cells in the visual cortex of the cat and monkey. *J. Comp. Neurol.* **195:**547–566.

Somogyi, P., and Smith, A. D., 1979, Projection of neostriatal spiny neurons to the substantia nigra. Application of a combined Golgi-staining and horseradish peroxidase transport procedure at both light and electron microscope levels, *Brain Res.* **178:**3–15.

Somogyi, P., Hodgson, A. F. and Smith, A. D., 1979, An approach to tracing neuron networks in the cerebral cortex and basal ganglia. Combination of Golgi staining, retrograde transport of horseradish peroxidase and arterograde degeneration of synaptic boutons in the same material, *Neuroscience* **4:**1804–1852.

Somogyi, P., Bolam, F. P., and Smith, A. D., 1981, Monosynaptic cortical input and local axon collaterals of identified striato-nigral neurons. A light and electron microscopic study using the Golgi-peroxidase transport-degeneration procedure, *J. Comp. Neurol.* **195:**567–584.

Stell, W. K., 1965, Correlation of retinal cytoarchitecture and ultrastructure in Golgi preparations, *Anat. Rec.* **153:**389–397.

Stell, W. K., 1967, The structure and relationships of horizontal cells and photoreceptor–bipolar synaptic complexes in goldfish retina, *Am. J. Anat.* **121:**401–424.

Stell, W. K., and Witkovsky, P., 1973, Retinal structure in the smooth dogfish, *Mustelus canis:* General description and light microscopy of giant ganglion cells, *J. Comp. Neurol.* **148:**1–31.

Stell, W. K., Lightfoot, D. O., Wheeler, T. G., and Leeper, H. F., 1975, Goldfish retina: Functional polarization of some horizontal cell dendrites and synapses, *Science* **190:**989–990.

Valverde, F., 1976, Aspects of cortical organization related to the geometry of neurons with intra-cortical axons, *J. Neurocytol.* **5:**509–529.

White, E. L., 1978, Identified neurons in mouse SmI cortex which are postsynaptic to thalamocortical axon terminals: A combined Golgi–electron microscopic and degeneration study, *J. Comp. Neurol.* **181:**627–661.

Fluorescence Histochemical Methods

13

Neurotransmitter Histochemistry

ROBERT Y. MOORE

I. INTRODUCTION

The methods that have commonly been employed to investigate the morphology of the central nervous system have been directed toward two major goals. The first is to identify groups of neurons with similar morphological features, and the second is to demonstrate the axonal projections of such groups of neurons to define the precise neuronal circuitry of the brain and spinal cord. The massive volume of information currently available on the organization of the central nervous system attests to the effectiveness with which these methods have been applied. More recently, new methods have been developed that enhance our capacity to define specific neuronal groups, and these have added to our further definition of the functional organization of the central nervous system. The purpose of this chapter is to describe one group of these techniques, the fluorescence histochemical methods for intracellular demonstration of biogenic amines, methods that permit the designation of groups of neurons, not on the basis of their cellular morphology, location, or axonal projection, but on the basis of their

ROBERT Y. MOORE • Department of Neurology, State University of New York at Stony Brook, Stony Brook, New York 11794.

production of a chemical that is utilized for the process of synaptic transmission. These methods have proven to be extraordinarily useful in the analysis of central nervous system structure (see Fuxe *et al.*, 1970; Björklund *et al.*, 1972b; Hökfelt and Ljungdahl, 1972a; Moore and Bloom, 1978, 1979, for reviews) and have led to discoveries of fundamental functional importance.

The basis for this methodology is twofold. First, it is generally accepted that synaptic transmission in the central nervous system is carried on almost exclusively by the presynaptic release of a chemical that either depolarizes or hyperpolarizes a postsynaptic membrane. Second, most central neurons would appear to conform to the view first proposed by Dale (1935) that each neuron produces a single neurotransmitter chemical and, thus, can be identified by its content of that chemical. Until recently, the identification of specific neurons in the central nervous system by histochemical methods has been limited to a specific class of neurons, those that use monoamines as neurotransmitter compounds (Fig. 1). In the mammalian central nervous system, as well as those of most vertebrates, the neurotransmitters that can be identified by histochemistry are the catecholamines, dopamine, norepinephrine, and epinephrine, and the indolamine, serotonin. Recently developed immunohistochemical methods (see Hökfelt *et al.*, 1973; Elde and Hökfelt, 1978, for review) have added greatly to this field and will be discussed in detail in Chapter 14. These methods depend on the production of an immune reaction to an enzyme in the biosynthetic sequence of a neurotransmitter or to a peptide hormone or transmitter. The antibody formed is used to show the enzyme or peptide by an antigen–antibody reaction. Those methods are now gaining wide use and provide powerful tools for demonstration of specific neuronal systems. The methods to be described in this portion of the chapter are applicable, at the present time, only of the compounds noted above, and groups of neurons are found in the mammalian central nervous system that produce each of the monoamines noted above.

The catecholamine-producing neurons fall into three specific groups. Neurons that contain the enzymes tyrosine hydroxylase and DOPAdecarboxylase but do not contain dopamine-β-hydroxylase (DBH) or phenylethanolamine-N-methyltransferase (PNMT) will produce dopamine as their neurotransmitter. Neurons that contain DBH as well as the first two synthetic enzymes but lack PNMT will produce norepinephrine as their neurotransmitter, and neurons that contain all of the synthetic enzymes will produce epinephrine as their neurotransmitter (Fig. 1). Indolamine neurons all appear to contain the same enzymatic steps in the mammalian nervous system and synthesize serotonin, 5-hydroxytryptamine, by hydroxylation of the amino acid tryptophan to produce 5-hydroxytryptophan which is decarboxylated to yield serotonin.

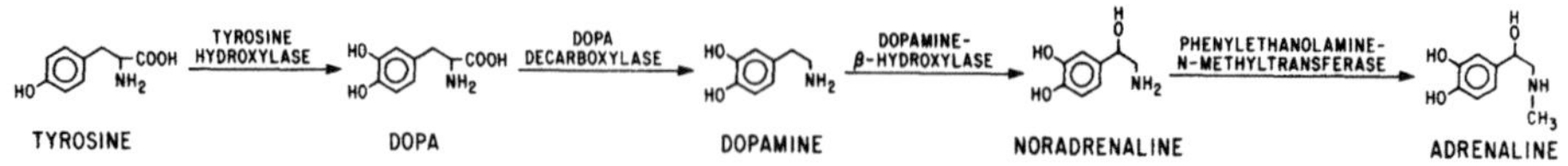

Figure 1. Metabolic pathway for the synthesis of catecholamines from tyrosine.

Although the methods for neurotransmitter histochemistry to be described in this chapter are limited in their application, they are extremely useful because they demonstrate restricted classes of neurons without showing other neuronal groups and, consequently, are extremely specific.

The history of the development of these methods has been reviewed previously (Corrodi and Jonsson, 1967) and will not be repeated here. The essential background for the development of the method was the establishment of the concept of neurochemical transmission and the identification of putative neurotransmitter agents in the central nervous system (see Werman, 1966). The history of the development of the concept of chemical transmission and its application to the peripheral nervous system is too well-known to be repeated in detail. Norepinephrine was demonstrated conclusively in the late 1940s to be the peripheral sympathetic neurotransmitter (von Euler, 1946), but it was it was not until 1954 that M. Vogt demonstrated the presence of epinephrine and norepinephrine in brain in a regional distribution independent of its sympathetic innervation (Vogt, 1954). Almost simultaneously, Twarog and Page (1953) and Amin *et al.* (1954) identified serotonin in brain. Subsequently, Bertler and Rosengren (1959) demonstrated the regional localization of dopamine within the neostriatum, and this was followed shortly thereafter by the striking clinical–pathological correlation of a markedly reduced dopamine content in the neostriatum of parkinsonian patients (Ehringer and Hornykiewicz, 1960). It was evident at this time, however, that chemical analysis of monoamines in brain was inadequate to demonstrate the specific neuronal systems with which they were associated, and a number of investigators, particularly Hillarp and his associates (see Corrodi and Jonsson, 1967, for review) undertook to develop a histochemical method for demonstrating monoamines in brain. Many early attempts to demonstrate monoamines and to maintain them in their *in vivo* location were frustrated because the monoamines are water-soluble and extremely labile, properties which provide formidable obstacles to the development of an adequate histochemical method.

However, the introduction by Falck, Hillarp and their co-workers (Falck and Torp, 1961; Falck, 1962; Falck *et al.* 1962) of a method based on the condensation of a biogenic amine with formaldehyde in a gas-phase reaction overcame these problems better than any prior techniques. The use of the gas-phase reaction on dried or freeze-dried tissue or in model systems in which the biogenic amine was dried in a protein film prevented the diffusion of the water-soluble biogenic amines and produced an intense fluorophore that could be readily identified in the fluorescence microscope. The application of the original Falck–Hillarp method (Falck, 1962; Falck and Owman, 1965; Björklund *et al.*, 1972b) led to extensive advances in our knowledge of the organization of the central monoamine neuron systems (cf. Dahlström and Fuxe, 1965a,b; Fuxe, 1965, Andén *et al.*, 1966; Fuxe *et al.*, 1968, 1970; Ungerstedt, 1971). Subsequent improvements in the fluorescence histochemical methodology, such as the Vibratome–formaldehyde method (Hökfelt and Ljungdahl, 1972b) and, particularly, the glyoxylic acid method (Björklund *et al.*, 1972a; Lindvall *et al.*, 1974; Lindvall and Björklund, 1974a), have led to further advances, so our knowledge of the organization of central catecholamine neuron systems probably approaches that of any central system save for the primary sensory and motor systems (see Lindvall and Björklund, 1974b, 1978; Moore and Bloom, 1978, 1979, for reviews).

The advantages of the neurotransmitter histochemical techniques currently available, particularly the glyoxylic acid method and its variants, are their selectivity and their sensitivity. There is general agreement that the available methods demonstrate only dopamine, norepinephrine, epinephrine, and serotonin neurons in the mammalian central nervous system. The sensitivity of the methods has been reviewed by Björklund *et al.* (1975) who conclude from information in the literature and from their own observations that the fluorescence histochemical techniques are sensitive to as little as 5 $\times 10^{-6}$ micromoles of norepinephrine or dopamine contained in a single nerve terminal. The glyoxylic acid technique is more sensitive than the original fluorescence histochemical techniques so that for dopamine, for example, it is calculated that 10^{-7} micromoles can readily be detected (Björklund *et al.*, 1975). In general, the methods are less sensitive for epinephrine and the indolamines, but these specific problems will be discussed in regard to the application of individual methods. It is worth noting, however, that the original methods (Falck, 1962; Falck *et al.*, 1962; Falck and Owman, 1965) were sufficiently sensitive to demonstrate the cell bodies and terminal plexuses of neurons containing catecholamines and, in some instances, serotonin, but did not demonstrate preterminal axons. The newer methods are sufficiently sensitive to demonstrate the entire neuron, cell body, dendrites, preterminal axon, and terminal axonal plexus, and thereby represent substantive improvements over prior methods.

At the present time there are two major variants of the neurotransmitter histochemical methods. Since all of these depend on the formation of a fluorophore from the reaction of the neurotransmitter with another compound, they will be referred to as fluorescence histochemical methods. A fluorophore is a compound that is selectively activated by one wavelength of light to emit light of a specific, higher wavelength that is characteristic of the compound. This property, exhibited by many compounds, provides the essential basis for a fluorescence histochemical method. The first method to be discussed will be the original Falck–Hillarp technique and its modifications, and the second will be the glyoxylic acid method. Each variant has its special uses and advantages, but it should be pointed out that this is a period of intense investigation in technology in this field, and very rapid advances are being made in improving the methods. These include the development of new methods as well as the improvement of existing ones.

II. CHEMICAL BASIS OF THE FLUORESCENCE HISTOCHEMICAL METHODS

A. Introduction

The original fluorescence histochemical method utilized the reaction of formaldehyde with a primary or secondary amine to form a fluorescent compound. Although this reaction had been known for many years and had been applied as the basis for a biochemical assay for monoamines (Hess and Udenfriend, 1959) and to demonstrate norepinephrine in adrenal medulla (Eränko, 1955), it was not until Falck and Hillarp and their associates introduced use of the formaldehyde reaction with dried tissue and gaseous formaldehyde that a specific and sensitive histochemical method could be developed (Falck *et al.*, 1962; Falck, 1962; Carlsson *et al.*, 1962). Subsequent work in Falck's laboratory led to the establishment of glyoxylic acid as a second compound that could be used to provide an equally specific and more sensitive method for histochemical demonstration of monoamines in tissue (Lindvall and Björklund, 1974a). The details of these procedures and their modifications will be described in subsequent sections of this chapter. The chemical bases for the methods will be reviewed below.

B. The Falck–Hillarp Method (Formaldehyde Condensation)

The condensation or cyclization of formaldehyde with a primary catecholamine or indoleamine to form a fluorophore has been known for a number of years (see Corrodi and Jonsson, 1967, for review). This reaction forms the chemical basis of the original Falck–Hillarp method. Shortly after the discovery of the Falck–Hillarp method, the chemical basis was substantiated by a series of studies principally carried out by Corrodi and Hillarp (see Corrodi and Jonsson, 1967; Björklund *et al.*, 1975, for reviews). As indicated above, it is based on the foundation of many years' work in the organic chemistry of cyclization reactions. The reaction of formaldehyde with a primary catecholamine or indolamine, termed a Pictet–Spengler reaction, is well-known in organic chemistry and has been the subject of extensive study (Whaley and Govindachari, 1951). As applied to the catecholamines, it is a general condensation method for reacting primary or secondary β-arylethylamines with a carbonyl compound to yield a tetrahydroisoquinoline (Fig. 2A). A 5-hydroxyindolylethylamine (e.g., serotonin) undergoes a similar condensation reaction to form a 6-hydroxy-1,2,3,4-tetrahydro-β-carboline (Fig. 2B). In each case the product of the condensation reaction is only weakly fluorescent. To form a sufficiently intense fluorophore for visualization in a fluorescence microscope, the reaction must take place in the presence of a protein. In general it will not occur in the absence of protein, but the substitution of peptides or even amino acids (only alanine and glycine are effective substitutes) is adequate to catalyze the reaction. The products, a 3,4-dihydroisoquinoline in the case of the catecholamines (Fig. 2) and a 6-hydroxy-3,4-dihydro-β-carboline in the case of the indolamine (Fig. 2), are intensely fluorescent. At neutral pH, as occurs in a tissue, these fluorophores are in their quinoidal forms, whereas the nonquinoidal forms predominate at lower pH. This pH-dependent tautomerism is reflected in characteristic changes in the spectral properties of the fluorophore and forms the basis by which the dopamine fluorophore can be distinguished from the norepinephrine fluorophore by acid treatment (Björklund *et al.*, 1975). For example,

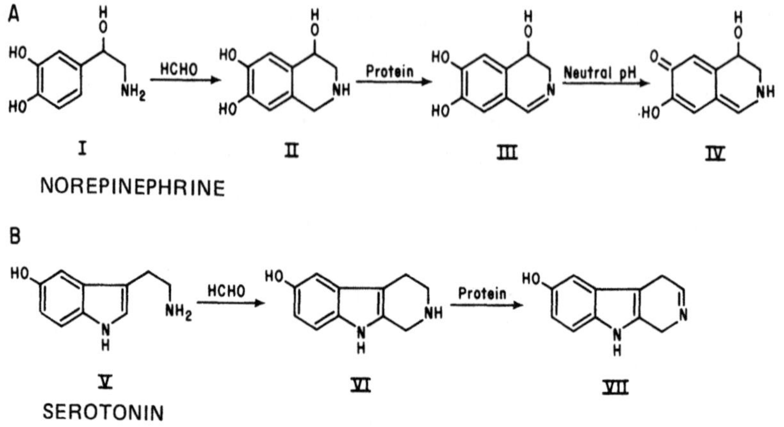

Figure 2. (A) Reaction between norepinephrine and formaldehyde in the Falck–Hillarp method. Norepinephrine (1) undergoes a cyclization (Pictet–Spengler reaction) with formaldehyde to form a 6,7-dihydroxy-1,2,3,4-tetrahydroisoquinolone (II). This is converted to a 3,4-dihydroisoquinoline (III) in a protein-promoted reaction. In the pH range 6–10 in the protein layer, the tautomeric quinoidal form (IV) predominates and is intensely fluorescent. (B) Reaction between serotonin (V) and formaldehyde in the Falck–Hillarp method. This, too, involves an initial cyclization reaction to form a 6-hydroxy-1,2,3,4-tetrahydro-β-carboline (VI) which is then dehydrogenated in a protein-catalyzed reaction to a 6-hydroxy-3,4-dihydro-β-carboline (VII). (See Corrodi and Jonsson, 1967; Björklund *et al.*, 1975, for reviews.)

both dopamine and norepinephrine have the same excitation maximum (that is, the wavelength of light maximally effective in producing fluorescence) after formaldehyde treatment. Because norepinephrine has a labile hydroxy group affected by acid, treatment with HCl vapor lowers the norepinephrine excitation wavelength and permits its discrimination from dopamine by microspectrofluorimetry (Björklund *et al.*, 1972b). At the present time, there is extensive information available on the chemistry of these reactions, including their mechanism specificity, sensitivity, factors influencing the spectral properties of fluorophores, and their potential application to other compounds (see Corrodi and Jonsson, 1967; Björklund *et al.*, 1972b; Björklund *et al.*, 1975, for reviews).

C. The Glyoxylic Acid Method

The formation of a fluorophore from primary and secondary phenylethylamines and indolylethylamines is quite similar in the glyoxylic acid method to that in the formaldehyde condensation method (Lindvall *et al.*, 1974; Björklund *et al.*, 1975). It should be noted, however, that the chemical basis of the glyoxylic acid method is not nearly as well understood as that for the formaldehyde condensation method (Björklund *et al.*, 1975). As it is now viewed, the reaction proceeds in two steps (Fig. 3). In the first step, a phenylethylamine or an indolylethylamine reacts with glyoxylic acid in an

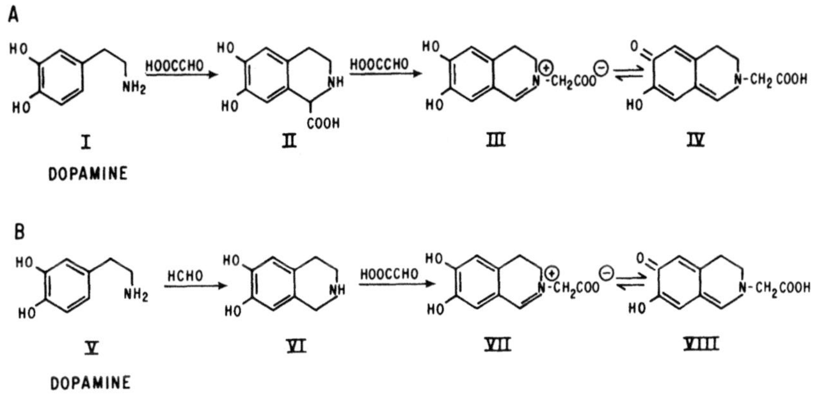

Figure 3. (A) Proposed reaction between dopamine and glyoxylic acid. Dopamine (I) reacts with glyoxylic acid in a Pictet–Spengler cyclization to form the weakly fluorescent 6,7-dihydroxy-1,2,3,4-tetrahydroisoquinolone-1-carboxylic acid derivative (II). Fluorophor formation then continues by an intramolecularly acid-catalyzed reaction with a second glyoxylic acid molecule to form a strongly fluorescent 2-carboxymethyl-6,7-dihydroxy-3,4-dihydroisoquinolinium compound (III). This fluorophor is in a pH-dependent equilibrium with its tautomeric quinoidal form (IV) (see Lindvall *et al.*, 1974; Björklund *et al.*, 1975, for details). (B) Proposed reaction for dopamine in the combined formaldehyde and glyoxylic reaction. The initial Pictet–Spengler cyclization is the same as in Fig. 1, and the final product is the same as in the glyoxylic acid reaction in (A).

acid-catalyzed Pictet–Spengler reaction to yield a 1,2,3,4-tetrahydroisoquinoline-1-carboxylic acid or 1,2,3,4-tetrahydro-β-carboline-1-carboxylic acid, respectively, via a Schiff's base. In the second step, these weakly fluorescent compounds may be transformed into strongly fluorescent products in two alternative ways. The first is via an autooxidative decarboxylation to the 3,4-dihydroisoquinoline or 3,4-dihydro-β-carboline, and the second through an intramolecularly, acid-catalyzed reaction with glyoxylic acid to the 2-carboxymethyl-3,4-dihydoisoquinolinium or 2-carboxymethyl-3,4-dihydro-β-carbolinium compound. A further decarboxylation to the 2 methylated compounds may occur. These glyoxylic acid-induced fluorophores also exhibit a pH-dependent tautomerism similar to that of the formaldehyde-induced fluorophores. At neutral pH, they will be in the quinoidal form, and, at acid pH, in the nonquinoidal forms. Typical excitation and emission spectra for a glyoxylic acid fluorophore are shown in Fig. 4. In addition to being specific for the glyoxylic acid fluorophores, these show what is meant by "fluorescence" and the general basis of the fluorescence histochemical methods. A compound formed by one of the reactions, such as that in Fig. 4, is activated by light of a specific band of wavelengths (excitation spectrum) to emit light with a higher band of wavelengths (emission spectrum) but with less energy.

Experiments with tissue and with model systems have demonstrated that glyoxylic acid yields a stronger fluorophore than does formaldehyde (Björklund *et al.*, 1975). This can be attributed to several phenomena including the formation of strongly fluorescent products by the intramolecular acid catal-

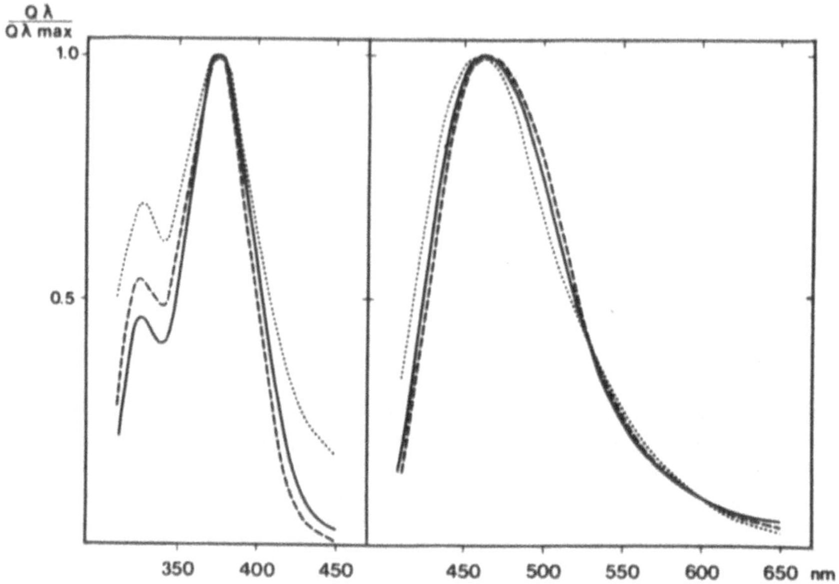

Figure 4. Excitation (left) and emission (right) spectra from glyoxylic acid-induced fluorescence of dopamine in dried albumin droplets (———), in most cells of the bovine splenic nerve (......), and of the authentic 2-carboxymethyl-6,7-dihydroxy-3,4-dihydroisoquinolinium compound in dried albumin droplet models (----). All specimens had been exposed to GA vapor (+100°C, 300 torr, 3 min). Concentration: 2.0×10^{-3} M. (From Lindvall *et al.*, 1974c)

ysis exerted by the carboxyl group on the 1-carbon of the tetrahydroisoquinoline or tetrahydro-β-carboline molecules, the promotion of the initial cyclization step because of the acid catalysis participating in both steps of the fluorophore formation, and the acidification which is known to increase the fluorescence intensity of some amine fluorophores. It should be noted that, although fluorophore formation occurs with a reaction between glyoxylic acid and the indolylethylamines in model experiments, this has not been applied successfully to tissue as a histochemical method for the indolylethylamines as yet. Consequently, the glyoxylic acid methods to be described in this chapter are only applicable to the study of catecholamine neuron systems in tissue with one exception. The exception is in instances in which methods incorporate the utilization of both a formaldehyde condensation reaction and glyoxylic acid. The chemical basis for this is shown in Fig. 2B. These methods result in the same fluorophore as is formed from the glyoxylic acid method itself and are coming into increasing use.

III. EQUIPMENT

The details of the equipment necessary to perform the fluorescence histochemical methods and their use are provided in the Appendix (Section VII) to this chapter. In general, these include at least two, and probably three, major items of equipment. The first is a freeze-dryer appropriate for the proper drying of tissue. A very suitable instrument developed by Olson and Ungerstedt (1970a) is commercially available or can be produced by suitably equipped shop at a price currently about $5000. A cryostat, usual cost around $5000, may be substituted if one of the cryostat section methods described below appears optimal. Similarly, a Vibratome, price about $2000, or a paraffin microtome, price about $2000, may be required. The major item of equipment needed, regardless of the fluorescence histochemical method employed, is a fluorescence microscope with a camera. Depending upon the company, the type of objective lenses desired, and the quality of the microscope, these range from about $7500 to $15,000. In addition to this equipment, the usual histological facilities and supplies are required as is access to a photographic darkroom.

IV. METHODS USING FORMALDEHYDE OR GLYOXYLIC ACID CONDENSATION

A. Introduction

The Falck–Hillarp method is comprised of three essential steps. The first is preservation of the tissue so that the cellular localization of monoamines is maintained. This is accomplished by one of three procedures, each directed to obtaining a dried tissue without monoamine diffusion. Freeze-drying is the most generally useful method for central nervous system. In this, pieces of fresh brain are rapidly frozen to the temperature of liquid nitrogen and dried in a special apparatus (see Section IV.B.2). A simpler method, applicable to thin sheets of tissue such as iris, is to make a stretch preparation of the tissue and quickly air-dry it prior to formaldehyde treatment. Finally, if precise tissue morphology is not required, tissue such as brain can be smeared on a glass slide, air-dried, and treated.

The second step of the Falck–Hillarp method is the treatment of the dried tissue with gaseous formaldehyde. The use of formaldehyde vapor was the critical step that allowed Falck and Hillarp to achieve cellular localization of monoamines in tissue.

The third step is fluorescence microscopy. This utilizes the fluorescence properties of the components formed by formaldehyde treatment of monoamine-containing tissue to permit visualization and localization of the monoamines. Specific methods will now be described in detail.

B. The Falck–Hillarp Method

1. Stretch and Smear Preparations

The original application of the Falck–Hillarp method was to model systems in which amines in a dry protein film or thin preparations of tissue were exposed to formaldehyde vapor (Falck and Torp, 1961; Falck *et al.*,

1962; Falck, 1962). The application of the method to preparations of thin sheets of fresh material continues to be useful for two purposes. First, it is easily performed and can serve the investigator to assess whether the method is working properly in his hands and to validate the funtion of the fluorescence microscope and photographic exposure times. Second, certain experimental situations only require stretch or smear preparations or, alternatively, can be adapted to employ these techniques. The tissue most often used for stretch preparation is the iris of the albino rat. This is particularly useful because it has an extremely rich innervation of sympathetic, norepinephrine-containing fibers. This preparation is made by removing the eye, incising the entire circumference of the globe with a small scissors (such as an iris scissors), removing the lens, and placing the anterior portion of the globe with the cornea down on a clean glass slide. It is then possible, using a small forceps under a dissecting microscope, to gently tease the iris away from its attachment at the ciliary body. The iris is then transferred to an adjacent portion of the glass slide where it is gently stretched to approximate its original shape. The remaining tissue is discarded, and excess fluid around the iris blotted away. The slide is then placed in an evacuated dessicator containing phosphorus pentoxide and dried overnight. The next day it is treated with formaldehyde vapor (the parameters of the formaldehyde vapor treatment will be discussed in Section IV.B.2) at 80° for 1 hr, covered with liquid paraffin, coverslipped, and viewed in the fluorescence microscope (Fig. 5). This technique is applicable not only to iris but to any thin tissue, such as meninges, mesentery, pleura, that can be stretched and is sufficiently thin to dry rapidly and allow visualization of discrete structures in the microscope. A typical iris stretch preparation is shown in Fig. 6B.

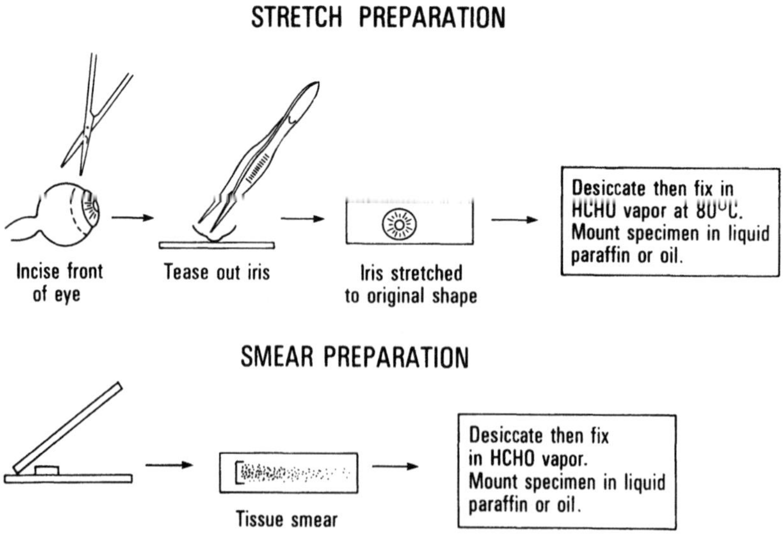

Figure 5. Diagram illustrating the method of preparing an iris stretch preparation and a tissue smear preparation. See text for description.

An alternative to the stretch preparation is the smear preparation first utilized by Olson and Ungerstedt (1970b). This has been applied to a variety of areas in the central nervous system but predominantly to cerebral cortex. It is identical to the stretch preparation except that the tissue to be examined is dissected so that a small piece, approximately 1 × 1 × 2 mm or less, is removed and placed on a clean glass slide. A second slide is then placed on edge at an acute angle to the first slide and gently pulled across the tissue so that a thin, even smear of the tissue is obtained (Fig. 5). This is identical to the technique of obtaining blood smears. As with the iris preparation, the slide is then placed in an evacuated dessicator containing phosphorus pentoxide overnight and treated the next day with formaldehyde vapor, coverslipped, and visualized in the fluorescence microscope. For most areas of brain, this method demonstrates fluorescent varicosities that are present in number approximately in proportion to the innervation of the area making up the tissue smeared in the preparation. As noted above, this is used typically for such tissues as cortex where the innervation tends to be homogeneous and in situations where one does not require particular information about the localization of the fluorescent material. In essence, it is a semiquantitative method for determining innervation by catecholamine neurons and does not provide information on localization. That is, the number and fluorescence intensity of varicosities observed in a tissue smear is roughly proportional to the density of innervation of the area smeared. It is not useful for demonstrating serotonin terminals. One primary use of the method has been as a screening procedure for rapid assessment of the effects of pharmacological manipulation of catecholamine pathways or of lesions that destroy catecholamine pathways.

2. The Falck–Hillarp Freeze-Dry Method

The methodology for the conventional Falck–Hillarp freeze-dry technique has been described on a number of occasions (see Falck and Owman, 1965; Eränko, 1967; Fuxe *et al.*, 1970; Björklund *et al.*, 1972b, for reviews) in great detail. While the basic procedure is similar in all laboratories, there are some individual variations in detail that will not be described here. In essence, each laboratory develops minor modifications that appear to suit that particular laboratory better than others. The procedure to be described is that used in the author's laboratory; the reader interested in variations and further details should consult the reviews cited above.

The technique used for catecholamines in this laboratory is as follows (Fig. 7). The animal is sacrificed by decapitation. With rats this is done directly, but larger animals are decapitated under anesthesia with pentobarbital (40 mg/kg). The brain is removed rapidly and dissected into pieces that may range in size from 2 mm in diameter (or smaller) to as large as 0.5 × 1 cm in diameter. It is possible to freeze-dry larger pieces of tissue, but the results for the larger pieces are not always satisfactory. It is our practice to

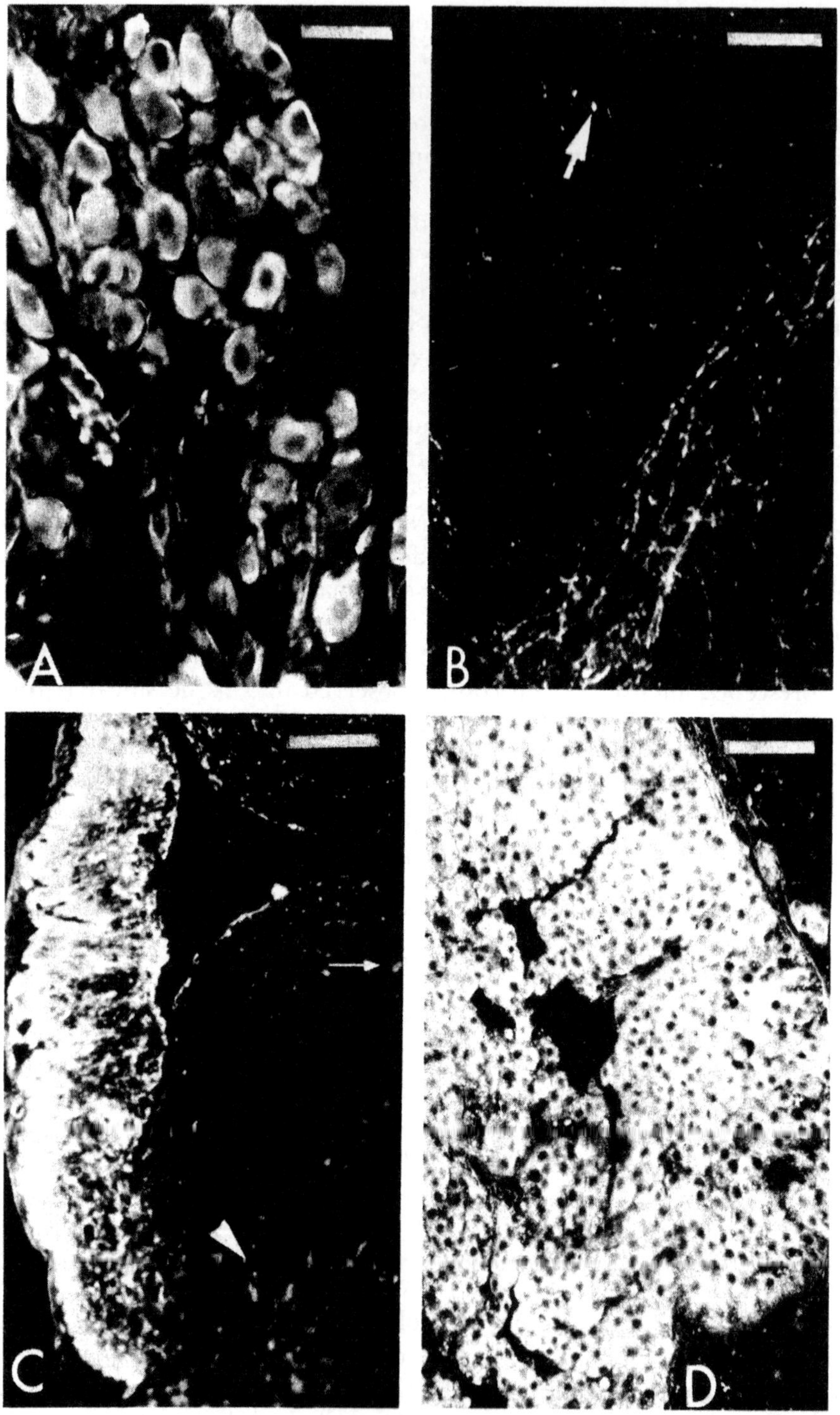

Figure 6. (A) Superior cervical ganglion demonstrating fluorescent cell bodies of peripheral, sympathetic norepinephrine-producing neurons. Note that the nucleus in the center of the cell body is not fluorescent. Freeze-dry Falck–Hillarp method. Bar, 30 μm. (B) Iris prepared

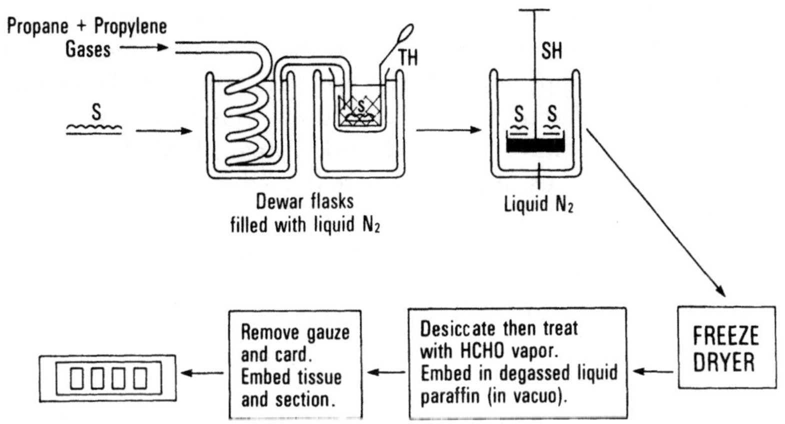

Figure 7. Diagram illustrating the Falck–Hillarp freeze–dry method. See text for description.

place the tissue on a rigid piece of paper (a cut-up index card works well) slightly larger than the specimen. Landmarks are readily lost in the freeze-dried tissue, so that the orientation of the block on the paper should be identified. It is our general practice to place the face of the block to be cut down on the paper. The back of the paper is numbered so that each specimen frozen can be recorded. After the tissue has been placed on the paper, it is covered with a fine-mesh cotton gauze. The gauze adheres readily to fresh tissue, and excess gauze extending beyond the borders of the paper is cut away. The tissue is then placed in a mixture of liquid propane and propylene. We use a combination of commercial propane and propylene that remains liquid when cooled to the temperature of liquid nitrogen. This is accomplished in practice by running the gaseous propane and propylene through coiled copper tubing that is immersed in liquid nitrogen in a Dewar flask. The end of the tubing is placed in a cylindrical cooper container (approximately 10×4 cm) immersed in liquid nitrogen in a second Dewar flask.

---

as a stretch preparation. The bundle of fluorescent axons running diagonally across the photograph is sympathetic innervation about a small blood vessel within the iris. The remainder of the fluorescent structures constitute the autonomic ground plexus of the iris. This is composed of varicose segements that represent terminals of the axons with thin, interposed non-varicose segments. A particularly intense varicosity is designated by an arrow. (C) Ventral hypothalamus showing the heavily innervated median eminence (ME) lying above the third ventricle (V). The lack of distinctness of fluorescence in this is characteristic of material prepared by the freeze-dry, Falck–Hillarp method. Another characteristic is the crack running vertically upward from the bottom of the photograph (large arrow). A portion of the arcuate nucleus with diffuse, varicose terminals and a dopamine-producing cell body (small arrow) is evident above the median eminence and lateral to the third ventricle. Falck–Hillarp method. Bar, 80 μm. (D) Intermediate lobe of the monkey pituitary gland exhibiting dense, yellow, indoleamine fluorescence in the cells of the intermediate lobe. The border of the intermediate lobe with the anterior lobe is designated by arrows. All of the cells of the intermediate lobe in this photograph demonstrate intense fluorescence throughout the cytoplasm, with the nucleus showing as a nonfluorescent spot in the center of the cytoplasm. Falck–Hillarp method. Bar, 80 μm.

Pure propane freezes at the temperature of liquid nitrogen, so it is necessary to add sufficient propylene to keep the mixture liquid. The copper container contains a copper-mesh basket with a handle, and the tissue pieces to be frozen are placed in this and then immersed rapidly in the propane–propylene mixture (see Falck and Owman, 1965; Björklund *et al.*, 1972b, for details). The tissue pieces freeze immediately and are transferred to the tissue plate of the freeze-dryer which is held by its handle and immersed in another Dewar flask containing liquid nitrogen. The tissue can be stored for long periods in liquid nitrogen without losing monoamines and, consequently, tissue may be collected for several days before a freeze-dry run is initiated. The use of the propane–propylene mixture is to allow the most rapid possible freezing of the tissue. It is evident from a number of reviews (cf. Björklund *et al.*, 1972b) that the critical point in the freezing is that at which the tissue passes between the temperatures of −30° and −40°C. If this is done rapidly, it minimizes ice-crystal artifact and provides the best tissue preservation. Liquid nitrogen itself does not give good preparations because of the rapid gas formation adjacent to the tissue which insulates it from the rapid cooling and consequently slows freezing through the critical temperatures. Isopentane, although used widely, is less satisfactory because it freezes at the temperature of liquid nitrogen and is difficult to keep in a liquid state. Rapid freezing of tissue as described above results in cracking of the tissue, and the gauze covering prevents loss of tissue pieces formed by cracks during the process of freezing.

After the tissue has been frozen and collected on the tissue plate, it is transferred to the freeze-dryer. The dessicant-containing baskets are then placed over the tissue, and the glass plate covering the cooling well is lightly covered around the edge with vacuum grease and placed over the o-ring. At this point the temperature in the alcohol bath should be at −35°C, and the tissue will rapidly equilibrate to this temperature. The mechanical vacuum pump is turned on and, as the initial evacuation of the chamber takes place, the pump makes a bubbling sound which gradually subsides if the vacuum seal is sound. Evacuation of the chamber should be allowed to continue to a vacuum of 10^{-1} torr when the oil diffusion pump can be turned on. For very small pieces of tissue, 4–5 days in the freeze-dryer are sufficient to assure good drying. Larger pieces may require as long as 10–14 days. At about 2 days, or with longer runs, at 3–4 days, the temperature in the alcohol bath is gradually raised to room temperature. At the conclusion of the period in the freeze-dryer, the oil diffusion pump is turned off and allowed to cool. At this point the mechanical pump can be turned off, and air is admitted to the chamber containing the tissue plate. If the humidity of the room air is high, this should be passed through a trap immersed in liquid nitrogen or obtained from a chamber containing a desiccant. Alternatively, pure nitrogen gas can be used as a substitute. The tissue basket is then removed rapidly and placed in a dessicator over phosphorus pentoxide, under vacuum, or treated immediately with formaldehyde vapor. Tissue kept in a dessicator prior to treatment can be held for several days without noticeable depreciation of fluorescence intensity.

The formaldehyde vapor treatment is carried out in a 1-liter glass vessel containing about 5 g of paraformaldehyde in an oven at 80°C for 1 to 2 hr. Paraformaldehyde is depolymerized during heating to form formaldhyde gas which undergoes the specific condensation reaction with primary and secondary amines. It should be noted that the reaction with secondary amines (e.g., epinephrine) is slower, and a 3-hr exposure is optimal for a demonstration of this compound. The paraformaldehyde is discarded after use. The water content of the paraformaldehyde is essential to the optimal formation of the fluorophore in the Falck–Hillarp freeze-dry method. If water content is too low, the fluorescent yield is low, and few amine-containing structures are visualized. On the other hand, if the water content is too high, the fluorescence yield is high, but the amines are so diffused that structural analysis is difficult. Indeed, in some preparations, diffusion may be so great as to make morphologic analysis impossible. The ideal situation is to have a paraformaldehyde with sufficient water content to produce an optimal yield of fluorescence with minimal diffusion (Falck and Owman, 1965; Hamburger *et al.,* 1965; Björklund *et al.,* 1972b). For reasons that are not entirely understood, the optimal water content appears to vary from time to time within the same laboratory and also from laboratory to laboratory. Standardization of the water content of paraformaldehyde was carried out by Hamburger *et al.* (1965) who introduced the technique of equilibrating paraformaldehyde to a constant humidity provided by solutions of sulfuric acid. In practice, small vials of paraformaldehyde are maintained in desiccators with varied solutions of concentrated sulfuric acid and water. These solutions produce, in a closed container, a constant relative humidity and, hence, a constant water content of the paraformaldehyde. In most laboratories a relative humidity between 50 and 90% will produce optimal results, but this can only be determined empirically. For catecholamines we routinely use paraformaldehyde equilibrated to a relative humidity between 50 and 70%. Figure 6A shows the norepinephrine containing neurons of the superior cervical ganglion that exhibit a high fluorescence yield and, at the same time, good morphological detail. This is exemplified by the lack of fluorescence over the cell nuclei and by the very distinct cell borders. In contrast, Fig. 6C from the arcuate nucleus–median eminence complex shows a very good fluoresence yield but with considerable diffusion as exhibited by the fluorescence over the nuclei, the indistinct cell borders, and the diffusion of varicosities within the arcuate nucleus.

Following the formaldehyde vapor treatment, the tissue block should be infiltrated as rapidly as possible in degassed paraffin. The infiltration is performed *in vacuo* to insure rapid and complete penetration of the dried tissue. The infiltration is carried out with the tissue still attached to the paper strips and with the gauze covering intact. After the initial infiltration the paper is removed, the gauze stripped off the tissue using a warm forceps, and the block embedded by any standard technique for sectioning in a paraffin microtome. Sections are cut on a rotary microtome at approximately 6–15 μm, mounted on clean glass slides in either Entellan,®* Fluoromount,®† liquid paraffin, or immersion oil. The sections are then gently warmed for a short time on a slide warmer at approximately 60°C to dissolve the paraffin into the mounting medium. They are then visualized in the fluorescence microscope. Central nervous system tissue shows stable fluorescence in paraffin blocks for approximately 3–6 months, but the deparaffinized sections show rapid fading over a few days associated with a significant increase in background fluorescence. Consequently, once sections are made, the material must be analyzed and photographed immediately. The fluorescence can be maintained somewhat longer by keeping the sections in the refrigerator, but standard Falck–Hillarp preparations can be used for only a short time.

The above methodology has been principally used for catecholamines and is much less satisfactory for serotonin-containing neurons. Indeed, with the methodology described above, most investigators have only been able to demonstrate serotonin cell bodies, with very few terminal areas visible (cf. Kuhar *et al.*, 1972). The visualization of serotonin cell bodies and terminals is improved by a pretreatment of the animal with a monoamine oxidase (MAO) inhibitor (e.g., nialamide, pargyline). This treatment significantly elevates the serotonin content of the brain and enhances the brightness of the fluorophore obtained from formaldehyde vapor treatment. Catecholamine fluorescence is not greatly affected by MAO inhibitors, as the principal enzyme for catecholamine degradation in brain is catechol-*O*-methyltransferase. In addition, Fuxe and Jonsson (1967) have reported that increased serotonin fluorescence is obtained using 1-hr exposures to formaldehyde vapor generated from paraformaldehyde equilibrated to a low relative humidity (60%) for the first hour and a high humidity (90%) for the second hour. In our experience, the most successful demonstration of brain serotonin neurons, including pathways and terminals, is obtained with the method of Aghajanian *et al.* (1973) in which the animals are treated with an MAO inhibitor (pargyline, 100 mg/kg, i.p.) followed 15 min later by L-tryptophan (100 mg/kg, i.p.). Two hours thereafter, the animals are decapitated, and the freeze-drying is carried out as detailed above, with the formaldehyde vapor treatment performed for 1.5 hr at 80°C using paraformaldehyde equilibrated to a relative humidity of 60%.

The interpretation of material prepared by the Falck–Hillarp method in the central nervous system has followed from information obtained from studies of the peripheral sympathetic system. It is well known that the peripheral sympathetic neuron (Geffen and Livett, 1971) conains a moderate level of catecholamine in the cell body, low levels of catecholamine in preter-

*E. Merck, Darmstadt, Federal Republic of Germany.
†E. Gurr, London, England.

minal axons, and very high levels in the axon terminals. The sensitivity of the standard Falck–Hillarp method is such that, in the peripheral nervous system, it is possible to visualize the cell bodies of sympathetic neurons (Fig. 6A) and the terminal axons (Fig. 6B). Some preterminal axons are visualized, but this is typically at the point near axon terminals. With the procedure outlined above and the fluorescence microscope equipped with filters as described previously, catecholamine neurons exhibit a green to green–yellow fluorescence. The method does not make a fluorescence microscopic distinction between dopamine and norepinephrine fluorophores, but this can be done by special treatment or by microspectrofluorometry (cf. Björklund *et al.,* 1975). The methods for distinguishing between dopamine and norepinephrine, particularly microspectrofluorometry, are cumbersome and require specialized, expensive equipment. With newer variants of the fluorescence histochemical methods (see Section IV.C), distinctions can be made on the basis of morphology alone. In contrast to the catecholamine fluorophores, the serotonin fluorophore has a distinctive yellow appearance. The sensitivity of the Falck–Hillarp method, however, is significantly less for serotonin than for the catecholamines and, as noted above, it is unusual to demonstrate serotonin axons or axon terminals in pharmacologically untreated material. Serotonin cell bodies are evident, presumably because the combination of the amount of serotonin present and the thickness of sections provides sufficient amine to obtain a visible fluorophore (Fig. 8A), whereas the axons and terminals are too small to be visible. A further problem with the serotonin fluorophore is that it undergoes a very rapid photodecomposition so that in a very few minutes of being exposed to the excitation light, the fluorophore is no longer visible. Both of these problems are overcome to some extent with the pharmacological manipulation of MAO inhibition and tryptophan loading, but not to the extent that the result is equivalent to that for catecholamines. Despite this, the conventional Falck–Hillarp method has remained the most reliable one for histochemically demonstrating serotonin neurons in the central nervous system.

The major difficulties in interpretation of Falck–Hillarp material are threefold. The first is the limited sensitivity of the method, which has been described above. The second is artifactual fluorescent material. An example of this is autofluorescent material seen in some neuronal perikarya. Frequently, autofluorescent material in neuronal cell bodies may be misinterpreted as serotonin fluorescence, but the distinction can be made readily. The autofluorescent material (cf. Fig. 9D) appears as orange–yellow granules in the neuronal cytoplasm which do not fade with exposure to light from the excitation filter. This autofluorescence is much more common in older than in young animals and probably represents the accumulation of lipochrome pigment within the neuronal cell bodies. In addition, when present it is apparently ubiquitously distributed and is independent of the formaldehyde vapor treatment. That is, control tissue can be run through the histochemical procedure, not treated with formaldehyde vapor, and show

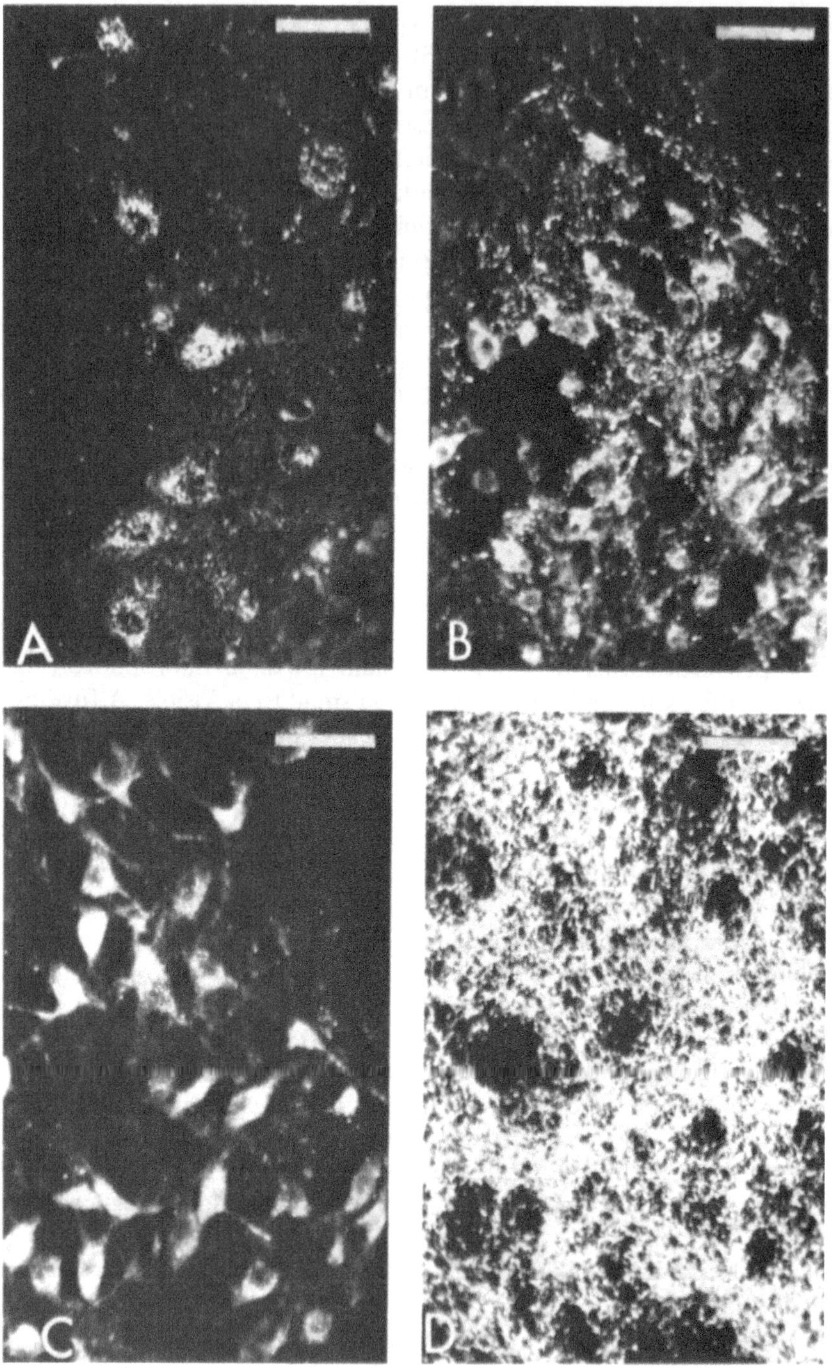

Figure 8. (A) Fluorescent cell bodies in the dorsal raphe nucleus of the midbrain. In the microscope these have the characteristic, yellow fluorescence and rapid photodecomposition of indoleamines. Falck–Hillarp method. Bar, 50 μm. (B) Nucleus locus coeruleus demonstrating

the same autofluorescence. In addition, the autofluorescence does not fade with exposure to ultraviolet light as the specific monoamine fluorescence does. A second source of artifact in the Falck–Hillarp material is the frequent cracking attendant upon the freeze-drying procedure. In some preparations this is minimal, whereas in others it is common, and it may occur in an area where it results in an unsightly gap through what would have made an excellent photograph. A third problem in interpretation of Falck—Hillarp material lies in the failure to demonstrate fluorescent material on the one hand or excessive diffusion of fluorescent material on the other. That is, the line between an optimal reaction and one that produces little fluorescence on the one hand, or excessive diffusion on the other, is a fine one. Even the most experienced laboratories have periodic, inexplicable failures with the method. A typical crack and some diffusion is evident in Fig. 6C. A last problem, but one that does not pertain to the central nervous system as far as is currently known, is shown in Fig. 6D. This, from the pars intermedia of the monkey pituitary, exhibits intense fluorescence within the endocrine cells of the pars intermedia which, in the fluorescence microscope, has the appearance of an indolamine. There has been work to establish, however, that this represents a cyclization reaction with a terminal tryptophan molecule on a peptide in the pars intermedia. The specific conformational arrangements of the peptide with the terminal tryptophan provide for the intense fluorescence. This is a type of result that must be carefully analyzed in studies of peripheral tissue.

Despite the difficulties with the Falck–Hillarp method, it continues to be valuable and extensively used for several reasons. First, it offers the simplicity of the stretch and smear techniques. Once established, it is easy to perform and, despite the ocasional failures, reasonably reliable. In addition, it has been, until recently, the only method available for the examination of large tissue blocks and for the study of serotonin neuron systems. If one is interested in studying the larger brains of animals such as cat or monkey, it is possible to take quite large blocks for processing, which allows serial sectioning through areas of the nervous system.*

*Recently a new modification (Ajelis *et al.*, 1979; Lorén *et al.*, 1980; Björklund *et al.*, 1980) has markedly enhanced the utility of the Falck–Hillarp method.

---

densely packed norepinephrine-containing neurons with a neuropil heavily innervated by catecholamine-containing terminals. Some of these are terminals of norepinephrine neurons, and others of epinephrine neurons. They cannot be distinguished in the Falck–Hillarp material. Vibratome–formaldehyde method. Bar, 80 μm. (C) Dopamine-containing cell bodies of the substantia nigra pars compacta. Occasional catecholamine-containing varicosities are scattered through the nucleus. Vibratome–formaldehyde method. Bar, 80 μm. (D) Caudate–putamen complex demonstrating dense innervation by dopamine-containing axons which form a meshwork of fine, preterminal axons and small varicose terminals. The open, nonfluorescent areas are zones where bundles of fibers of the internal capsule traverse the caudate–putamen complex. Glyoxylic acid method. Bar, 80 μm.

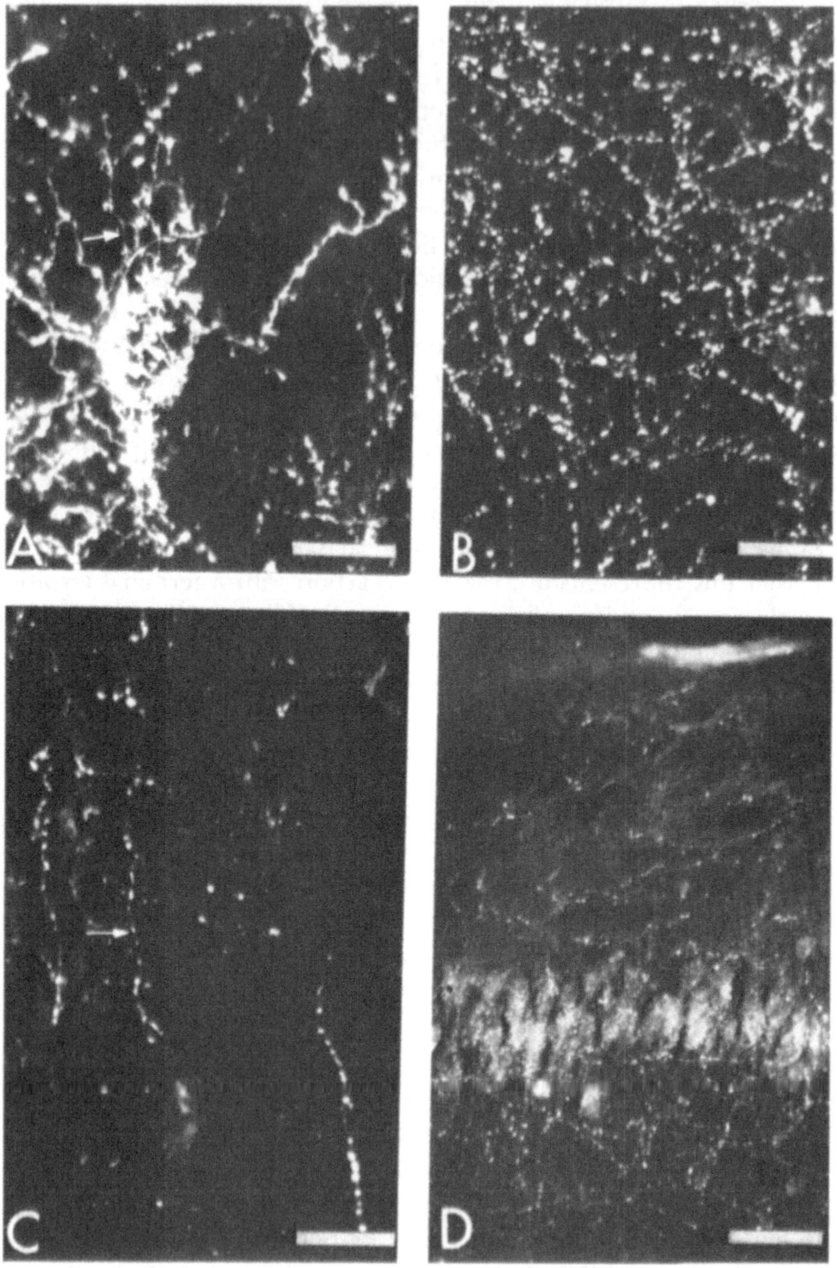

Figure 9. (A) Lateral septal nucleus demonstrating a neuron densely innervated by dopamine axons originating in the ventral tegmental area. The fine dopamine axons (small arrow) form an intensely fluorescent, dense plexus of axons and terminals about the lateral septal nucleus neuron and its proximal dendrites. Additional axons are present'in the adjacent neuropil. This form of dopamine innervation is quite distinct from that shown in Fig. 5D. Glyoxylic acid method. Bar, 35 μm. (B) Hippocampus (stratum radiatum, CA3) showing typical norepinephrine innervation arising from locus coeruleus. The axons show fairly uniform, regularly

3. The Vibratome–Formaldehyde Technique

The original Falck–Hillarp method was modified by Hökfelt and Ljungdahl (1972b) in two significant ways. First, they eliminated freeze-drying and introduced the use of perfused material. Second, they introduced the Vibratome* to fluorescence histochemical methodology. The instrument is described in Section VII.C briefly and will not be commented upon further except in regard to its use in this method.

The Vibratome–formaldehyde method utilizes two solutions. The first of these is a buffered formaldehyde solution made by combining 83 ml of a solution of monobasic sodium phosphate (2.26 g/100 ml) with 17 ml of a sodium hydroxide solution (2.52 g/100 ml), adding 4 g paraformaldehyde, and heating to 60°C to dissolve the paraformaldehyde. The pH is then adjusted to 7.2–7.4 with sodium hydroxide. The second solution is a calcium-free Tyrode solution (sodium chloride, 0.8 g; potassium chloride, 0.02 g; magnesium chloride, 0.01 g; dibasic sodium phosphate, 0.05 g; monobasic sodium phosphate, 0.1 g; and dextrose, 0.1 g added to 100 ml distilled water and the pH adjusted to 7.2–7.4).

The procedure for the Vibratome–formaldehyde method is as follows. The animal is anesthetized with pentobarbital (40 mg/kg, i.p.). The thoracic cavity is opened, and the descending aorta is clamped. The right auricle or vena cava is then incised, and a second incision is placed in the left ventricle, through which a cannula is placed into the ascending aorta. We have found it most satisfactory to attach the cannula to a large syringe. The perfusion may be carried out at room temperature, but results are often improved by chilling the anesthetized animal in an ice bath so that the perfusion is carried out under hypothermia. As with other perfusion techniques, the solution should be carefully filled to the tip of the cannula at the start of the perfusion so as to avoid air embolism occluding cerebral vessels and resulting in incomplete brain perfusion. The perfusion is carried out over 2–10 min with 100–200 ml of the buffered formaldehyde solution at 4°C. The volume of perfusion solution is determined empirically and depends on the size of the rat. The brain is then immediately removed and stored in ice-cold, buffered formaldehyde solution for approximately 20 min. Following this postperfusion fixation, the brain is dissected into pieces approximately 4 mm in diameter which are then mounted for sectioning. Other pieces may be main-

*Oxford Instruments, San Mateo, California.

spaced varicosities with short intervaricose segments. Vibratome–formaldehyde method. Bar, 50 μm. (C) Rostrum of the corpus callosum exhibiting fluorescent preterminal axons of the locus coeruleus type (arrows). The photograph is oriented so that the corpus callosum is running in the vertical direction. The fibers show thin intervaricose segments with small, fusiform varicosities. Vibratome–formaldehyde method. Bar, 50 μm. (D) Hippocampus (stratum oriens, SO; stratum radiatum, SR on either side of the CA1 stratum pyramidale) showing intense autofluorescence of the stratum pyramidale neurons, typical locus coerulus norepinephrine axon innervation of the dorsal stratum radiatum, and a similar sparse innervation of the stratum oriens. Vibratome–formaldehyde method. Bar, 80 μm.

tained for several hours in the fixative and give satisfactory results when subsequently sectioned. In our laboratory sections are usually cut at 10–25 μm as measured by the Vibratome scale. This is generally performed with the vibration rate at 6–9 scale units and a feeding speed of 1–3 scale units. For sectioning, the trough of the Vibratome is filled with Tyrode solution kept at or below 5°C. We accomplish this by transferring large copper bars between the Tyrode solution and a mixture of dry ice and acetone and monitoring the temperature in the trough. The copper bars are transferred any time the temperature rises above 5°C. Immediately after being cut, the sections are placed on a clean glass slide, blotted around the edges to remove excess Tyrode solution, dried under the warm air stream of a hair dryer for 15 min, and placed in a desiccator overnight under vacuum and over phosphorus pentoxide. The following day they are treated for 1 hr at 80°C in an oven with paraformaldehyde equilibrated to 70–80% relative humidity. The sections are mounted in liquid paraffin, Entellan,® or immersion oil and examined in the fluorescence microscope. Sections prepared by the Vibratome–formaldehyde method are more durable than those prepared by the conventional Falck–Hillarp method. If kept refrigerated between use, the sections may be examined without significant loss of fluorescence quality for 1–2 weeks. Examples of material prepared by the Vibratome–formaldehyde method are shown in Fig. 8B,C and Fig. 9B–D.

It should be added at this point that the examination of central nervous system material by dark-field microscopy is not an easy task even for the individual experienced in neuroanatomy. Landmarks are difficult to visualize, and the fluorescent axons and cell bodies may be difficult to localize. In this respect, the Vibratome–formaldehyde method offers a significant advantage over the original Falck–Hillarp method or the glyoxylic acid method which will be described below. The sections adhere well to the slide and, following visualization in the fluorescence microscope, the cover slip may be readily removed and the section cleared through xylene and graded alcohols to water, stained with cresyl violet, dehydrated, and again coverslipped for viewing by routine light microscopy. Thus, if drawings or photographs are made from the material in the fluorescence microscope, exact verification of their location can be obtained in the stained section. This can be applied, of course, to the Falck–Hillarp or glyoxylic acid material, but the histologic quality of the material is so much poorer with these methods that exact confirmation of tissue localization is not as readily carried out.

At the present time, the Vibratome–formaldehyde method is routinely used in our laboratory because of its reliability, the ease of cutting Vibratome sections from formaldehyde-fixed material, and the quality of the histological material. It does have several disadvantages, however, and the most significant of these is that serotonin neurons are not shown by this method. In addition, the axonal and terminal morphology of Vibratome–formaldehyde material is not comparable in quality to that of good sections prepared by the glyoxylic acid method (Lindvall and Björklund, 1974a) to be described in Section IV.C, and, finally, the necessity of cutting Vibratome sections limits the amount of material that can be prepared from a single brain in 1 day. An individual with patience and skill in using the Vibratome will rarely cut more than 15–20 sections in a single day so that only small amounts of tissue can be prepared for study. Further, only small tissue blocks can be cut so that, if rapid screening of a large area is important, the original Falck–Hillarp freeze-dryer method or one of the glyoxylic acid methods may be preferable. There is some evidence that material prepared by the Vibratome–formaldehyde method may be significantly improved by taking the cut sections and immersing them for 1–2 min in a solution of Tyrode solution containing 5 g magnesium sulfate/100 ml solution. This appears to greatly enhance fluorescence and axonal morphology (Lorén *et al.*, 1977).

C. Methods Using Glyoxylic Acid

1. The Vibratome–Glyoxylic Acid Method

The use of glyoxylic acid (GA) as a reagent to form a fluorophore with biogenic amines was introduced by Björklund, Lindvall, Falck and Svensson in a series of papers (see Björklund *et al.*, 1972a, 1975; Lindvall *et al.*, 1974, for reviews). These studies established the basic reactions of biogenic amines with glyoxylic acid and the application of the Vibratome provided the basis for a viable histochemical technique (Lindvall *et al.*, 1973; Lindvall and Björklund, 1974a). This method will be described in detal as it has provided the most powerful histochemical method for the analysis of the organization of catecholamine neuron systems in the central nervous system. The technique to be described is essentially that of Lindvall and Björklund (1974a).

The Vibratome–GA method utilizes two primary solutions. The first is a 2% GA solution in a Krebs–Ringer bicarbonate buffer. The buffer composition is as follows (in g/l): sodium chloride, 6.923; potassium chloride, 0.354; calcium chloride ($CaCl_2 \cdot 2H_2O$), 0.165; monobasic potassium phosphate, 0.162; magnesium sulfate ($MgSO_4 \cdot 7H_2O$), 0.294; sodium bicarbonate, 2.100; glucose, 1.8. The buffer is approximately pH 7.0. After it is made up, it is saturated with a mixture of 95% O_2 and 5% CO_2. Glyoxylic acid is then added to a concentration of 2% (20 g/1). In this laboratory the monohydrate of glyoxylic acid is routinely used. This compound is somewhat unstable and, when received in the laboratory in 100-g lots, individual containers (e.g., scintillation vials) are filled with small amounts of GA which approximate those that might be used in a day (5–10 g). These are sealed, placed in a dessicator over phosphorus pentoxide, and kept in a freezer. They are removed individually for use. Ater the GA is added to the buffer, the solution is acidic, and the pH is adjusted to 7.0 with 10 N sodium hydroxide. The second solution is Krebs–Ringer bicarbonate buffer, pH 7.0, prepared as described above including the saturation with 95% O_2 and 5% CO_2. The histological procedure is essentially that used for the

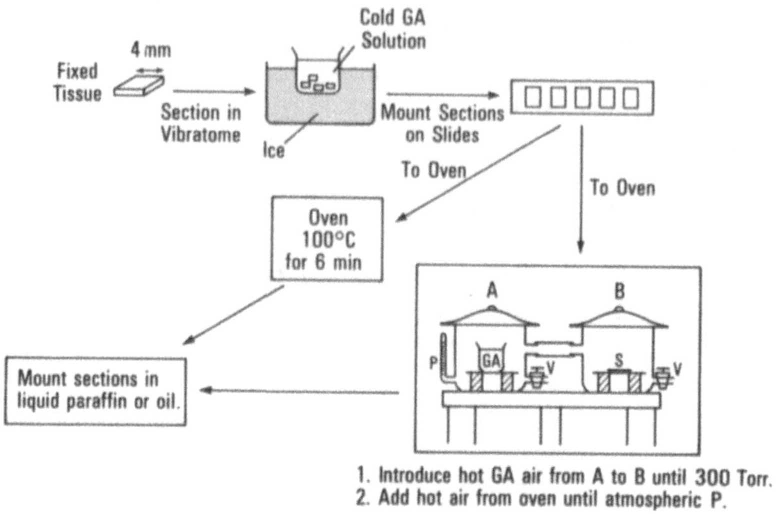

Figure 10. Diagram illustrating the glyoxylic acid method. See text for description.

Vibratome–formaldehyde method. The animal is anesthetized and the head perfused with GA solution through a cannula placed in the ascending aorta. Fo the GA method, 150 ml of 2% GA solution is perfused rapidly (in less than 2 min). This amount is for adult rats (approximately 200–250 g); larger amounts of perfusion solution would be necessary for larger animals, and proportionally smaller amounts for smaller ones.

Following perfusion the brain is removed rapidly, cooled in buffer, and dissected to provide the desired specimen which should be no greater than 5 mm thick in any dimension. This is then glued to the Vibratome specimen holder with a rapidly hardening adhesive (Loctite Quickset Adhesive 404®*). The specimen and holder are immediately immersed in cold buffer, transferred to the Vibratome, and submerged in ice-cold Krebs–Ringer bicarbonate buffer in the Vibratome trough. As with the Vibratome–formaldehyde method, the buffer is kept at 0.5°C through the sectioning which is performed at a vibration rate usually in the range of 6–7 units and a feeding speed of 1–3 scale units. Unlike the specimens prepared by formaldehyde perfusion, GA-perfused specimens are quite soft, and it is extremely difficult to cut good sections. With experience, however, most individuals can cut acceptable sections at 25–35 μm. The tissue piece will usually be maintained in good condition for 3–4 hr, during which an experienced individual can obtain at least ten acceptable sections. Some areas are much easier to cut than others. Septum, caudate, thalamus, and hypothalaus are much easier to cut than, for example, brainstem or cerebral cortex. The Vibratome–GA method is presented schematically in Fig. 10.

*Loctite Corp., Newington, Connecticut.

After a section is cut, it is transferred with a blunt glass rod or a fine brush to a container of ice-cold GA solution (a small beaker or staining dish in an ice bath will suffice) for 3–5 min. While it is in the GA solution, the section must be kept submerged below the surface of the solution throughout the treatment. It is then floated onto a clean glass slide, excess solution is blotted away, and the section is dried under the warm air stream of a hair dryer for 15 min. The section is then transferred to a desiccator where all of the sections collected from that day are kept in a slide rack over phosphorus pentoxide. When all the sections have been collected, the desiccator is evacuated and kept in the dark overnight. The development of the fluorophore can be accomplished by one of two methods, further GA vapor treatment or heating. These have been described in detail (Lindvall and Björklund, 1974a; Björklund *et al.* 1975).

Glyoxylic acid treatment requires two vessels of about 1 liter capacity. In this laboratory we use small desiccators. One vessel, the reaction vessel, consists of a desiccator that has three outlets. One of these is attached to a manometer. The second is connected by a piece of tubing to the second vessel, and the third is formed by a valve that can be opened to the outside atmosphere. The second vessel, the GA vessel, is another small desicator equipped with a valve and the aforementioned connection to the reaction vessel. The treatment is carried out as follows. GA (2 g) is dried in a desiccator over phosphorus pentoxide for 24 hr. placed in the GA vessel, and heated at 100°C for 1 hr. The two vessels are then placed together in the oven at 100°C. The microscope slides containing the sections are then placed in the oven for about 3 min and transferred to the reaction vessel. The reaction vessel is evacuated with a vacuum pump, and hot, GA-saturated air from the GA vessel is introduced into the hot reaction vessel to a pressure of 300 torr. Hot air from the oven is then allowed into the reaction vessel until atmospheric pressure is attained. The sections are treated in a reaction vessel for 2 min and then removed. For the alternative method of developing the fluorophore, reaction by heating, the slides containing the sections are placed in the slide rack and heated at 100°C for 6 min. In general, the fluorescence yielded from the heating treatment is less than that from the GA vapor treatment, but heating alone produces less background fluorescence than the GA vapor treatment.

Following the treatment, whether by heating or GA vapor treatment, the sections are mounted in liquid paraffin or immersion oil, coverslipped, and examined in the fluorescence microscope. Glyoxylic acid-treated material is extremely stable over time and can be kept, with excellent fluorescence, for several weeks if the slides are placed in a freezer at −20°C when not being examined. Over time GA sections tend to have increasing background fluorescence but, when kept cold, this is decreased while the specific fluorescence is maintained. Examples of GA method material are shown in Figs. 8D, 9A, and 11.

The principal advantage of the GA method is the extremely discrete morphology revealed in good preparations and the marked sensitivity of the

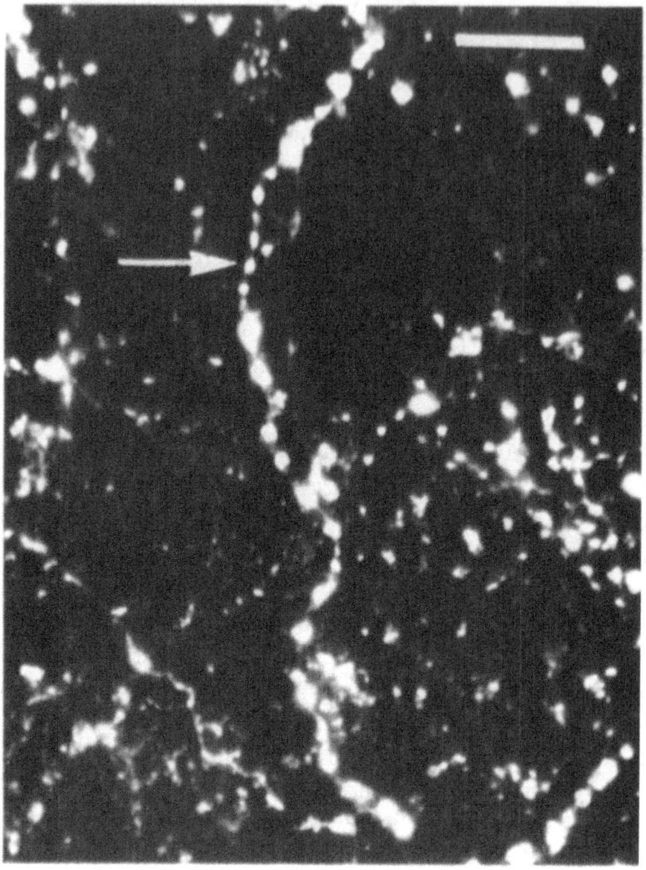

Figure 11. Lateral septal nucleus at high magnification showing a typical norepinephrine-containing axon of brain stem origin (arrow). The axon runs vertically across the photograph with large varicosities and small varicosities irregularly spaced and very thin intervaricose segments. Glyoxylic acid method. Bar, 20 μm.

method as compared to previous fluorescence histochemical methods. Preterminal axons are routinely observed and, as Lindvall and Björklund (1974b) have shown, it is possible to identify the preterminal axons of dopamine neurons as distinct from norepinephrine neurons. In general, the axons of dopamine neurons are thin and, in the preterminal segments, contain no varicosities. At sites of termination they collateralize into two forms of terminal networks. The first is into dense plexuses of fine, highly collateralized, intertwined axons with small varicosities, as in the neostriatum (Fig. 8C), or into more discrete, less collateralized axons with larger varicosities, as in the septum (Fig. 9A). As shown in Fig. 9A, it is typical for such dopamine axons to form dense pericellular arrays of axons in terminals. In contrast to this, preterminal norepinephrine axons, particularly of locus coeruleus origin, are thin but have fusiform varicosities (Fig. 9C). In areas of terminal distribution, norepinephrine axons collateralize into a plexus of fine axons with regularly spaced varicosities. These are shown well in Vibratome–formaldehyde material as in the GA material (Figs. 9B,D). In general, the varicosities of norepinephrine axons of locus coeruleus origin are smaller than those arising from other brainstem norepinephrine cell groups and have a more regular distribution in the terminal field. Brainstem axons tend to have large varicosities but with considerable variance in morphology in the terminal field (Fig. 11). It is not possible at this time to make any statement about preterminal axons of brainstem norepinephrine cell group origin as there appears to be no particular rule about their appearance. The varicosities observed in GA material occurring along preterminal axons are conventionally viewed only as areas of somewhat higher local neurotransmitter contact. Those observed in terminal areas are interpreted, as are those in Falck–Hillarp material, as terminals of axons that would contain vesicles and make synaptic contacts. Certainly, the high level of fluorescence in the varicosities within terminal areas implies a high neurotransmitter content.

Despite the morphological advantages of the GA method, it has a number of disadvantages. First, it does not demonstrate serotonin as currently applied. Second, sectioning is quite difficult, and only a relatively few sections can be obtained from a small piece of tissue in a full day. Third, the method is not applicable to large blocks of tissue, and the cytology of the material obtained is quite poor in comparison to that from Vibratome–formaldehyde preparations. There are several recent variations of the Vibratome–GA method that offer considerable promise in obviating some of these disadvantages. These will be described below. One simple but very useful modification is that of Berger *et al.* (1976) in which magnesium chloride is substituted for calcium chloride at the same molarity in the perfusion solution, and the sections are treated with formaldehyde vapor generated from paraformaldehyde equilibrated to 50% humidity for 80°C for 1 hr.

2. The Cryostat GA Methods

At the present time, three cyrostat GA methods have been described (Watson and Barchas, 1975, 1977; Battenberg and Bloom, 1975; Bloom and Battenberg, 1976; de la Torre and Surgeon, 1976). Two of these methods will be described here.

a. de la Torre and Surgeon Modifications (de la Torre and Surgeon, 1976). This is the simplest of the GA methods. The only special equipment required is the cryostat. One solution is utilized, a 1% GA solution in phosphate buffer. This is made up as follows: GA monohydrate, 2.5 g; sucrose, 17.1 g; KH_2PO_4, 7.5 g; distilled water, 250 ml. This 1% GA solution is brought to pH 7.4 with 1 N sodium hydroxide.

The procedure is as follows. The animal is sacrificed by decapitation. Fresh brain is removed rapidly, and the appropriate area is blocked and placed on a cryostat chuck at −30°C. It is attached to the chuck with a thin film of water. Several blocks can be taken from a single brain and kept in the cryostat chamber until cut. Sections are cut as 15 μm (with a range from 6–30 μm) at −30°C. The sections are mounted directly on a clean glass slide. They are then removed from the cryostat and dipped three times, approximately 5 sec for each dipping, in the 1% GA solution. Excess Ga solution is blotted from around the section, and the sections are then dried for 3 min under the warm air stream from a hair dryer. Following drying they are placed in any oven at 80°C for 5 min. The timing of this step is critical; longer times in the oven increase background significantly. Following removal from the oven the section is covered with liquid paraffin and coverslipped. It is then placed on a slide warmer at a surface temperature of 85° for 90 sec. This equivalent to the heat treatment in the original Vibratome–GA method. This method has recently been improved by de la Torre (1980).

b. Cryostat Formaldehyde–GA Method. This combination formaldehyde and GA method for use on cryostat sections was developed by Battenberg and Bloom (1975; Bloom and Battenberg, 1976). Two solutions are used. The first is a combined paraformaldehyde–GA perfusion solution. This is made of two stock solutions prepared no more than 4 hr prior to use. Paraformaldehyde is depolymerized in double-strength phosphate buffer (0.5 M, pH 7.4, 65°C) in a concentration of 1 g/100 ml; the solution is then chilled to 2–4°C prior to use. Glyoxylic acid monohydrate is dissolved in mammalian Ringer's solution at room temperature in a concentration of 4 g/100 ml. The precooled paraformaldehyde and GA solutions are combined in equal volumes to derive a final solution containing 0.5% paraformaldehyde and 2% GA in phosphate-buffered Ringer's solution (approximately 450 milliomolar). The pH is then adjusted to 7.4 with 10 N sodium hydroxide, and the solution chilled to 2–4°C prior to use.

The procedure is carried out as follows. The animal is anesthetized, the thorax opened, and the descending aorta clamped. The right vena cava is cut, and a small incision is made in the left ventricle. The head is immersed in crushed ice. The paraformaldehyde–GA perfusion solution is perfused through the ascending aorta under pressure so that the entire volume of fixative passes through within 90 sec (a total volume of 250 ml is optimal for rats in the weight range 100–400 g). The brain is rapidly removed, dissected into 3- to 5-mm slabs in any desired plane of section, and these are immediately frozen onto brass cryostat chucks inserted into powdered dry ice. The chucks are then placed in the cryostat chamber. After the chucks have equilibrated to the temperature of the cryostat chamber (−16° to −20°C), sections are cut at 8–36 μm thickness, transferred to a glass slide with a chilled brush, and thawed onto the slide with finger pressure to the back of the slide while the slide is maintained within the cryostat chamber. The slides are immediately immersed in the second solution, an ice-cold 2%

glyoxylic acid solution in phosphate-buffered mammalian Ringer's solution (pH 7.4), for 11 min and then dried in the warm air stream from a hair dryer at approximately 37°C for 5 min. The slides are then placed in a preheated, covered Coplin jar at 100°C for 10 min in an oven. This, again, is equivalent to the heat development in the Vibratome–GA method. Following heat treatment the slides are removed, covered with liquid paraffin, coverslipped, and observed in the fluorescence microscope.

As with the previous method, this is a newly developed method which has been in use for only a short time. In our experience it has not provided the morphological detail of the Vibratome–formaldehyde method, but we cannot state that this is more than a lack of experience with its use. We have attained material with excellent visualization of cell bodies and some visualization of terminal areas, but it will require the test of a number of laboratories in order to determine if this method will be as useful as the original Vibratome–GA method. It has the undeniable advantage of much better tissue preservation because of the utilization of formaldehyde in the original perfusion solution and the quickness of cutting cryostat sections. Further, larger blocks of tissue can be utilized in the cryostat than in the Vibratome.

c. Freeze-Dry Formaldehyde–GA Method. Recently a method has been described combining formaldehyde–GA perfusion with freeze-drying and subsequent formaldehyde treatment for neurotransmitter histochemistry. (Lorén *et al.*, 1976a,b, 1977). Although this method is new, it shows considerable promise and has many advantages over previously described methods, so that it warrants description.

One solution is used for the method. It must be made fresh each day, no more than 4 hr prior to use. Either sodium glyoxylate or GA monohydrate may be used. When sodium glyoxylate is used, a 0.5% formaldehyde solution is prepared by dissolving paraformaldehyde (0.75 g/150 ml solution) in 0.1 M phosphate buffer, pH 7.0, and putting the formaldehyde-containing mixture in a boiling water bath to dissolve the paraformaldehyde. This is then cooled with crushed ice. When the formaldehyde solution has reached room temperature, sodium glyoxylate is added to a final concentration of 2.5% (to yield a glyoxylic acid concentration of 2%). At that point 40 g magnesium sulfate is added, and the pH is adjusted to between 4 and 5 with 6 M HCl. When Ga monohydrate is used, 0.75 g paraformaldehyde and 40 g magnesium sulfate are dissolved in 75 ml 0.1 M phosphate buffer, pH 7.0, as described above. Another solution is made with 3 g GA monohydrate dissolved in 75 ml 0.1 M phosphate buffer, pH 7.0, and the pH is then adjusted to 4.8 to 5.0 with solid sodium hydroxide. The solutions are then mixed to give a final concentration of 0.5% formaldehyde and 2% GA. If necessary, pH is adjusted to between 4 and 5 with 6 M HCl. The procedure is carried out as follows. The animal is deeply anesthetized with a suitable agent (e.g., pentobarbital, 40 mg/kg). The chest is opened as described for the previous perfusion methods, and the descending aorta clamped. The animal is perfused via the ascending aorta with 150 ml ice-cold perfusion solution with the clamp on the descending aorta removed after the first 30-ml perfusion. The perfusion should be carried out rapidly, requiring no more than 1–2 min. The brain is then removed immediately, and desired regions dissected. These brain pieces are then prepared exactly as for the original Falck–Hillarp procedure. That is, they are placed on pieces of paper, covered with gauze, frozen in a liquid propane–propylene mixture, freeze-dried, and treated with formaldehyde vapor. They are then embedded in paraffin, sectioned, mounted, and observed in the fluorescence microscope described above.

This method has significant advantages over the other GA method. First, it can be used for large pieces of brain. Second, in animals pretreated with an MAO inhibitor and tryptophan, as described above, serotonin can be demonstrated. For optimal demonstration of serotonin, a two-step treatment with formaldehyde vapor is recommended in which the specimens are treated for 1 hr at 80°C in the presence of paraformaldehyde equilibrated to a relative humidity of 50% and then for a second hour in the presence of paraformaldehyde equilibrated to a 70–90% relative humidity. A further advantage of this method is that it can be applied to fetal or newborn material as the freeze-drying procedure maintains morphological integrity that cannot be maintained with either Vibratome or cryostat sectioning. Although this method is new, our experience with it has been quite good and, in the view of this investigator, it will become an important tool in neurotransmitter histochemistry. It has some of the disadvantages of the original Falck–Hillarp method in that tissue cracking may be a problem and, occasionally, diffusion may occur, but it has the advantage of the GA method in providing great sensitivity and in preserving excellent axonal morphology. Further examples of preparations using glyoxylic acid methods are shown in Fig. 12.

Another recently described method offers the possibility of combining fluorescence histochemistry for monoamines with ultrastructural studies (Furness *et al.*, 1978). The method utilizes perfusion with a solution of 4% formaldehyde and 0.5–1% glutaraldehyde in 0.1 M cacodylate buffer (pH 7.0) 200 μl at a pressure of 120 mm Hg. Vibratome sections are cut, placed in the formaldehyde–glutaraldehyde solution for 15 min, transferred to slides, mounted in paraffin oil, and examined in the fluorescence microscope. Areas exhibiting specific catecholamine fluorescence are dissected and prepared by usual techniques for electron microscopy.

Figure 12. Illustrations of the glyoxylic acid method.(A) Dopaminergic cells and axons in the substantia nigra pars compacta; glyoxylic acid method. (B) Terminal plexus of the nigrostriatal dopamine system in the rat neostriatum; glyoxylic acid method. (C) Motor trigeminal nucleus showing numerous green norepinephrine-containing terminals surrounding the neuronal perikarya of the nucleus. These are visible by their orange–yellow autofluorescence; method of Lorén *et al.* (1976a). (D) Raphe nucleus containing yellow fluorescent serotonin-containing neurons; method of Lorén *et al.* (1976a).

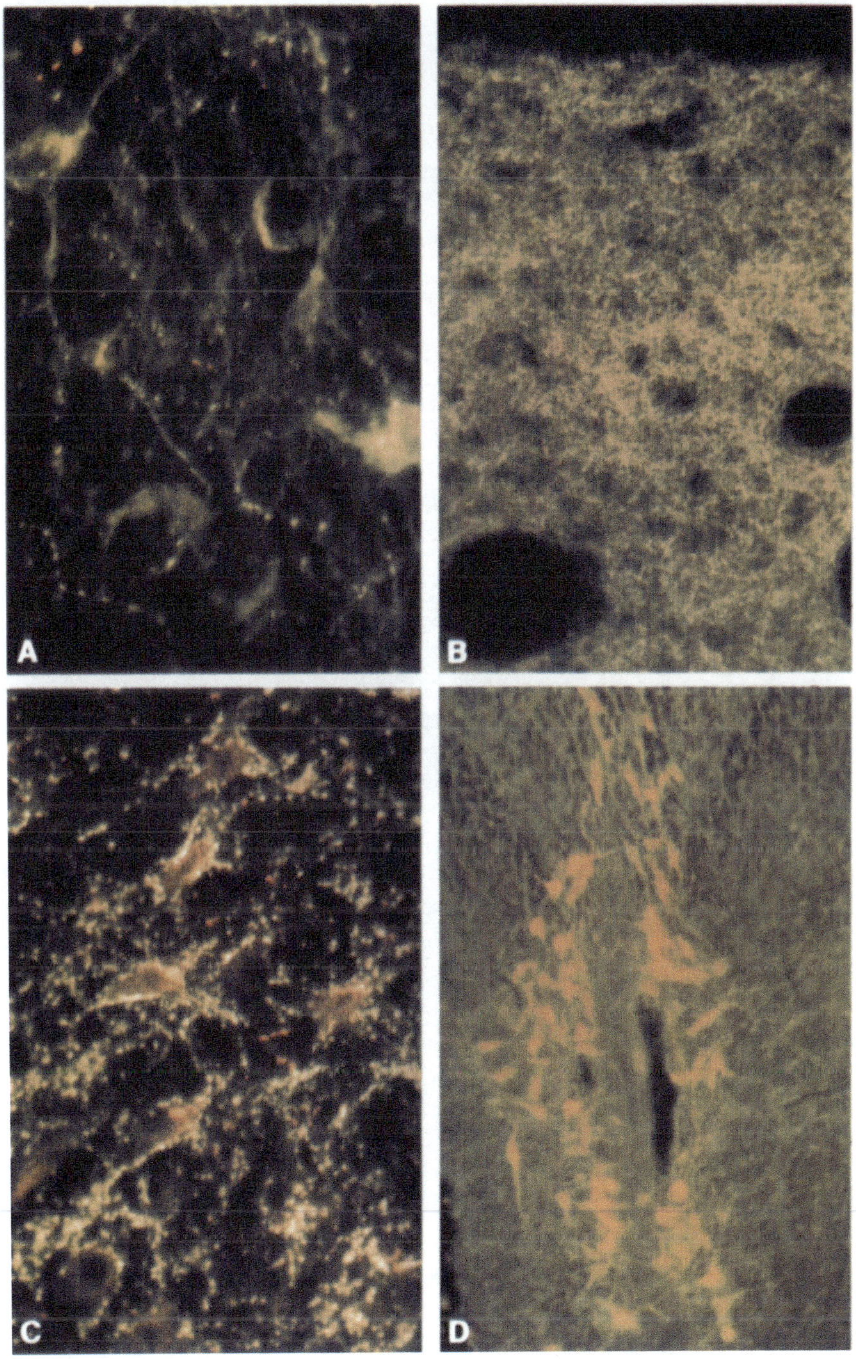

Figure 12

V. THE SELECTION OF FLUORESCENCE HISTOCHEMICAL METHODS

The selection of a fluorescence histochemical method is dependent upon several factors. First, for the investigator who wishes to carry out extensive studies of the morphology of monoamine neuron systems, it would be appropriate to set up a fully equipped laboratory that would include a freeze-dryer, a cryostat, and, perhaps, a Vibratome. This permits maximum flexibility in selecting a method. Second, for the investigator who has only limited or intermittent use for fluorescence histochemistry, one of the cryostat methods is probably optimal. A cryostat is a common piece of equipment in anatomical laboratories; the methods are quick and relatively uncomplicated and, if they are performed properly, consistent, reliable results can be obtained. The restriction of methods is that they are generally limited to relatively small areas of the nervous system. A further aspect of selection of an appropriate fluorescence histochemical method is the monoamine or monamines to be studied and the sensitivity required by the proposed experiment. As noted above, the original Falck–Hillarp method is the only one that reliably shows serotonin neuron cell bodies and terminals. Its sensitivity is limited even with respect to cell bodies (Beaudet and Descarries, 1979), and the rapid photodecomposition of terminals, even after pharmacological enhancement of the fluorophore, limits anatomical analysis. Serotonin neuron cell bodies are demonstrated by the Lorén *et al.* (1976a, 1977) method in a manner that appears as sensitive as the Falck–Hillarp method.

Since all of the fluorescence histochemical methods demonstrate catecholamines in tissue, the major factors that would influence selection of one or another method would be the requirements for specificity, sensitivity, and preservation of morphological integrity. All of the methods would appear to have approximately the same specificity. That is, all show norepinephrine, dopamine, and, probably, epinephrine in tissue. It has become evident, however, that the glyoxylic acid methods offer a much greater sensitivity than the original Falck–Hillarp method, and they also provide the best preservation of morphological integrity. This is particularly true of the original glyoxylic acid–Vibratome method (Lindvall and Björklund, 1974a). In the author's experience, when optimal preparations are obtained with the glyoxylic acid–Vibratome method, catecholamine neurons are shown with greater sensitivity and more exquisite preservation of morphology than are obtainable with any of the other methods. With such material, and with material prepared by the methods of Lorén *et al.* (1976a, 1977), it is possible, to a large extent, to categorize the majority of preterminal axons as arising from either dopamine or norepinephrine cell groups and to characterize the terminal axonal arborizations as arising from either lateral tegmental norepinephrine groups, locus coeruleus, or mesencephalic dopamine cell groups (Moore, 1978; Levitt and Moore, 1979). The glyoxylic acid method provides significant problems, however, in that it is extremely difficult to obtain good sections on a reliable basis. Indeed, some individuals never master the cutting of good Vibratome sections on glyoxylic acid-perfused brain, and even a highly skilled technician can produce no more than 10–20 usable sections in a day. For that reason the extreme sensitivity and preservation of the catecholamine neurons offered by the glyoxylic acid–Vibratome method is generally employed only on special occasions such as when preparations are made to obtain photographs for publication.

The method of Lorén *et al.* (1976a, 1977) offers nearly the same sensitivity and preservation of morphology but is more reliable and technically much easier. Further, it permits examining large blocks of tissue; the only limitation is the size of a block that will permit adequate freeze-drying. The major problem encountered with this method is the fracturing that inevitably occurs upon quenching the tissue in liquid nitrogen. Since this frequently will occur across an area of special interest, a number of preparations are usually required. The cryostat methods (de la Torre and Surgeon, 1976; Bloom and Battenberg, 1976) provide the advantages of speed and a technical ease in carrying out the procedure but are somewhat less sensitive than the methods noted above. In addition, they are somewhat capricious, and good preparations are not always obtained readily. A new modification of the Falck–Hillarp method in which freeze-drying is preceded by perfusion with solutions containing a high concentration of aluminum (A. Björklund, personal communication) offers the promise of sensitivity equivalent to the glyoxylic acid methods with the simplicity of the original Falck–Hillarp method. The Vibratome–formaldehyde method (Hökfelt and Ljungdahl, 1972b) is probably the most reliable and generally useful fluorescence histochemical method when it is not necessary to examine large areas of nervous system in a single brain. Tissue preservation is generally excellent, but the sensitivity of the method and preservation of catecholamine neurons are not quite as good as with the glyoxylic acid methods. Nevertheless, this method is routinely used for screening in this laboratory because of its reliability. The method of Berger *et al.* (1976) shares many of the advantages of the Vibratome–formaldehyde method but is somewhat less useful because the sectioning is more difficult and the method somewhat more capricious.

VI. ADVANTAGES AND LIMITATIONS OF THE FLUORESCENCE HISTOCHEMICAL METHOD

A. Advantages

The principal advantage of the fluorescence histochemical methods used in neurotransmitter histochemistry is that they permit the selective demonstration of neurons containing a known neurotransmitter. The available methods are sensitive and reliable, particularly for catecholamine neuron systems. Demonstration of indolamine neuron systems has been accomplished, but not with equivalent sensitivity. Used in combination with other experimental procedures, such as drugs or lesions, or in combination with other neuroanatomical methods such as autoradiography, the autoradiographic and retrograde tracing methods, and ultrastructural analysis, the fluorescence histochemical methods have added utility in the analysis of the organization of monoamine neuron systems.

B. Limitations

The application of the fluorescence histochemical methods has three major limitations. First, the methods at present are only applicable to monoamine neuron systems and, as these represent only a very small proportion of all neuron systems, one can only use them for a limited analysis of neuronal organization. Second, the methods provide a cumbersome problem of interpretation. Because unstained sections are examined with dark-field illumination, and only fluorescent material is distinctly visible, that is, catecholamine and indoleamine neuronal perikarya and processes, few landmarks are available for reference. Consequently, unless the investigator is thoroughly familiar with the area under study and sections are cut in a familiar plane, interpretation of the exact localization of fluorescent structures may be difficult. Third, it must be emphasized that the apparent absence of an innervation in fluorescence histochemical material does not necessarily mean that it does not exist. Such a judgment must be based upon knowledge of the method employed, whether the material is optimal for that method, and, in the best of circumstances, verification by another neuroanatomical method.

C. Conclusion

Neurotransmitter histochemistry applicable to the central nervous system was developed more than 15 years ago with the advent of the Falck–Hillarp histochemical method (Falck and Torp, 1961; Falck, 1962; Falck *et al.*, 1962). Since the development of the original Falck–Hillarp method, there have been a number of significant technical improvements in methodology, particularly the introduction of the glyoxylic acid method. These have led to significant advances in our understanding of the organization of neuron systems that are characterized by transmitter content, and there is now a large literature extant on the morphology of monoamine neuron systems in the nervous systems of invertebrates, lower vertebrates, and mammals (see Moore and Bloom, 1978, 1979, for review). This information has led, in turn, to a vast number of studies of the function of monoamine neuron systems. These have demonstrated that the monoamine neuron systems participate importantly in a variety of functions such as the regulation of sleep–waking behavior, feeding behavior, motor regulation, neuroendocrine regulation, central nervous system reward mechanisms, and some aspects of higher brain functions. As technical advances have occurred, the methods of neurotransmitter histochemistry are becoming accessible to larger groups of investigators, and the future should provide even further advances which will contribute to our general understanding of the organization and function of the central nervous system.

VII. APPENDIX

A. Fluorescence Microscopy and the Fluorescence Microscope

The fluorescence microscopy requisite for the histochemical methods to be described below can be carried out with an ordinary light microscope suitably equipped for fluorescence microscopy. There are now a great variety of such microscopes available, utilizing either transmitter light or incident (epi-) illumination, and the individual investigator will select the one that is either available to him or appears most suitable for his purposes. The fluorescence microscope to be described below is that in use in this laboratory and in the laboratory of Falck in Lund and has proven satisfactory for high-quality fluorescence microscopy over the years. Microscopes suitable for use are produced by virtually all major manufacturers, and standard laboratory microscopes can be modified by adding attachments so that they may serve both in routine microscopy and in fluorescence microscopy. It should be noted that transmitted light is usually more advantageous for work with low-power objectives, whereas incident light illumination is particularly advantageous for use with high-power objectives with a numerical aperture greater than 0.75. The differences are not sufficiently great, however, that one would necessarily recommend one form of illumination over the other, and there are currently considerable advances being made in incident light illumination technology to suggest that this will become a more widely used tool for fluorescence microscopy. The microscope in use in our laboratory is an ordinary Zeiss fluorescence microscope which is fixed to a stand that has an attached lamp housing and camera holder. The lamp housing holds a 200-watt mercury vapor lamp (Osram HBO 200) which requires a special power supply. The light emitted from the lamp passes through a heat absorption filter (Schott KG1) and through suitable primary filters that select activation wavelengths as close as possible to the activation maximum (absorption maximum) of the fluorophores and that are sufficiently thick to minimize unwanted light. The excitation maxima for the biogenic amines that can be visualized in the central nervous system by the fluorescence histochemical methods are in the range of 390–410 nm, and the excitation or primary filter commonly used is a Schott BG-12. On our microscope we have four holders for excitation filters that can be placed in or out of the light path. These contain BG-12 filters of 2-mm, 3-mm, and 4-mm (2) thickness, respectively, so that a combination of filters can be chosen to minimize background illumination and maximize specific fluorescence. The secondary, or barrier, filter should have a high absorption below 500 nm, and an example of such a filter is a Schott OG4. This excludes the blue component of the excitation light and, since the catecholamine fluorophores have an emission maximum of about 475 nm, the combination of BG-12 primary filters and an OG-4 secondary filter results in catecholamines appearing green to greenish-yellow in the fluorescence microscope. Serotonin, which has an emission maximum of approximately 525 nm, will appear yellow with this combination of filters. The fluorescence microscope should be set up so that the light emitted from the lamp passes through the heat absorption and primary filters in a direct path to a silvered mirror that directs it onto the specimen. The fluorescence emitted by the specimen then goes through the objective lens of the microscope directly to the eye pieces. It should be emphasized that the fluorescence intensity usually achieved with the fluorescence histochemical methods is low and, consequently, no beam splitter or prism should be in the light path, as it will reduce the light sufficiently to make even extremely good specimens very difficult to utilize. For this reason, and because the electronics of the instruments are, for the most part, inadequate to assess the low levels of illumination obtained as fluorescence in the fluorescence histochemical methods, we have not found any of the automatic microscopes we have tried to be applicable to photomicroscopy with these methods. And, since most of these microscopes utilize a beam splitter, they are difficult to use even for routine microscopy.

The essence of the fluorescence microscope then is a simple setup (Fig. 13) that contains a light source (either a 100- or a 200-watt mercury vapor lamp or, in special instances, a xenon lamp), a cooling filter, primary filters,

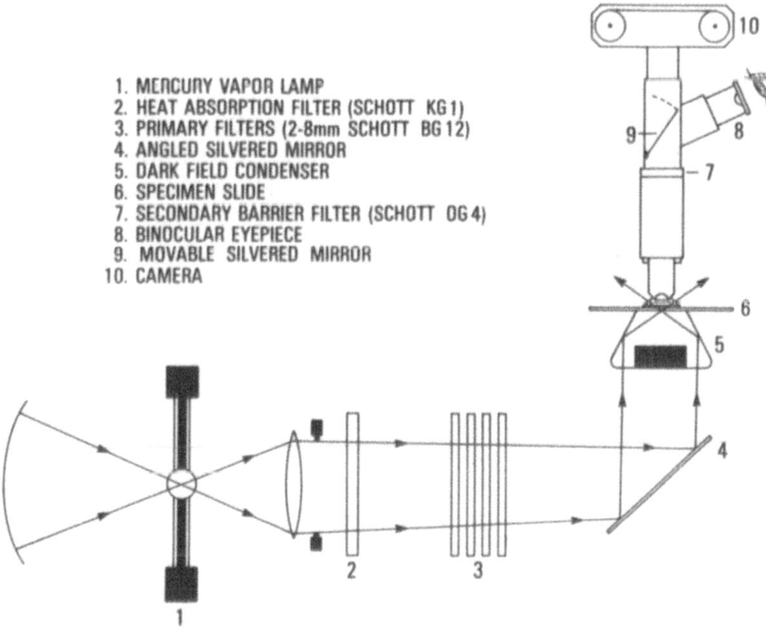

Figure 13. Diagram of a fluorescence microscope using transmitted light as described in Section VII.A.

a dark-field condenser, the specimen, objective lenses, a secondary or barrier filter, and eye pieces. It should be noted that we use only condensers that require oil; the dry dark-field condensors do not provide sufficient illumination for good visualization of fluorescence histochemical material or photography. The trinocular head of our microscope contains a removable mirror which directs all of the light to the eye pieces or, when removed, to the camera. There is an objective lens in the trinocular head between the objective lens and the camera back, which is a simple, 35-mm back attached to a connector between the microscope and the camera. The connector is held on a horizontal rod attached to a vertical rod that arises from the microscope stand. The connector has a viewing lens, and the light can be directed either into the viewing lens or into the camera, but, again, all of the light goes one direction or the other. To give the reader some idea of the levels of illumination that are obtained with this methodology, it is useful to make some comments on photomicrography. If we wish merely to record the appearance of material for a permanent record but not publication, we commonly employ Kodak Tri-X film with exposures of 15 to 30 sec. The exposure times are determined empirically. If we take photographs for publication, we use either Kodak Panatomic-X or Kodak SO-115. Exposure times for Panatomic-X are from 2–10 min, depending on the specimen and the objective lens. The lower the objective lens magnification, the shorter the exposure time required. For SO-115, which is less sensitive in the green ranges of the visual spectrum, exposures of 6–12 min are used for the usual material. The latter film has the advantage, however, for the patient microscopist of offering superb contrast. For color photography we use Kodak Hi-Speed Ektachrome with exposures in the range of 3–10 min. Again, it should be emphasized that exposure times for photomicrography are determined empirically for each microscopic setup.

One further practical point that should be made about fluorescence microscopy concerns the life-span of mercury vapor lamps. Our experience has been that these commonly are useful and give good fluorescence intensity for approximately 150–200 hr. They usually do not "burn out" in the conventional sense, but, rather, slowly give less and less intense light. This is difficult for the individual to perceive on many occasions, and it often becomes evident only when exposure times for photography that had been routinely satisfactory result in underexposed film. For that reason, we monitor the exposure times carefully and keep a running record of the length of time that a particular mercury vapor lamp has been in use and replace any lamp which has been in use more than 200 hr.

B. Freeze-Dryers

The principle of freeze-drying is to obtain the rapid removal of water from frozen tissues by sublimation at a temperature below the freezing point of the tissue. This is necessary to preserve localization of the monoamines which are water soluble and tend to diffuse from their normal cellular location if the tissue is not properly treated. Freeze-drying is carried out by keeping the tissue under vacuum to promote the vaporization of the water and trapping water vapors so formed so that all residual water is removed from the tissue. There are two principal types of freeze-dryer currently in use for the fluorescence histochemical techniques, one in which a trap for water vapor is provided by a cold finger and one in which the trap is a desiccant chemical (see Falck and Owman, 1965; Eränko, 1967; Björklund *et al.*, 1972b, for reviews). There are a number of commercial models available that are adequate for freeze-drying for the fluorescence histochemical methods, but, as Björklund *et al.* (1972b) emphasized, simple and inexpensive freeze-dryers can be constructed in nearly any laboratory. In this laboratory, we employ a commercial freeze-dryer designed by Olson and Ungerstedt (1970a; Figs. 14 and 15). This is produced by the Bergman and Beving Company, Stockholm, Sweden, and is a desiccant-type freeze-dryer that has the advantages of being compact, easily used, reliable, and has a high tissue capacity. The instrument is easily placed on a table or laboratory bench (Fig. 14). The principal components are as follows: vacuum gauge, vacuum pump, freezing compartment (this contains a compressor and a metal cabi-

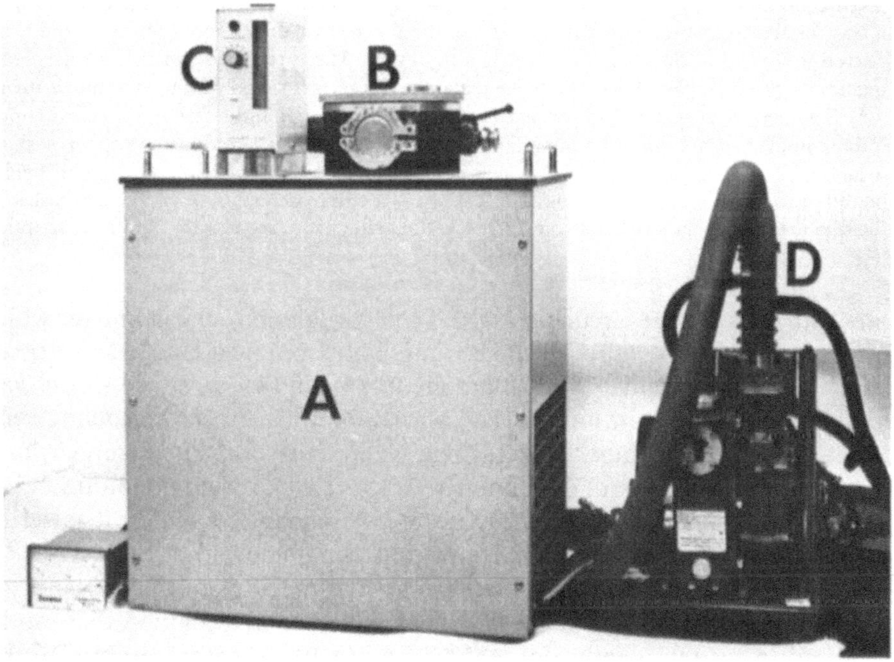

Figure 14. Photograph of the Olson–Ungerstedt (1970a) freeze-dryer. The large cabinet (A) is a refrigeration unit with a container filled with methyl alcohol in which the tissue chamber (B) is immersed. The temperature–thermometer unit (C) can be set to maintain a constant temperature. A two-stage vacuum pump (D) is also shown.

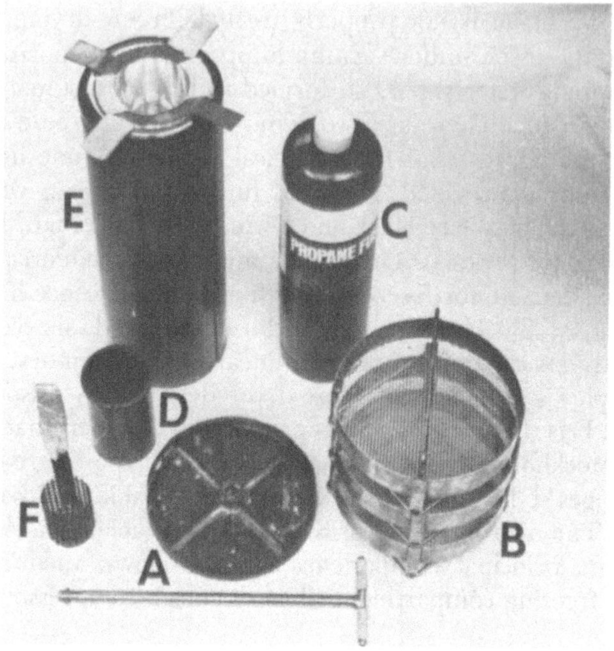

Figure 15. Photograph of the tissue plate (A), phosphorous pentoxide trays (B), and the equipment for freeze-drying. Tissue is collected in the tissue plate under liquid nitrogen and transferred to the tissue chamber in the refrigeration unit. The three trays containing phosphorus pentoxide are then placed over the tissue plate, the chamber is closed, and the vacuum is turned on. Commercial propane (C) is cooled to its liquid form running it through a coiled copper tubing immersed in liquid nitrogen. The end of the tubing is inserted in the copper flask (D) which itself is immersed in liquid nitrogen in a Dewar flask (E). The tissue plate (A) is held in liquid nitrogen in a large Dewar flask. Pieces of tissue are frozen in the small copper basket (F) held in the copper flask (D) and transferred to the tissue plate. See text for further description.

net with coils for the circulation of the coolant solution), thermometer which regulates the temperature in the cooling bath from +20°C to −35°C, tissue and dessicant container. The components of the tissue and dessicant container are as follows: cylindrical metal chamber that sits in the cooling solution, base of the chamber which is flat, clamp over port into chamber, connection to vacuum gauge, connector with valve to vacuum pump, valve which can be opened to let in air, tissue-holding plate, handle for lowering tissue-holding plate into chamber, baskets for holding desiccant. Thus, in this freeze-dryer the tissue is placed on a flat metal plate that rests in the bottom of a cylindrical metal well that is immersed in a solution that can be lowered to the temperature of −35°C; we use 100% methyl alcohol. In this situation the temperature of the plate will adjust rapidly to the temperature of the solution. After the tissue is placed in the plate and into the metal well, the metal mesh baskets, which contain granular phosphorus pentoxide (Granusec®, Baker) placed on filter paper, are placed over the tissue plate.

The top of the cylinder is covered by a glass plate which rests on an O-ring. In our laboratory we use a two-stage vacuum pump, a mechanical and oil diffusion pump. This generates a vacuum in excess of 10^{-3} torr in the freeze-drying chamber. The alcohol bath is continuously circulated so that the temperature remains constant.

The other type of freeze-dryer, the cold-finger type, employs the principle that vaporized water from the tissue is trapped by a cold finger that rests a short distance above the tissue. The details of cold-finger type freeze-dryers have been thoroughly reviewed by Björklund *et al.*, :1972).

C. Vibratome

The Vibratome is a special type of microtome (manufactured by the Oxford Instrument Company, San Mateo, California) useful in fluorescence histochemical techniques because the brain tissue must be prepared with minimal fixation; the fixation method will result in diffusion of monoamines and, hence, failure to accomplish their cellular localization. Fresh or fixed tissue is placed on a chuck that is fixed to the Vibratome as in any microtome with the exception that it is in a trough containing a cold solution. The Vibratome operates by moving a sharp blade in short, rapid horizontal strokes while also moving it forward. The instrument is designed to allow small blocks of relatively soft tissue to be cut into sections sufficiently thin and of sufficiently good histological quality to be useful. The instrument is made so that the rate of horizontal movement, the rate of foreward movement, and the thickness of section can be controlled independently. These parameters will vary for different tissues, and the use of the Vibratome is empirical with respect to them. Successful use of the Vibratome takes considerable practice, significantly more than for routine microtomy, and cutting good sections from fresh tissue or that prepared with glyoxylic acid perfusion is often difficult even for the expert. The details of the use of the Vibratome are discussed in Sections IV.B.3 and IV.C.1.

D. Cryostat

Methods for the application of the fluorescence histochemical method to cryostat material are described in Section IV.C.2 These methods can be applied to sections obtained from any good, commercially available cryostat. However, a freezing microtome is not an adequate substitute, and the methods will only work on cryostat sections. The basis for this undoubtedly is that the cooling chamber of the cryostat keeps the sections cold after they are cut prior to mounting. This, then, retains the easily diffusible amines in their original, *in vivo* location.

ACKNOWLEDGMENTS. The preparation of this review was supported in part by USPHS Grant NS 16303 from the National Institutes of Health.

REFERENCES

Aghajanian, G. K., Kuhar, M. J. and Roth, R. H., 1973, Serotonin-containing neuronal perikarya and terminals: Differential effects of *p*-chlorophenylalanine, *Brain Res.* **54**:85–101.

Ajelis, V., Björklund, A., Falck, B., Lindvall, O., Lorén, I., and Walles, B., 1979, Application of the aluminum–formaldehyde histoflourescence method for demonstration of periferal stores of catecholamines and indoleamines in freeze-dried, paraffin-embedded tissue, cryostat sections and whole-mounts, *Histochemistry* **65**:1–15.

Amin, A. H., Crawford, T. B. B., and Gaddum, J. N., 1954, The distribution of substance P and 5-hydroxytryptamine in the central nervous system of the dog, *J. Physiol. (Lond.)* **126**:596–618.

Andén, N.-E., Dahlström, A., Fuxe, K., Larsson, K., Olson, L., and Ungerstedt, U., 1966, Ascending monoamine neurons to the telencephalon and diencephalon, *Acta Physiol. Scand.* **67**:313–326.

Battenberg, E. L. F., and Bloom, F. E., 1975, A rapid, simple and more sensitive method for the demonstration of central catecholamine-containing neurons and axons by glyoxylic acid induced fluorescence. I. Specificity, *Psychopharmacol. Commun.* **1**:3–13.

Beaudet, A., and Descarries, L., 1979, Radioautographic characterization of a serotonin-accumulating nerve cell group in adult rat hypothalamus, *Brain Res.* **160**:231–243.

Berger, B., Thierry, A. M., Tassin, J. P., and Moyne, M. A., 1976, Dopaminergic innervation of the rat prefrontal cortex: A fluorescence histochemical study, *Brain Res.* **106**:133–145.

Bertler, A., and Rosengren, E., 1959, Occurrence and distribution of dopamine in brain and other tissues, *Experientia* **15**:10–11.

Björklund, A., Lindvall, O., and Svensson, L.-Å., 1972a, Mechanisms of fluorophore formation in the histochemical glyoxylic acid method for monoamines, *Histochemie* **32**:113–131.

Björklund, A., Falck, B., and Owman, Ch., 1972b, Fluorescence microscopic and microspectrofluorometric techniques for the cellular localization and characterization of biogenic amines, in: *Methods of Investigative and Diagnostic Endocrinology, Vol. I., The Thyroid and Biogenic Amines* (J. E. Rall and I. J. Kopin, eds.), pp. 318–378, North-Holland, Amsterdam.

Björklund, A., Falck, B., and Lindvall, O., 1975, Microspectrofluorometric analysis of cellular monoamines after formaldehyde or glyoxylic acid condensation, in: *Methods in Brain Research* (P. B. Bradley, ed.), pp. 249–294, Wiley, London.

Björklund, A., Falck, B., Lindvall, O., and Lorén, I., 1980, The aluminum–formaldehyde (ALFA) histoflourescence method for improved visualization of indoleamines and catecholeamines. 2. Model experiments. *J. Neurosci. Methods* **2**:301–318.

Bloom, F. E., and Battenberg, E. L. F., 1976, A rapid, simple and sensitive method for the demonstration of central catecholamine-containing neurons and axons by glyoxylic acid-induced fluorescence. II. Detailed description of methodology, *J. Histochem. Cytochem.* **24**:561–571.

Carlsson, A., Falck, B., and Hillarp, N.-Å., 1962, Cellular localization of brain monoamines, *Acta Physiol. Scand.* [*Suppl.*] **196**:1–27.

Corrodi, H., and Jonsson, G., 1967, The formaldehyde fluorescence method for the histochemical demonstration of biogenic monoamines, *J. Histochem. Cytochem.* **15**:65–78.

Dahlström, A., and Fuxe, K., 1965a, Evidence for the existence of monoamine-containing neurons in the central nervous system. I. Demonstration of monoamines in the cell bodies of brain stem neurons, *Acta Physiol. Scand.* [*Suppl.*] **232**:1–55.

Dahlström, A., and Fuxe, K., 1965b, Evidence for the existence of monoamine neurons in the central nervous system. II. Experimentally induced changes in the intraneuronal amine levels of bulbospinal neuron systems, *Acta Physiol. Scand.* [*Suppl.*] **247**:1–36.

Dale, H. H., 1935, Pharmacology and nerve endings, *Proc. R. Soc. Med.* **28**:319–332.

de la Torre, J. C., 1980, An improved approach to histofluorescence using the SPC method for tissue monoamines, *J. Neurosci. Methods* **3**:1–5.

de la Torre, J. C., and Surgeon, J. W., 1976, A methodological approach to rapid sensitive monoamine histofluorescence using a modified glyoxylic acid technique: The SPG method, *Histochemistry* **49:**81–93.

Ehringer, H., and Hornykewicz, O., 1960, Verteilung von Noradrenalin und Dopamin (3-Hydroxytyramin) in behirn des Menschen und ihr Verhalten bei Erkrankungen des Extrapyramidalen Systems, *Klin. Wochenschr.* **38:**1236–1239.

Elde, R., and Hökfelt, T., 1978, Distribution of hypothalamic hormones and other peptides in the brain, *Front. Neuroendocrinol.* **5:**1–34.

Eränko, O., 1955, Histochemistry of noradrenaline in the adrenal medulla of rats and mice, *Endocrinology* **57:**363–368.

Eränko, O., 1967, The practical histochemical demonstration of catecholamines by formaldehyde-induced fluorescence, *J. R. Microsc. Soc.* **87:**257–276.

Euler, U. S. von, 1946, Specific sympathomimetric ergone in adenergic nerve fibers (Sympathin) and its relations to adrenaline and noradrenaline, *Acta Physiol. Scand.* **12:**73–97.

Falck, B., 1962, Observations on the possibilities of the cellular localization of monoamines by a fluorescence method, *Acta Physiol. Scand.* [*Suppl.*] **197:**1–25.

Falck, B., and Owman, Ch., 1965, A detailed methodological description of the fluorescence method for the cellular demonstration of biogenic monoamines, *Acta Univ. Lund.* [*Sec. II*] **7:**1–23.

Falck, B., and Torp, A., 1961, A new evidence for the localization of noradrenaline in adrenergic nerve terminals, *Med. Exp.* **5:**169–197.

Falck, B., Hillarp, N.-Å., Thieme, G., and Torp, A., 1962, Fluorescence of catecholamines and related compounds condensed with formaldehyde, *J. Histochem. Cytochem.* **10:**348–354.

Furness, J. B., Health, J. W., and Costa, M., 1978, Aqueous aldehyde (Faglu) methods for the fluorescence histochemical localization of catecholamines and for ultrastructural studies of central nervous tissue, *Histochemistry* **57:**285–295.

Fuxe, K., 1965, Evidence for the existence of monoamine neurons in the central nervous system. IV. Distribution of monoamine nerve terminals in the central nervous system, *Acta Physiol. Scand.* [*Suppl.*] **247:**39–85.

Fuxe, K., and Jonsson, G., 1967, A modification of the histochemical fluorescence method for the improved localization of 5-hydroxytryptamine, *Histochemie* **11:**161–166.

Fuxe, K., Hökfelt, T., and Ungerstedt, U., 1968, Localization of indolealkylamines in CNS, *Adv. Pharmacol.* **6A:**235–251.

Fuxe, K., Hökfelt, T., Jonsson, G., and Ungerstedt, U., 1970, Fluorescence microscopy in neuroanatomy, in : *Contemporary Research Methods in Neuroanatomy* (W. J. H. Nauta and S. O. E. Ebbesson, eds.), pp. 275–314, Springer-Verlag, New York.

Geffen, L. B., and Livett, B. F., 1971, Synaptic vesicles in sympathetic neurons, *Physiol. Rev.* **51:**98–157.

Hamburger, B. T., Malmfors, T., and Sachs, C., 1965, Standardization of paraformaldehyde and of certain procedures for the histochemical demonstration of catecholamines, *J. Histochem. Cytochem.* **13:**147 150.

Hess, S. M., and Udenfriend, S., 1959, A fluorimetric procedure for the measurement of tryptamine in tissues, *J. Pharmacol. Exp. Ther.* **127:**175 179.

Hökfelt, T. G. M., and Ljungdahl, A. S., 1972a, Histochemical determination of neurotransmitter distribution, *Res. Pub. Assoc. Res. Nerv. Ment. Dis.* **50:**1–24.

Hökfelt, T., and Ljungdahl, A., 1972b, Modification of the Falck–Hillarp formaldehyde fluorescence method using the Vibratome: Simple, rapid, sensitive localization of catecholamines in sections of unfixed or formalin fixed brain tissue, *Histochemie* **29:**324–393.

Hökfelt, T., Fuxe, K., Goldstein, M., and Joh, T. Y., 1973, Immunohistochemical localization of those catecholamine synthesizing enzymes: Aspects on methodology, *Histochemie* **33:**231–254.

Kuhar, M. J., Aghajanian, G. K., and Roth, R. M., 1972, Tryptophan hydroxylase activity and synaptosomal uptake of serotonin in discrete brain regions after midbrain raphe lesions: Correlations with serotonin levels and histochemical fluorescence, *Brain Res.* **44:**165–176.

Levitt, P., and Moore, R. Y., 1979, Origin and organization of brainstem catecholamine innervation in the rat, *J. Comp. Neurol.* **186:**505–528.

Lindvall, O., and Björklund, A., 1974a, The glyoxylic acid fluorescence histochemical method: A detailed account of the methodology for the visualization of central catecholamine neurons, *Histochemistry* **39:**97–127.

Lindvall, O., and Björklund, A., 1974b, The organization of the ascending catecholamine neuron systems in the rat brain, *Acta Physiol. Scand.* [*Suppl.*] **412:**1–48.

Lindvall, O., Björklund, A., 1978, Organization of catecholamine neurons in the rat central nervous system, in: *Handbook of Psychopharmacology, Vol. 9* (S. D. Iversen and S. H. Snyder, eds.), pp. 139–231, Plenum Press, New York.

Lindvall, O., Björklund, A., Hökfelt, T., and Ljungdahl, Å., 1973, Application of the glyoxylic acid method to Vibratome sections for improved visualization of central catecholamine neurons, *Histochemie,* **35:**31–38.

Lindvall, O., Björklund, A., and Svensson, L.-Å., 1974, Fluorophore formation from catecholamines and related components in the glyoxylic acid fluorescence histochemical method, *Histochemistry* **39:**197–227.

Lorén, I., Björklund, A., Falck, B., and Lindvall, O., 1976a, An improved histofluorescence procedure for freeze-dried paraffin-embedded tissue based on combined formaldehyde–glyoxylic acid perfusion with high magnesium content and acid pH, *Histochemistry* **49:**177–192.

Lorén, I., Björklund, A., and Lindvall, O., 1976b, The catecholamine systems in the developing rat brain; improved visualization by a modified glyoxylic acid–formaldehyde method, *Brain Res.* **117:**313–318.

Lorén, I., Björklund, A., and Lindvall, O., 1977, Magnesium ions in catecholamine fluorescence histochemistry, *Histochemistry* **52:**223–239.

Lorén, I., Björklund, A., Falck, B., and Lindvall, O., 1980, The aluminum–formaldehyde flourescence (ALFA) method for improved visualization of catecholeamines and indoleamines. 1. A detailed account of the methodology for central nervous system tissue using paraffin, cryostat or vibratome sections, *J. Neurosci. Methods* **2:**277–300.

Moore, R. Y., 1978, Catecholamine innervation of the basal forebrain. I. The septal area, *J. Comp. Neurol.* **177:**665–684.

Moore, R. Y., and Bloom, F. E., 1978, Central catecholamine neuron systems: Anatomy and physiology of the dopamine systems, *Annu. Rev. Neurosci.* **1:**129–169.

Moore, R. Y., and Bloom, F. E., 1979, Central catecholamine neuron systems: Anatomy and physiology of the norepinephrine and epinephrine systems, *Annu. Rev. Neurosci.* **2:**113–168.

Olson, L., and Ungerstedt, U., 1970a, A simple high capacity freeze-drier for histochemical use, *Histochemie,* **22:**8–19.

Olson, L., and Ungerstedt, U., 1970b, Monoamine fluorescence in CNS smears: Sensitive and rapid visualization of nerve terminals without freeze-drying, *Brain Res.* **17:**343–347.

Twarog, B. M., and Page, I. H., 1953, Serotonin content of some mammalian tissues and urine, *Am. J. Physiol.* **175:**157–161.

Ungerstedt, U., 1971, Stereotaxic mapping of the monoamine pathways in the rat brain, *Acta Physiol. Scand.* [*Suppl.*] **367:**1–48.

Vogt, M., 1954, The concentration of sympathin in different parts of the central nervous system under normal conditions and after the administration of drugs, *J. Physiol.* (*Lond.*) **123:**451–481.

Watson, S. J., and Barchas, J. D., 1975, Histofluorescence in the unperfused CNS by cryostat and glyoxylic acid: A preliminary report, *Psychopharmacol. Commun.* **1:**523–531.

Watson, S. J., and Barchas, J. D., 1977, Catecholamine histofluorescence using cryostat sectioning and glyoxylic acid in unperfused frozen brain: A detailed description of technique, *Histochem. J.* **9:**183–195.

Werman, R., 1966, Criteria for identification of a central nervous system transmitter, *Comp. Biochem. Physiol.* **18:**745–766.

Whaley, W. M., and Govindachari, T. R., 1951, The Pictet–Spengler synthesis of tetrahydroisoquinolones and related compounds, in: *Organic Reactions, Vol. 6* (R. Adams, ed.), pp. 151–206, Wiley, New York.

Immunocytochemical Methods

14

VIRGINIA M. PICKEL

I. INTRODUCTION

A few essential features of immunology form the basis of immunocytochemical methods used for the demonstration of chemically distinct neuronal pathways in the central nervous system. The immonologic concepts of antigens and antibodies and their *in vitro* reaction in tissue sections are discussed briefly as is the rationale for choosing the immunocytochemical method for mapping neuronal pathways.

A. Immunologic Basis

Many macromolecules, particularly proteins and polysaccharides, that occur naturally in neural tissues can be extracted and used as antigens or substances that, on introduction into a suitable host, give rise to the formation of antibodies. In addition to the naturally occuring antigens, other small polypeptides, either isolated from neural tissues or synthesized in the labo-

VIRGINIA M. PICKEL • Laboratory of Neurobiology, Department of Neurology, Cornell University Medical College, New York, New York 10021.

ratory, can be made antigenic by coupling them to a carrier compound that might be a protein or a polysaccharide (Weir, 1973). These antigenic substances, whether natural or produced in the laboratory, are characterized by their capacity to induce antibody formation (immunogenicity) and by their specificity for combining with the formed antibody. The antibody, an immunoglobulin (IgG) composed of polypeptide chains, usually has one or more combining sites for the specific antigen (Sternberger, 1974; Weir, 1973). The antigen used in the production of the primary antibody can be demonstrated subsequently in homogenates or sections of tissue by the observable combination with the antibody. The antigen–antibody reaction may be shown in homogenates of tissue by the formation of a precipitate or an agglutinate. In sections of tissue, the reaction can be visualized by labeling the antibody with a marker compound (Weir, 1973). The *in vitro* immunologic reaction using marker compounds constitutes the basis for the localization of antigens in neuronal pathways by immunocytochemical methods.

The value of the immunocytochemical technique is directly related to the specificity of the antibody for the antigen that induced its formation.

The specificity of the antiserum produced against large enzymes can be assessed by immunoelectrophoresis and by enzyme inhibition (Joh *et al.*, 1963). Immunoelectrophoresis or double immunodiffusion of the antibodies, run against partially or completely purified enzyme, yields a single precipitant arc. No precipitant arcs are formed when the antibodies are run against enzymes other than those producing the antibody. The specific antibody also inhibits the activity of only its parent enzyme. In contrast to larger enzymes, the specificity of the antibodies for the the small peptides is largely based on cross-reactivity between the formed antibody and its parent peptide (Pickel *et al.*, 1979).

B. Rationale

The choice of the immunocytochemical technique for demonstrating neuronal pathways is dependent on the availability of a purified antigen and its specific antibody as well as on the necessity for demonstrating the chemical nature of a pathway as opposed to anatomical connections. The purification procedures for most neuron-specific antigens, such as the neurotransmitter-synthesizing enzymes, are difficult and laborious (Gibbs *et al.*, 1967; Joh *et al.*, 1963). The antibodies to only a small number of these compounds are available and these in a relatively few laboratories.

II. TYPES OF IMMUNOCYTOCHEMICAL TECHNIQUES

A number of recent reviews have dealt with the use of immunocytochemical techniques for the localization of cellular antigens (Nairn, 1969; Sternberger, 1974; Tanaka, 1968; Weir, 1973). In the present chapter, a brief survey of the general types of immunocytochemical techniques serves as an introduction to the more detailed description of the use of the peroxidase–antiperoxidase (PAP) method (Sternberger, 1974) for the localization of neurotransmitter-synthesizing enzymes and peptides in neurons of the central nervous system.

The two major types of immunocytochemical techniques can be distinguished on the basis of whether the marker compound used for identification of the site of the antigen–antibody reaction is attached *directly* to the antibody or *indirectly* to an anti-intermediary immunoglobulin (IgG).

A. Direct Method

One of the first methods used for the immunocytochemical localization of antigens was the direct immunofluorescence technique developed by Coons and Kaplan (1950). The direct antibody-labeling technique is based on the finding that antibodies can be conjugated with simple compounds such as fluorescein isothiocyanate (FITC) without destroying their capacity to bind with the specific antigen in the tissues. Following the development of the direct immunofluorescence technique for light microscopy, other direct labeling methods were developed for electron microscopy. These methods employ either compounds that can be made electron-opaque, such as uranium, or compounds with structured electron opacity, such as ferritin (Sternberger, 1974). As shown in Fig. 1A, the immunocytochemical labeling of antigenic sites in sections of tissues by the direct immunocytochemical method involves only one step, the application of an antiserum (A) previously bound to an appropriate marker (F) (e.g., FITC or ferritin).

The direct labeling method greatly reduces the reactivity of the antibodies and thus is unable to demonstrate cellular constituents that are present in low concentrations in tissue (Sternberger, 1974). The direct antibody-labeling method is also limited by the fact that a different labeled antibody preparation is needed for each antigen to be localized. The procedure thus generally requires a large amount of antiserum and is not convenient for antigens that are difficult to purify or are present in small quantities.

B. Indirect Methods

More sensitive indirect immunocytochemical labeling methods employ an anti-IgG intermediary to couple the antibodies to the marker compound (Avrameas, 1969). The anti-IgG is usually either goat or sheep antiserum prepared against the IgG of the species (e.g., rabbit) used for the production of the primary antiserum. The marker compounds include many of the same substances used for direct immunocytochemical labeling, such as FITC, ferritin, or specific enzymes like horseradish peroxidase (Avrameas, 1969). The indirect immunocytochemical methods may be classified into two

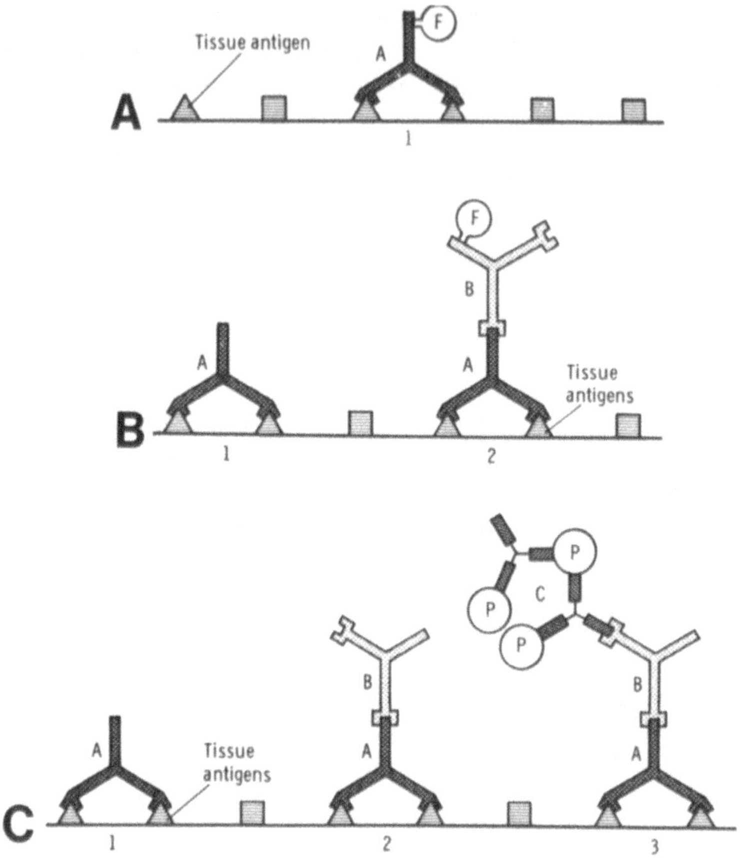

Figure 1. Immunocytochemical technique. (A) *Direct.* The specific antigens in the tissue sections are localized in a one-step procedure consisting of the application of a primary antibody (A) prepared against the specific tissue antigen. Antiserum (A) is produced in a suitable species such as rabbits and subsequently bound covalently to a marker compound, FITC or ferritin (F). (B) *Indirect immunocytochemical technique using covalent bonding.* The immunocytochemical labeling procedure involves two steps. (1) Apply antiserum (A) prepared against the specific tissue antigen. Antiserum (A) is produced in one species such as rabbit. (2) Apply anti-IgG (B) produced in a different species such as sheep or goat against the rabbit immunoglobulin. The anti-IgG has previously been covalently bound to a marker compound such as FITC or ferritin (F). (C) *Indirect immunocytochemical technique: immunologic bonding using the PAP technique.* The indirect immunocytochemical procedure employing PAP consists of three steps. (1) Apply antiserum (A) prepared against the specific tissue antigen. Antiserum (A) is produced in a species such as rabbit. (2) Apply anti-IgG (B) produced in a second species (i.e., sheep or goat) against rabbit IgG. (3) Apply the PAP (C) solution. The antiperoxidase in the PAP was produced in the same species (i.e., rabbit) as antiserum (A).

groups based on whether the bonding between the anti-IgG and the marker compound is covalent or immunologic (Sternberger, 1974; see Fig. 1B,C).

The *covalent* bonding of anti-IgG to FITC depends on the introduction of reactive groups into the molecule. For ferritin or horseradish peroxidase, this bonding is achieved by the use of bifunctional reagents, such as toluene

2,4-diisocyanate, 4,4′-difluoro-3,3′-dinitrophenylsulfone (Modesto and Pesce, 1971), or glutaraldehyde (Avrameas, 1969). As shown in Fig. 1B, immunocytochemical labeling by the indirect covalent bonding method involves two steps: the application of an unlabeled antibody (A) produced against a specific tissue antigen and the application of an anti-IgG (B) which is covalently bonded to the marker compound (F). The most frequently used direct methods employing covalent bonding of a marker compound to anti-IgG are immunofluorescence and immunoenzyme techniques.

For light microscopy, the indirect immunofluorescent technique has wide applicability. The technique has, for example, been used extensively for the immunocytochemical localization of the catecholamine-synthesizing enzymes (Goldstein *et al.*, 1971; Hartman, 1973; Swanson and Hartman, 1975). The enzyme used in each step in the biosynthesis of catecholamines is shown in Fig. 2. The localization of tyrosine hydroxylase would identify both noradrenergic and dopaminergic neurons, whereas the localization of dopamine β-hydroxylase is specific for noradrenergic neurons. The immunofluorescent localization of dopamine-β-hydroxylase by Swanson and Hartman is shown in Fig. 3. In this photomicrograph, the fluorescent product clearly demonstrates the localization of the enzyme in neuronal perikarya and processes of noradrenergic neurons in the locus coeruleus and in axon terminals in the anteroventral nucleus of the thalamus. A more extensive mapping of the distribution of the noradrenergic neurons in the brain has been prepared by the same investigators (Swanson and Hartman, 1975).

The immunoenzyme techniques, which are applicable for both light and electron microscopy, were introduced by Nakane and Pierce (1966) and have since been used for the localization of pituitary hormones and other antigens (Avrameas, 1969). The most commonly used marker enzyme, horseradish peroxidase, is localized in tissue sections by the formation of a brown precipitate when reacted with the substrate, hydrogen peroxide, in the presence of

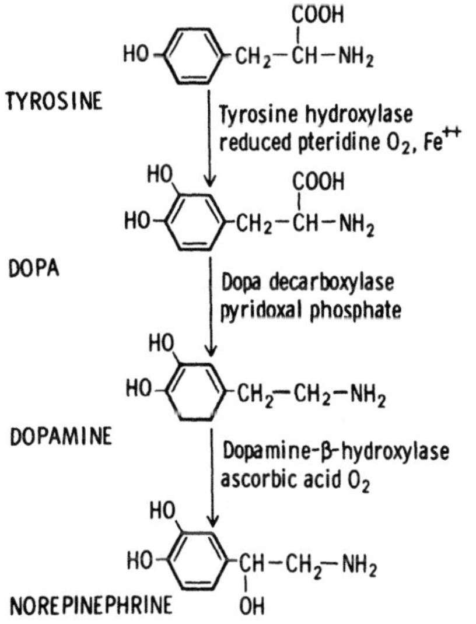

Figure 2. The enzymatic steps in the synthesis of norepinephrine from tyrosine. (From Weiner, 1970.)

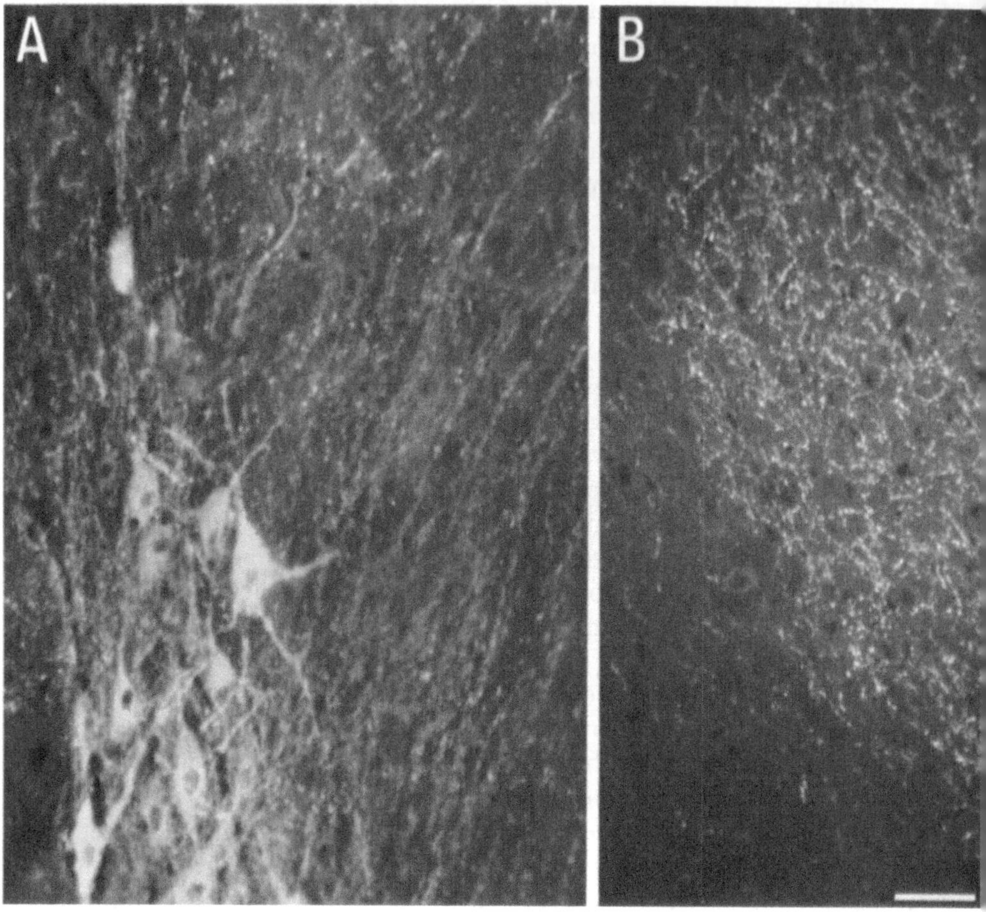

Figure 3. Indirect immunofluorescence localization of dopamine-β-hydroxylase. (A) A horizontal section through the rostral part of the locus coeruleus stained with antiserum to dopamine-β-hydroxylase to show the morphology of its cell bodies and the efferent fibers (upper right) that leave the nucleus to innervate wide parts of the central nervous system. Rostral direction is at the top of the figure. (B) The anteroventral nucleus of the thalamus as viewed in a frontal section stained with antiserum to dopamine-β-hydroxylase. Note that the sharp border between this nucleus (upper right) and the ventral nuclear complex (lower left) is clearly demonstrated by the presence of adrenergic innervation density in the anteroventral nucleus. Both photomicrographs are from 10-μm frozen sections of the rat. Bar, 100 μm. (Courtesy of L. W. Swanson and B. K. Hartman.)

the capturing agent, 3,3′-diaminobenzidine (Graham and Karnovsky, 1966). The brown reaction product can be made electron-dense by the application of osmium tetroxide. The major limitations of this method are the reduced specificity of the anti-IgG produced by the conjugation procedure and the relatively low yield of peroxidase-conjugated antibodies produced by the existing methods (Sternberger, 1974).

In the *indirect immunologic labeling techniques*, many of the problems encountered with the convalently labeled IgG were overcome by the immunologic bonding of the marker compound to anti-IgG. The peroxidase–antiperoxidase (PAP) technique developed by Sternberger (1974) relies exclusively on immunologic bonding.

Immunocytochemical labeling by the PAP method involves essentially three steps (Fig. 1C): the application of a specific antibody (A) against the antigen in the tissues, the application of anti-IgG (B), and the application of the PAP molecule (C). The PAP molecules are pentagonal structures having a ratio of three molecules of peroxidase to two molecules of antiperoxidase (Sternberger, 1974). The antibodies to the tissue antigen and to horseradish peroxidase are made in the same species (i.e., rabbit), whereas the anti-IgG is prepared in sheep or goats against rabbit IgG.

In addition to being highly sensitive (Burns, 1975; Moriarty *et al.*, 1973), the PAP method is especially useful for electron microscopy, since it can be used either before or after the tissue is embedded in plastic. The initial PAP technique, as described by Sternberger (1974), was used exclusively as a postembedding procedure. However, the preembedding procedure is used for the localization of neurotransmitter-synthesizing enzymes (Pickel *et al.*, 1975a,b, 1979). The reasons for using the preembedding procedure are twofold: embedding in plastic reduces the antigen–antibody reaction for catecholamine-synthesizing enzymes, and anatomical landmarks necessary for tracing neuronal pathways are more easily visualized in the large sections used for the preembedding labeling technique.

C. Summary

1. Direct Method

Presently, the direct method is used to detect only those antigens that are present in large concentrations in tissues.

2. Indirect Methods

The covalent variation includes three different techniques: immunofluorescence (used for light microscopy of many peptides and enzymes), immunoferritin (used for electron microscopy), and immunoenzyme (used for light and electron microscopy of certain specific antigens. The immunologic variation includes the PAP technique, which is popularly used for light and electron microscopy.

III. THE PEROXIDASE–ANTIPEROXIDASE (PAP) TECHNIQUE

The PAP technique described here is suitable for the light and electron microscopic localization of the catecholamine-synthesizing enzymes (Pickel *et al.*, 1975a, 1979) as exemplified by tyrosine hydroxylase and at least two peptides, substance P (Chang and Leeman, 1971) and [Met^5]-enkephalin (Hughes *et al.*, 1975; Pickel *et al.*, 1979). The steps involved in the use of the PAP technique for the cellular localization of these antigens in rat brain are described.

A. Fixation

The brains are fixed by vascular perfusion for 6 min with either 4% paraformaldehyde in phosphate buffer (0.1 M, pH 7.3) for light microscopy, or with 4% paraformaldehyde and 0.2% glutaraldehyde in phosphate buffer for electron microscopy. The brains are then removed, sliced with a razor blade into 1-mm coronal or sagittal sections, and placed in a fresh solution containing the same fixatives for an additional 20 min.

B. Sectioning

For both light and electron microscopy, the fixed brain slices are usually sectioned at 15–20 μm on a vibrating microtome (Hökfelt and Ljungdahl, 1972). This instrument has been described by Moore in Chapter 13. However, paraffin-embedded or frozen sections can also be used for light microscopy. (See Section IV.B for choice of sectioning method.)

C. Immunolabeling

The sections are subsequently processing through the following series of steps.

1. Tris–Saline Wash

Paraffin or frozen sections are mounted on glass slides with gelatin. The paraffin sections are deparaffinized with xylene and hydrated through graded ethanol to water prior to being placed in a Tris–saline solution. Frozen sections are placed directly in the Tris–saline. Vibratome sections are not mounted at this time. Instead, each section is transferred individually with a wooden stick to Porcelain Spot Test Plates* containing Tris–saline. To prepare Tris–saline, add 9.0 g NaCl to 1 liter of Tris–water. Tris–water is produced by the addition of concentrated HC1 to 0.5 M solution of Trizma® base† until a final pH of 7.6 has been obtained.

*A. H. Thomas Co., Philadelphia, Pennsylvania.
†Sigma Chemical Co., St. Louis, Missouri.

2. Triton X-100

The sections are transferred for 15 min to a 0.2% solution of a detergent, Triton X-100,* in Tris-saline.

3. Goat Serum

Following two 8-min washes in Tris–saline, the sections are transferred for 30 min to a diluted (1:30) normal goat serum.*

4. Antiserum

Following two 8-min washes in Tris–saline, the sections are transferred for 24 hr at 4°C to appropriately diluted solutions of antiserum or control serum (described in Section V). Under these conditions, we find the optimal dilution of our preparation of tyrosine hydroxylase antiserum to be 1:500. For substance P and [Met^5]-enkephalin antiserum, the dilution is 1:400. The dilutions of the antisera are prepared with Tris–saline containing 1% normal goat serum.

5. Anti-IgG

After two 8-min washes in Tris–saline containing 1% goat serum, the sections are transferred for 30 min to a 1:10 dilution of goat-anti-rabbit IgG.†

6. PAP

The sections are washed twice (8 min each) in Tris–saline containing 1% goat serum and then transferred for 30 min to a solution of 1:30 diluted PAP. The antiperoxidase is produced in rabbits that are injected with Sigma type VI horseradish peroxidase. PAP complex is prepared by combining the specific antiserum with horseradish peroxidase according to the method of Sternberger (1974).

7. Diaminobenzidine

After another two 8-min washes in Tris–saline, the sections are transferred for 6 min to a solution of 3,3′-diaminobenzidine and hydrogen peroxide composed of:

*Grand Island Biological Co., Grand Island, New York.
†Miles Laboratories, Kankakee, Illinois.

Tris–water (0.5 M, pH 7.6)	350 ml
3,3′-Diaminobenzidine	44 mg
30% Hydrogen peroxide	0.04 ml

8. Distilled Water

The sections are transferred to three 5-min washes in distilled water.

D. Light Microscopy

For light microscopy, the Vibratome sections are mounted on glass slides and dried with a hair dryer after completion of the immunocytochemical labeling procedure. All sections (Vibratome, paraffin, and frozen) are then dehydrated and mounted with a coverslip using a mounting resin. The slides are examined with the light microscope using either bright-field, phase-contrast, or Nomarski optics. The Nomarski optics usually give the best visualization of both the immunocytochemically labeled neurons and the unlabeled cells.

E. Electron Microscopy

For electron microscopy, the areas of labeling in the Vibratome sections are visualized by the use of a dissecting microscope and selectively removed with a razor blade. The small pieces of tissue are subsqeuently fixed for 2 hr in 2.0% osmium tetroxide in 0.1 M phophate buffer. The osmium-fixed tissues are then washed in phosphate buffer, dehydrated through graded ethanols, and embedded in Epon 812. The embedding technique is a flat-embedding procedure using a plastic slide and coverslip (Hendrickson, 1975). Ultrathin sections cut from the surface of each block show a progression through three layers: an outer layer which contains considerable nonspecific labeling and poor cellular morphology, a middle layer which contains specific labeling, and an inner layer which contains no labeling because of insufficient penetration of antisera.

IV. VARIATIONS IN PAP TECHNIQUE

Variation either in the methods of tissue preparation, such as fixation and sectioning, or in the reagents used in the PAP-labeling procedure may be necessary in order to achieve selective localization of antigens other than those described in the previous section.

A. Fixation

Preservation of cellular morphology for both light and electron microscopy usually requires the use of chemical fixatives such as formaldehyde and/or glutaraldehyde (Ellar *et al.*, 1970). These fixatives may either diminish or abolish the antigen–antibody reactions that constitutes the primary step in the immunocytochemical labeling procedure. The extent of the reduction in antigen–antibody binding that is induced by the chemical fixation depends on the properties of the antigen. For the localization of tyrosine hydroxylase as well as for many other enzymes and peptides, the concentration of the aldehydes and the duration of the fixation time must be made as low as is compatible with maintaining cytological details in the labeled cells (Pickel *et al.*, 1975b).

B. Sectioning

Numerous techniques are available for the preparation of sections of fixed tissues for light microscopic immunocytochemical localization of antigens. These techniques include Vibratome sectioning of unembedded tissue, microtome sectioning of tissues embedded in paraffin, and crytostat sectioning of frozen tissues. All of these methods are suitable for the localization of tyrosine hydroxylase and neuropeptides in brain. The method of choice depends on whether the preservation of cellular morphology or the production of thin serial sections is of primary importance.

1. The Vibratome Method

This method offers the best preservation of cellular morphology. The tissues cut on the vibrating microtome are continuously surrounded by phosphate buffer (0.1 M) and are not damaged by a paraffin-embedding procedure or freezing conditions. The major limitation of this method is that the sections are relatively thick (15–20 μm) and difficult to collect in a serial manner. For electron microscopic localization of antigens in tissue, the Vibratome method has been used most extensively for sectioning tissues prior to labeling with the PAP method. The Vibratome sections have adequate preservation of ultrastructural morphology and are sufficiently thin for light microscopy. The light microscopic analysis is useful in the identification of selectively labeled neuronal pathways that are then embedded in plastic for electron microscopy.

2. The Microtome Method

Microtome sections of paraffin-embedded tissues offer the best method of obtaining thin (4–6 μm) serial sections as are required in experiments involving the quantitative determination of the number of labeled cells within a specific nuclear region. The major limitation of the paraffin-embedding technique is the reduction in the intensity of the antibody labeling that is produced by the organic solvents and/or high temperatures of the embedding procedure.

3. The Cryostat Method

Cryostat sections of frozen tissues are thinner (6–8 μm) than Vibratome sections, and the labeling is not reduced by the embedding procedure. Frozen sections are especially useful for localizing antigens in fetal brains, which are poorly labeled after paraffin embedding and are difficult to section on a vibrating microtome. The major limitation in the use of frozen sections is the formation of ice-crystal artifacts that are produced by slow freezing with dry ice. The ice-crystal formation can partially be eliminated by the incubation of the tissue with 30% sucrose at 4°C for 24 hr prior to freezing. Alternatively, the tissues may be frozen rapidly in freon or a similar quenching agent that produces fewer ice crystals. However, quick freezing often fractures large blocks of tissue.

C. Reagents

Modifications in molarity, concentration, or duration of incubation with various reagents may or may not significantly alter the immunocytochemical labeling.

The *molarity of the Tris buffer* may range from 0.05 to 0.5. The 0.5 M Tris buffer gives better penetration of the antibody and less nonspecific background labeling. However, the lower molar concentrations have been used successfully for many antibodies (Sternberger, 1974). For the final diaminobenzidine reaction, Weir (1973) found that the ionic strength of the buffer had no appreciable effect on the specificity or sensitivity of the method.

Detergents such as Triton X-100 facilitate the penetration of antiserum. The use of these agents is especially necessary for the immunocytochemical localization of membrane-bound enzymes such as a dopamine-β-hydroxylase (Hartman, 1973) or tyrosine hydroxylase in distal axons and axon terminals. Since Triton X-100 dissolves membranes, this reagent should always be used at the lowest concentration and shortest incubation periods that are compatible with the immunocytochemical labeling. For example, using 0.25% Triton X-100 for longer than the described 15 min produces poor membrane preservation by electron microscopy.

The optimal *dilution of the specific antiserum* depends on the characteristics of the antibody as well as on the time and temperature of the incubation. For tyrosine hydroxylase, an overnight incubation with a dilute antiserum

(1:500) at 4°C gives more selective localization than incubation with a concentrated antiserum (1:20) for 1–2 hr at room temperature. Extended incubation periods of up to 48 hr give selective labeling at even higher (1:1000) dilutions, but the ultrastructural morphology begins to deteriorate after the long incubations with the buffer solutions. The dilutions of the anti-IgG and PAP are similarly related to time and temperature of the incubation. However, the exact dilutions appear to be less critical as long as these agents are present in quantities sufficiently high to bind to all of the antibody-labeled sites. Although the 1:10 dilution of the anti-IgG and 1:30 dilution of the PAP are used routinely, dilutions as high as 1:50 and 1:100, respectively, have been used with no noticeable changes in the immunocytochemical labeling.

The time of *incubation with 3,3′-diaminobenzidine* and hydrogen peroxide may range from 1–30 min. The specific brown product formed by the reaction of horseradish peroxidase with hydrogen peroxide and diaminobenzidine is visible in less than 1 min. However, longer exposure times of up to 10 min may increase the intensity of the labeling as a result of greater penetration of the reagents. After 10 min, the nonspecific background labeling begins to increase, and the tissue may appear damaged when examined by electron microscopy. The ultrastructural damage is probably a consequence of exposure to the hydrogen peroxide which can produce holes in the tissue at the site of the reaction.

Numerous investigators have sought to find a *substitute for the 3,3′-diaminobenzidine,* which is apparently carcinogenic (Hanker *et al.*, 1977; Hardy and Heimer, 1977). Other less carcinogenic chromogens that have been used for the detection of horseradish peroxidase include tetramethylbenzidine (TMB) (Hardy and Heimer, 1977) and paraphenylene diamine pepocatechol, the Hanker–Yates reagent (Hanker *et al.*, 1977).* Although these reagents are reasonably good substitutes for diaminobenzidine for light microscopy, they are not very suitable for electron microscopic immunocytochemistry. Thus, at the present time, the primary agent used for electron microscopic localization of horseradish peroxidase is 3,3′-diaminobenzidine, which should be used with considerable safety precautions.

V. SPECIFICITY OF THE PAP TECHNIQUE

A variety of procedures is available to test the specificity of the immunocytochemical labeling method. If the specific antiserum is replaced by normal rabbit serum, infinitely high dilutions of the specific antiserum, or blocked antiserum (Petrusz *et al.*, 1976; Pickel *et al.*, 1975b), then there should be an absence of specific labeling. Replacement of the specific antiserum with normal rabbit serum collected from the rabbits prior to immunization (preimmune) is often not a particularly good method for determin-

*Polysciences Inc., Warrington, Pennsylvania.

ing the specificity of the reaction, since intrinsic labeling is often present in normal serum.

The second method, using a dilution series that ranges from 100–0% antiserum, should show a gradual reduction in antibody labeling at the higher dilutions until, finally, a complete absence of labeling is present when the antibody is at 0%. This method has the added advantage of demonstrating the concentration of antiserum that gives optimal specific labeling.

The most critical test of specificity is obtained by use of blocked antiserum. The blocked antiserum demonstrates whether or not the labeling is inhibited by the addition of effective concentrations of the antigen to the working solution. The effective concentrations of tyrosine hydroxylase, for example, would be the amount of enzyme sufficient to remove reactivity by immunoelectrophoresis, enzyme inhibition, or immunostaining capability. The immunoabsorption method is limited by the purity of the antigen. Impurities in the antigen preparation serve to induce the formation of antibodies that could also be selectively blocked (Weir, 1973). Also, immunoabsorption of the antiserum with other antigens thought to be present in the histological preparation will test for crossreactivity. One or more of these three control procedures is often necessary to determine the specificity of the immunocytochemical method.

VI. USE OF THE PAP TECHNIQUE IN THE DEMONSTRATION OF CATECHOLAMINERGIC NEURONS

The immunocytochemical localization of tyrosine hydroxylase serves as an example of the results that can be obtained by the use of the PAP technique. The method can be used for the light microscopic demonstration of neuronal pathways containing a specific neurotransmitter-synthesizing enzyme and for the electron microscopic identification of perikarya, processes, and terminals containing the specific enzyme and the ultrastructural localization of the enzyme with respect to specific cellular organelles.

A. Light Microscopic Pathways

Tyrosine hydroxylase catalyzes the first step in the biosynthesis of catecholamines (Fig. 2) and is thus selectively localized to neuronal pathways that have been previously characterized as catecholaminergic by fluorescence histochemistry (Dahlström and Fuxe, 1964). One such *dopaminergic pathway* is the nigrostriatal projection (Andén *et al.*, 1965). The dopaminergic perikarya in the pars compacta of the substantia nigra show dark brown reaction product in the cytoplasm of neurons selectively labeled for tyrosine hydroxylase (Fig. 4A). The dendrites of the dopaminergic neurons in the pars reticulata of the substantia nigra and the dopaminergic axons that ascend through the lateral hypothalamus are also selectively labeled for the enzyme (Fig. 4B). In the passage through the lateral hypothalamus, the bundles of

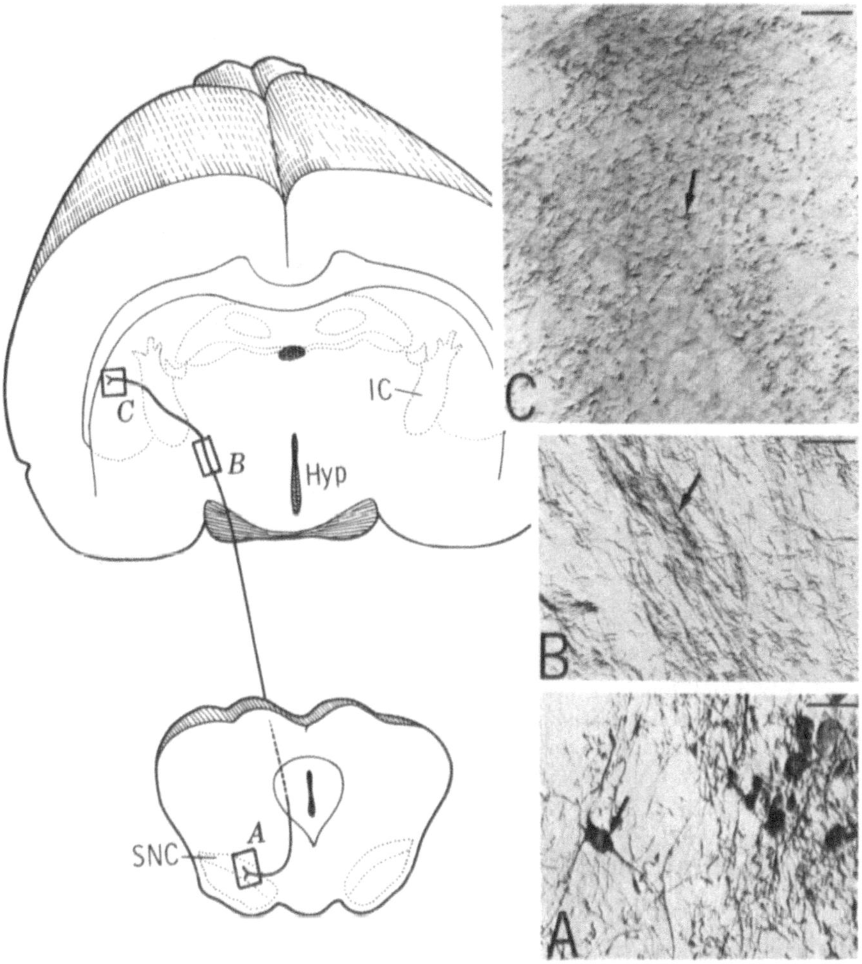

Figure 4. Light microscopic localization of tyrosine hydroxylase in dopaminergic neurons of the nigrostriatal system using the PAP method. The nigrostriatal system is schematically illustrated in the left portion of the figure. Neurons in the pars compacta of the substantia nigra (SNC) in the midbrain (A) send efferent axons through the lateral hypothalamus (B). These axons continue through the internal capsule (IC) to terminate in the caudate nucleus (C). (A) Photomicrograph showing the peroxidase reaction product (arrow) in the cytoplasm of neuronal perikarya located in SNC, labeled area A of schematic drawing. (B) Photomicrograph showing the peroxidase reaction product (arrow) in axons passing through the lateral hypothalamus and identified as area B in the drawing. (C) Photomicrograph showing the peroxidase reaction product (arrow) in terminal varicosities in the caudate nucleus taken from the area labeled C in the schematic diagram. 20-μm Vibratome sections and Nomarski optics throughout. (Bar, 50 μm throughout)

labeled axons divide into numerous fine axonal branches. The terminal varicosities of these axons form a plexus within the caudate nucleus (Fig. 4C). The varicosities usually surround unlabeled neuronal perikarya.

Another example of the use of the PAP technique is the localization of tyrosine hydroxylase in a *noradrenergic pathway* such as the projections from

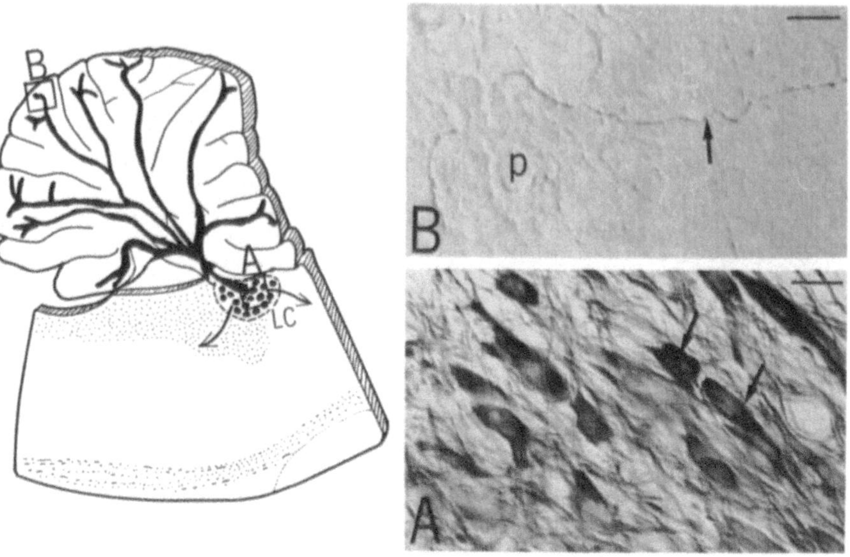

Figure 5. Light microscopic immunocytochemical localization of tyrosine hydroxylase in the noradrenergic pathway from locus coeruleus to cerebellum using the PAP method. The pathway, as illustrated in the schematic drawing in the left portion of the figure, consists of the neuronal perikarya of noradrenergic neurons located in area A in the locus coeruleus (LC) and the efferent axons from these neurons which ascend through the superior cerebellar peduncle to terminate in the cerebellar cortex located in area B. (A) Photomicrograph showing peroxidase reaction product for tyrosine hydroxylase in neuronal perikarya (arrows) from the area labeled A in the schematic. (B) Light micrograph showing the peroxidase reaction product for tyrosine hydroxylase in axons (arrows) passign through the purkinje cell layer (P) and into the molecular layer, from the area labeled B in the schematic drawing. Nomarksi optics and 20-μm Vibratome sections. Bar, 50 μm.

the locus coeruleus to the cerebellum (Dahlström and Fuxe, 1964). The labeling for tyrosine hydroxylase in the locus coeruleus is restricted to the cytoplasm of neuronal perikarya and proximal axons and dendrites (Fig. 5A). Labeled axons can be traced through the medial border of the superior cerebellar peduncle to the cerebellar cortex. The axons pass from the white matter of the cerebellum into the granule, Purkinje, and molecular layers. The most extensive terminal ramifications are present in the molecular layer (Fig. 5B).

B. Ultrastructural Localization of Tyrosine Hydroxylase

The perikarya and processes of neurons containing tyrosine hydroxylase can be identified by the electron-dense material formed by the combination of osmium tetroxide with the peroxidase reaction product (Fig. 6).

Within *perikarya* of noradrenergic neurons, the reaction product for tyrosine hydroxylase is widely distributed throughout the cytoplasm of selec-

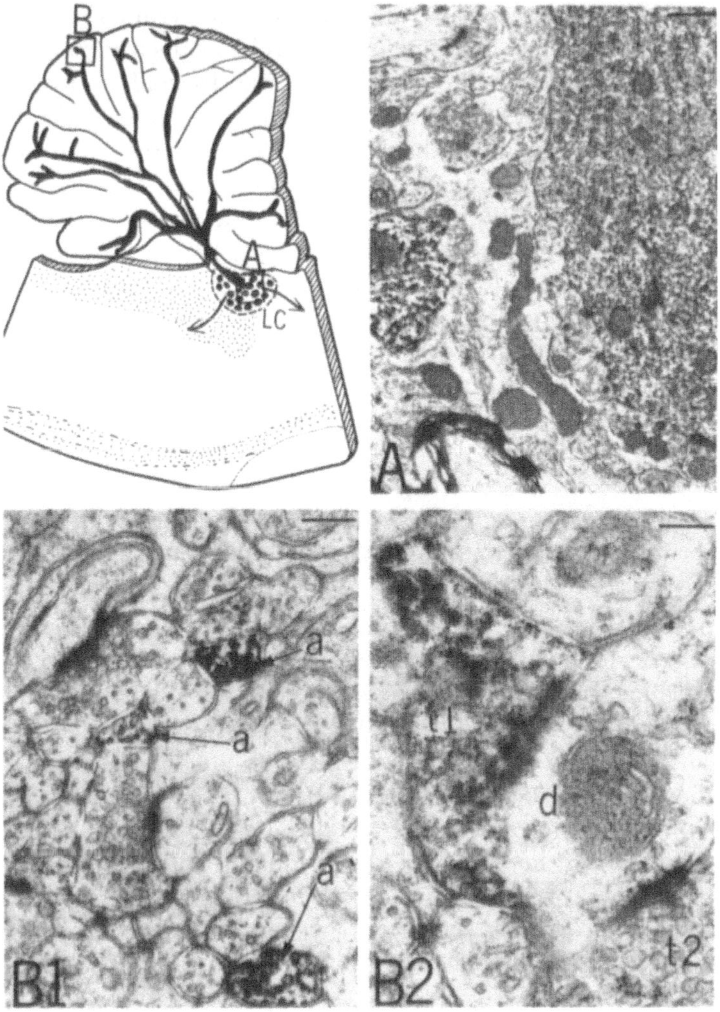

Figure 6. Electron microscopic immunocytochemical localization, using the PAP method, of tyrosine hydroxylase in noradrenergic neurons in the pathway from locus coeruleus to cerebellum. The pathway schematically illustrated in the upper left portion of the figure shows the neurons located in area labeled A in the locus coeruleus (LC) projecting to the cerebellar cortex in the area labeled B. (A) An electron micrograph, taken from area A of the schematic drawing, which shows the localization of tyrosine hydroxylase in a portion of the cytoplasm of neuronal perikarya (right side) and proximal processes of noradrenergic neurons. (B1) Electron micrograph from the molecular layer of the cerebellar cortex that is identified as area B in the schematic drawing. Selective localization of the reaction for tyrosine hydroxylase is found in unmyelinated axons (a). (B2) The ultrastructural labeling for the enzyme is demonstrated in an axon terminal (t1) in the molecular layer of the cerebellum. The labeled axon terminal forms an asymmetric synapse with an unlabeled dendrite (d). Note that a second terminal (t2) on the same dendrite is unlabeled. Bar, 0.1 μm.

tively labeled neurons (Fig. 6A). However, there is slightly more intensive labeling associated with the endoplasmic reticulum, Golgi apparatus, and microtubules (Pickel *et al.*, 1975b). In *axons and dendrites*, the reaction product is primarily associated with round subcellular organelles or longitudinally oriented fibers. These organelles are probably microtubules since they are similar in diameter (22 nm) and distribution. In small unmyelinated axons in the cerebellar cortex (Fig. 6B1), the microtubules labeled for tyrosine hydroxylase are slightly larger than the unlabeled microtubules in adjacent axons. This is presumably because of the accumulation of reaction product. In *axon terminals*, the reaction product for tyrosine hydroxylase is associated with small (40–60 nm) vesicles and an occasional large vesicle (75–100 nm). The labeled axon terminals form asymmetric junctions with unlabeled dendrites (Fig. 6B2). Noradrenergic terminals have previously been demonstrated in the cerebellar molecular layer by uptake of [^{3}H]-norepinephrine (Bloom, 1973).

VII. USE OF THE PAP TECHNIQUE IN THE LOCALIZATION OF NEUROPEPTIDES

Neuropeptides with possible neurotransmitter or neuromodulatory functions have recently been isolated from neural tissues. Examples of these peptides include substance P (Chang and Leeman, 1971) and the morphine-like pentapeptides, [Leu^5]- and [Met^5]-enkephalin (Hughes *et al.*, 1975). Although both groups of peptides are widely distributed in the central nervous system, they are particularly prevalent in the pathways related to pain and analgesia and in areas containing catecholaminergic neurons (Hökfelt *et al.*, 1977c). Immunocytochemical labeling techniques offer the only method presently available for the cellular localization of these neuropeptides. The PAP techniques provides a method for determining the light microscopic distribution of selectively labeled neuronal perikarya and terminals, the ultrastructure of the terminals containing the peptides, and synaptic interactions between peptidergic and nonpeptidergic neurons.

A. Light Microscopy

The perikarya of neurons containing substance P or enkephalin are difficult to demonstrate by the PAP technique without the use of drugs, such as cholchicine or vinblastine, that block axonal transport (Hökfelt *et al.*, 1977a,b). However, in a few regions of the brain, e.g., the perifornical area, perikarya of neurons containing enkephalin can be lightly labeled with the PAP method without the use of special blockers of axonal transport (Fig. 7A).

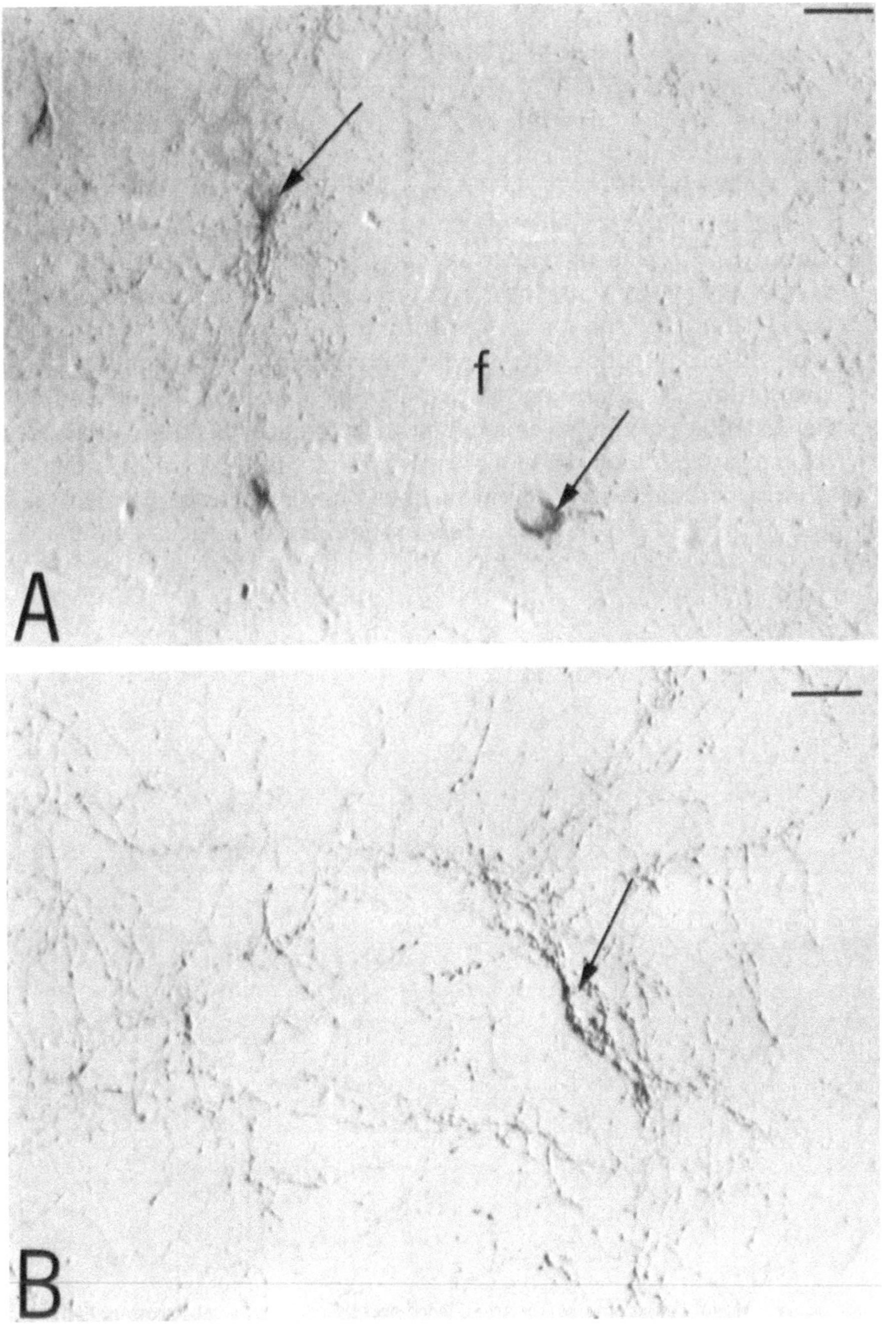

Figure 7. Light microscopic immunocytochemical localization by the PAP method of [Met^5]-enkephalin in perikarya and proximal processes and in distal axon terminals. (A) Photomicrograph showing the reaction product (arrows) for enkephalin in the cytoplasm of neuronal perikarya located adjacent to the fibers of the fornix (f) in the posterior hypothalamus. (B) Photomicrograph showing the reaction product for enkephalin in axon terminals that outline the perikarya and dendrites of a neuron (arrow) in the cerebral cortex. Nomarski optics and 20-μm Vibratome sections throughout. Bar, 50 μm.

The *axon terminals* labeled with these peptides are more easily demonstrated by the PAP method. For example, in certain areas, such as the lateral septal nucleus or cerebral cortex (Fig. 7B), the number of enkephalin-containing terminals is so dense that the entire perikarya and dendrites of receptive neurons are outlined. The distribution of the axon terminals containing substance P or enkephalin are similar (Hökfelt *et al.*, 1977c). They are both extensively localized in regions of the dorsal fasciculus of the spinal cord, the paratrigeminal nucleus (Chan-Palay, 1978), the nucleus tractus solitarius (NTS), and in regions containing catecholamines such as substantia nigra, caudate nucleus, locus coeruleus, and A2 regions. The labeling for enkephalin and substance P within the A2 region is shown in Fig. 8. Whereas both peptides are present within many of the same regions, differences in the distribution of labeled axon terminals are found in these areas. For example, in the lower medulla, enkephalin labeling is localized to fine axonal varicosities in the more ventral portion of the paratrigeminal nucleus, whereas substance P is found in fiber bundles entering the dorsal surface and in processes within the dorsal portion of this nucleus (Fig. 9).

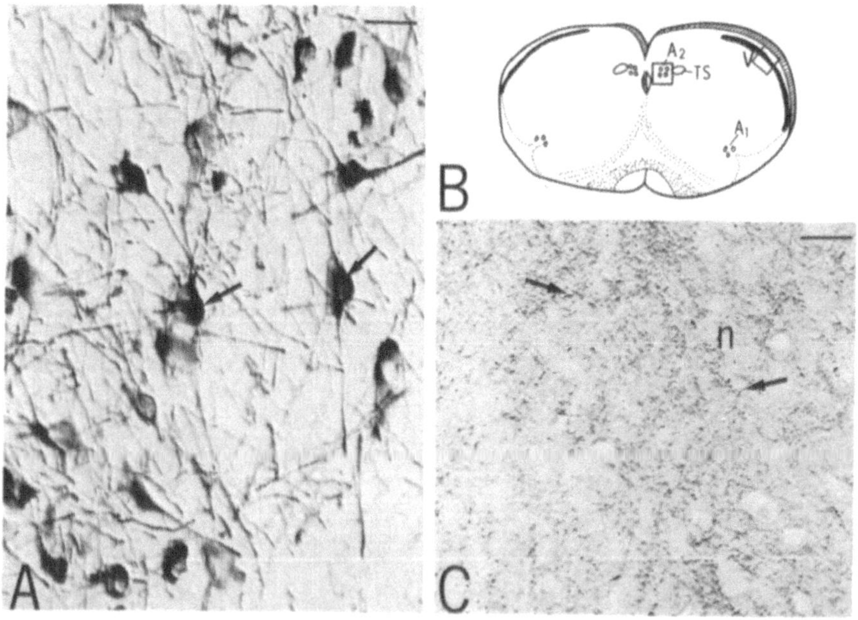

Figure 8. Light microscopic immunocytochemical localization of tyrosine hydroxylase and [Met⁵]-enkephalin in A2 region by the PAP method. (A) Immunocytochemical localization of the reaction product for tyrosine hydroxylase in neuronal perikarya (arrows) and proximal processes in the A2 region which is shown in the schematic drawing. (B) Schematic drawing showing a cross section of the medulla at the level of the obex. V, partrigeminal; NTS, nucleus tractus solitarius; A1 and A2, catecholaminergic nuclei. (C) Photomicrograph showing the location of enkephalin in terminal varicosities (arrows) surrounding unlabeled neurons (N). 20-μm Vibratome sections and Nomarski optics in A and C. Bar, 50 μm.

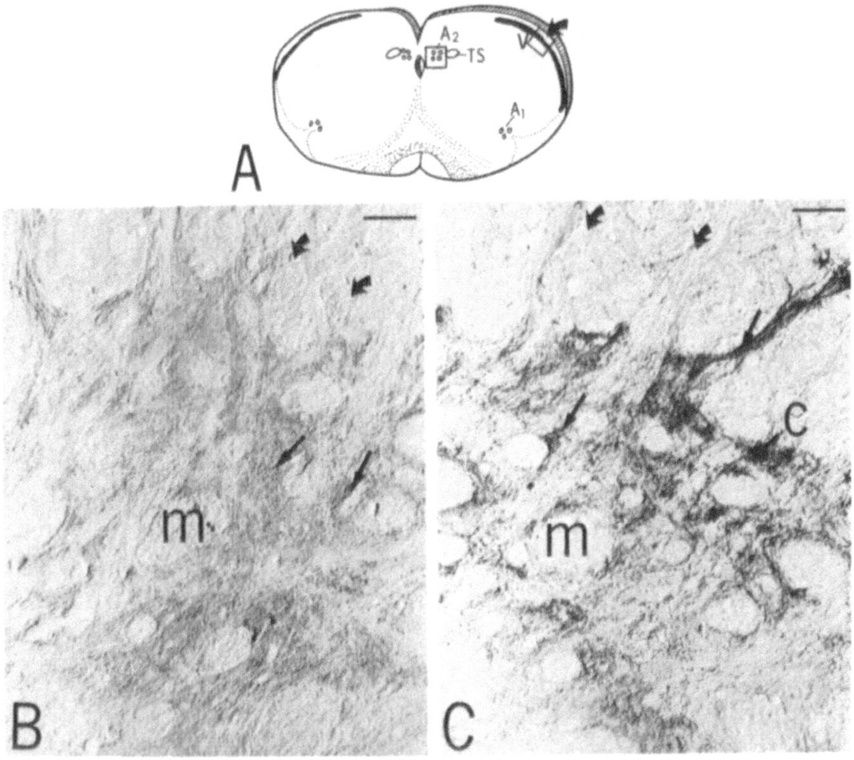

Figure 9. Light microscopic immunocytochemical localization of [Met[5]]-enkephalin and substance P in the paratrigeminal nucleus in the lower medulla. (A) Schematic illustration of a cross section of the medulla oblongata at the level of the paratrigeminal nucleus (V). Curved arrow illustrates dorsal surface of brain at the point of entering substance-P-containing axons. The catecholaminergic neurons in A1 and A2 are also indicated. A2 is located in the medial portion of the nucleus tractus solitarius (TS). (B) Micrograph showing the reaction product (arrows) for enkephalin that is contained within beaded varicosities surrounding neurons in the ventral portion of the paratrigeminal nucleus. Many myelinated bundles (m) are unlabeled. (C) Photomicrograph showing the localization of substance P in neurons of the more dorsal portion of the paratrigeminal nucleus (arrows) that enter from the dorsal brain surface (curved arrows). Round dense structures (c) could either be labeled perikarya or cross sections of densely labeled bundles of axons. Unlabeled bundles of myelinated axons (m) are also present. Nomarski optics and 20 μm Vibratome sections in both B and C. Bar, 50 μm.

B. Ultrastructural Localization of Peptides in Axon Terminals

The axon terminals labeled for substance P and enkephalin in locus coeruleus and A2 region are morphologically similar. The labeled axon terminals contain both large dense vesicles (80–100 nm) and small clear vesicles (40–60 nm) and form asymmetric synaptic contacts with unlabeled dendrites (Fig. 10). The reaction product extends throughout the large dense vesicles (Fig. 10B,C) and forms a rim of labeling around smaller vesicles.

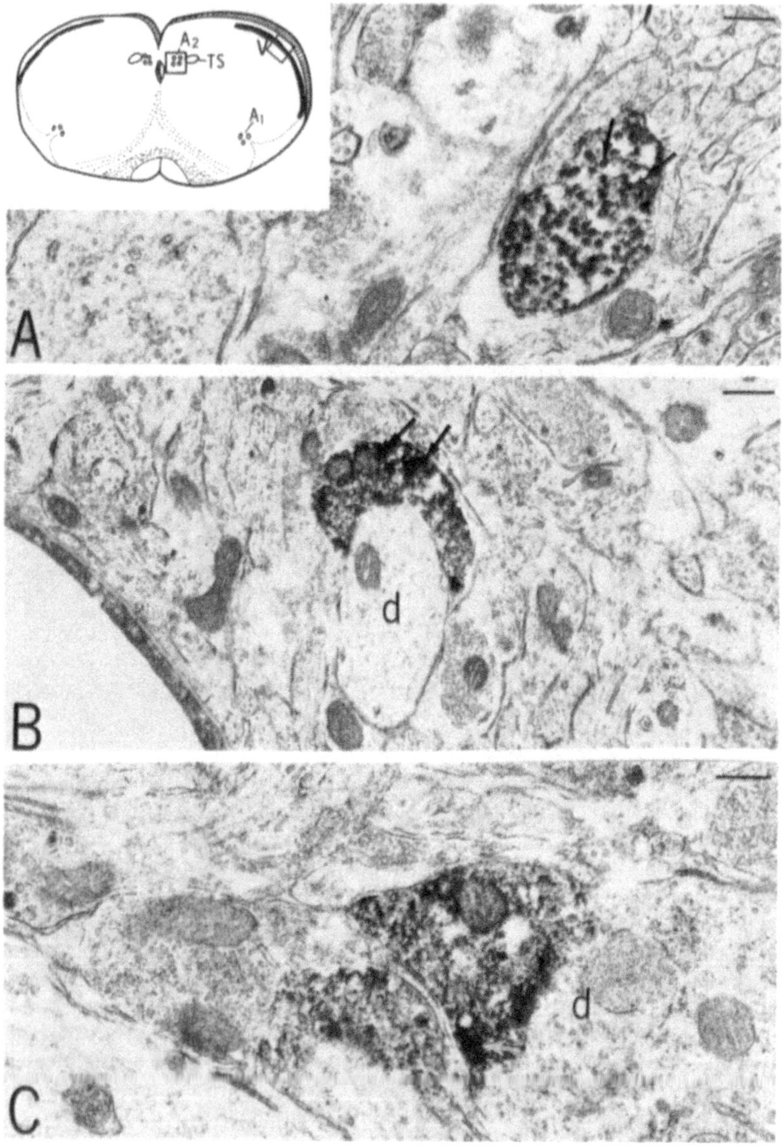

Figure 10. Electron microscopic localization of tyrosine hydroxylase, [Met⁵]-enkephalin, and substance P in the A2 region by the PAP method. (A) Electron micrograph of a dendrite showing reaction product for tyrosine hydroxylase associated with microtubules (arrow), taken from the A2 region shown in the inserted schematic drawing. Abbreviations for insert as in legend to Fig. 9. (B) Electron micrograph of an axon terminal showing reaction product for [Met⁵]-enkephalin associated with both the large dense vesicles (arrows) and the outer rim of smaller vesicles. The labeled terminal forms a synaptic contact with the unlabeled dendrite (d) in the same A2 region (see Schematic drawing in A). (C) Electron micrograph of an axon terminal labeled with substance P in the same A2 region. The reaction is associated with both large and small vesicles in a terminal that forms a synaptic contact with dendrite (d). Bar, 0.1 μm.

C. Synaptic Interactions between Peptidergic Axons and Catecholaminergic Neurons

The light and electron microscopic localization of neurotransmitter-synthesizing enzymes and neuropeptides by the PAP technique provides a method for examining possible synaptic interactions that may exist between catecholaminergic and peptidergic neurons (Pickel *et al.*, 1979). The localization of tyrosine hydroxylase, substance P, and [Met^5]-enkephalin in the A2 region of the medulla serves to exemplify the use of the immunocytochemical technique for determining synaptic interactions. By light microscopy, tyrosine hydroxylase is localized to perikarya and dendrites of intrinsic neurons in the A2 region, whereas substance P and enkephalin are localized to beaded varicosities surrounding the unlabeled neuronal perikarya (Fig. 10). Electron microscopic examination shows that tyrosine hydroxylase is distributed in the perikarya and dendrites of selectively labeled neurons, a distribution similar to that found in the locus coeruleus (Fig. 10A). The dendrites labeled for tyrosine hydroxylase form synapses with terminals containing both large and small vesicles (Pickel *et al.*, 1979). These terminals are similar to the axon terminals labeled for substance P and enkephalin (Fig. 10B,C). This type of analysis suggests that a synaptic interaction exists between the axon terminals containing the peptides and the catecholaminergic dendrites. However, a more critical evaluation could be determined by double-labeling techniques (Gonatas *et al.*, 1974). For example, such techniques might include peptides tagged with a radioisotope and localized by autoradiography coupled with immunocytochemical localization of tyrosine hydroxylase. Such methods have not been possible until recently, since radiolabeled peptides have not been available.

VIII. SUMMARY OF ADVANTAGES AND LIMITATIONS OF THE PAP TECHNIQUE

A. Advantages

1. With proper attention to fixation and controls, the PAP technique can be used to identify specific neuronal antigens. At present this identification is limited to qualitative assessment, although as the methodology improves, quantitative analysis will also become feasible.
2. The marker compound associated with the identified antigen is visualized in direct topographical relationship to the cellular structure(s) containing that antigen.
3. The PAP technique is, perhaps, the most sensitive method for localization of antigens in sections of tissue.
4. This technique can be used at both the light and electron microscope levels. In fact, the same piece of tissue examined with the light mi-

croscope can subsequently be analyzed at the electron microscope level.

5. This method can often be combined with other techniques (e.g., autoradiography, lesion-induced degeneration, fluorescence microscopy) to provide simultaneous characterization of two substances or structures.

B. Limitations

The following limitations are inherent in the use of the immunocytochemical technique.

1. The localization of any antigen is entirely dependent on the specificity of the antibody and on the specificity of the immunocytochemical technique.
2. Given proper controls, positive immunocytochemical labeling demonstrates that a given antigen is present in a section of the tissue; however, negative immunocytochemical labeling does not necessarily indicate that the antigen is absent. The antigen may be present, but undetected, because of technical limitations such as the effect of fixation on the antigen–antibody reaction or insufficient penetration of the antiserum.
3. Identification of labeled subcellular organelles is frequently difficult because of the accumulation of peroxidase reaction product on the surface or in the lumen of the labeled organelle. In this case, reliance must be placed on the relative size and distribution of the organelles under study as a means of identification.
4. The localization of a reaction product in association with a specific organelle indicates a close proximity between the labeled antigen and the organelle, but it does not necessarily indicate biochemical binding. The observed localization may be the result of crosslinking between adjacent proteins induced by glutaraldehyde.

IX. CONCLUSIONS

The advent of immunocytochemical techniques for the localization of specific antigens in tissues provides a method for the light and electron microscopic identification of chemically distinct neurons within the central nervous system. Highly sensitive indirect immunocytochemical labeling methods such as the PAP method demonstrate the cellular localization of both the large neurotransmitter-synthesizing enzyme, tyrosine hydroxylase, and the small peptides, substance P and enkephalin. The details of the immunocytochemical procedure used to localize these compounds may not be entirely applicable to the localization of all other antigens in neurons. However, the described protocol does serve as a useful starting point for the development of procedures that may be useful for the localization of other enzymes and peptides.

ACKNOWLEDGMENTS. The studies upon which this account is based were done in collaboration with T. H. Joh and D. J. Reis (Cornell University Medical College), Susan Leeman (Harvard Medical School), and Richard Miller (University of Chicago), and were supported by grants from NIH (MH24285, NS06911, and HL18974) and NASA (NGR 33-019-179). In addition, I am recipient of a Research Career Development Award from NIMH (MH00078). I would also like to express my appreciation to Dr. Ludwig Sternberger and Dr. Marc Nathan for helpful suggestions on the manuscript.

REFERENCES

Andén, N.-E., Dahlstrom, A., Fuxe, K., and Larsson, K., 1965, Further evidence for the presence of nigro–neostriatal dopamine neurons in the rat, *Am. J. Anat.* **116**:328–334.

Avrameas, S., 1969, Coupling of enzymes to proteins with glutaraldehyde. Use of the conjugates for the detection of antigens and antibodies, *Immunochemistry* **6**:43–52.

Bloom, F. E., 1973, Ultrastructural identification of catecholamine-containing central synaptic terminals, *J. Histochem. Cytochem.* **21**:333–348.

Burns, T., 1975, Background staining and sensitivity of the unlabeled antibody–enzyme (PAP) method comparison with the peroxidase labeled antibody sandwich method using formalin-fixed paraffin embedded material, *Histochemistry* **43**:291–294.

Chan-Palay, V., 1978, The paratrigeminal nucleus. I. Neurons and synaptic organization and II. Identification and interrelations of catecholamine, indoleamine axons, and substance P immunoreactive cells in the neuropil, *J. Neurocytol.* **7**:405–442.

Chang, M. M., and Leeman, S. E., 1970, The isolation of a sialogogic peptide from bovine hypothalamic tissue and its characterization as substance P, *J. Biol. Chem.* **245**:4784–4790.

Coons, A. H., and Kaplan, M. H., 1950, Localization of antigens in tissue cells. II. Improvement in a method for the detection of antigen by means of fluorescent antibody, *J. Exp. Med.* **91**:1–14.

Dahlström, A., and Fuxe, K., 1964, Evidence for the existence of monoamine containing neurons in the central nervous system. I. Demonstration of monoamines in cell bodies of brain stem neurons, *Acta Physiol. Scand.* [*Supp.*] **232**:1–55.

Ellar, D. J., Nunoz, E., and Salton, M. R. D., 1970, The effects of low concentrations of glutaraldehyde on *Micrococcus lysodeikticus:* Changes in the release of membrane associated enzymes and membrane structure, *Biochim. Biophys. Acta* **255**:140–149.

Gibbs, J. W., Spector, B. S., and Udenfriend, S., 1967, Production of antibodies to dopamine-β-hydroxylase of bovine adrenal medulla, *Mol. Pharmacol.* **3**:473–478.

Goldstein, M., Fuxe, K., Hökfelt, T., and Joh, T. H., 1971, Immunohistochemical studies on phenylethanolamine-N-methyltransferase, DOPA-decarboxylase and dopamine-β-hydroxylase, *Experientia* **27**:951–952.

Gonatas, N. K., Stieber, A., Gonatas, J., Gambetti, P., Antoine, J. C., and Avrameas, S., 1974, Ultrastructural autoradiographic detection of intracellular immunoglobulins with iodinated Fab fragments of antibody. The combined use of ultrastructural autoradiography and perioxidase cytochemistry for the detection of two antigens (double labeling), *J. Histochem. Cytochem.* **22**:999–1009.

Graham, R. C., Jr., and Karnovsky, M. J., 1966, The early stages of absorption of injected horseradish peroxidase in the proximal tubules of mouse kidney: Ultrastructural cytochemistry by a new technique, *J. Histochem. Cytochem.* **14**:291–302.

Hanker, J. S., Yates, P. E., Metz, C. B., and Rustioni, A., 1977, A new specific, sensitive and non-carcinogenic reagent for the demonstration of horseradish peroxidase, *Histochem. J.* **9**:789–792.

Hardy, H., and Heimer, L., 1977, A safer and more sensitive substitute for diaminobenzidine in the light and electron microscopic demonstration of retrograde and anterograde axonal transport of HRP, *Neurosci. Lett.* **5**:235–240.

Hartman, B. K., 1973, Immunofluorescence of dopamine-β-hydroxylase. Application of improved methodology to the localization of the peripheral and central noradrenergic nervous system, *J. Histochem. Cytochem.* **21**:312–332.

Hendrickson, A., 1975, Technical modification to facilitate tracing synapses by electron microscopic autoradiography, *Brain Res.* **85**:241–247.

Hökfelt, T., and Ljungdahl, A., 1972, A modification of the Falck–Hillarp formaldehyde fluorescence method using the Vibratome: Simple, rapid and sensitive localization of catecholamines in sections of unfixed brain tissue, *Histochemie.* **29**:352–353.

Hökfelt, T., Elde, R., Johansson, O., Terenius, L., and Stein, L., 1977a, The distribution of enkephalin-immunoreactive cells bodies in the central nervous system. *Neurosci. Lett.* **5**:25–31.

Hökfelt, T., Johansson, O., Kellerth, J. O., Ljungdahl, A., Nilsson, G., Nygards, A., and Pernow, B., 1977b, Immunohistochemical distribution of substance P, in: *Substance P*, (U.S. von Euler and B. Pernow, eds.), pp. 117–146, Raven Press, New York.

Hökfelt, T., Ljungdahl, A., Terenius, L., Elde, R., and Nilsson, G., 1977c, Immunohistochemical analysis of peptide pathways possibly related to pain and analgesia: Enkephalin and substance P, *Proc. Natl. Acad. Sci. USA* **74**:3081–3085.

Hughes, J., Smith, T. W., Kosterlitz, H. W., Fothergill, L. A., Morgan, B. A., and Morris, H. R., 1975, Identification of two related pentapeptides from the brain with potent opiate agonist activity, *Nature* **258**:577–579.

Joh, T. H., Gegham, C., and Reis, D. J., 1963, Immunochemical demonstration of increased tyrosine hydroxylase protein in sympathetic ganglia and adrenal medulla elicited by reserpine, *Proc. Natl. Acad. Sci. USA* **70**:2767–2771.

Modesto, R. R., and Pesce, A. J., 1971, The reaction of 4,4′-difluoro-3.3′-dinitro diphenyl sulfone with gamma globulin and horseradish peroxidase, *Biochm. Biophys. Acta* **229**:384–395.

Moriarty, G. C., Moriarity, C. M., and Sternberger, L. A., 1973, Ultrastructural imunocytochemistry by unlabeled antibodies and the peroxidase–antiperoxidase complex (PAP). A technique more sensitive than radioimmunoassay, *J. Histochem. Cytochem.* **21**:825–833.

Nairn, R. C., (ed.), 1969, *Fluorescent Protein Tracing* (3rd ed.), E. M. S. Livingston, Ltd., Edinburgh and London.

Nakane, P. K., and Pierce, G. B., Jr., 1966, Enzyme-labeled antibodies: Preparation and application for the localization of antigens, *J. Histochem. Cytochem.* **14**:929–931.

Petrusz, P., Sar, M., Ordronneau, P., DiMeo, P., 1976, Specificity in immunocytochemical staining, *J. Histochem. Cytochem.*, **24**:1110–1112.

Pickel, V. M., John, T. H., Field, P. E., Becker, C. G., and Reis, D. J., 1975a, Cellular localization of tyrosine hydroxylase by immunocytochemistry, *J. Histochem. Cytochem.* **23**:1–13.

Pickel, V. M., Joh, T. H., and Reis, D. J., 1975b, Ultrastructural localization of tyrosine hydroxylase in noradrenergic neurons of brain, *Proc. Natl. Acad. Sci. USA* **72**:659–663.

Pickel, V. M., Joh, T. H., Reis, D. J., Leeman, S. E., and Miller, R. J., 1979, Electron microscopic localization of substance P and enkephalin in axon terminals related to dendrites of catecholaminergic neurons, *Brain Res.* **160**:387–400.

Sternberger, L. A., 1974, *Immunocytochemistry*, Prentice–Hall, Englewood Cliffs, New Jersey.

Swanson, L. W., and Hartman, B. K., 1975, The central adrenergic system an immunofluorescence study of the location of cell bodies and their efferent connections in the rat utilizing dopamine-β-hydroxylase as a marker, *J. Comp. Neurol.* **163**:467–506.

Tanaka, H., 1968, The ferritin-labeled antibody method: Its advantages and disadvantages. A methodological review, *Acta Haematol. Jpn.* **31**:125–137.

Weiner, N., 1970, Regulation of norepinephrine biosynthesis, *Annu. Rev. Pharmacol.* **10:**273–290.

Weir, D. M., (ed), 1973, *Handbook of Experimental Immunology,* Second Edition, Blackwell Scientific Publications, Oxford.

Weir, E. E., Pretlow, T. G., Pitts, A., and Williams, E. E., 1974, A more sensitive and specific histochemical peroxidase stain for the localization of cellular antigens by the enzyme–antibody conjugate method, *J. Histochem. Cytochem.* **22:**1134–1140.

The 2-Deoxyglucose Method

15

PETER J. HAND

I. INTRODUCTION

The [^{14}C]-2-deoxy-D-glucose (2DG) method, developed by Sokoloff *et al.* (1977) for the purpose of studying local cerebral glucose metabolism *in vivo*, can provide information concerning the functional organization of larger arrays of neurons. For example, it is possible to examine functional connectivity of an entire sensory pathway, such as that of the somesthetic system, within an individual animal (e.g., Hand *et al.*, 1977; Fig. 1). Such mapping is made possible because of the close link between brain metabolism, which in the adult is almost entirely dependent upon glucose, and neural function (Sokoloff *et al.*, 1977; Sokoloff, 1977). For example, recent work of Yarowsky *et al.* (1979) and Miyaoka *et al.* (1979) has demonstrated a linear relationship between the amount of peripheral stimulation and the values of increased glucose metabolism in such structures as the superior cervical ganglion following electrical stimulation of its afferents and the dorsal lateral geniculate nucleus following photic stimulation of the retina.

PETER J. HAND • Laboratories of Anatomy, Department of Animal Biology, School of Veterinary Medicine, and Institute of Neurological Sciences, University of Pennsylvania, Philadelphia, Pennsylvania 19104.

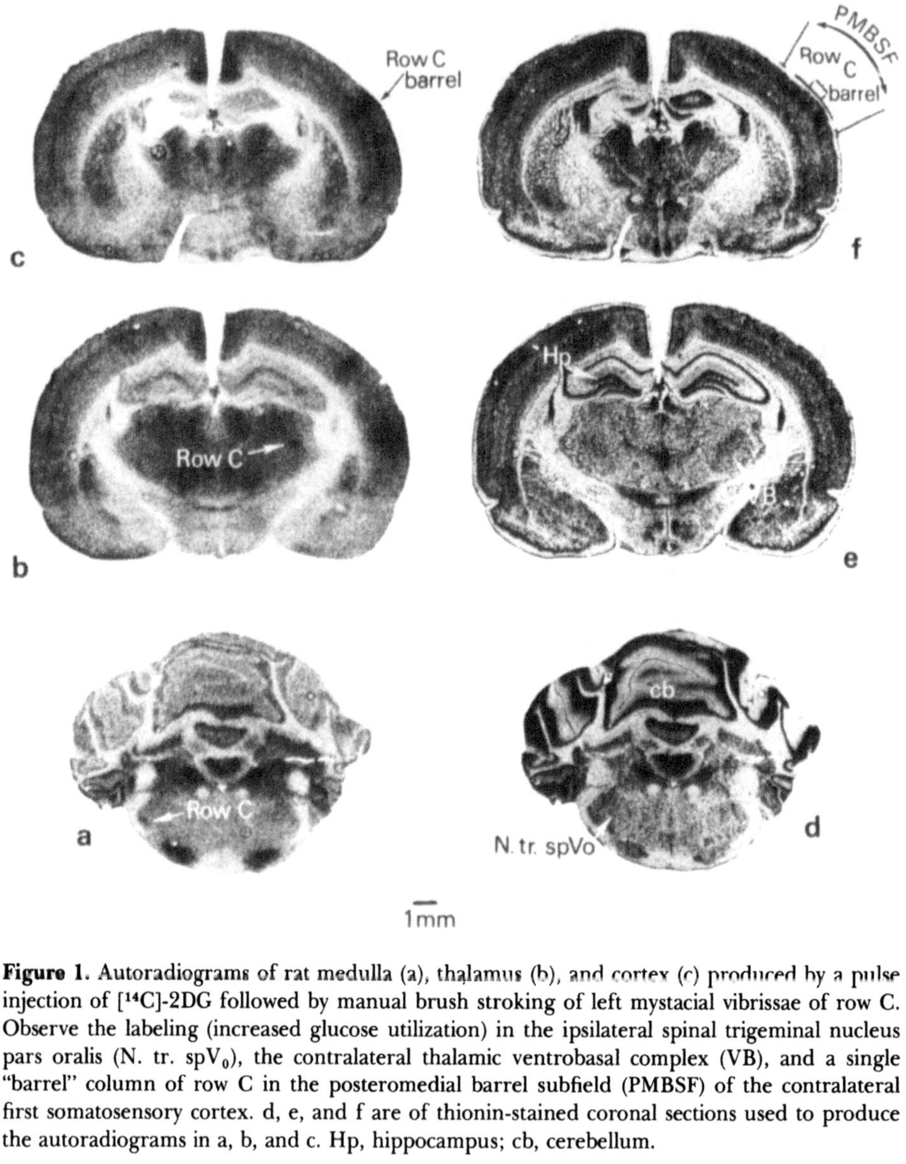

Figure 1. Autoradiograms of rat medulla (a), thalamus (b), and cortex (c) produced by a pulse injection of [^{14}C]-2DG followed by manual brush stroking of left mystacial vibrissae of row C. Observe the labeling (increased glucose utilization) in the ipsilateral spinal trigeminal nucleus pars oralis (N. tr. spV$_0$), the contralateral thalamic ventrobasal complex (VB), and a single "barrel" column of row C in the posteromedial barrel subfield (PMBSF) of the contralateral first somatosensory cortex. d, e, and f are of thionin-stained coronal sections used to produce the autoradiograms in a, b, and c. Hp, hippocampus; cb, cerebellum.

II. BASIC PRINCIPLES OF THE METHOD

The reader is referred to the paper by Sokoloff *et al.* (1977) for details of the theory of the 2DG method. Briefly, an analogue of glucose, 2-deoxy-D-glucose, injected intravenously is transported from arterial plasma to the brain as well as to other tissues of the body by the carrier that transports glucose (Fig. 2). When it has reached the brain, 2DG competes with glucose for the same enzyme, hexokinase, which phosphorylates them to 2DG-6-

phosphate (2DG-6-PO_4) and glucose-6-phosphate, respectively. It is at this stage that the altered structure of 2DG, in which a hydroxyl group on the second carbon atom is replaced by a hydrogen atom, prevents the 2DG-6-PO_4 from being converted to fructose-6-PO_4 by phosphohexoisomerase in order to continue in the glycolytic cycle to become CO_2 and H_2O as does glucose-6-phosphate (Wick *et al.*, 1957; Sokoloff *et al.*, 1977; Fig. 2). Although pharmacological doses of deoxyglucose can competitively inhibit glucose-6-phosphate metabolism at the phosphohexoisomerase step and produce hypoglycemia (Wick *et al.*, 1967; Tower, 1968, Horton *et al.*, 1973), dosages commonly used in the [^{14}C]-2DG experiments are insufficient to have any significant effect on glucose metabolism. Also, 2DG-6-PO_4 does not appear to be a substrate for the enzyme glucose-6-phosphate dehydrogenase, which is the initial step in the hexose-monophosphate shunt, an alternative pathway of glucose metabolism. Additionally, 2DG-6-phosphatase activity, which could hydrolyze 2DG-6-PO_4 to free 2DG, is not present in the brain in sufficient amounts to negate the 2DG model. Thus 2DG-6-PO_4 becomes trapped within the cell (half-life in gray matter, 7.7 hr; in white matter, 9.2 hr; Sokoloff *et al.*, 1977). If 2DG is labeled with a radioactive tracer such as ^{14}C, tritium, or, in the case of positron-emission tomography in man, ^{18}F, the loci of 2DG uptake can be revealed pictorially using an autoradi-

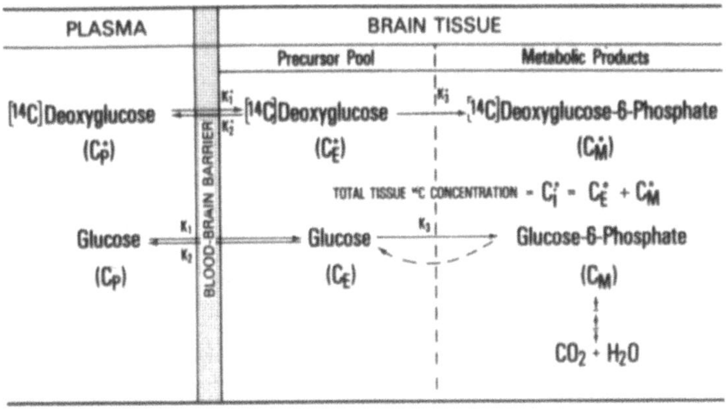

Figure 2. Diagrammatic representation of the theoretical model taken from Fig. 1 of Sokoloff *et al.* (1977). C^* represents the total ^{14}C concentration in a single homogeneous tissue of the brain. C_p^* and C_p represent the concentrations of [^{14}C]-deoxyglucose and glucose in the arterial plasma, respectively; C_E^* and C_E represent their respective concentrations in the tissue pools that serve as substrates for hexokinase. C_M^* represents the concentration of [^{14}C]-deoxyglucose-6-phosphate in the tissue. The constants K_1^*, K_2^*, and K_3^* represent the rate constants for carrier-mediated transport of [^{14}C]-deoxyglucose from plasma to tissue, for carrier-mediated transport back from tissue to plasma, and for phosphorylation by hexokinase, respectively. The constants K_1, K_2, and K_3 are the equivalent rate constants for glucose. [^{14}C]-deoxyglucose and glucose share and compete for the carrier that transports both between plasma and tissue and for hexokinase which phosphorylates them to their respective hexose-6-phosphates. The dashed arrow represents the possibility of glucose-6-phosphate hydrolysis by glucose-6-phosphatase activity, if any. (Reproduced by permission of the authors and Pergamon Press.)

ographic method (Sokoloff *et al.*, 1977; Sejnowski *et al.*, 1979; Basinger *et al.*, 1979; Reivich *et al.*, 1979a,b).

In the case of mapping studies, a system or group of systems is activated using natural, electrical, or chemical stimulation in the restrained or actively behaving animal (Hand *et al.*, 1977; Hubel and Wiesel, 1977; Gallistel *et al.*, 1977; Durham and Woolsey 1977; Stewart *et al.*, 1979). The increased neural activity produces an increased demand for, and uptake of, glucose as well as [^{14}C]-2DG. Thus, structures of the activated neural systems become more densely labeled than those of nonstimulated systems.

Because the 2DG method is dependent on glucose metabolism for its effects, and neurons of the central nervous system possess a significant basal metabolic rate for glucose, it has a lower signal-to-noise ratio of autoradiographic labeling than do the axoplasmic flow types of tracer studies. Thus, the 2DG method produces the most readily identifiable labeling where a functional system has dense, localized, unilateral connections, such as in the rodent vibrissal–first somatosensory cortical "barrel" system (Figs. 1, 3, and 4; Hand *et al.*, 1977; Durham and Woolsey, 1977). Unilateral systems are useful models to study because the same animal can serve as its own control for basal levels of 2DG labeling. Functional systems with less dense and localized connections as well as systems with bilateral terminations (e.g., Mehler *et al.*, 1960; Edwards, 1975) can also be investigated with the 2DG technique, but these require more exacting quantification of the local rates of glucose utilization and a more careful selection of controls to differentiate activated from nonactivated regions.

Although some of the most impressive labeling is observed in pathways possessing dense and highly localized terminations, the 2DG method, by nature of its global labeling characteristics, appears to be most ideally suited to the study of functional systems related to complex behavior (e.g., Gallistel *et al.*, 1977; Jarvis *et al.*, 1978). The use of classical electrophysiological methods would be highly impractical in such investigations.

Because metabolic rates are much lower in white matter than gray matter (Sokoloff *et al.*, 1977), the 2DG method is not well suited to the study of the functional organization of fiber tracts themselves. One of the major limitations of the 2DG method, furthermore, is that the mechanisms underlying the relationship between functional and metabolic activity is unknown, as is the localization of the 2DG labeling to specific tissue components such as neuronal perikarya, dendrites, synaptic endings, and glial cells.

Some recent studies have provided information that may help resolve the localization and mechanism issues (Des Rosiers and Descarries, 1978; Basinger *et al.*, 1979; Schwartz *et al.*, 1979; Mata *et al.*, 1980). A definitive answer to the localization of the 2DG uptake, however, must await the development of a modification of the 2DG method that is suitable for use with the electron microscope. Such a method, based on *in situ* perfusion–fixation and embedding, is presently being developed by Des Rosiers and Descarries (1978). Their preliminary electron micrographs of normal spinal gray matter show 2DG labeling over all portions of the neuropil, including neuronal and glial perikarya.

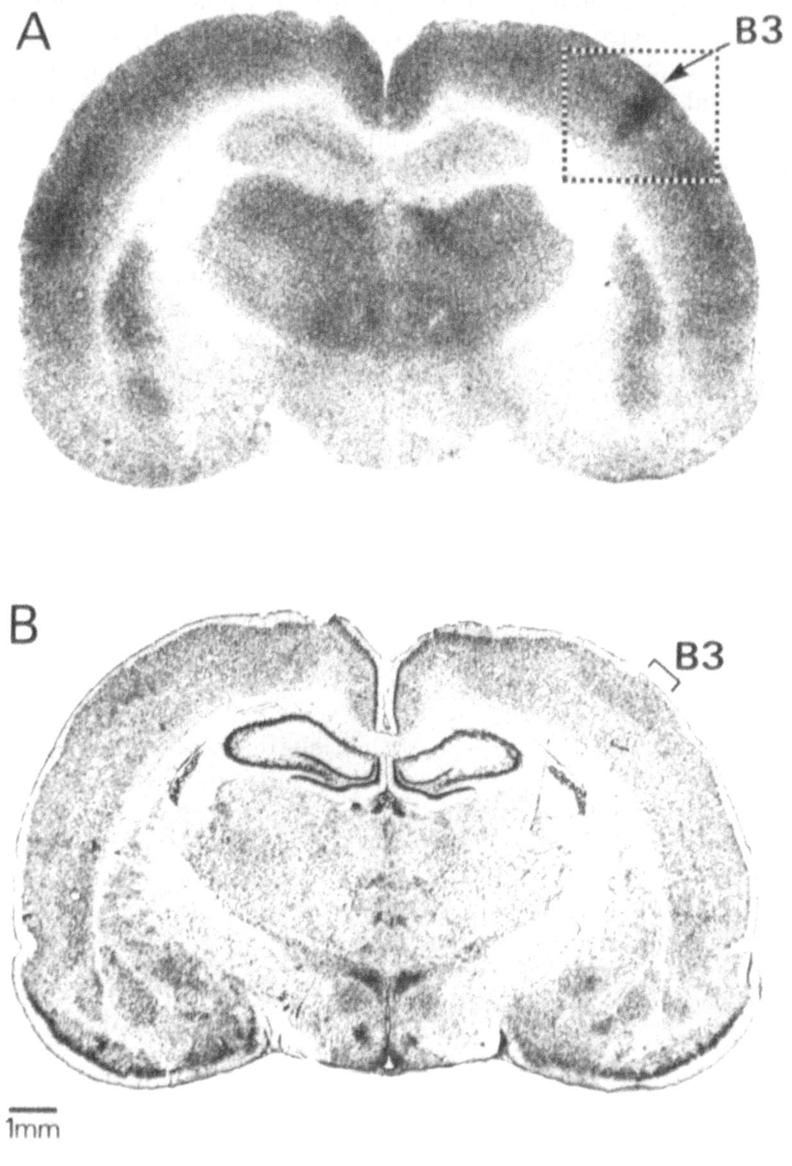

Figure 3. (A) Autoradiogram of [^{14}C]-2DG labeling in a coronally-sectioned rat "barrel" column B3 of the contralateral first somatosensory cortex. Labeling was produced by brush stroking of the left mystacial vibrissa No. 3 of row B. The insert represents the area scanned and reconstructed in color in Fig. 4. (B) A thionin-stained section that was used to produce the autoradiogram in (A).

Basinger *et al.* (1979) demonstrated that specific cell types as well as localized regions of the neuropil of isolated and incubated goldfish retina could be labeled differentially with [^{3}H]-2DG following dark-adaptation or red or white light stimulation.

A recent [^{14}C]-2DG study by Schwartz *et al.* (1979) showed increased labeling in the posterior pituitary of the rat following physiological stimulation of the hypothalamo–posterior hypophyseal system by salt-loading of the animals. The labeled region possessed only axon terminals, whereas nuclear regions containing the cells of origin of this activated system, i.e., the supraoptic and paraventricular nuclei, were not labeled. The study by Mata *et al.* (1980) extends that of Schwartz and co-workers. Using an *in vitro* [14C]-2DG method in which they electrically stimulated an incubated rat posterior pituitary, they determined that increases in energy metabolism are a function of the increased activity of the sodium pump. The application of veratridine, an alkaloid that increases activity of the sodium conductance mechanism in the cellular membrane, produced increased 2DG labeling in the posterior pituitary, whereas electrical stimulation in the presence of ouabain, which inhibits the sodium pump and Na^+, K^+-ATPase, prevented increased labeling.

The findings of Mata and co-workers indicate that the major energy consumption of activated neural tissue is in the ion pumping by the sodium pump that accompanies depolarization of the membrane. They suggest that, since increases in energy utilization are a function of the surface-to-volume ratio of the tissue (Greengard and Ritchie, 1971), and since the highest surface-to-volume ratios are found in nerve endings and dendrites, the greatest relative increases in glucose utilization should be present in regions of the brain richest in these structures.

In contrast, stimulation of the facial sinus hairs of the rat following a pulse injection of [^{14}C]-2DG produces labeling in the rostrally located first somatosensory cortical "hollow barrels." These barrels are cylindrical aggregations of layer IV neurons related on a one-to-one basis to input from facial sinus hair receptors (Welker, 1976). The region of the barrel most densely labeled was that of the cell-dense wall (Woolsey and Van der Loos, 1970), whereas the cell-sparse barrel center, rich in axon terminals (Woolsey and

---

Figure 4. Computer-enhanced pseudocolor reconstructions of [^{14}C]-2-deoxyglucose (2DG) labeling of a coronally sectioned single cortical barrel column (B3) in the posteromedial barrel subfield (PMBSF) of S-I in the same unanesthetized, restrained rat as in Fig. 3. (A) Low-power (×4.5) reconstruction included for orientation purposes. (B) Higher-magnification (×32) reconstruction. The metabolism in deep lamina I through upper layer VI of the column was increased above background levels by the manual brush-stroking of vibrissa No. 3 of row B (B3). The color spectrum on the right side of the figure represents different rates of glucose utilization in μmoles/100 g per min.

Compare these computer-enhanced color images of the B3 barrel column with that in the standard autoradiogram of Fig. 3. Note that the enhanced color image reveals details of a heterogeneity of 2DG labeling within the barrel column. (The laminae are labeling differentially—layer IV labels most densely, layers I and VI most lightly.) In addition, the color image reveals the presence of increased metabolic labeling extending tangentially from the barrel column as irregular fingerlike projections. Observe that the extent of tangential labeling is not equal for all layers. Such detailed information concerning the pattern of increased local glucose utilization is not possible using standard autoradiograms.

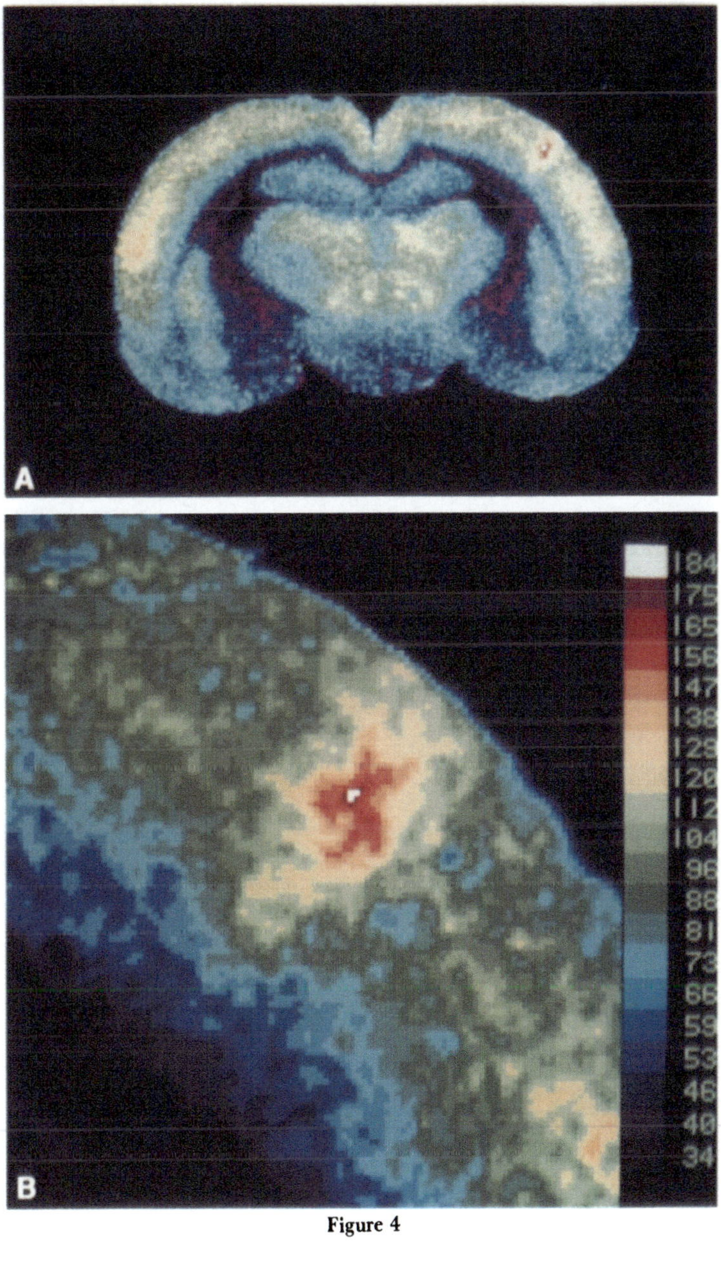

Figure 4

Van der Loos, 1970; Killackey and Leshin, 1975; Donaldson *et al.*, 1975) was lightly labeled (Hand, unpublished observations).

These conflicting results suggest that different cellular components may be labeled in different functional systems.

In spite of the problem of labeling localization, the 2DG method has proved invaluable in determining the pattern of functional activity in such systems as the rat and monkey somatosensory cortices (Hand *et al.*, 1977, 1978, 1979a,b; Juliano *et al.*, 1979). This would not have been possible using classical anatomical and functional methods.

III. GENERAL APPLICATIONS OF THE METHOD

The 2DG method has been utilized to map the functional organization of portions of the visual (Hubel and Wiesel, 1977, Hubel *et al.*, 1978; Schwartz and Gainer, 1977; Jarvis *et al.*, 1978), auditory (Silverman *et al.*, 1977), somatosensory (Hand *et al.*, 1977, 1978, 1979a,b; Juliano *et al.*, 1979; Durham and Woolsey, 1977), gustatory (Orlandi *et al.*, 1977), olfactory (Stewart *et al.*, 1979), and selfstimulation (Gallistel *et al.*, 1977) system. More recently, the use of the 2DG method has been extended to the study of (1) the functional development of portions of a system such as the cerebral cortical columns (Des Rosiers *et al.*, 1978; Hand *et al.*, 1979b), (2) alterations of cortical activity following ablations of peripheral receptor organs in the neonatal and adult rat (Hand *et al.*, 1978; Kossut *et al.*, 1979), (3) the effects of selective lesions on central nervous system functions (Wooten and Collins, 1979; Ferron *et al.*, 1979), (4) effects on neural functions of experimentally produced ischemia (Ginsberg *et al.*, 1977) and seizure activity (Collins, 1978), and (5) cerebral metabolic effects of various anesthetics (Sokoloff *et al.*, 1977; Shapiro *et al.*, 1978) and drugs (Weinberger *et al.*, 1979; Brown and Wolfson, 1979).

Although the 2DG method has been used largely to study metabolism (function) within the central nervous system of vertebrates, recent reports in the literature suggest that it may now have a much wider application. For example, Rapoport *et al.* (1978) and Toop *et al.* (1979) have used the 2DG method to identify skeletal muscle fibers activated by electrical stimulation to either the femoral nerve or a single motor unit. Successful extension of the 2DG technique to invertebrates was demonstrated by Buchner *et al.* (1979) who mapped movement-specific functional activity in one of the visual ganglia of Drosophila.

Although the 2DG method was originally developed for the purpose of investigating glucose metabolism *in vivo,* two recent investigations have shown that the 2DG technique can be used *in vitro* and that such studies are important to our understanding of what neuropil structures may be labeled (Basinger *et al.*, 1979; Mata *et al.*, 1980).

IV. METHODOLOGY FOR [^{14}C]-2DG

The basic method to be described here has been developed for the rat. When using other species, especially larger animals such as the cat or the monkey, one will have to make adjustments, particularly concerning methods of restraint.

A. Injection of Deoxyglucose and Methods for Determining Metabolic Rates of Glucose

For quantitative studies, halothane anesthesia is used. The animal's femoral arteries are catheterized—one for sampling arterial plasma levels of deoxyglucose and glucose, and the other for monitoring blood pressure. Additionally, one (or both) femoral vein(s) is (are) catheterized for the purpose of injecting the deoxyglucose, saline flushes, anticoagulants, if such substances are used, and the overdose of barbiturate anesthesia to sacrifice the animal. The surgical incisions are infiltrated with a long-acting local anesthetic such as Marcaine® HCl,* 0.25%. The animal is lightly restrained by taping it to a padded block of wood and allowed to recover from the anesthesia for approximately 3 hr. This recovery period is necessary, since anesthetics have been shown to significantly depress glucose metabolism and, hence, labeling, particularly in cortex and thalamus (Sokoloff *et al.,* 1977; Shapiro *et al.*, 1978). Body temperature is maintained at 38°C with a servo-controlled heat lamp. The stimulus is begun about 2 min prior to the injection of [^{14}C]-2DG in order to stabilize the blood flow and metabolism of the system under investigation, and it is continued for the full 45-min postinjection period. For a 300 g rat, 30 μCi of 2-deoxy-D-[^{14}C]glucose (specific activity, 50–56 mCi/mmole)* in 0.5 ml saline is injected. The isotope is supplied dissolved in an ethanol solution and must be dehydrated by evaporation and reconstituted with normal saline, as is done with other autoradiographic methods (Chapter 5). Arterial blood samples are taken at prescribed intervals for analysis of plasma levels of 2DG and glucose. In order to capture the rapidly occuring peak of the arterial curve as well as the more slowly changing values of the latter part of the arterial curve, the samples are taken at the following times: 15, 30, and 45 sec, and 1, 2, 3, 5, 7, 10, 15, 20, 25, 30, 40, and 45 min. Larger samples are taken at 15 sec and 10, 20, 30, and 45 min so that glucose analyses can be done on a portion of the plasma obtained. These values are required for the operational equation for quantitation of local cerebral metabolic rates of glucose. Sampling vials are heparinized prior to the experiment. The vials must be air-dried or warmed in an oven so that the heparin, or other anticoagulant, dries to a powder. In this way the volume of the blood sample is not altered. The

*Bupivacaine HCl, Breon Laboratories, Inc., New York, New York.
†New England Nuclear Corporation, Boston, Massachusetts.

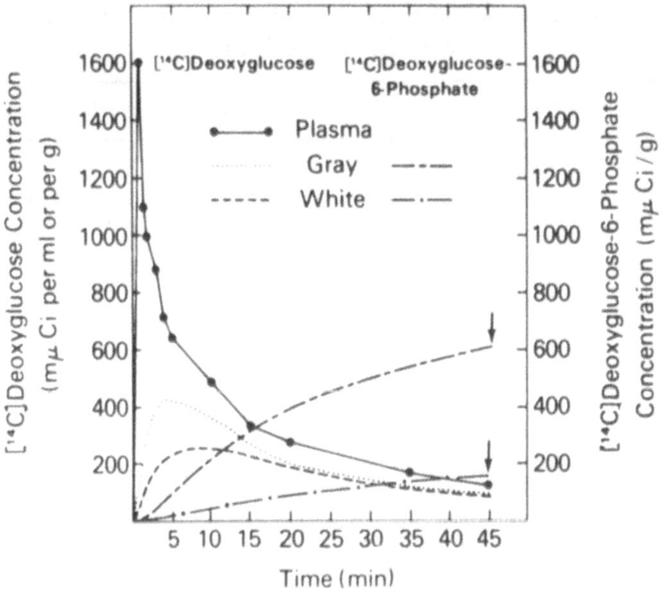

Figure 5. Graphical representation, taken from Figure 5A of Sokoloff *et al.* (1977), of the significant variables in the equation used to calculate local cerebral glucose utilization. Shown are the time courses of [^{14}C]-deoxyglucose concentrations in arterial plasma and in average gray and white matter and [^{14}C]-deoxyglucose-6-phosphate concentrations in average gray and white matter following an intravenous pulse injection of 50 μCi of [^{14}C]-deoxyglucose concentration. For the method of determination of the tissue concentrations of [^{14}C]-deoxyglucose and [^{14}C]-deoxyglucose-6-phosphate, see Sokoloff *et al.* (1977). The arrows point to the concentrations of [^{14}C]-deoxyglucose and [^{14}C]-deoxyglucose-6-phosphate in the plasma and tissues at the time of sacrifice; the autoradiographic technique measures the total ^{14}C content (i.e., the sum of these concentrations) at that time. Note that at the time of death, the total ^{14}C content represents mainly [^{14}C]-deoxyglucose-6-phosphate concentration, especially in gray matter. (Reproduced by permission of the authors and Pergamon Press.)

blood samples are centrifuged within 1 min of sampling to prevent coagulation. Measured amounts (10–30 μl) of plasma are placed in liquid scintillation vials for counting, and these are also used for glucose analyses on the Beckman glucose analyzer.* Both procedures are necessary for the proper quantification of local cerebral glucose utilization rates. The animal is sacrificed with an overdose of pentobarbital sodium at 45 min post-pulse-injection. At 45 min, the brain tissue concentration of [^{14}C]-deoxyglucose-6-phosphate trapped within the cells is at its highest value, while the "free" [^{14}C]-deoxyglucose concentration in brain tissue and arterial plasma is at its lowest value (Fig. 5). Thus, the subsequent autoradiographic labeling which represents the total ^{14}C content ([^{14}C]-2DG-6-PO_4 and [^{14}C]-2DG) is largely that of [^{14}C]-2DG-6-PO_4 trapped within the cells. The quantitative method has been discussed in detail by Sokoloff *et al.* (1977).

*Beckman Instruments, Fullerton, California.

Table I. Rates of Local Cerebral Glucose Utilization (LCMRG) in Lamina IV of Different Regions of the Neocortex of Two Rats (RP2 and RP3) as a Function of Perfusion[a]

	Visual	Auditory	Barrel	Parietal	Frontal
RP3					
Nonperfused	68.8 ± 15 (78)	115.5 ± 21.1 (58)	109.4 ± 20 (15)	66.6 ± 7.2 (82)	61.6 ± 7.2 (79)
Perfused	72.2 ± 8.8 (73)	97.7 ± 11.6 (48)	122 ± 13.8 (16)	65.5 ± 9.4 (84)	54.4 ± 7.2 (79)
RP2					
Nonperfused	82.7 ± 15 (68)	149.4 ± 20 (64)	181.1 ± 10.5 (14)	88.8 ± 5 (81)	85.5 ± 6.6 (72)
Perfused	101.6 ± 9.4 (83)	157.7 ± 10 (67)	186.1 ± 11.6 (17)	93.8 ± 6.1 (80)	92.7 ± 11.6 (83)

[a]In each animal a single facial vibrissa (No. 3 of row C; Woolsey and Van der Loos, 1970) was bilaterally stroked manually for 45 min following a pulse injection of [^{14}C]-2DG. The rats were then sacrificed by an overdose of pentobarbital sodium. The calvarium was rapidly removed and one nonperfused (NP) hemicortex removed by blunt dissection. The remaining portions of the CNS, including the contralateral hemicortex, were quickly perfused with 3.3% formalin, PO_4-buffered, pH 7.4 using the method of Collins (1978). The perfused hemicortex (P) was then removed by blunt dissection and, together with the unfixed cortex, gently flattened in the tangential plane, frozen in freon, serially sectioned in that plane, and exposed to X-ray film. The autoradiograms of the labeled cortical regions, including the activated C3 barrel (barrel), were carefully read using a mânually operated optical densitometer, and local glucose rates determined. The number of readings/area are shown in parentheses. Data are mean ± S.D. Note that the glucose rates in the perfused and nonperfused cortices are essentially the same.

B. Fixation and Sectioning

1. Fixation

The animal is perfused intracardially with 3.3% formalin, pH 7.4 (Collins, 1978; see Section VIII.A for the formula). Such treatment does not alter the 2DG labeling qualitatively or quantitatively (Table I), and it improves brain removal, cryostat sectioning, and staining of the sections. Furthermore, there seems to be an improvement in the signal-to-noise ratio of the 2DG labeling as a result of the perfusion fixation with the abovementioned formalin. The basis for such an improvement is unknown. Other fixatives, such as the standard 4% paraformaldehyde–5% glutaraldehyde mixture, may also be used, but only in qualitative studies as they may alter 2DG labeling. For example, use of the standard paraformaldehyde–glutaraldehyde fixative produces a 14.5% reduction in 2DG labeling of the single barrel column (unpublished observations). One should also be aware that these fixatives may alter the sectioning characteristics of the tissue, thereby producing more artifacts.

Following the perfusion, the brain is removed carefully. Experience has shown that it is not necessary to remove the brain quickly. Both we and others (e.g., Sokoloff, unpublished observations) have removed labeled brains from the skull and allowed them to air dry for 1 hr before freezing without any detectable movement of the isotope with subsequent loss of resolution.

2. Sectioning

Prior to freezing the brain, it is a good idea to photograph it for orientation purposes, particularly if one studies the more complex brain of a carnivore or a primate. Such a procedure facilitates reconstruction of labeling on the lateral surface of the brain as well as in complex sulcal areas. Frequently the brain is blocked into smaller segments, not only to improve cryostat sectioning but also to orient blocks of tissue for planes of section that best reveal the pattern of 2DG labeling in the region of the brain to be studied. In addition, photographs of the blocking procedure aid in the later identification and orientation of the brain blocks in the cryostat prior to sectioning. Should three-dimensional reconstructions of autoradiograms be contemplated, reference marks can readily be placed in the unfrozen tissue at this time.

The brain blocks are then placed on standard histological slides of appropriate size and lowered—using a gauze sling or similar device—into a Dewar flask containing Freon XXII at approximately −45° to −50°C. Other brain-freezing procedures may be used. Sokoloff *et al.* (1977) report the use of Freon XII chilled to −75°C in liquid nitrogen, whereas Stewart *et al.* (1979) use isopentane cooled to dry ice temperature. Kennedy *et al.* (1978), however, suggest that freezing at temperatures lower than −50°C results in cracking of the brain, a problem particularly common with larger pieces of brain tissue. We have experimented with both warmer (−40°C) and colder (approximately −75°C) temperatures and have found that cracking will occur at either of these extremes. The cracking during the freezing procedure appears to be related to differences between the expansion rates of the cerebrospinal fluid and the brain tissue and possibly also between those of white and gray matter.

The placement of the relatively soft brain block on a histological slide appears to be an improvement over the earlier method of placing the brain directly into a gauze sling and lowering it into the freon. The earlier procedure resulted in the gauze sticking to the surface of the brain and leaving behind a marring imprint. In addition, distortion of the brain frequently resulted as it conformed to the shape of the gauze sling. An alternative method is to mount the brain block onto a cryostat specimen holder using appropriate embedding matrix and to lower both into the freezing agent, using various devices. This is particularly useful when dealing with small pieces of tissue, such as a ganglion, where the normal freezing procedure, which is followed by a brief partial thawing during the mounting of the tissue onto the frozen cryostat chuck, produces movement of the isotope and a blurring of the radiographic image. With standard size tissue blocks, this is not a significant problem.

The mounting of unfrozen tissue blocks on histological slides can be modified to include the placement of a second slide on top of the block for the purpose of gently flattening a piece of tissue for sectioning in a particular plane. Labeling in the barrel columns of laminae I–VI in S-I of the rat, for instance, is best revealed by tangential sectioning of the flattened cortex (Hand *et al.*, 1977).

As one lowers the tissue and slide into the freezing agent, long forceps may be helpful to push them under the surface so that they are completely submerged. When the bubbling of the freon around the immersed brain ceases, the brain is removed and quickly placed into a labeled plastic bag that has previously been cooled in dry ice in order to prevent warming of the brain tissue. The bag is quickly sealed and placed into dry ice (−70°C) or an electric deep freeze (−70° to −80°C) for storage. The tissue may be stored at this temperature until cryostat sectioning. The length of time that one may store tissue in the deep freeze before freezer burn and other forms of deterioration occur is unknown. We have not observed alterations in the 2DG labeling or in the histology of sections of brains that have been stored in the freezer for up to 2 years. Increases in ambient temperature have a more profound effect on the quality of the 2DG labeling. For example, brain tissue stored in a cryostat at the usual cryostat temperature of −14° to −22°C for over a day loses some localization of its labeling and visual cortical columns begin to appear fuzzy. After 3 days of cryostat storage, visual cortical columns are no longer visible (Sokoloff, unpublished observations).

On the day of sectioning, the tissue block is removed from the dry ice or deep freeze storage and placed in the cryostat and allowed to equilibrate to cryostat temperature over a period of approximately 45 min. This is necessary to prevent knife "chattering" marks on the tissue. The temperature of the cryostat is adjusted to between −14° and −18°C. Unfixed tissue appears to section best at the higher temperature (−14°C). For a discussion of general cryostat technique and associated problems, the reader is referred to Chapter 1 and to a standard textbook on histological techniques (e.g., Bancroft and Stevens, 1977). While the tissue is equilibrating, the specimen holder is cooled rapidly by being placed face down on dry ice. After approximately 2 or 3 min this supercooled holder is placed in the cryostat, and room temperature embedding matrix* is placed on the specimen holder. As the matrix becomes partially frozen ("slushy"), the tissue block is carefully oriented on the holder using long forceps that have been cooled to dry ice temperature to prevent any softening and subsequent damage to the tissue. In this manner, tissue thawing and resultant isotope diffusion around the edges and surface of the tissue contacting the matrix are minimized. Additional small amounts of matrix may need to be dribbled around the base of the tissue to aid in sealing it to the frozen matrix. Such additions should be minimal to prevent isotope spread and loss of clarity of labeling. The use of cold matrix applied to the supercooled specimen holder has not proved sat-

*M-1 Embedding Matrix, Lipshaw Manufacturing Co., Detroit, Michigan, or an equivalent.

isfactory because it freezes so quickly that the matrix does not reach the depths of the grooves in the holder, and thus does not anchor the brain block firmly; this can result in the brain block and matrix separating from the holder during sectioning. The final orientation of the tissue to the knife edge can be achieved by adjusting the fine cryostat controls, if such are available. Routinely, sections are cut to 20 μm. We have found that thicker sections (30 μm) do not section as nicely as those at 20 μm.

3. Sectioning Artifacts

Cryostat sectioning has a tendency to produce artifacts that may obscure the labeling. However, many of the artifacts can be eliminated, or at least minimized, through experience. Some of the most important factors contributing to the production of good sections are use of (1) a sharp microtome knife, (2) a knife edge kept free of tissue fragments and ice buildup, (3) proper angulation of the knife blade in relation to the face of the tissue block, (4) proper alignment of the guide plate roll bar to the knife edge, (5) proper sectioning speed, (6) correct cryostat temperature for a particular tissue block, (7) careful flattening of the histological sections on the knife prior to pick up with two small brushes that have been cooled to cryostat temperature, and (8) proper placement of the coverslip on the section and subsequent removal of the coverslip with attached section from the knife blade.

The accurate placement and removal of the coverslip can be accomplished by the aid of a suction device*; this tends to smooth out potential tissue folds and generally improves the quality of the mounted sections.

Despite all these precautions, two common artifacts—small tissue folds and air bubbles (Fig. 6)—remain and may be accentuated under certain conditions. For example, these artifacts are more pronounced in animals with larger brains, presumably as the result of the larger surface area on the tissue block. The increased surface area provides increased opportunities for sectioning distortions to occur. The blocking of the larger brains into small segments though creating more tissue blocks to section greatly reduces these sectioning artifacts. Accentuation of the fold and air-bubble artifacts seems also to depend on three other factors: the increased heterogeneity of the tissue block, the shape of the block, and the presence of sulci. The greatest number of artifacts occurs in irregularly shaped blocks possessing many sulci and in blocks containing large amounts of white matter.

C. Preparation of Autoradiograms

Following sectioning and the careful flattening of the tissue section on the microtome knife, a No. 2 coverslip at room temperature and of suitable

*Such as that from SLEE International, New York.

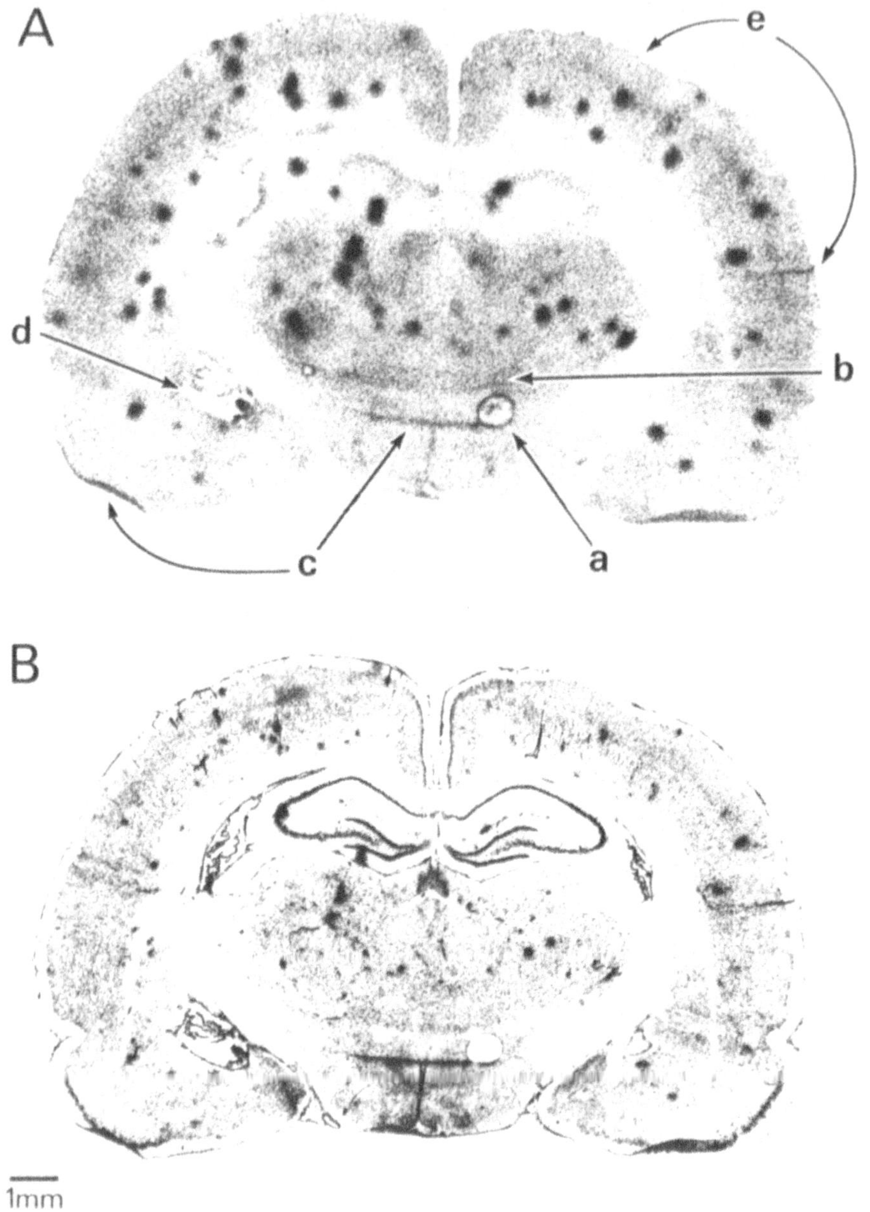

Figure 6. (A) Autoradiogram of a coronal section of the rat brain included to demonstrate common tissue artifacts (arrows): (a) air bubble; (b) tissue wrinkles; (c) tissue fold; (d) tissue tear; and (e) isotope spread at the edge of the tissue produced by room temperature embedding matrix contacting and thawing the surface of the brain. Note the blurriness of the autoradiographic image. This was the result of a poor contact between the film emulsion and the tissue surface. In addition, the large black dots were produced by a protozoan infection known as Nosematosis (Basson *et al.*, 1966). (B) A thionin-stained section that was used to produce the autoradiogram in (A).

size is gently dropped onto the tissue. The manner in which the coverslip is dropped is an individual matter. In the past, the method of choice was to place one edge of the coverslip on the microtome knife and release the opposite edge to allow the coverslip to fall onto the tissue in an oblique manner. This appears to alleviate some of the tissue-fold and air-bubble problems. However, the use of a suction pick-up device that places uniform pressure over the entire surface of the section is now the preferred method. It is the most effective method we have yet discovered to minimize folds and bubbles.

Using a No. 2 rather than a No. 1 coverslip also assists in reducing artifacts, because its greater thickness causes it to freeze at a slower rate on the knife, thus permitting some of the bubbles and folds to resolve themselves. As soon as the coverslip and the attached histological section have refrozen on the microtome knife—a matter of a few seconds—it is removed and placed, tissue side up, on a standard slide-warming tray at 65°C. This procedure rapidly dehydrates and "bakes" the section onto the coverslip. Such sections are routinely left on the tray for approximately 2–15 min. The longer times do not appear to affect the quality of the autoradiograms or the cytoarchitecture as seen in the thionin-stained sections.

The coverslip with the dehydrated sections is then glued* onto 10 in. × 12 in. four-ply posterboard in serial order. Important information is written on the posterboard itself: experiment number, identification of the particular brain block, orientation of the block, section thickness, frequency of sections (e.g., "every fifth," "serial"), and individual section numbers. Later, this information is transferred to the exposed and developed autoradiograms. In the center of the posterboard, a small rectangular section is cut out for the inclusion of a set of [^{14}C]methyl methacrylate standards† which are attached to a strip of paper. Because the standards are thicker than the No. 2 coverslips, an effort is made to establish a condition in which the histological section and the standards make the same close contacts with the X-ray film emulsion. This is accomplished by mounting the standards on paper that is thinner than the posterboard. Thus, the thinner paper compensates for the thicker standards. If this correction is not made, the standards will raise the film surrounding them and prevent close contact of the X-ray film emulsion with the adjacent labeled section, producing blurring of the autoradiogram (Fig. 6). The inclusion of these known standards permits the quantification of the amount of ^{14}C in structures of interest, or, in conjunction with known arterial plasma values of 2DG and glucose, the actual metabolic rate of glucose in a given brain region.

The posterboard with the attached sections and methacrylate standards is placed in a nonintensifying-screened standard 10 in. × 12 in. X-ray cas-

*Duco Cement, Du Pont Co., Wilmington, Delaware.

†New England Nuclear Corporation, Boston, Massachusetts; Amersham/Searle Corporation, Arlington Heights, Illinois.

sette. In the darkroom, and under a safelight,* a sheet of Kodak SB-5 special blue-sensitive medical X-ray film† is placed, with its emulsion side (gray, nonshiny) down, in direct contact with the dehydrated 2DG-labeled brain sections and the methacrylate standards. The cassette is then closed and stored in a cool, dry, and dark place. Storage in excessively warm and damp places is to be avoided, as these conditions produce increased levels of background labeling, fading of the latent image, and subsequent reduction in clarity of the autoradiographic image (Rogers, 1973). The storing of the cassettes in a dark environment reduces the problem of light leaks, which occur frequently when older cassettes are used. The X-ray film is then exposed to the sections for the period of from 1 to 6 weeks, depending on the amount of isotope and the species used. For example, in a 300-g rat injected with 50 μCi of [^{14}C]-2DG, the exposure time is typically 10 days. In larger animals, such as the monkey, which has a metabolic rate one-third to one-half that of the rat (Kennedy *et al.*, 1978) and where, for reasons of economy, relatively less isotope per body weight is injected, the exposures are longer, usually from 4 to 6 weeks. Because of the variability in the larger animals, we normally run a series of test exposures to determine the most ideal exposure time. This can be done by cutting the 10 in. × 12 in. posterboard and X-ray film into four strips of approximately equal size. Several noncritical sections, attached to coverslips, are glued to each of the four posterboard strips. In the darkroom, the X-ray film strips are taped, with emulsion-side down, to the sections on the posterboard strips. Taping prevents shifting of the film–posterboard pairs during subsequent handling of the individual pairs. The test cassette is then stored with the others that were loaded on the same day with the same brain tissue type. At selected time intervals, e.g., 2, 3, 4, and 5 weeks, a test strip of film is removed and developed to determine the ideal exposure time for the remaining films in that series. Should the films be exposed for an insufficient length of time, be developed improperly, or be of poor quality as the result of fogging or light leaks in the cassettes, the sections can be reexposed using new film.

Following the appropriate exposure time, the X-ray films are developed according to the instructions supplied with the film. Manual tanks stocked with standard X-ray developer and fixer solutions and, ideally, maintained at 68°F are used. The commercial rapid processing machines are not suitable for use with the SB-5 film. For general principles of X-ray film development, the reader is referred to a standard radiography manual such as *Fundamentals of Radiography*.‡

Other films can be utilized for improved resolution, but these require longer exposure times. For example, Lo Doz film§ has an exposure time

*Such as Thomas Duplex-super safe light, Thomas Instrument Company, Charlottesville, Virginia.

†Eastman Kodak Company, Rochester, New York.

‡12th edition, Eastman Kodak Co., Rochester, New York, 1980.

§Du Pont Co., Wilmington, Delaware.

three times as long as that of the SB-5 film, and Panatomic X* has an exposure time approximately 18 times that of the SB-5. Mammography film, such as Kodak MR-1, is one of the finest-grained films; it has an exposure time approximately three times that of the SB-5.

Following film development, information about the experimental procedure, e.g., brain orientation, experiment identification, etc., is transferred from the posterboard to the nonemulsion side of the film using a standard wax pencil or acetate pen. Because the developed emulsion can be damaged by repeated handling of the original autoradiograms, it is advisable to have the original films copied, using a standard X-ray film-duplicating process. The duplicate can then be used as a working copy, thus saving the original for critical analyses.

The developed X-ray films may be viewed on a standard light box or X-ray viewer. The use of a magnifying glass facilitates the localization of labeling in small structures. The analysis of the 2DG labeling can then proceed along qualitative or quantitative lines as discussed in Section VI.

The histological sections that produced the autoradiographic images can be stained with a cell body stain such as thionin (Section VIII.B). The importance of this step to the accurate localization of 2DG labeling in specific nuclear groups or in specific cytoarchitectonic areas cannot be overemphasized. For example, the use of a thionin stain on 2DG sections has permitted us to localize labeling to individual laminae of a single somatosensory cortical column of the rodent (Hand *et al.*, 1978, 1979b) and to individual cytoarchitectural areas of the first somatosensory cortex of the monkey (Hand *et al.*, 1979a; Juliano *et al.*, 1979).

V. METHODOLOGY FOR [^{3}H]-2DG

Tritiated 2DG, although commercially available, has been less commonly used than [^{14}C]-2DG. Tritium is a weaker beta emitter than carbon-14—a spot source of ^{3}H travels 2 μm at best, whereas 90% of the silver grains exposed by a spot source of ^{14}C are found within 25 μm of this source (Rogers, 1973). Therefore, the resolution of labeling using ^{3}H should be greatly improved, permitting analysis at the cellular level. One of the major limitations of the [^{3}H]-2DG method as it is presently used is that it usually requires the use of a liquid nuclear emulsion that, when it contacts the water-soluble 2DG-6-PO_4, causes it to move from its area of localization. This movement of the isotope can be prevented by rapid dehydration of the emulsion by warming it or blowing air over it. This procedure, however, creates increased levels of background labeling caused by stresses on the emulsion (Chapter 5).

Although a number of [^{3}H]-2DG modifications exist, they all involve the

*Eastman Kodak Co., Rochester, New York.

precoating of histological slides with a nuclear emulsion such as NTB2 or NTB3.* Cryostat sectioning and mounting of the frozen sections onto either room temperature or frozen emulsion-coated slides is done in the dark room.

When the sections are allowed to thaw on room temperature emulsion-coated slides and are then dried slowly, the [^{3}H]-2DG moves from its locus. The movement, however, appears to be towards the emulsion itself and not in the tangential plane (Darcy Kelley and Oswald Stewart, unpublished observations). Thus, in 5-μm thin sections, the developed silver grains of the NTB emulsion essentially underlie the metabolically active neurons.

If frozen emulsion-coated slides are used, and frozen sections are mounted on them, the problem of isotope movement is avoided, but the sections may not contact the emulsion uniformly, producing tissue folds and wrinkles or poor tissue adherence. Improper contact of the tissue with the emulsion produces blurring of the image and loss of resolution (Rogers, 1973). The procedures of storage, exposure, and development of the [^{3}H]-2DG slides are similar to those used for standard [^{3}H]amino acid–axoplasmic flow preparations (Chapter 5).

The possibility that some lateral movement of the [^{3}H]-2DG occurs during the thawing and slow drying process of the tissue suggests that caution should be exercized before a definitive statement is made concerning the localization of the labeling. The use of other dehydration, fixation, or embedding methodologies—such as freeze-substitution coupled with epon embedding, or periodate–lysine–paraformaldehyde fixation which binds carbohydrate moieties to the protein in cell membranes (McLean and Nakane, 1974)—may be required to determine the localization of 2DG labeling. Another alternative would be to perfuse, fix, and epon-embed the brain *in situ* for either light or electron microscopy (Des Rosiers and Descarries, 1978). The more frequent use of the ^{3}H modification of the 2DG method must await its further development.

VI. DATA ANALYSIS

A. Qualitative

The use of the qualitative 2DG method may suffice for the determination of the general pattern of labeling in some functional systems. For example, this type of study is best done in systems that possess dense and localized connections such as those of the rodent somatosensory cortical "barrels" (Durham and Woolsey, 1977; Hand *et al.*, 1977) or the first somatosensory and visual cortices of the monkey (Hand *et al.*, 1979a; Juliano *et al.*, 1979; Hubel and Wiesel, 1977). The presence of dense and localized

*Eastman Kodak Co., Rochester, New York.

connections facilitates visualization of increased 2DG labeling produced by their activation.

In addition, the qualitative method may be the only one feasible in certain types of experiments, such as those requiring a freely moving animal and those using very small animals. In behavioral studies requiring the use of a freely moving animal, arterial catheterization for quantitative studies may not be practical. Chronic arterial catheters frequently become clotted with blood and, in addition, must be of sufficient length to permit free movement of the animal. The extended length of the arterial catheter with its increased "dead space" presents problems of accurate glucose and deoxyglucose determinations during the rapid arterial sampling procedure required for the determination of the local cerebral metabolic rates of glucose. In addition, the arterial sampling procedure itself may disrupt the behavior of the animal.

Another use of the qualitative method is in studies of very small animals such as the mouse or baby rat (Durham and Woolsey, 1977; Hand *et al.*, 1979b). In these cases, reliable arterial catheterization would be difficult at best, and the amount of arterial blood to be taken for quantitative purposes produces a significant lowering of blood pressure and might result in the death of the animal.

Once it has been determined that the qualitative method is applicable to the study in question, data analysis begins with an examination of the X-ray film containing the autoradiographic images. A standard X-ray viewer and magnifying glass are used during this procedure. Areas that appear darker on the autoradiograms are those that are more densely labeled; that is to say, they have higher metabolic rates. In order to establish that a pattern of increased labeling in a single autoradiogram is real and not artifactual in nature, it is important to examine a number of adjacent autoradiographic sections. The autoradiograms can then be photographed using black-and-white film, and 8 × or 10 × magnification prints can be made for the purpose of publication. Reproducing high quality autoradiographic images for publication is one of the major problems of the [^{14}C]-2DG method. This limitation, however, has largely been resolved by the use of image-enhancement techniques (Goochee *et al.*, 1980); these will be discussed in Section VI.B.2.

B. Quantitative

Quantification of local glucose utilization is an important procedure if one wishes to do more than map dense [^{14}C]-2DG labeling of a functional system that has been activated by a given stimulus. For example, quantification of the local cerebral metabolic rates of glucose (LCMRG) permits direct comparisons of the amount of metabolism (function) occurring simultaneously within a number of structures, such as individual laminae of a cortical column or several columns (Hand *et al.*, 1978; Kossut *et al.*, 1979; Fig.

4) in a single animal or within different animals. Such quantitative comparisons permit careful analyses of the responsiveness of different structures to different experimental situations, e.g., different stimulus parameters (Yarowski *et al.*, 1979; Miyaoka *et al.*, 1979).

Quantification is also of great importance when attempting to determine if the observed 2DG labeling is significantly elevated above normal background levels and reflects a true metabolic activation of that structure or functional system, for example, the tangential labeling emanating from a single barrel column (Fig. 4). Such minimal metabolic changes might be expected to occur in functional systems that possess diffuse, sparse, and bilateral connectivity (Mehler *et al.*, 1960; Edwards, 1975).

In order to determine the plasma concentrations of [^{14}C]-2DG and glucose, which are necessary to quantify local glucose utilization of the neural tissue, blood samples are withdrawn from the femoral arterial catheter at prescribed intervals (see Section IV.A). The blood samples are centrifuged immediately, and 30 μm plasma is pipetted off, added to deionized water in scintillation vials, and counted in Bray's liquid scintillation cocktail solution* or equivalent, in a Tri-Carb® liquid scintillation counter.† Ten microliters of plasma are pipetted from designated samples to be analyzed for glucose concentrations using a Beckman glucose analyzer. These values, in conjuction with the optical densitometric readings of the autoradiograms, are then used in the formula for the quantification of LCMRG (Sokoloff *et al.*, 1977).

1. Manual Optical Densitometry

The densitometric readings of structures on the autoradiograms are accomplished using either a manually operated optical densitometer‡ with a 100 μm spot size and interfaced with a PDP11-34 computer§ or an automated scanning image-enhancement–color display system (Goochee *et al.*, 1980).

When the manual system is used, areas of the X-ray film between the autoradiographic images (background regions) as well as the [^{14}C]methyl methacrylate standards are each "read" 40 times by the densitometer. To minimize the contributions of such artifacts as variations in section thickness, tissue folding, tissue tears, and air bubbles, as many of the labeled brain sections as possible are read.

The manually operated system, however, is quite tedious and time-consuming to use, particularly when reading numerous brain structures on numerous autoradiographic sections. For example, we routinely read each of seven cortical structures of interest approximately ten times on approximately 100 serial tangential sections, a total of 7000 readings. Such readings

*New England Nuclear Corporation, Boston, Massachusetts.
†Hewlett–Packard Co., King of Prussia, Pennsylvania.
‡Gamma Scientific Inc., San Diego, California.
§Digital Equipment Co., Maynard, Massachusetts.

produce large volumes of numerical data that become difficult to manipulate and to present in publication form.

In addition, the attempt at manually positioning the 100-μm spot accurately over small regions, such as a single cortical column or a single cortical lamina within a column, can be extremely difficult and may result in inaccurate readings. Conversely, readings over large regions of the brain such as the reticular formation cannot be done in as consistent a manner as with an automated optical densitometer. Consequently, significant variations in optical density readings are frequently observed among individual sections in a given animal and in different animals.

2. Automated Image-Enhancement Technique

The use of an automated scanning system with its precise densitometric readings of the autoradiographic sections and color display capabilities eliminates inaccurate readings and significantly reduces the extensive numerical data to a relatively few optical density ranges. These optical densities are then converted to local cerebral metabolic rates of glucose (LCMRG), which are associated with the enhanced color images of the brain (Fig. 4).

Details concerning the use of the automated scanning system may be found in an article by Goochee *et al.* (1980). Briefly, the system is comprised of an Optronics® optical scanner* which is interfaced to a PDP11-34 computer†, a black-and-white monitor, a color monitor with color tables, a display refresh memory, and two joysticks enabling on-line interaction with the displayed images as well as the computer. The Optronics scanner is a rotary drum scanner with a light source external to the drum on which the X-ray film is mounted. The light source is directed radially into the center of the drum where a phototube measures the transmitted light intensity. As the cylinder spins, the light source–phototube assembly moves axially and measures the transmitted light along parallel bands (25 μm–200 μm, selectable) of the film. Although the aperture size of the scanner (spot size) is switch-selectable, a setting of 100 μm represents a good compromise with the limited resolution of the film and isotope.

The automated analysis of the data proceeds along the following lines. An entire X-ray film containing approximately 100 autoradiographic images of brain sections is scanned very coarsely (100-μm aperture reading every 1000 μm), thus permitting display of the entire X-ray film. The image of each brain section is poor but sufficient for identification of a given autoradiogram. Using the joysticks, a variable size rectangle is positioned over an autoradiographic section of interest, and the computer instructs the scanner to rescan the X-ray film over the area covered by the rectangle. This scan is

*Model P 1000, Optronics International Inc., Chelmsford, Massachusetts.

†The interface permits both transfer of data from the scanner to the computer as well as computer control of the scanner.

made with the same 100-μm spot, but takes data every 100, 50, or 25 μm (selectable) and produces an image of high resolution and good quality. The image is displayed on both the black-and-white and the color monitor. A simple transformation converts the optical density reading to one in which the shade of gray or the color is proportional to optical density and finally to one in which the gray scale or color is directly proportional to cerebral metabolic rate (Fig. 4). Placing a labeled gray scale or labeled color scale beside the image allows direct quantification of the metabolic rate from the displayed image. The color tables enable the color monitor to assign different colors to different metabolic rates. Additionally, under software control, a number of data manipulations can be applied to the image. The color assignments (i.e., which range of metabolic rate is assigned to which color) can be manipulated in order to accentuate certain structures of interest. Such a manipulation can be very helpful in highlighting trends but must be interpreted carefully with full appreciation of the exaggeration of effects that may result. The system also has the facility for enlarging any part of the image to visualize more clearly the brain structures of interest. This is particularly helpful in studying the uptake of [^{14}C]-2DG in very small regions such as individual cortical laminae (Fig. 4), thalamic or brainstem nuclei, and spinal gray. The displayed image is then photographed to save the data for later analysis.

3. Resolution

Using the automated scanner image-enhancement–color display system, in conjunction with the [^{14}C]-2DG method permits a resolution of approximately 100 μm (Goochee *et al.*, 1980). Recent work (unpublished observations) on a rostrally located S-I "hollow-core" barrel, which is related to a single facial sinus hair (Welker, 1976), can be used as an example. The S-I cortex is an ideal brain region in that it possesses dense and localized connectivity and has cell-dense and cell-sparse portions in direct apposition. [^{14}C]-2DG labeling approximately 70 μm in diameter was detected in the cell-dense wall of this barrel following brush stroking of the sinus hair. Thus, the resolution of the [^{14}C]-2DG technique permits analysis approaching that of the single cell. When [^{3}H]-2DG is used, differential labeling of individual neurons is possible (Basinger *et al.*, 1979).

VII. ADVANTAGES AND LIMITATIONS

A. Advantages

The 2DG method pictorially demonstrates global functional activity related either to passively applied stimuli (natural, electrical, or chemical) or to an active behavioral situation. In this manner, functional relationships of an entire system or group of systems can be revealed in an individual animal (e.g., Fig. 1). The method is not dependent on axonal degeneration or axoplasmic flow mechanisms for its effects, and it is not restricted by localized placements of micro- or macroelectrodes.

In addition, the 2DG technique can be utilized either to add new information about systems whose connections and general functions are known or to demonstrate, as an initial survey method, heretofore unknown functional relationships. For example, the rodent mystacial vibrissal–S-I cortical barrel system, the connections of which have been established (Woolsey and Van der Loos, 1970; Killackey and Leshin, 1975; Donaldson *et al.*, 1975; Welker, 1976), has been utilized as a model for a 2DG study of the effects of peripheral receptor organ damage on cortical functioning (Hand *et al.*, 1978; Kossut *et al.*, 1979). In the latter type of study, Gallistel *et al.* (1977), using [^{14}C]-2DG and electrical stimuli controlled by the animal itself, labeled functional systems related to rewarding brain stimulation.

Finally, the [^{14}C]-2DG method is a relatively rapid technique usually requiring about 10 days for a rat and 4–6 weeks for a monkey. In addition, the use of X-ray film autoradiography permits a rapid survey of 2DG labeling in an entire brain.

B. Limitations

One of the major problems is that the exact structure(s) taking up [^{14}C]-2DG is (are) presently unknown. However, the increased use of the [^{3}H]-2DG light and electron microscopic modifications should soon provide the information needed to resolve this issue, thereby making functional–anatomical correlations more exact.

Any stimulus, intentional or incidental, may evoke enough neural activity and metabolism to produce detectable increases in labeling. Thus, the mere positioning or restraining of the animal during an experiment or, perhaps, the attempt by a paralyzed, awake animal to initiate a response can produce incidental labeling. Such labeling can create a problem in the interpretation of the observed pattern. Careful and extensive controls must be used in order to rule out any incidental labeling.

Another problem arises when anesthetics are required to immobilize an animal for the precise control of stimulus parameters. Anesthetics suppress the metabolism and dramatically reduce the extent of 2DG labeling, particularly within the cerebral cortex (Sokoloff *et al.*, 1977; Shapiro *et al.*, 1978). This reduction in metabolism results in a loss of the general heterogeneity of 2DG labeling in the brain and, thus, a lack of definition in the labeling of individual structures.

An alternative approach is to use the paralyzed, unanesthetized preparation, in which case one must take precautions to infiltrate any and all surgical wounds with a long-acting local anesthetic. Because of this limitation, certain 2DG studies may not be able to be conducted as readily as those on the rat vibrissal–cortical barrel system (Hand *et al*, 1977, 1978, 1979b). For example, investigations of "pain" systems should not be conducted in unanesthetized animals unless a hypnotic drug such as nitrous oxide is used. However, nitrous oxide has been shown to selectively suppress brain metabolism by 12 to 18%, primarily affecting the cerebral cortex but not thalamic projection nuclei (Sokoloff, unpublished observations). Thus, some difficulty could be encountered, for example, in determining the site and pattern of cortical representation of noxious information.

Finally, because glucose metabolism is much lower in white matter than in gray matter, the 2DG technique is not a good method for tracing functional pathways within the white matter.

VIII. APPENDIX

A. Fixative

The fixative used is a solution of 3.3% formalin in 0.12 M PO_4 buffer, pH 7.4 (Collins, 1978). The phosphate buffer solution contains sodium phosphate, monobasic, 3.18 g, and potassium phosphate, dibasic, 16.8 g, dissolved in enough distilled water to make 1000 ml; the pH is then adjusted to pH 7.4.

Add 33 ml of a stock solution of formalin (37–40% formaldehyde by weight) to each 967 ml of the pH 7.4 phosphate buffer solution.

B. Thionin Stain for 2DG Sections*

Prepare a stock solution by combining 100 ml of 1 M acetic acid (29.25 g = 27.8 ml in 500 ml distilled water) with 38 ml of 1 M sodium hydroxide (25 g in 500 ml distilled water).

Dilute to 500 ml with distilled water. Adjust to pH 4.6 with 1 M sodium hydroxide. Heat this solution to steaming, add 5 g thionin, and stir occasionally for 1 hr. Filter. Stain at room temperature.

C. Staining Procedure for [^{14}C]-2DG Sections*

1. Using a scalpel or a small pair of scissors, cut coverslip containing the desired histological sections from the 10 in. × 12 in. posterboard. Do not attempt to remove the piece of posterboard that is glued to the back of the coverslip as this usually results in cracking of the fragile coverslip and the accompanying tissue section. Immersion of the coverslip in the xylene will dissolve the glue, and the posterboard

*Staining protocol is after LaVail and LaVail (1979). For staining of emulsion-coated slides, see Chapter 5.

backing can be removed following the soak in distilled water prior to the staining step.

2. The coverslips may be individually carried through the staining procedure, or they may be carried through as a group in a coverslip holder.
3. Coverslips are placed in the following sequence of solutions.

100% Alcohol	2 min
Xylene	3 min
100% Alcohol	1 min
100% Alcohol	1 min
95% Alcohol	2 min
70% Alcohol	2 min
Distilled water	2–5 min

Gently remove the posterboard backing. Soak in thionin for 5–30 sec, depending on species, age of animal, type of fixative used (or unfixed tissue), and region of the nervous system. Run a test section first. Carefully agitate the section(s) by dipping it (them) up and down. Then place into, in sequence, 70% alcohol, 95% alcohol, and 100% alcohol for as long or as short a time interval as necessary to remove excess thionin and for proper differentiation. Most differentiation will occur in 70%; least in 100%. Then place in 100% alcohol for 2 min and xylene for at least 2 min.

4. Mount coverslips (tissue-side down) on histological slides of appropriate size using mounting medium.

ACKNOWLEDGMENTS. I wish to thank Dr. Louis Sokoloff for his informative discussions concerning the use of the [^{14}C]-2DG method, and for his generosity in permitting me to use the automated scanning–pseudocolor display equipment housed at the Laboratory of Cerebral Metabolism, NIMH. In addition, I wish to thank Dr. Martin Reivich for the extensive use of his facilities, and Dr. Joel Greenberg, Mr. Chuck Goochee, and Mrs. Carol Hand for their technical assistance. This work was supported by NIH grants NS 06716, NS 14935-03, and BMRSG 07083-14 sb. 25.

REFERENCES

Bancroft, J. D., and Stevens, A. (eds.), 1977, *Theory and Practice of Histological Techniques*, Churchill Livingstone, New York.

Basinger, S. F., Gordon, W. C., and Lam, C. M. K., 1979, Differential labelling of retinal neurons by ^{3}H-2-deoxyglucose, *Nature* **280**:682–684.

Basson, P. A., McCully, R. M., and Warnes, W. E. J., 1966, Nosematosis: Report of a canine case in the Republic of South Africa, *J. S. Afr. Vet. Med. Assoc.* **37**:3–8.

Brown, L. L., and Wolfson, L. I., 1979, Autoradiographic deoxyglucose studies of apomorphine effects in halothane anesthetized rats, *Neurosci. Abtr.* **5**:67.

Buchner, E., Buchner, S., and Hengstenberg, R., 1979, 2-Deoxy-D-glucose maps movement-specific nervous activity in the second visual ganglion of Drosophila, *Science* **205**:687–688.

Collins, R. C. 1978, Use of cortical circuits during focal penicillin seizures: An autoradiographic study with (^{14}C) deoxyglucose, *Brain Res.* **150**:487–501.

Des Rosiers, M. H., and Descarries, L., 1978, Adaptation de la méthode au désoxyglucose à l'èchelle cellulaire: Préparation histologique du système nerveux central en vue de la radioautographie á haute résolution, *C. R. Acad. Sci. [D] (Paris)* **287**:53–56.

Des Rosiers, M. H., Sakurada, O., Jehle, J., Shinohara, M., Kennedy, C., and Sokoloff, L., 1978, Functional plasticity in the immature striate cortex of the monkey shown by the (^{14}C) deoxyglucose method, *Science* **200**:447–449.

Deuel, R. K., Collins, R. C., Dunlop, N., and Caston, T. V. 1979, Recovery from unilateral neglect: Behavioral and functional anatomic correlations in monkey, *Neurosci. Abstr.* **5**:624.

Donaldson, L., Hand, P. J., and Morrison, A. R., 1975, Cortical–thalamic relationships in the rat, *Exp. Neurol.* **47**:448–458.

Durham, D., and Woolsey, T. A., 1977, Barrels and columnar cortical organization: Evidence from 2-deoxyglucose (2DG) experiments, *Brain Res.* **137**:169–174.

Edwards, S. B., 1975, Autoradiographic studies of the projections of the midbrain reticular formation: Descending projections of nucleus cuneiformis, *J. Comp. Neurol.* **161**:341–358.

Ferron, A., Des Rosiers, M. H., de Montigny, C., Bosler, O., Reader, T. A., and Descarries, L., 1979, Effects of unilateral destruction of the nigro–striatal dopamine system on local cerebral glucose utilization in adult rat, *Neurosci. Abstr.* **5**:70.

Gallistel, C. R., Karreman, G. A., Reivich, M., 1977, (^{14}C)-2-deoxyglucose uptake marks systems activated by rewarding brain stimulation, *Brain Res. Bull.* **2**:149–152.

Ginsberg, M. D., Reivich, M., Giagiandomenico, A., and Greenberg, J. H, 1977, Local glucose utilization in acute focal cerebral ischemia: Local dysmetabolism and diaschisis, *Neurology* **27**:1042–1048.

Goochee, C., Rasband, W., Sokoloff, L., 1980, Computerized densitometry and color coding of (^{14}C) deoxyglucose autoradiographs *Ann. Neurol.* **7**:359–370.

Greengard, P., and Ritchie, J. M., 1971, Metabolism and function in nerve fibers in: Handbook of Neurochemistry Vol. V.A. (A. Lajtha, ed.), pp. 317–335, Plenum Press, New York.

Hand, P. J., Miselis, R. R., and Reivich, M., 1977, A (^{14}C)-2-deoxyglucose metabolic mapping study of the rat posteromedial barrel subfield, *Neurosci. Abstr.* **3**:483.

Hand, P. J., Greenberg, J. H., Miselis, R. R., Weller, W. L., and Reivich, M., 1978, A normal and altered cortical column: A quantitative and qualitative (^{14}C)-2-deoxyglucose (2DG) mapping study, *Neurosci. Abstr.* **4**:553.

Hand, P. J., Juliano, S., Whitsel, B., and Goochee, C., 1979a, A (^{14}C)-deoxyglucose study of the modular organization in somatosensory cortices of the cynomolgous monkey, *Anat. Rec.* **193**:560–561.

Hand, P. J., Kossut, M., Patel, U., and Goochee, C., 1979b, The development of the cortical column: A (^{14}C)-2-deoxyglucose study in the rat, *Neurosci. Abstr.* **5**:162.

Horton, R. W., Meldrun, B. S., and Bachelard, H. S., 1973, Enzymic and cerebral metabolic effects of 2-deoxy-D-glucose, *J. Neurochem.* **21**:507–520.

Hubel, D. H., and Wiesel, T. N., 1977, Functional architecture of Macaque monkey visual cortex, *Proc. R. Soc. Lond. B.* **198**:1–59.

Hubel, D. H., Wiesel, T. N., and Stryker, M. P., 1978, Anatomical demonstration of orientation columns in Macaque monkey, *J. Comp. Neurol.* **177**:361–380.

Jarvis, C. D., Mishkin, M., Shinohara, M., Sakurada, O., Miyaoka, M., and Kennedy, C., 1978, Mapping the primate visual system with the (^{14}C)-2-deoxyglucose technique, *Neurosci. Abstr.* **4**:632.

Juliano, S., Hand, P. J., Whitsel, B., Goochee, C., Karp, P., and Bajcsy, R., 1979, A (^{14}C)-deoxyglucose study of somatosensory and associated cortical areas in the monkey, *Neurosci. Abstr.* **5**:117.

Kennedy, C., Sakurada, O., Shinohara, M. D., Jehle, J., and Sokoloff, L., 1978, Local cerebral glucose utilization in the normal conscious Macaque monkey, *Ann. Neurol.* **4**:293–301.

Killackey, H. P., and Leshin, S., 1975, The organization of specific thalamo–cortical projections to the posteromedial barrel subfield of the rat somatic–sensory cortex, *Brain Res.* **86**:469–472.

Kossut, M., Hand, P. J., Greenberg, J., Sylvestro, A., Goochee, C., and Reivich, M., 1979, Plasticity of a somatosensory cortical column: A comparison of the effects of neonatal and adult receptor ablations in the rat using the (^{14}C)-2-deoxyglucose technique, *Neurosci. Abstr.* **5**:630.

La Vail, J., and La Vail, M., 1979, *Basic Neuroanatomical Methods,* Cold Spring Harbor Workshop on Neurobiology.

Mata, M., Fink, D. J., Gainer, H., Smith, C. B., Davidsen, L., Savaki, H., Schwartz, W. J., and Sokoloff, L., 1980, Activity-dependent energy metabolism in rat posterior pituitary primarily reflects sodium pump activity, *J. Neurochem.* **34**:213–215.

McLean, I. W., and Nakane, P. K., 1974, Periodate–lysine–paraformaldehyde fixative: A new fixative for immunoelectron microscopy, *J. Histochem. Cytochem.* **22**:1077–1083.

Mehler, W. R., Feferman, M. E., and Nauta, W. J. H., 1960, Ascending axonal degeneration following anterolateral cordotomy. An experimental study in the monkey, *Brain* **83**:718–752.

Miyaoka, M., Shinohara, M., Batipps, M., Pettigrew, K. D., Kennedy, C., and Sokoloff, L., 1979, The relationship between the intensity of the stimulus and the metabolic response in the visual system of the rat, *Neurosci. Abstr.* **5**:411.

Orlandi, M. A., Erickson, R. P., Johnson, A. E., and Skeen, L. C., 1977, Deoxyglucose and electrophysiological mapping of chorda tympani activity in rat nucleus solitarius, *Neurosci. Abstr.* **3**:81.

Rapoport, S. I., Ohno, K., and Schwartz, W. J., 1978, Activity-related regional uptake of (^{14}C) deoxyglucose by rat quadriceps femoris muscle, *Exp. Neurol.* **60**:168–174.

Reivich, M., Kuhl, D., Wolf, A., Greenberg, J., Phelps, M., Ido, T., Casella, V., Fowler, J., Hoffman, E., Alavi, A., and Sokoloff, L., 1979a, The ^{18}F-fluorodeoxyglucose method for the measurement of local cerebral glucose utilization in man, *Circ. Res.* **44**:127–137.

Reivich, M., Greenberg, J., Alavi, A., Christman, D., Fowler, J., Hand, P. J., Rosenquist, A., Rintelmann, W., and Wolf, A., 1979b, The use of the ^{18}F-fluorodeoxyglucose technique for mapping of functional neural pathways in man, *Acta Neurol. Scand.* [*Suppl.*] **60**:198–199.

Rogers, A. W., 1973, *Techniques of Autoradiography,* 2nd Edition, Elsevier, Amsterdam.

Schwartz, W. J., and Gainer, H., 1977, Suprachiasmatic nucleus: Use of the ^{14}C-labeled deoxyglucose uptake as a functional marker, *Science* **197**:1089–1091.

Schwartz, W. J., Smith, C. B., Davidsen, L., Savaki, H. Sokoloff, L., Mata, M., Fink, D. J., and Gainer, H., 1979, Metabolic mapping of functional activity in the hypothalamo–neurohypophyseal system of the rat, *Science* **205**:723–725.

Sejnowski, T. J., Kelley, D. B., Paton, J. A., and Yodlowski, M. L., 1979, Quantitative (^{3}H) and (^{14}C)-2-deoxyglucose mapping of the auditory central nervous system in *Xenopus laevis, Neurosci. Abstr.,* **5**:30.

Shapiro, H. M., Greenberg, J. H., Reivich, M. Ashmead, G., and Sokoloff, L., 1978, Local cerebral glucose uptake in awake and halothane anesthetized primates, *Anesthesiology* **48**:97–103.

Silverman, M. S., Hendrickson, A. E., and Clopton, B. M., 1977, Mapping of the tonotopic organization of the auditory system by uptake of radioactive metabolites, *Neurosci. Abstr.* **3**:11.

Sokoloff, L., 1977, Relation between physiological function and energy metabolism in the central nervous system, *J. Neurochem.* **29**:13–26.

Sokoloff, L., Reivich, M., Kennedy, C., Des Rosiers, M. H., Patlak, C. S., Pettigrew, K. D., Sakurada, O., and Shinohara, M., 1977, the (^{14}C) deoxyglucose method for measurement of local cerebral glucose utilization: Theory, procedure, and normal values in the conscious and anesthetized albino rat, *J. Neurochem.* **28**:897–916.

Stewart, W. B., Kauer, J. S., and Shepherd, G. M., 1979, Functional organization of rat olfactory bulb analyzed by the 2-deoxyglucose method, *J. Comp. Neurol.* **185**:715–734.

Toop, J., Burke, R. E., Dum, R. P., O'Donovan, M. J., and Smith, C. B., 1979, 2-Deoxy-(^{14}C)glucose as a marker for identifying acutely active skeletal muscle fibers in individual motor units, *Neurosci. Abstr.* **5**:388.

Tower, D. B., 1958, The effect of 2-deoxy-D-glucose on metabolism of slices of cerebral cortex incubated *in vitro, J. Neurochem.* **3**:185–205.

Weinberger, J., Greenberg, J. H., Waldman, M., Sylvestro, A., and Reivich, M., 1979, The effect of scopolamine on local glucose metabolism in rat brain, *Brain Res.* **177**:337–345.

Welker, C., 1976, Receptive fields of barrels in the somatosensory neocortex of the rat, *J. Comp. Neurol.* **166**:173–189.

Wick, A. N., Drury, D. R., Nakada, H. I., and Wolfe, J. B., 1957, Localization of the primary metabolic block produced by 2-deoxyglucose, *J. Biol. Chem.* **224**:963–969.

Woolsey, T. A., and Van der Loos, H., 1970, The structural organization of layer IV in the somatosensory region (SI) of mouse cerebral cortex. The description of a cortical field composed of discrete cytoarchitectonic units, *Brain Res.* **17**:205–242.

Wooten, G. F., and Collins, R. C., 1979, Regional brain glucose utilization following intrastriatal injections of kainic acid in rats, *Neurosci. Abstr.* **5**:578.

Yarowsky, P., Jehle, J., Ingvar, D. H., and Sokoloff, L., 1979, Relationship between functional activity and glucose utilization in the rat superior cervical ganglion *in vivo, Neurosci. Abstr.* **5**:421.

Epilogue

Some General Advice to the Young Investigator

The success of a neuroanatomical research project depends to a large extent on the techniques that are used and how effectively they are applied. The purpose of this book is to introduce the student to the major tract-tracing methods and to emphasize the importance of selecting the procedure that is most appropriate for solving a particular neuroanatomical question.

Success in research, however, is not only dependent on technical matters. Motivation and a clear intellect are basic prerequisites, and, as in many other fields of human endeavor, interpersonal relations play a most important role. Many scientists have written about these matters with the intention of advising and encouraging young investigators, and these nontechnical matters will therefore be considered only briefly.

It is difficult to define precisely the combination of personal qualities that characterize a scientist and compel him or her to engage in scientific activities. However, a few of the ingredients necessary for a successful and fulfilling scientific career can be identified. Foremost, there must be curiosity and eagerness to solve a problem. This means that the person should have a certain amount of compulsion and be irritated with unanswered questions.

Although scientists occasionally gain national or international recognition for their efforts, the individual who embarks on a scientific career to garner fame and fortune is likely to be frustrated. For those who eagerly await the clinical application of their discoveries, some solace must be taken from the certainty that the more we discover about the circuitry of the nervous system, the better prepared we will be to understand human behavior, and the more effectively we will be able to manage neurological and mental disorders.

Although practical laboratory work can at times be tedious, it should basically be a gratifying experience, and new preparations and experiments should be greeted with enthusiasm. If you frequently feel scientific work to be tiring and monotonous, or if you find yourself intentionally or unconsciously diverting yourself to other activities, you may wish to reassess your career choice. It is not a tragedy to discontinue a scientific career; rather, the tragedy is to continue a career that does not give personal satisfaction.

The student–supervisor relationship is an important factor in the initial stages of a research career. Certainly, there is no standard or "ideal" arrangement for this relationship, and it will vary considerably depending on the interacting personalities. Yet, some general comments may be made. If no results are forthcoming in spite of great investment of time and effort, there may be reasons to take a closer look at the situation in the laboratory. Occasionally, a supervisor's dominance and rigid attitudes may stifle the initiative of a younger colleague, or, at the other extreme, the supervisor may offer little or no guidance, merely allowing the student to drift. The faculty member may at times be so preoccupied with administrative or teaching responsibilities that adequate guidance cannot be given. Or he may be out of date with respect to techniques and equipment. Such a supervisor may nevertheless be the one providing the all-important economic basis for the institution. Resourceful students, however, can usually solve even these difficult problems with initiative and a positive approach. They can undertake technical challenges on their own, using the published literature as a springboard, and write letters to well-regarded scientists in other institutions when specific problems arise.

Startling discoveries have been made by scientists who have followed a seemingly farfetched idea with great boldness, persistence, and independence, and it is important that students with bold, original ideas carry out their plans. However, the apprentice should also be cautioned against continuing a project for a protracted period of time if it seems hopeless, and especially if more experienced colleagues have warned of its difficulty and uncertainty. Such warnings are usually well founded, because scientists are hesitant to criticize colleagues for excessive independence, especially as experience has taught them that a seemingly worthless experiment can lead to valuable results.

Many people make up the research team, some with sophisticated academic training and years of experience and some with either one or the other. The student can benefit enormously by developing a good relationship with laboratory technical personnel, many of whom may have rich practical experience. Technicians are vital to the productivity of the laboratory; students and supervisors should treat them with respect, admire their achievements when deserved, and—when necessary—demonstrate in a polite and friendly way how their techniques might be improved.

Author Index

Subject Index

A bold page number indicates that the subject is treated in the photomicrograph or diagram on the page cited.

GPSR Compliance

The European Union's (EU) General Product Safety Regulation (GPSR) is a set of rules that requires consumer products to be safe and our obligations to ensure this.

If you have any concerns about our products, you can contact us on ProductSafety@springernature.com

In case Publisher is established outside the EU, the EU authorized representative is:

Springer Nature Customer Service Center GmbH
Europaplatz 3
69115 Heidelberg, Germany

Batch number: 10838062

Printed by Printforce, the Netherlands